T0391092

Springer Transactions in Civil and Environmental Engineering

Springer Transactions in Civil and Environmental Engineering (STICEE) publishes the latest developments in Civil and Environmental Engineering. The intent is to cover all the main branches of Civil and Environmental Engineering, both theoretical and applied, including, but not limited to: Structural Mechanics, Steel Structures, Concrete Structures, Reinforced Cement Concrete, Civil Engineering Materials, Soil Mechanics, Ground Improvement, Geotechnical Engineering, Foundation Engineering, Earthquake Engineering, Structural Health and Monitoring, Water Resources Engineering, Engineering Hydrology, Solid Waste Engineering, Environmental Engineering, Wastewater Management, Transportation Engineering, Sustainable Civil Infrastructure, Fluid Mechanics, Pavement Engineering, Soil Dynamics, Rock Mechanics, Timber Engineering, Hazardous Waste Disposal Instrumentation and Monitoring, Construction Management, Civil Engineering Construction, Surveying and GIS Strength of Materials (Mechanics of Materials), Environmental Geotechnics, Concrete Engineering, Timber Structures.

Within the scopes of the series are monographs, professional books, graduate and undergraduate textbooks, edited volumes and handbooks devoted to the above subject areas.

More information about this series at https://link.springer.com/bookseries/13593

Krishna R. Reddy · Rathish Kumar Pancharathi ·
Narala Gangadhara Reddy · Suchith Reddy Arukala
Editors

Advances in Sustainable Materials and Resilient Infrastructure

Editors
Krishna R. Reddy
Department of Civil, Materials, and Environmental Engineering
University of Illinois
Chicago, IL, USA

Narala Gangadhara Reddy
Department of Civil Engineering
Kakatiya Institute of Technology and Science
Warangal, India

Rathish Kumar Pancharathi
Department of Civil Engineering
National Institute of Technology Warangal
Warangal, India

Suchith Reddy Arukala
Department of Civil Engineering
Kakatiya Institute of Technology and Science
Warangal, India

ISSN 2363-7633 ISSN 2363-7641 (electronic)
Springer Transactions in Civil and Environmental Engineering
ISBN 978-981-16-9746-3 ISBN 978-981-16-9744-9 (eBook)
https://doi.org/10.1007/978-981-16-9744-9

This Springer imprint is published by the registered company Springer Nature Singapore Pte Ltd.
The registered company address is: 152 Beach Road, #21-01/04 Gateway East, Singapore 189721, Singapore

Preface

The edited book on *Advances in Sustainable Materials and Resilient Infrastructure* stemmed out from the idea of creation and development of sustainable and resilient infrastructure and comprises of invited book chapter contributions from a domain of experts of repute. The book covers the most critical and emerging topics for creating sustainable solutions in the construction industry, promoting technologies, and monitoring methods for a resilient infrastructure with an aim to deliver high-quality end solutions in the field of civil engineering.

This book provides a path for recent developments and advances in utilizing materials, techniques, and methods with the goal of achieving sustainability. This book provides knowledge-based information for readers to assess, monitor, measure, and practice sustainability for resilient infrastructure. The content in the book is a blend of academic research work and industry-based case studies covering the use of sustainable materials like Lime–Pozzolana Binders, bio-polymers, lignosulphonate-based materials, lightweight aggregates made from fly ash, calcined clay, paper ash, and limestone as amendments/ameliorators for soil remediation, focuses on the development of neo-construction materials and composites for civil engineering applications. Topics related to advances in characterization of structural and geomaterial, design of innovative pavements using alkali activation, and pervious concrete for sustainable infrastructure are also presented. Some of the contents provide tools, indicators, techniques, and performance-based design for energy-efficient buildings. Few chapters covered the role of Civil Engineers in achieving UN sustainable development goals, effect of COVID-19 on public transportation, promoting climate change design for urban landscapes, and modeling building energy demand. Special emphasis is laid on the valorization of industrial, C&D waste, recycle and reuse of wastes and life cycle analysis of materials.

Towards the potential areas of growth and vision for future of sustainability some of the book chapters summarize the guiding concepts to reduce the carbon footprint, development of low carbon materials, durability, and service life assessment, MCDM methods for selecting materials, promoting sustainable materials, techniques, and construction, methodologies for synthesis and design of novel composite material, performance-based design for resilient infrastructures, Durability and Service life improvement of buildings, waste minimization, and material efficiency for circular ecology materials help the reader to learn more concepts on reuse, replace, and recycle, tools and technologies for health monitoring of structures and construction sites.

Finally, this book is framed to address the principles and practice from the point of view of geoenvironmental, sustainable construction materials, low carbon materials, energy efficiency, and waste management, designed to create, evaluate, and integrate engineering approaches in the built environment to incorporate sustainability strategies in building resilient infrastructure.

This book derives the attention of the people who intend to contribute to environmental, social, and economic solutions (triple bottom line). This not only focuses on the needs of a designer but also its concerns on the strategies to improve the design and construction, eliminating the uncertainty in decision making at each stage of infrastructure development. Some practical cases are also addressed, which could be a valuable reference for faculty, researchers, field experts, scientists, and practising engineers. Overall, the book is versatile in presenting all-round developments and state of the art in the thrust area of sustainability.

Chicago, USA Krishna R. Reddy
Warangal, India Rathish Kumar Pancharathi
Warangal, India Narala Gangadhara Reddy
Warangal, India Suchith Reddy Arukala

Contents

Editors and Contributors

About the Editors

Dr. Krishna R. Reddy is a professor of civil and environmental engineering, the director of the Sustainable Engineering Research Laboratory (SERL), and also the director of the Geotechnical and Geoenvironmental Engineering Laboratory (GAGEL) in the Department of Civil and Materials Engineering at the University of Illinois at Chicago (UIC). His research expertise includes (1) environmental remediation of soils, sediments, groundwater, and stormwater; (2) solid and hazardous waste management and landfill engineering; (3) engineering applications of waste/recycled materials; (4) life cycle assessment and sustainable and resilient engineering; and (5) geotechnical engineering. He is the author of four major books: *(1) Geoenvironmental Engineering: Site Remediation, Waste Containment, and Emerging Waste Management Technologies, (2) Sustainable Engineering: Drivers, Metrics, Tools, and Applications, (3) Sustainable Remediation of Contaminated Sites, and (4) Electrochemical Remediation Technologies for Polluted Soils, Sediments and Groundwater*. He is also author of **264** journal papers, **27** edited books/conference proceedings, **22** book chapters, and **226** full conference papers (with h-index of **66** with over 14,000 citations). He has served or currently serves as an associate editor or editorial board member of over 10 different journals. He has also served on various professional committees, including the Geoenvironmental Engineering Committee and Technical Coordinating Council of the Geo-Institute (GI) of the American Society of Civil Engineers (ASCE) and the Environmental Geotechnics Committee of the International Society for Soil Mechanics and Geotechnical Engineering (ISSMGE). He has received several awards for excellence in research and teaching, including the ASTM Hogentogler Award, the UIC Distinguished Researcher Award, the University of Illinois Scholar Award, and the University of Illinois Award for Excellence in Teaching. He is a fellow of the American Society of Civil Engineers (FASCE), a Diplomate of Geotechnical Engineering (DGE), and a Board Certified Environmental Engineer (BCEE). He is also a registered Professional Civil Engineer (PE) and an Envision™ Sustainability Professional (ENV SP).

Dr. Rathish Kumar Pancharathi is currently Professor and Head in the Department of Civil Engineering, National Institute of Technology Warangal. He obtained his Ph.D. from the NIT Warangal and **Doctor of Engineering & Post-Doctoral research from Japan**. His major areas of interest include new/alternate/supplementary cementitious materials, special concretes, rehabilitation of structures, construction technology and management, structural health monitoring, earthquake engineering and sustainable materials and technologies. He has published **223 technical articles** in various international and national journals and conferences of repute including **15 Book Chapters** and authored 1 Book. He has guided **7 Doctoral students and 13 students** are presently working with him. He has guided **66 Master's students** and **35 Bachelor's students**. He is a recipient of several awards including **Aftab Mufti Medal, prestigious Monbusho and JSPS Scholarships of the Japanese Government, Best Engineering Researcher Award, Heritage Scholarship to pursue research at IST Portugal, Danish Government Scholarship, Italy Government Post Doctoral Scholarship, Heritage Scholarship under Erasmus Mundus, AIT Fellowship, ASEM-Duo India Fellowship, Slovakian Government Scholarship, Jawaharlal Nehru memorial fellowship and Distinguished Alumnus award**. Prof Kumar is a member of the Research Advisory Council member of the National Council for Cement and Building Materials (NCCBM), Hyderabad for the past six years. He represents several Bureau of **Indian Standard (BIS) committees like CED-4 i.e. Lime and Gypsum Products of the BIS, Member of the Preparation of Handbook of lime for the BIS, Convenor for revision of all parts of Indian Standard Code IS 2542-1978 including tests for Gypsum, Plaster, Concrete and Mortars, Member of Revision of Indian Standard Code IS 712:1984 i.e. Specification for building limes.** He is on the editorial board of Journal of Facta Universtatis, Journal of Cement Wapno Beton among others and is a reviewer of several International Journals published by Elsevier, Thomas Telford, Taylor and Francis etc. He has 28 Years of Teaching and Research experience and 17 Years of administrative experience. He has worked as Associate Dean Admissions, Planning and Development, Head of Structures Division, training and placement activities, warden, faculty in charge of the project engineer's office and member of the master planning committee and DPR for NIT AP. He does lot of active consultancy in designs and stability checks and auditing of buildings and bridges.

Dr. Narala Gangadhara Reddy is an Assistant Professor in the Department of Civil Engineering at Kakatiya Institute of Technology and Science, Warangal, where he teaches and conducts research. Dr. Reddy also worked as a Geotechnical Engineer in Mumbai, India. He obtained his Ph.D. from IIT Bhubaneswar, India and his Master's degree from NIT Bhopal, India. After completion of Ph.D., he worked as a research associate at Shantou University, China and as a visiting research fellow to Nanjing University, China. He is recipient of a Chinese government scholarship award (CSC-2019). His major areas of research include geotechnical and geoenvironmental engineering and pavement geotechnics, with special focus on the characterization of problematic soils and waste materials. He has research experience on sustainable materials such as biochar and biopolymer-amended soils and wastes for mitigating of soil and industrial waste problems. He has published 32 technical articles in various journals and conferences of repute including 3 book chapters. He is a reviewer of several International Journals published by Elsevier, Springer, Thomas Telford etc. He has organized several short term training programs, seminars, and workshops.

Dr. Suchith Reddy Arukala is a Chartered Engineer by qualification in civil engineering and he is an expert in the areas of sustainable construction, materials, technologies, and management, Multi-Criteria Decision Making Techniques, Fuzzy Logic, and Life Cycle Assessment of buildings materials. He has published more than 30 papers in his area in reputed international journals. He also acts as a reviewer for some of the SCI and Scopus indexed journals like Taylor and Francis, ICE virtual, Springer and Elsevier in his area of specialization.

He is recognized as a member of ASCE, MIE, MCIOB, MISTE, and Honorary Member of London Journal Press. He has completed his Masters from the UK in 2010 and a Ph.D. from NIT WARANGAL, India. Before Research and Teaching experience, Dr. Suchith has worked for giant MNC's like Hindustan Construction Company, Madhucon Projects, SVEC Construction Ltd., Indu Projects. His vast experience in teaching and research along with an intention and motivation for sustainable development is recognized and awarded him as an Earth Leader by Council for Green Revolution and KPR foundation. He is currently working as a faculty in the prestigious institute, KITS Warangal one of the top institutes in India. He has good collaborations with the National Academy of Construction and industry sector and initiated and conducted programs to bridge the gap between academic and industry.

Contributors

P. Teja Abhilash Kakatiya Institute of Technology and Science, Warangal, Telangana, India

Lavanya Addagada CSIR- National Environmental Engineering Research Institute Nagpur, Nagpur, Maharashtra, India

Francisco Agrela Universidad de Córdoba, Córdoba, Spain

Shamshad Alam Chandigarh University, Mohali, India

Anshuman Department of Civil Engineering, Birla Institute of Technology and Science, Pilani, Rajasthan, India

B. R. Anupam School of Infrastructure, Indian Institute of Technology, Bhubaneswar, India

Rizwan Anwar National Council for Cement & Building Materials, Ballabgarh, Haryana, India

Dali Naidu Arnepalli Department of Civil Engineering, Indian Institute of Technology Madras, Chennai, Tamilnadu, India

Suchith Reddy Arukala Department of Civil Engineering, Kakatiya Institute of Technology & Science, Warangal, Telangana, India

Sumanth Kumar Bandaru Kakatiya Institute of Technology and Science, Warangal, India

Patrick Amoah Bekoe Senior Adjunct Lecturer, Department of Civil Engineering, Kwame Nkrumah University of Science and Technology, Kumasi, Ghana

Sriman Pankaj Boindala Civil and Environmental Engineering, Technion—Israel Institute of Technology, Haifa, Israel

Venkata Joga Rao Bulusu Department of Civil Engineering, IIT Kharagpur, Kharagpur, West Bengal, India

Manuel Cabrera Universidad de Córdoba, Córdoba, Spain

Suzane A. V. Carneiro University of Illinois, Chicago, USA

Sarath Chandra Department of Civil Engineering, CHRIST (Deemed To Be University), Bangalore, India

Yu Chen School of Highway, Chang'an University, Xi'an, China

Chinchu Cherian Faculty of Applied Science, School of Engineering, The University of British Columbia, Kelowna, British Columbia, Canada

Jyoti K. Chetri University of Illinois, Chicago, USA

Bajaya K. Das KIIT University, Bhubaneswar, India

Sarat Kumar Das Indian Institute of Technology (ISM), Dhanbad, India

Rajan Gandhimathi Department of Civil Engineering, National Institute of Technology, Tiruchirappalli, Tamilnadu, India

Udaya B. Halabe Department of Civil and Environmental Engineering, West Virginia University, Morgantown, WV, USA

Nabil Hossiney Department of Civil Engineering, CHRIST (Deemed To Be University), Bangalore, India

Nauman Ijaz Key Laboratory of Geotechnical and Underground Engineering of Ministry of Education, College of Civil Engineering, Tongji University, Shanghai, China

Zain Ijaz Key Laboratory of Geotechnical and Underground Engineering of Ministry of Education, College of Civil Engineering, Tongji University, Shanghai, China

Abin Joy ASIET, Kalady, Kerala, India

G. Kannan School of Civil Engineering, SASTRA Deemed University, Thanjavur, Tamil Nadu, India

Chandra Sekhar Karadumpa Department of Civil Engineering, National Institute of Technology, Warangal, Telangana, India

Arkamitra Kar Department of Civil Engineering, BITS-Pilani, Hyderabad Campus, Hyderabad, Telangana, India

Puneet Kaura National Council for Cement & Building Materials, Ballabgarh, Haryana, India

Swarna Swetha Kolaventi NIT Warangal, Warangal, India

Ashutosh Kumar Indian Institute of Technology Mandi, Kamand, Mandi, India

Ayush Kumar Indian Institute of Technology Mandi, Kamand, Mandi, India

Degloorkar Nikhil Kumar National Institute of Technology Warangal, Warangal, Telangana, India

S. Anandha Kumar Department of Civil Engineering, Aditya Engineering College (Autonomous), Surampalem, Andhra Pradesh, India

Sonu Kumar Indian Institute of Technology Mandi, Kamand, Mandi, India

Sudhakar Reddy Kusam Department of Civil Engineering, IIT Kharagpur, Kharagpur, West Bengal, India

Kundan Meshram Guru Ghasidas Vishwavidyalaya (A Central University), Bilaspur, India

Anshumali Mishra Indian Institute of Technology (ISM), Dhanbad, India

Ramya Sri Mullapudi Civil Engineering Department, Indian Institute of Technology, Hyderabad, India

Manu S. Nadesan ASIET, Kalady, Kerala, India

Vishnu Sai Nagavelly Department of Civil Engineering, School of Engineering and Technology, Central Queensland University, Melbourne Campus, Australia

P. N. Ojha National Council for Cement & Building Materials, Ballabgarh, Haryana, India

Rathish Kumar Pancharathi Department of Civil Engineering, National Institute of Technology, Warangal, Telangana, India

Vikas Patel National Council for Cement & Building Materials, Ballabgarh, Haryana, India

S. Prusty Palladium Consulting India Pvt. Ltd, Pune, India

Shashi Ram Department of Civil Engineering, N.I.T. Warangal, Telangana, India

Kruthi Kiran Ramagiri Department of Civil Engineering, BITS-Pilani, Hyderabad Campus, Hyderabad, Telangana, India

Srikrishnaperumal T. Ramesh Department of Civil Engineering, National Institute of Technology, Tiruchirappalli, Tamilnadu, India

K. Rangaswamy National Institute of Technology Calicut, Calicut, India

Dwarika N. Ratha Department of Civil Engineering, Thapar Institute of Engineering and Technology, Patiala, Punjab, India

P. Mani Rathnam Department of Civil Engineering, N.I.T. Warangal, Telangana, India

Indrajit Ray Department of Civil and Environmental Engineering, University of West Indies, St. Augustine, Trinidad and Tobago

Krishna R. Reddy University of Illinois, Chicago, USA

Narala Gangadhara Reddy Kakatiya Institute of Technology and Science, Warangal, India

T. V. G. Reddy National Council for Cement & Building Materials, Ballabgarh, Haryana, India

Mercedes Regadío The University of Sheffield, Sheffield, UK

Zia ur Rehman Department of Civil Engineering, University of Engineering and Technology, Taxila, Pakistan

Jaqueline R. Robles University of Illinois, Chicago, USA

Julia Rosales Universidad de Córdoba, Córdoba, Spain

Prangya Ranjan Rout Department of Biotechnology, Thapar Institute of Engineering and Technology, Patiala, Punjab, India

Umesh Chandra Sahoo School of Infrastructure, Indian Institute of Technology, Bhubaneswar, India

Bhaskar Sangoju CSIR-Structural Engineering Research Centre, Chennai, India

Rajiv Satyakam Netra, National Thermal Power Corporation Limited, Noida, India

P. V. V. Satyanarayana Andhra University, Visakhapatnam, Andhra Pradesh, India

Hima Kiran Sepuri Department of Civil Engineering, CHRIST (Deemed To Be University), Bangalore, India

D. Rama Seshu National Institute of Technology, Warangal, India

Sumi Siddiqua Faculty of Applied Science, School of Engineering, The University of British Columbia, Kelowna, British Columbia, Canada

Brijesh Singh National Council for Cement & Building Materials, Ballabgarh, Haryana, India

S. B. Singh Department of Civil Engineering, Birla Institute of Technology and Science, Pilani, Rajasthan, India

M. V. N. Siva Kumar NIT Warangal, Warangal, India

S. Smitha National Institute of Technology Calicut, Calicut, India

Evangelin Ramani Sujatha Centre for Advanced Research On Environment, School of Civil Engineering, SASTRA Deemed University, Thanjavur, Tamil Nadu, India

M. B. Sushma Department of Civil Engineering, Kakatiya Institute of Technology and Science, Warangal, India

Tezeswi Tadepalli NIT Warangal, Warangal, India

Naveen Thakur Department of Civil Engineering, Birla Institute of Technology and Science, Pilani, Rajasthan, India

K. Tharani National Institute of Technology, Warangal, Telangana, India

Steven F. Thornton The University of Sheffield, Sheffield, UK

Avinash Unnikrishnan Department of Civil and Environmental Engineering, Portland State University, Portland, OR, USA

M. Vishweswaran School of Civil Engineering, SASTRA Deemed University, Thanjavur, Tamil Nadu, India

Anush K. Chandrappa School of Infrastructure, Indian Institute of Technology, Bhubaneswar, India

Chapter 1
Tiered Quantitative Assessment of Life Cycle Sustainability and Resilience (TQUALICSR): Framework for Design of Engineering Projects

Krishna R. Reddy, Jaqueline R. Robles, Suzane A. V. Carneiro, and Jyoti K. Chetri

1.1 Introduction

Recent advancements in the field of engineering have led to exponential growth in resource consumption. The manifold increase in infrastructure projects, such as high-rise buildings, multi-lane highways, airports, and bridges, has led to significant increase in consumption of raw materials, energy usage, waste generation, and harmful emissions. Engineering designs are normally based on technical performance and construction costs; hence, little consideration is given to the broader environmental and socio-economic impacts associated with the project (Reddy et al. 2019). In recent years, nevertheless, a focus on sustainability has gained momentum as a global trend, and it is increasingly being adopted in various disciplines of engineering. Organizations such as the American Society of Civil Engineers (ASCE) have started promoting sustainability in engineering projects by introducing policies and guidelines, such as ASCE Policy 418, which requires engineering designers and planners to assess the net environmental, social, and economic benefits, also referred to as the triple bottom line (TBL), and to consider sustainability in various life cycle stages of a project (Reddy and Kumar 2018a).

K. R. Reddy (✉) · J. R. Robles · S. A. V. Carneiro · J. K. Chetri
University of Illinois, Chicago, USA
e-mail: kreddy@uic.edu

J. R. Robles
e-mail: jrojas21@uic.edu

S. A. V. Carneiro
e-mail: svieir2@uic.edu

J. K. Chetri
e-mail: jkc4@uic.edu

K. R. Reddy et al. (eds.), *Advances in Sustainable Materials and Resilient Infrastructure*, Springer Transactions in Civil and Environmental Engineering,
https://doi.org/10.1007/978-981-16-9744-9_1

Various guidelines have been developed to encourage incorporation of sustainability and resilience principles in engineering, such as green and sustainable buildings (ASTM 2019) and green and sustainable cleanups (ASTM 2020, ASTM 2016, ITRC 2011). Various rating tools have also been developed to promote sustainable infrastructure, among which Leadership in Energy and Environmental Design (LEED) and Envision^{TM} are frequently used for sustainability assessment and certification. In recent years, the life cycle or cradle-to-grave approach has gained importance in applications toward sustainability considerations. Life cycle assessment (LCA) is considered a comprehensive tool to quantify environmental sustainability considering environmental impacts from various life cycle stages, from material acquisition to waste disposal (Reddy and Kumar 2018a). Although LCA is an excellent tool to evaluate environmental sustainability, it does not consider the other two pillars of the triple bottom line: social and economic. For the overall sustainability of a system or a project, it is of the utmost importance to consider net environmental, economic, and social benefits during the life cycle of a project (Damians et al. 2016; Reddy and Kumar 2018b; Reddy et al. 2019).

In the past few years, more attention has been given to resilience of the built environment (Burroughs 2017; Ayyub 2020), which is often assessed from a technical view and limited to physical criteria. A general definition of resilience is the measure of the ability of a system to protect, sustain, and restore the functionality after exposure to sudden and damaging actions or events (Reddy et al. 2019; Kumar and Reddy 2020). There is a greater need to holistically integrate sustainability and resilience, as the sustainability of a system will be challenged in the absence of a resilient design. However, whenever resilience is incorporated into a project, it is usually done from an environmental perspective. Social resilience and economic resilience are either not included or are considered at later phases of the design and selection process, at which point efforts may be of limited value. Therefore, to effectively address the triple bottom line of resilience, it should be incorporated from an early stage in the design and construction of a system to prepare for an effective recovery from any disruptive social, economic, or environmental events.

Lastly, many of the existing sustainability assessment frameworks are better suited for broader or larger scale applications. For example, the United Nations' Sustainable Development Goals (SDGs) are designed for global application, and Envision^{TM} is designed for projects with relatively larger scope (e.g., regional stormwater management, or airport infrastructure). In addition, many existing frameworks require use of advanced, costly, and time-consuming tools, limiting their frequent use to incorporate sustainability aspects in engineering projects. Hence, there is a need for a sustainability and resilience framework of simple implementation that can be used even at a project component assessment level (e.g., foundation choice for a building, or technology option for groundwater cleanup).

A framework called Quantitative Assessment of Life Cycle Sustainability (QUALICS) was previously developed based on life cycle triple bottom line sustainability assessment (Reddy et al. 2019). The QUALICS framework used advanced quantitative tools, such as LCA, and specialty tools, such as Spreadsheets for Environmental Footprint Analysis (SEFA) and SiteWise^{TM}, for environmental impact assessment, direct and indirect costs for economic assessment, Social Sustainability Evaluation Matrix (SSEM) for social sustainability assessment, and integrating these triple bottom line indicators through multi-criteria decision analysis (Trentin et al. 2019; Reddy et al. 2019). Although such tools are exceptional in quantifying the sustainability, their application can be challenging due to limited available input data. Moreover, QUALICS did not consider resilience. To provide greater flexibility and breadth, and to integrate resilience and sustainability, a new framework called Tiered Quantitative Assessment of Life Cycle Sustainability and Resilience (TQUALICSR) is presented in this chapter. TQUALICSR uses a tiered approach to assess the environmental, economic, and social impacts that incorporates resilience considerations, depending on the level of data and information available on the project.

1.2 Framework (TQUALICSR) Basis

The TQUALICSR framework preserves the approach of integrating the triple bottom line of sustainability as demonstrated in our previously reported QUALICS framework. A key goal of TQUALICSR is to give a wide group of users a flexible framework to perform a holistic, integrated sustainability and resilience assessment at a project scale. Key features of the TQUALICSR framework include the following: (1) integration of resilience and sustainability into a unified assessment framework, including a triple bottom line approach to the resilience criteria; (2) flexible, tier-based selection of tools to assess the environmental, economic, and social impacts and resilience of a project based on the information, data, and resources available; (3) integration of interdependent relationships between the technical, environmental, social, economic, and resilience dimensions into a simple yet comprehensive model; and (4) applicability to any life cycle stage of an engineering project of any size, from planning to decommissioning.

The TQUALICSR framework is one of few attempts to integrate sustainability and resilience in a unified framework. Throughout the TQUALICSR framework, resilience is treated as an input toward sustainability-related steps. The guiding principle for this relationship as incorporated in the TQUALICSR framework is that "resilience is a precondition for sustainability and sustainable development" (Roostai et al. 2019). To achieve a holistic, integrated approach, considerations of interdependent relationships are included. The first interdependent relationship addressed is between technical aspects (design or construction), resilience, and sustainability. TQUALICSR requires the user to consider more than the technical constraints of the project and project site, which are often prioritized over sustainability and resilience considerations. For the development of sustainable and resilient designs, TQUALICSR incorporates technical, sustainability, and resilience constraints. The second interdependent relationship addressed by the TQUALICSR framework is the consideration of connections between the three dimensions of sustainability, in which each dimension is dependent on the others, and they all affect each other. For example, environmental impacts, such as air pollution, can be used as inputs to the social

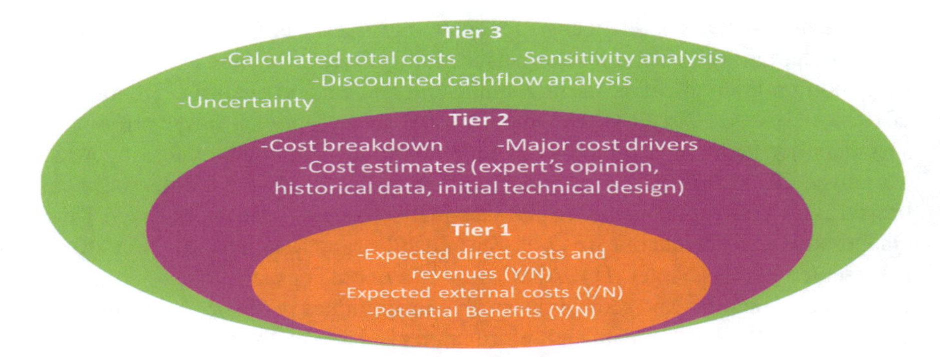

Fig. 1.1 Example sustainability and resilience economic assessment tiers

sustainability assessment of individual health and well-being, and environmental and social impacts can be monetized into economic impacts and benefits.

The sustainability and resilience assessment tools in the TQUALICSR framework are categorized into three tiers. Tier 3 has the highest ranking, followed by Tier 2 and then Tier 1. The rankings are based on the degree of quantitative nature, time required, and accuracy involved. Fig. 1.1 shows tiers and typical tools to assess economic aspects. The categorization of the tools into tiers makes it possible to perform an integrated sustainability and resilience assessment that considers the amount and type of data, information, and quantitative tools available to the user.

1.2.1 Tier 1 Tools and Indicators

Tier 1 tools and indicators encompass best management practice (BMP)-based tools. Indicators are assigned a value from a yes or no (Y/N) approach, using the basic knowledge of the project to convert the qualitative responses into semi-quantitative. A value of 1 is attributed if a BMP is incorporated in the project to address the impact being assessed or to bring additional benefits; otherwise, a value of 0 is given if the BMP is not addressed or incorporated.

1.2.2 Tier 2 Tools and Indicators

Tier 2 tools and indicators encompass rating system-based tools. Indicators are assigned a value using a number relative to a numerical scale, such as 1–4, where 4 can be used to indicate the most positive, and 1 is equivalent to the least positive. The assessment is based on preliminary information, basic technical design, and expert knowledge, and the results are considered semi-quantitative.

1.2.3 Tier 3 Tools and Indicators

Tier 3 tools and indicators encompass advanced quantification tools. Indicators are assigned a value by utilizing specific quantitative data and methods. The data can be analyzed using specially designed software or statistical and mathematical models to assign a numerical value to each indicator. Information and data should be as accurate as possible and often involves a detailed analysis of the technical design and expert knowledge to generate reliable quantitative results.

1.3 Framework (TQUALICSR) Description

The TQUALICSR framework is divided into four consecutive phases, with data and results from the previous phase feeding into the next one, as shown in Fig. 1.2. Phase 1 involves the definition of the project scope, technical and sustainability constraints based on inputs from stakeholders, exposure and vulnerability assessment, and calculating resilience index using multi-criteria decision analysis (MCDA) tools. Then, the different designs/alternatives are selected in Phase 2 for the sustainability resilience assessment. The criteria and indicators for the triple bottom line requirements are defined, and weightages are assigned to each indicator, criterion and requirement involving stakeholders' engagement. In Phase 3, the integrated sustainability and resilience assessment is performed using the framework to assign values for each indicator and to calculate a sustainability resilience index using MCDA. Different weights are assigned to each indicator based on the level of tools and data/information available, and sensitivity and uncertainty analyses are performed to establish a confidence level for the final index. At last, Phase 4 is the decision-making phase, where the sustainability resilience indexes are evaluated, and the best design/alternative is recommended.

1.3.1 Phase 1: Design Constraints, Exposure and Vulnerability Assessment

The first phase aims to provide a structured flow to include more than technical considerations early in the design process and to help identify resilience goals, and constraints in an informed manner. Users begin by defining the project goal and scope. Stakeholders' engagement is sought in the early design to incorporate their preferences in the technical design and resilience constraints. Then technical design is developed incorporating stakeholders' preferences and resilience consideration. As a simple example of a technical constraint for an infrastructure project, consider the maximum allowable operating pressure for a pipeline. Establishing technical specifications to abide by this constraint does not require consideration of sustainability

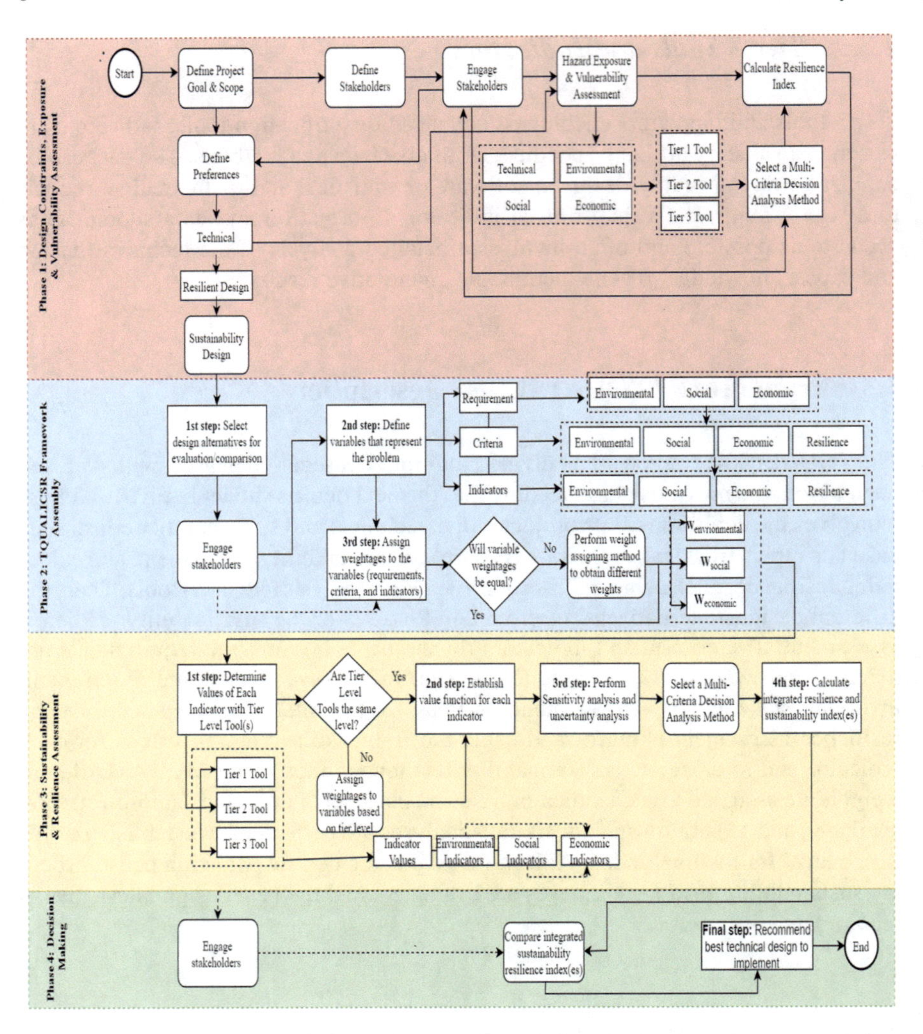

Fig. 1.2 Schematic of tiered quantitative assessment of life cycle sustainability and resilience (TQUALICSR) framework

or resilience constraints. However, if sustainability and resilience considerations are not included early in the project, the frequency and/or magnitude of negative environmental, social, and economic consequences can increase. Using the same pipeline as an example, disregard of durable and/or environmentally friendly materials may lead to the release of toxic chemicals into the environment, resulting in both environmental and social impacts, as well as high mitigation, repair, and replacement costs. An exposure and vulnerability assessment of the technical design is performed to calculate the resilience index. The vulnerability is assessed for the four criteria: technical, environmental, social, and economic. Vulnerability indicators for each criterion are identified based on the exposure conditions. The vulnerability indicators under each criterion are quantified using tiered approach. Resilience index is calculated following a multi-criteria decision analysis (similar approach explained in Phase 3 Sect. 1.3.3).

To foster proactive sustainable and resilient engineering designs, users are encouraged to define sustainability constraints, resilience goals, resilience criteria, and mitigation strategies before they can optimize the design. This is achieved in the TQUAL-ICSR framework by performing an exposure and vulnerability assessment commonly used in disaster management. Using the viewpoint that systems are often interdependent and have social and economic connections, social and economic vulnerabilities are considered in the assessment along with technical and environmental. The calculated technical, environmental, social, and economic vulnerabilities can then be used to inform the stakeholders to shape their preferences and modify technical design to incorporate resilience goals and constraints, and finally develop resilient design.

1.3.1.1 Hazard Exposure and Vulnerability Assessment Tiered Tools

In considering exposure and vulnerability assessment tools, Tier 1 tools could follow BMP considerations, such as those included in the Envision™ resilience section scorecard or the RELi™ rating guidelines from the U.S. Green Building Council (USGBC) for resilient design and construction. Envision™ treats resilience more broadly and looks more at the resilience associated with technical performance than social and economic consequences. RELi™ has a more extensive set of resilience considerations that tackle the interdependent relationship between the technical, sustainability, and resilience concepts. More social and economic facets are included in RELi™. As demonstrated by the Tier 1 examples, considerations could be assessed using a Y/N approach to determine whether the project is vulnerable to such risk events/hazards and whether social and economic features of the society are vulnerable. For Tier 2, the considerations could be extended to a ratings system, such as through use of a scoring matrix. Finally, for Tier 3, peer-reviewed vulnerability/risk quantitative assessments may be considered. Comprehensive standard quantitative tools integrating resilience and sustainability along with interdependent relationships of the environmental-social and environmental-economic are not established. Therefore, the reader is encouraged to reference the literature for strategies specific to the context of the project. The outcomes of the exposure and vulnerability assessment could be used to set a baseline for the sustainability and resilience assessment in Phase 2. However, the key goal of Phase 1 is to identify resilience requirements that can be incorporated in the technical design process for a more holistic, sustainable design choice development.

Before Phase 1 can be completed, optimization of the design must be finalized. Pondering that resilience is considered as a precondition for sustainability, the resilience-related definitions and strategies are incorporated into the "Sustainability Design" step. With rating tools like LEED, users often utilize sustainability ratings/credits as constraints to optimize the design. However, it has been pointed out that resilience is not well integrated in the sustainability frameworks (Roostai et al. 2019). As an effort to address this matter, the TQUALICSR framework facilitates resilient design by considering resilience goals, resilience criteria, and mitigation strategies before initiating sustainability design. This allows users to fully define sustainability constraints that consider resilience aspects; thereby integrating sustainability and resilience. With technical, sustainable, and resilience aspects amassed, users can proceed to optimizing the technical designs and subsequently proceed to Phase 2 of TQUALICSR.

1.3.2 Phase 2: TQUALICSR Framework Assembly

Phase 2 is composed of three main sequential steps to build the integrated sustainability and resilience assessment framework: selection of the design alternatives, definition of qualitative and quantitative variables, and assignment of weights to the variables.

1.3.2.1 Step 1: Selection of the Design Alternatives

In this step, final technical design alternatives developed in Phase 1 are selected. A design identified as a sustainable choice at this phase should have incorporated the resilience requirements outlined in Phase 1, and therefore, should have features that address both resilience and sustainability. The user may have various sustainable designs that fall under this interdependent sustainability–resilience relationship and can include as many in the integrated assessment framework. However, the entire environmental, social, economic, and resilience assessment needs to be completed for each design. The proposed resilient-sustainable design could also be evaluated to the no-change or default solution (when nothing is done or modified) or compared to a design that is not resilient–sustainable. In order to have a baseline, the proposed sustainable solution should be compared to at least the "do-nothing" alternative.

1.3.2.2 Step 2: Defining Qualitative and Quantitative Variables

The second step of the framework focuses on establishing qualitative and quantitative variables that closely represent the environmental, social, and economic impacts of the various life cycle stages of the project. The framework emphasizes on stakeholders' engagement in defining variables for sustainability assessment. Following the integrated view of sustainability and resilience, resilience is integrated into each triple bottom line requirement. This is done by adding environmental resilience as part of the environmental requirement, social resilience as part of the social requirement, and economic resilience as part of the economic requirement. The variables

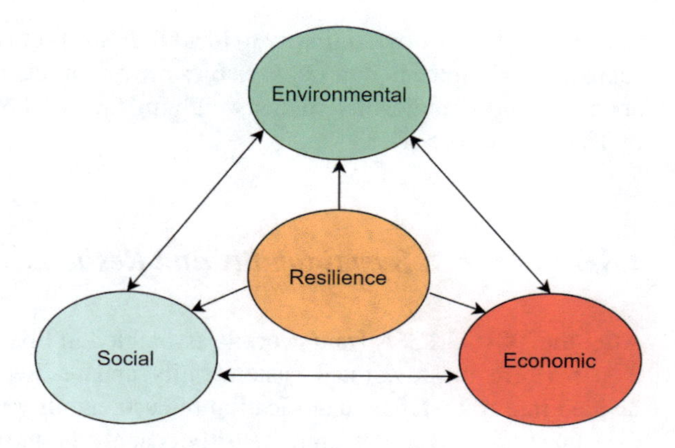

Fig. 1.3 Schematic of the interconnections of the criterion variables under each requirement level

are divided into three main categories, including requirement, criterion, and indicator. The variables in the requirement level are primarily based on the three pillars of sustainability; environmental, economic, and social. Each requirement variable is further forked into variables termed as criteria, with at least one obligatory resilience criterion for each requirement. In addition, each requirement can also have environmental, economic, and social criteria, as they are interconnected, as illustrated in Fig. 1.3. One example is the social impact criterion of jobs created due to a new project developed. A social criterion that accounts for resilience will not only measure the number of jobs created but also the longevity and resilience of the jobs created as a result of the new project being developed. Each criterion is further subdivided into a set of variables called indicators. The indicator-level variables represent the variables that have direct impact on the upper levels of the hierarchy, such as requirement level. For example, greenhouse gas emissions that have direct impact on the criterion "air", which in turn impacts the "environment" requirement (Reddy et al. 2019). Examples of variables based on the triple bottom line requirements and available tier tools are discussed in Phase 3.

1.3.2.3 Step 3: Assigning Weightages to the Variables

The third step of Phase 2 is to assign weights to the variables under each category. The stakeholders' can also be engaged in this step, however that is optional. This step needs expert judgement and a standardized approach to minimize biases. One such approach is the analytic hierarchy process (AHP) proposed by Saaty (2008). AHP is used to define the relative importance of the variables under each category and determine the weights of the requirements, criteria, and indicators (W_{req}, W_{cri}, and W_{ind}) (Trentin et al. 2019). AHP utilizes a pairwise comparison of the variables against each other to establish the priorities, which is then used to derive the weights. The process is explained in detail in Saaty (2008). Trentin et al. (2019) explains in detail its application with respect to the QUALICS framework. Other multi-criteria decision analysis tools for definition of weights include TOPSIS (Technique for

Order of Preference by Similarity to Ideal Solution) (Lai et al. 1994), and a weighted score method. Since higher tier variables provide more accurate information or data about the impact indicators, higher weighting factors should be provided in the order of Tier 3 > Tier 2 > Tier 1.

1.3.3 Phase 3: Sustainability and Resilience Assessment

After the TQUALICSR framework is assembled in Phase 2, users proceed to Phase 3 to perform the integrated sustainability and resilience assessment. Phase 3 is divided into four steps: quantification of sustainability indicators, determination of the value function for indicators, sensitivity analysis, and calculation of the integrated sustainability and resilience index(es).

1.3.3.1 Step 1: Quantifying Sustainability Indicators

The first step in Phase 3 is to quantify the variables in the indicator level. Quantifying variables can be daunting if the users do not have access to advanced impact assessment tools or mathematical and statistical models. Hence, TQUALICSR uses two additional tiers of indicators to facilitate easier quantification of the impacts, which utilize simple considerations and a rating system based on the indicator type and available information about the indicator. It uses a multidisciplinary approach, with greater emphasis placed on the contribution of experts from different areas and backgrounds mainly in Tier 1 and Tier 2. The indicators in Tier 1 will have values of "1" and "0" based on Y/N answers. Indicators in Tier 2 will typically have a rating system from 1 to 4, based on the level of potential impact to environmental, societal, and economic requirements. Finally, indicators in Tier 3 will have absolute quantities derived from quantitative tools, e.g., LCA, SEFA, Life Cycle Costing (LCC), and social factsheets. Different indicators within the same criterion and/or requirement can be estimated or calculated using different tier tools, as illustrated in Fig. 1.4.

1.3.3.2 Step 2: Determining Value Function for Indicators

Establishing value functions for each indicator is a crucial step in the TQUALICSR framework, as it allows normalization of the quantities with variable units into common, dimensionless values scaled 0–1, where 0 represents minimum satisfaction (S_{min}) and 1 represents maximum satisfaction (S_{max}) (Alarcon et al. 2011). The framework uses multi-criteria decision analysis (MCDA) methods, such as the Integrated Value Model for Sustainability Assessment (MIVES) model, for the standardization of the values of each impact indicator.

There are four basic steps to determine the satisfaction value: (1) determine tendencies of the value function (either increasing or decreasing), (2) define points of

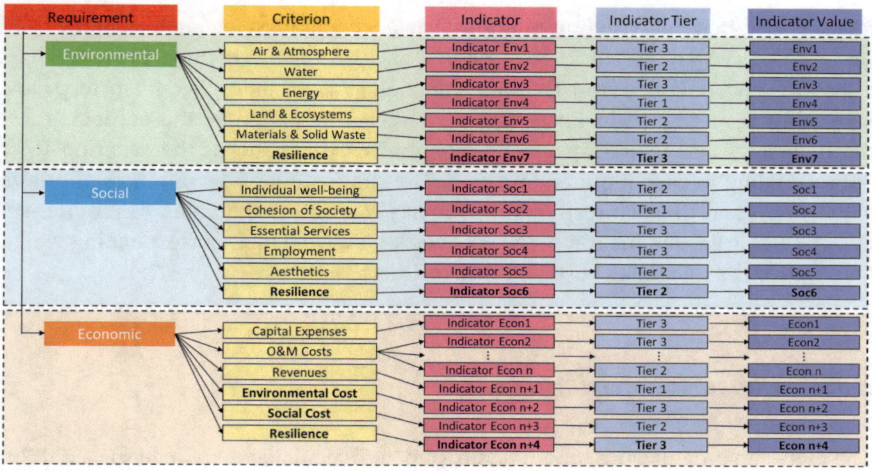

Fig. 1.4 Example of the tiered hierarchy for the TQUALICSR sustainability assessment. *Note* Criteria in bold are mandatory in the framework. Other criteria and the indicator tiers are illustrative, assigned based on project team's experience and subjected to change according to the project specific

minimum and maximum satisfaction (S_{min} and S_{max}), (3) choose a form for the value function (e.g., linear, concave, convex, S-shaped), and (4) establish a relevant mathematical relationship for evaluating the value function (Alarcon et al. 2011; Reddy et al. 2019). Determining value functions is not just limited to using MIVES methodology, any other tools/models that can normalize the indicator values can be used. The main objective is to convert indicators with different units into a common unit that allow for integration of the impact assessment results.

1.3.3.3 Step 3: Sensitivity Analysis

This step focuses on identifying the factors outweighing the influence of other parameters or impact categories on the overall sustainability of the project. A sensitivity analysis is performed by varying the identified parameter to evaluate the effect of the parameter on the overall sustainability assessment. For example, variability in transportation distances or different material costs from different suppliers could be explored in the sensitivity analysis. Uncertainty of any input values should also be assessed.

1.3.3.4 Step 4: Determining Integrated Sustainability Resilience Index(es)

The fourth step is determination of the sustainability resilience index for each design alternative considered in the sustainability assessment. The sustainability resilience index is also derived based on multi-criteria decision analysis, such as the MIVES methodology (Alarcon et al. 2011; Trentin et al. 2019). In this method, the normalized indicator value obtained from the value function (V_{ind}) is multiplied by its respective weight (W_{ind}) assigned in Phase 2, step 3. The summation of the products of all the indicators and their respective weights gives the value of the criterion (V_{cri}) corresponding to each requirement level. The summation of the products of criteria value (V_{cri}) and their corresponding weight (W_{cri}) gives the value of requirement (V_{req}). Lastly, the summation of the products of V_{req} and their corresponding weight (W_{req}) give the integrated sustainability resilience index (V_{final} or *SRI*).

1.3.4 Phase 4: Decision-Making

The final step is decision-making. In this step, the calculated integrated sustainability resilience indicators for the design(s) assessed are compared so that a decision can be made to implement the best design option. It is recommended that stakeholders' views be considered whenever possible regarding the results of the sustainability resilience indices for the different design options. This would allow for a participatory and comprehensive review of the designs before a choice is made. However, not all projects may require such thorough review, or it may not be feasible/manageable to do so. If that is the case, the users/evaluators can directly compare the integrated sustainability resilience index(es) and proceed to choosing the best technical design option to implement. It should be noted that the "best" option may be subject to external project considerations like budget and social acceptance of the solution. The final decision is sometimes subjected to the stakeholders' preference and relative importance of the environmental, economic, and social aspects for the project.

1.4 Sustainability Indicators and Tools

1.4.1 Environmental Requirements

The environmental requirement mainly focuses on assessing the environmental consequences of the project or activity. Environmental impacts differ from one process/activity/project/industry to another. One of the commonly used variables for assessing environmental impact is greenhouse gas (GHG) emissions, as they are more readily quantified than other impacts (Khasreen et al. 2009). However, there are a range of variables, such as ozone depletion, water consumption, eutrophication, and land use, which gauge the environmental impacts of project activities. Environmental impact assessment variables are mainly those that capture the consequences and effects of natural processes and human activities on the environment (UNEP 2015). There are various tools that have been developed to assess environmental impacts, such as the Environmental Impact Assessment (EIA) tool (IISD 2016), which are based on concepts of sustainable development and focus on preservation of natural resources and environment for future generations. Other tools include World Wide Fund for Nature and the American Red Cross's Environmental Stewardship Review for Humanitarian Aid (ESR) tool designed to evaluate environmental impacts of humanitarian aid projects; Rapid Environmental Impact Assessment in Disaster (REA) tool designed to assess potential environmental impacts in disaster situations (Kelly 2005); and LCA, which traces a product's or activity's environmental impacts from resource extraction through waste disposal and assesses both energy usage and pollution generation (UNEP 1996); SEFA, which focuses on analyzing environmental footprint for environmental cleanup (USEPA 2019); and SiteWise^{TM}, which assesses the environmental footprints of contaminated site remediation alternatives (Bhargava and Sirabian 2011). Each tool has its own set of variables to assess the environmental impacts; for example, SEFA uses variables such as on-site NOx, SOx and PM_{10} emissions, total GHG emissions, on-site water use, on-site refined materials use, etc. In addition, most of these tools are quantitative, meaning the impacts can be quantified in absolute numbers.

Each requirement is subdivided into criterion, and each criterion into indicator, as shown in Fig. 1.4. The previous version of our framework QUALICS derived the environmental impact variables from the well-established tools such as LCA. Although these tools are exceptional in quantifying the environmental impacts, their accessibility and availability of necessary databases are the main hurdles to the users. To address this challenge and make the framework accessible to a wide group of users, the TQUALICSR used the tiered approach described earlier in the framework description. Tier 1 comprises a set of indicator variables, which are based on simple considerations and provides a checklist similar to the Envision^{TM} self-assessment checklist (ASCE 2020).

The indicator variables in Tier 1 are chosen based on simple considerations which can be answered as "Yes" or "No" questions. The variables under criteria and indicators can vary depending upon the type of projects. For example, the variables for a structural construction project may be entirely different from a contamination remediation project. Hence, the variables are chosen which best define the project and its life cycle stages. The variables can be derived from the already available tools and frameworks, if needed. For example, for environmental remediation projects, the variables can be derived from resources such as the Standard Guide for Greener Cleanups (ASTM E2983-16), and ITRC's Green and Sustainable Remediation Framework (GSR-1, ITRC 2011), SEFA, and SiteWise^{TM}, which are readily available to users.

Tier 2 is designed to assess environmental impacts in a relatively more comprehensive manner than Tier 1 and hence requires a better understanding of the life cycle stages of the project. For example, regarding GHG, if the project involves vehicular activity, then ratings can be developed based on the CO_2 emission potential of fuel alternatives. Fuels with high CO_2 emissions potential, such as gasoline (USDE 2021) can be given the lowest rating, as it will have the highest negative environmental impact, and vehicles using green fuels, such as all-electric power, can be given the highest rating.

Tier 3 indicators focus on absolute quantification of the environmental impacts of the life cycle stages. Tier 1 and Tier 2 focus more on qualitative variables, while Tier 3 focuses on quantitative variables and hence uses various advanced tools to define and quantify environmental variables, such as LCA framework (ISO 14040), Sitewise™, SEFA (USEPA 2019), and Streamlined LCA (Todd and Curran 1999), for environmental impact quantification. There are many LCA-based tools in use for environmental impacts assessment. Some of the currently available LCA tools include SimaPro, GaBi, oneClickLCA, and openLCA. In addition, there are other tools for environmental impact measurement, such as Ecochain, SEFA, and SiteWise™.

One additional criterion of "Resilience" has been included in each criterion category to holistically incorporate resilience into sustainability. Like other environmental indicator variables, a tiered approach should be adopted to quantify resilience variables.

1.4.2 Social Requirements

Normally, society and humans are the direct recipients of environmental and economic consequences of any project, and yet, they are rarely considered in the decision-making process (Atanda 2019; Climent Gill et al. 2018). For a system to be sustainable, it is of the utmost importance to ensure sustainability within the social requirement, and not just the environmental and/or the economic requirement. Moreover, impacts are all interlinked, meaning environmental impacts will affect social aspects, which, in turn, will lead to economic impacts. In the TQUALICSR framework, social sustainability is incorporated through social requirements, which are subdivided into criteria and each criterion into indicators. Indicators can be grouped under Tier 1, Tier 2, or Tier 3 based on the quantification approach or information available.

It is crucial to identify social sustainability indicators in order to quantify them. There have been several attempts to quantify social sustainability (Reddy et al. 2014a; b; Popovic et al. 2017; Hossain et al. 2018; Atanda 2019). The social criteria should be chosen based on the type of the project, which reflects the socio-environmental and socio-economic impacts in a broader sense. Tier 1 social impact indicators can be quantified from simple considerations as done in environmental impacts. Tier 2 social indicators are quantified based on the rating system depending on the level of impact the project has on the society. For example, if the project does not affect the housing needs of the society, then it can be assigned a rating of 4; if housing needs are neglected, it may be assigned a rating of 1. Tier 3 social indicators will be quantified based on surveys (e.g., Qualtrics, Google form) among experts and stakeholders or information available from established factsheets. For example, if a project involves relocating people from the project location, then the number of people relocated may be used as an indicator of the negative social impact.

1.4.3 Economic Requirements

The economic sustainability assessment must consider both direct and indirect costs and benefits associated with the project, with a LCC perspective. LCC is applied to a specific project boundary, according to the goal and scope of the economic sustainability assessment, which can be the same used in the LCA. For semi-quantitative and quantitative assessments, the cost should be normalized using the same functional unit as the LCA, and an appropriate discount rate should be applied over the project's lifetime. When applied properly during the technical design stage of the project, LCC can lead to significant total cost savings (Frank 1997).

The economic indicators are broken down into two major types: direct cost and benefits, and indirect costs and benefits. The direct cost and benefits categories can be divided into research and development (R&D), capital, operational and maintenance (O&M), transportation, and end-of-life costs, and revenues. The indirect costs, in this context, are external costs and potential benefits, including technological, environmental, and social impacts, such as the social cost of carbon, willingness-to-pay, and potential cost savings. The indirect costs may also include the monetized values of the environmental impacts or monetary value of the non-marketed goods, such as payment for ecosystem services, and emission permits (USGAO 2020; Arendt et al. 2020; Pizzol et al. 2015). Other indirect costs may include costs for land use, embedded energy, employment opportunities, and time optimization. In the TQUALICSR framework, environmental and social costs should always be accounted as indirect costs. The economic resilience criterion is also mandatory in the analysis, as presented in Fig. 1.4. Such resilience indicators can include the impact of external funding support, inflation, land and real estate costs, community buying power, utilities, and transportation costs.

Economic sustainability indicators are also evaluated using the tiered approach as illustrated in Fig. 1.1. Tier 1 economic sustainability assessment can be performed in the early stage of the project to evaluate design options when cost information is not yet available. The incurrences of expected types of costs during all life cycle stages are assessed providing yes or no answers, with attributed values of 1 and 0, respectively. In Tier 2, the cost breakdown is determined, and a semi-quantitative or major cost-drivers analysis is performed, such as the streamlined life cycle costing (SLCC) presented by Roh et al. (2018). Tier 2 assessments can use qualitative data and simplifying assumptions based on expert opinions and historical data to attribute ratings for each cost input (even in terms of relative cost), and it can be very beneficial at the planning and developing stages to evaluate project alternatives. Finally, a Tier 3 economic sustainability assessment uses cost data and mathematical models to develop a discounted cash flow analysis for LCC, using an adequate discount rate (e.g., loan interest and inflation rates) to calculate the Net Present Value (NPV) or the Equivalent Annual Annuity (EAA) $\pm$ uncertainties. Indirect costs can be calculated using available data for the scenario analyzed. For example, a marketing analysis or survey in the region where the project will be implemented can determine the stakeholders' willingness to pay (Zalejska-Jonsson et al., 2020; Thormann and

Wicker, 2021). The social cost of carbon for a project in the United States can be calculated using the value for social cost of carbon estimated by the USEPA (2017) multiplied by the carbon emission output from any carbon footprint analyzer or LCA analysis. Monetary valuation methods, such as Stepwise2006 (Weidema 2009) and LIME (Life-cycle Impact Assessment Method based on Endpoint modeling) (Itsubo et al. 2004) can be used with caution as the base model to calculate social and environmental costs, by understanding and adapting the method's assumptions, simplifications, regional specificities, and the calculations adopted, as explained by Arendt et al. (2020) and Pizzol et al. (2015).

1.5 Conclusion

The tiered analysis can be applied to different project complexities and different stages of the project. This framework is being applied to different case studies, and results will be presented in future publications. Although the framework provides a comprehensive approach to assess sustainability including resilience considerations, there are some challenges associated with the availability of and variability in data, methodologies, and assumptions. There is not a standardized tool to quantify the triple bottom line resilience of a system; thus, qualitative, and semi-quantitative approaches based on rating systems are usually employed. Rating systems results, however, may vary depending on the user's assumptions, expertise, and biases. Similarly, the integrated sustainability resilience index value can be sensitive to the indicators and weightages assigned to the various impact categories in the framework; therefore, it is important to take into consideration the results from the risk assessment and execute this step with input from all stakeholders and multidisciplinary experts. When comparing different projects or solutions, it is recommended that the same user or group of users use the framework to assess different alternatives to avoid biases associated with the assumptions. In addition, social impacts are more abstract, and quantifying these impacts involve expert judgement, which may again have some biases that need to be accounted for in the analysis. Nevertheless, this framework provides a rational approach to considering life cycle triple bottom line environmental, social, and economic impacts complemented with resilience, which otherwise would have been generally ignored in the standard engineering practice.

References

Alarcon B, Aguado A, Manga R, Josa A (2011) A value function for assessing sustainability: application to industrial buildings. Sustainability 3(1):35–50. https://doi.org/10.3390/su3010035

Arendt R, Bachmann TM, Motoshita M, Bach V, Finkbeiner M (2020) Comparison of different monetization methods in LCA: a review. Sustainability 12(4):10493. https://doi.org/10.3390/su122410493

ASCE (American Society of Civil Engineers) (2020) Envision: Checklist/Scorecard. https://www.asce.org/envision/

ASTM (American Society for Testing and Materials) (2016) Standard guide for greener cleanups. ASTM International E2893-16e1, West Conshohocken, PA. http://www.astm.org

ASTM (American Society for Testing and Materials) (2019) Standard guide for general principles of sustainability relative to buildings. ASTM International E2432-19, West Conshohocken, PA. http://www.astm.org

ASTM (American Society for Testing and Materials) (2020) Standard guide for integrating sustainable objectives into cleanup. ASTM International E2876-13, West Conshohocken, PA. http://www.astm.org

Atanda JO (2019) Developing a social sustainability assessment framework. Sustain Cities Soc 44:237–252. https://doi.org/10.1016/j.scs.2018.09.023

Ayyub BM (2020) Infrastructure resilience and sustainability: definitions and relationships. ASCE-ASME J Risk Uncertain Eng Syst A Civ Eng 6(3):02520001. https://doi.org/10.1061/AJRUA6.0001067

Bhargava M, Sirabian R (2011). SiteWise™ version 3 user guide. NAVFAC Engineering Service Center, Port Hueneme UG-NAVFAC-EXWC-EV-1302

Burroughs S (2017) Development of a tool for assessing commercial building resilience. Proc Eng 180:1034–1043. https://doi.org/10.1016/j.proeng.2017.04.263

Climent Gill E, Aledo A, Vallejos A (2018) The social vulnerability approach for social impact assessment. Environ Impact Assess Rev 73:70–79. https://doi.org/10.1016/j.eiar.2018.07.005

Damians IP, Bathurst RJ, Adroguer EG, Josa A, Lloret A (2016) Sustainability assessment of earth-retaining wall structures. Environ Geotech 5(4):187–203. https://doi.org/10.1680/jenge.16.00004

Frank M (1997) Life cycle cost estimate. Cost estimating guide, chap 23. U.S. Department of Energy DOE G 430.1-1. 03-28-97

Hossain MU, Poon CS, Dong YH, Lo IM, Cheng JC (2018) Development of social sustainability assessment method and a comparative case study on assessing recycled construction materials. Int J Life Cycle Assess 23(8):1654–1674. https://doi.org/10.1007/s11367-017-1373-0

IISD (International Institute for Sustainable Development) (2016) Environmental impact assessment training manual. International Institute for Sustainable Development, IISD.org. https://www.iisd.org/learning/eia/wp-content/uploads/2016/06/EIA-Manual.pdf

ISO 14040 (2006) Environmental management–life cycle assessment–principles and framework: International Organization for Standardization

ITRC (Interstate Technology & Regulatory Council) (2011) Green and sustainable remediation: a practical framework. GSR-2. Interstate Technology & Regulatory Council, Green and Sustainable Remediation Team, Washington, DC. http://www.itrcweb.org

Itsubo N, Sakagami M, Washida T, Kokubu K, Inaba A (2004) Weighting across safeguard subjects for LCIA through the application of conjoint analysis. Int J Life Cycle Assess 9:196–205. https://doi.org/10.1007/BF02994194

Kelly C (2005) Guidelines for rapid environmental impact assessment in disasters. Benfield Hazard Research Centre, University of London and CARE International. https://www.humanitarianlibrary.org/sites/default/files/2013/06/Benfield,%20UCL,%20CARE,%20Guidelines%20for%20REIA%20in%20Disasters.pdf

Khasreen MM, Banfill PF, Menzies GF (2009) Life-cycle assessment and the environmental impact of buildings: a review. Sustainability 1(3):674–701. https://doi.org/10.3390/su1030674

Kumar G, Reddy KR (2020) Addressing climate change impacts and resiliency in contaminated site remediation. J Hazard Toxic Radioact Waste 24(4):04020026. https://doi.org/10.1061/(ASCE)HZ.2153-5515.0000515

Lai YJ, Liu TY, Hwang CL (1994) TOPSIS for MODM. Eur J Oper Res 76(3):486–500. https://doi.org/10.1016/0377-2217(94)90282-8

Pizzol et al (2015) Monetary valuation in life cycle assessment: a review. J Clean Prod 86:170–179. https://doi.org/10.1016/j.jclepro.2014.08.007

Popovic T, Kraslawski A, Barbosa-Póvoa A, Carvalho A (2017) Quantitative indicators for social sustainability assessment of society and product responsibility aspects in supply chains. J Int Stud 10(4). https://doi.org/10.14254/2071-8330.2017/10-4/1

Reddy KR, Kumar G (2018a) Addressing sustainable technologies in geotechnical and geoenvironmental engineering. In: Geotechnics for natural and engineered sustainable technologies. Springer, Singapore, pp 1–26. https://doi.org/10.1007/978-981-10-7721-0_1

Reddy KR, Kumar G (2018b) Green and sustainable remediation of polluted sites: new concept, assessment tools, and challenges. In: Proceedings of the 16th Danube—European conference on geotechnical engineering, vol 2(2–3), pp 83–92. https://doi.org/10.1002/cepa.663

Reddy KR, Sadasivam BY, Adams JA (2014a) Social sustainability evaluation matrix (SSEM) to quantify social aspects of sustainable remediation. In: International conference on sustainable infrastructure, Long Beach, CA, November 6–8, ASCE, Reston, VA. https://doi.org/10.1061/978 0784478745.078

Reddy KR, Sadasivam BY, Adams JA (2014b) Social sustainability evaluation matrix (SSEM) to quantify social aspects of sustainable remediation. In: ICSI 2014: Creating infrastructure for a sustainable world, pp 831–841. https://doi.org/10.1061/9780784478745.078

Reddy KR, Kumar G, Chetri JK (2019) Quantitative assessment of life cycle sustainability (QUALICS): application to engineering projects. In: Indian geotechnical conference: geotechnics for infrastructure development & urbanization (GeoINDUS). Springer Nature, pp 111–125. https://doi.org/10.1007/978-981-33-6590-2_9

Roh S, Tae S, Kim R (2018) Development of a streamlined environmental life cycle costing model for buildings in South Korea. Sustainability 10(6):1733. https://doi.org/10.3390/su10061733

Roostaie S, Nawari N, Kibert CJ (2019) Sustainability and resilience: a review of definitions, relationships, and their integration into a combined building assessment framework. Build Environ 154:132–144. https://doi.org/10.1016/j.buildenv.2019.02.042

Saaty TL (2008) Decision making with the analytic hierarchy process. Int J Serv Sci 1(1):83–98. https://doi.org/10.1504/IJSSCI.2008.017590

Thormann TF, Wicker P (2021) Willingness-to-pay for environmental measures in non-profit sport clubs. Sustainability 13(5):2841. https://doi.org/10.3390/su13052841

Todd JA, Curran MA (1999) Streamlined life cycle assessment: a final report from the Setac-North America streamlined LCA workgroup. In: Society of environmental toxicology and chemistry. SETAC Press, Pensacola, FL, 6/99, p 31

Trentin AWS, Reddy KR, Kumar G, Chetri JK, Thome A (2019) Quantitative assessment of life cycle sustainability (QUALICS): framework and its application to assess electrokinetic remediation. Chemosphere 230:92–106. https://doi.org/10.1016/j.chemosphere.2019.04.200

UNEP (United Nations Environment Program) (1996) Life cycle assessment: what it is and how to do it. United Nations Environment Program Industry and Environment, Paris, France, 1996, p 23. http://www.sciencenetwork.com/lca/unep_guide_to_lca.pdf

UNEP (United Nations Environment Program) (2015) An introduction to environmental assessment. United Nations Environment Program World Conservation Monitoring Centre (UNEP-WCMC), Cambridge, UK. https://www.wedocs.unep.org/bitstream/handle/20.500.11822/9731/-An_int roduction_to_environmental_assessment-2015UNEP-An_introduction_To_environmental_A ssessme.pdf.pdf?sequence=3&isAllowed=y

USDE (US Department of Energy) (2021) Alternative fuel data center. https://www.afdc.energy.gov/vehicles/electric_emissions.html. Accessed 18 May 2021

USEPA (United States Environmental Protection Agency) (2017) The social cost of carbon: estimating the benefits of reducing greenhouse gas emissions. https://www.19january2017snapshot.epa.gov/climatechange/social-cost-carbon_.html. Accessed 18 May 2021

USEPA (United States Environmental Protection Agency) (2019) Spreadsheets for environmental footprint analysis (SEFA): green remediation focus. https://www.clu-in.org/greenremediation/SEFA/. Accessed 18 May 2021

USGAO (US Government Accountability Office) (2020) Social cost of carbon: identifying a federal entity to address the national academies' recommendations could strengthen regulatory analysis.

A report to congressional requesters, June 2020, GAO-20-254. https://www.gao.gov/assets/gao-20-254.pdf
Weidema BP (2009) Using the budget constraint to monetarise impact assessment results. Ecol Econ 68:1591–1598. https://doi.org/10.1016/j.ecolecon.2008.01.019
Zalejska-Jonsson A, Wilkinson SJ, Wahlund R (2020) Willingness to pay for green infrastructure in residential development—a consumer perspective. Atmosphere 11(2):152. https://doi.org/10.3390/atmos11020152

Chapter 2
The Effect of COVID-19 on Public Transportation Sectors and Conceptualizing the Shifting Paradigm: A Report on Indian Scenario

M. B. Sushma and S. Prusty

2.1 Introduction

The transportation system is the vein of the country's economy, and transportation serves as a vital component to the economic sector by allying with it either directly or indirectly. The transportation system, mainly the public transportation sector acts as an essential aspect in economic growth. It provides people with mobility and access to various essential and recreational opportunities in communities. Public transportation systems (PTS) include various transit options such as buses, light rail, and subways. Public transportation options aids in broader economic, land-use planning, promote business opportunities and create a sense of community through transit-oriented development. With a mass population of around 1.3 billion people, India has one of the world's largest and most diverse transport sectors. India boasts of having such transport sectors as to cater to the needs of its population. For the financial year (fiscal year) of 2018–2019, the transport sector's overall contribution, including storage, communication, and service-related broadcasting, was about 8.84% to India's gross value added (GVA) at the current price (MOSPI 2019). In this, road transportation had a significant share with 4.17% of the overall share, followed by railways (0.91%), waterways (0.13%), and airways (0.09%) (Statista Research Department 2021). During the fiscal year 2020, the gross value added by road transportation services in India was the highest at approximately 58.2 billion US dollars. This was an increase from the previous FY.

As per the country's economy is considered, two-thirds of India's population lives in poverty, i.e., 68.8% of the people in India survive with less than $2 a day (NITI Aayog 2020). Thus, this shows that a large proportion of the population depends

M. B. Sushma (✉)
Department of Civil Engineering, Kakatiya Institute of Technology and Science, Warangal, India

S. Prusty
Palladium Consulting India Pvt. Ltd, Pune, India

K. R. Reddy et al. (eds.), *Advances in Sustainable Materials and Resilient Infrastructure*, Springer Transactions in Civil and Environmental Engineering,
https://doi.org/10.1007/978-981-16-9744-9_2

on the public transportation service. Besides, this tourism sector has a massive dependency on the public transportation service. In 2019, the contribution of travel and tourism was 277.1 billion US dollars to India's economy. However, due to the outbreak of the COVID-19 (Cui et al. 2019), there is a significant change in the travel scenarios. Presently, public transportation services face a tremendous economic crisis and need to implement various changes to cope with these challenging pandemic situations (TERI 2020).

The world has already faced pandemic situations in 2003 for SARS-CoV, in 2012 for MERS-CoV in 2012 (Chen et al. 2020; Lai et al. 2020; Shereen et al. 2020). But the effect of COVID-19 is most hazardous, and it has not only hindered public lives but had a significant negative impact on the world's economy. It had shut down the lives of many by sweeping of their jobs, impacting various economic sectors, and bringing the long-term behavioral effect on people's lives. Due to COVID, many social policies such as social distancing and lockdowns had affected the demand–supply or production–consumption chain for an indefinite period. It had paused various retail and hospitality sectors (Chauhan 2020). Along with this, the shared mobility like Ola, Uber, Meru, etc., considered as the component of the public transportation system, is also affected severely.

This study aims to understand the financial and operational impacts on the Indian Public transport sector due to the outbreak of the novel coronavirus. This study also provides a report on the choice of public transport during and after the lockdown period. The paper also discusses the people's consequences and changes in travel behavior patterns, which will help the public transport service providers understand the challenges and prepare to address them while resuming their services. The study is purely based on the primary data collected from the surveys, done through a predesigned Google form. Through the Google form survey, a total of 178 participants had shared their views. These participants were of various ages, gender, and professions from various parts of the country. The secondary data was collected from the information provided by the International Association of Public Transport (UITP), World Health Organization, Ministry of Health and Family Welfare (India), and the secondary information was collected from various news reports and research articles, and social media platforms. Thus, the primary objective of this study is to support the transportation providers in improving the transport scenario to a sustainable level post-COVID-19.

2.2 Level of Lockdown and Unlock Phases in India

The lockdown strategy in India led to a complete shutdown of the entire transportation system that included airports, Indian Railways, Metros, buses, and all forms of private vehicular movements. The nationwide lockdown was much tighter and strict than anywhere in the world. Table 2.1 distinctly refers to each lockdown period defined by the Government of India (GoI). The unlocking phase aims to aid the Indian economy to revive, which has registered a negative trend of 23.9%. It is expected

Table 2.1 National-level lockdown Phases

Phases of lockdown		
Phase 1	Nationwide lockdown	25 March–14 April 2020 (21 days)
Phase 2		15 April–3 May 2020 (19 days)
Phase 3		4 May–17 May 2020 (14 days)
Phase 4		18 May–31 May 2020 (14 days)

Source Ministry of Home Affairs (MHA), GoI

that the reopening of various economic sectors with the unlock phase would help the nation's economy to revive gradually to pre-COVID levels in the immediate future.

As we approach the end of July 2021, some restrictions continue in several states on various economic sectors, and it varies depending on the COVID-19 scenario. Such as still large gathering is prohibited. In some places, tourist destinations are being reopened in phases by the respective states depending on the spread of COVID-19. Social/religious/political functions shall be allowed with a maximum of 500 persons subjected to government norms. Wearing a mask, maintaining social distance, and provision of thermal scanning and hand sanitizer were mandatory.

2.2.1 *Lockdown-Through Lens of an Optimist*

In an attempt to curb the spread of the novel virus, the government had imposed a stringent and mandatory policy of 'stay at home. However, this lockdown policy positively affected the people, their happiness, and the environment. Restricting people's mobility has aided in reducing the rate of accidents (Kumar and Sankar 2020). The rate of road accidents per day has reduced drastically compared to the pre-lockdown period, as shown in Fig. 2.1. In quarterly data provided by the states and UTs to the Supreme Court committee on road safety reported that there is a fall of 68% in the death cases caused by road accidents between 24 March and 31 May 2020, as

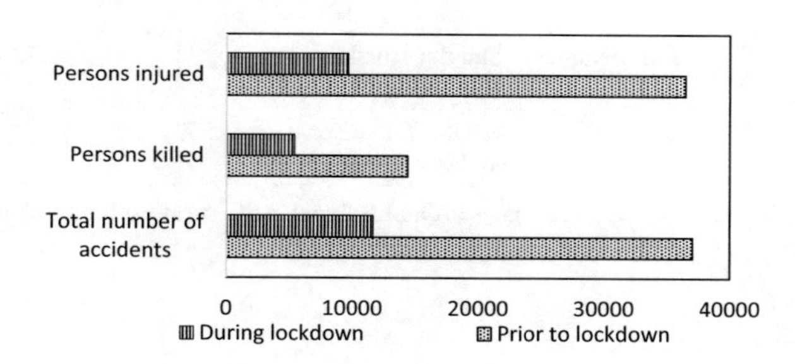

Fig. 2.1 Number of road accidents, injuries, and deaths during lockdown (24 March and 31 May 2020) and before lockdown (24 March and 31 May 2019) period

compared to the same period in 2019, i.e., the deaths due to road accidents have reduced to 11,645 from 37, 018 (TOI 2021).

Besides this, there is a considerable change in the environment, as the transportation sector vastly contributes to a significant reduction in the average emission level. It is noticed that the air pollution level has dropped to levels in the last few decades. There is a sharp decline in $PM_{2.5}$, PM_{10}, NO_2, and CO levels during the lockdown period in significant cities like Delhi and Mumbai due to strict transport restrictions, construction works, and industries. Table 2.2 shows the air quality index (AQI) for various locations for pre and post-lockdown. It states that the lockdown had helped the environment to revive from the critical pollution issues. Thus, the pandemic could be considered a blessing in disguise. However, the relaxations in the unlock phase have accelerated the levels significantly (Radhakrishnan et al. 2020).

Table 2.2 Weekly AQI data (before and during lockdown period)

Sl. no.	State name	Station name	AQI before lockdown	AQI after the declaration of lockdown			
			14 January 2020	24 March 2020	31 March 2020	7 April 2020	14 April 2020
1	Andhra Pradesh	Anand Kala Kshetram—APPCB	216	62	69	29	37
2	Bihar	Collectorate, Muzaffarpur—BSPCB	367	266	134	65	84
3	Chandigarh	Sector-25, Chandigarh—CPCC	75	35	38	41	45
4	Delhi	IHBAS, Dilshad Garden, Delhi—CPCB	301	99	24	37	53
5	Haryana	Sector-16A, Faridabad—HSPCB	315	151	121	121	78
6	Karnataka	Peenya, Bengaluru—CPCB	143	105	50	57	63
7	Maharashtra	Chhatrapati Shivaji Intl. Airport, Mumbai—MPCB	122	94	68	55	67
8	Punjab	Model Town, Patiala—PPCB	106	49	26	43	51
9	Tamil Nadu	Alandur Bus Depot, Chennai	172	42	40	29	27
10	Telangana	ICRISAT Patancheru, Hyderabad—TSPCB	130	64	64	47	77

Source NAQI (National Air Quality Index)—Central Pollution Control Board (CPCB) 2020

2.3 Economic Sector During Lockdown Period

One of the most affected economic sectors is the transportation service, experiencing a demand-cum-supply shock in this pandemic situation. Cities across the country had to impose massive restrictions on the travel of the public to control the transmission of this coronavirus. Thus, this significantly affected the count of ridership in public transportation sectors. The impacts of COVID-19 have surged over the transportation activities very severely. There is a significant reduction in travel and commercial activities. In India, the GVA from transportation sectors is 5.3%, where road transportation has a significant share with 4.17% of the overall share, followed by railways (0.91%), waterways (0.13%), and airways (0.09%). Motor transport accounts for employment of 1.3 million, i.e., 8.7% of the Industrial Employment in 2017 (Kumar and Sankar 2020).

This chapter focuses on the mass transportation system, which primarily depends on the people's demand, unlike the freight transportation system, which depends on goods and commodities. The transmission rate in mass transportation is higher than that of transportation systems carrying the goods or freight transportation system, as the virus transmission is significant through people's interactions. As per WHO, the virus transmits if a person directly contacts the infected person's saliva droplets (WHO 2020). Thus, there is a higher risk of transmission of the virus through human contact, and PTS tries to accommodate the public in a space where the area is confined and has limited ventilation. Moreover, there is no such infrastructural system where the access of the infected person can be controlled or limited. It is also exceedingly difficult to control the transmission as this system has a range of joint surfaces to contact, such as windows, seats, receipts, doorknobs, etc. As public transport is a primary service provider for mobility, it also played a significant role in pandemics by providing access to healthcare facilities (Leigh 2020). Thus, the key aspects that need to be focused on to continue the operation of public transportation services are focusing on their pandemic plan efforts on staff and safety and identifying employees who could act as backup for critical positions (UITP 2020). However, the exact span and the severity of COVID-19 are unpredictable at the current position. The country's economic status is greatly affected by the aversion behavior of the people in this pandemic. This behavior captures people's choice for travel and other social activities like preferences and choice of travel. To prevent getting in contact with the virus infections, people also have no alternative but to limit the travel for essentials only, resulting in a shortfall in revenue in various sectors, including transportation. Some potential economic channels affected by the pandemic outbreak for low- and middle-income group countries are provided by Evan and Over (2020), as shown in Fig. 2.2. The government also mandates the lockdown, social distancing policy, and some institutional protective decisions to minimize the spread of the infection that results in a shortfall of the revenue collection from different economic sectors, including manufacturing, retail services, trade, and transport.

The negative impact of COVID-19 has shown its devastating effect in many developing countries and so in India. Almost all the economic sectors in India have suffered

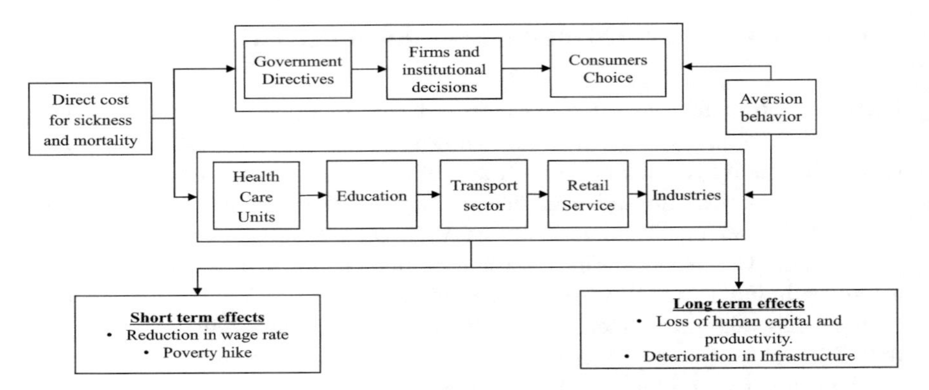

Fig. 2.2 Effect of a pandemic on the broad channels of the economy (*Source* Evan and Over 2020)

severely due to the pandemic, and the transportation sector is not an exception. India has long been struggling with the negative environmental impact, noise pollution, and carbon emission caused by the transportation system. These critical issues are due to the inefficiencies of the transportation system, inadequately handling the congestion level, inadequate public transportation system, inefficient integration of the multimodal system, deteriorated footpaths, and an exponential increase in the number of public vehicles. The imposition of lockdown has brought down the rate of pollution and helped nature to revive. Besides this, the rate of human mortality due to accidents and mishaps has also reduced drastically. However, the revenue from the transportation system had to compromise significantly. So, in this pandemic situation government is trying to improve the condition by implementing different development measures that can make the transportation system more efficient and sustainable. The outbreak of this pandemic situation had created much more significant challenges in the transportation sector, especially for those cities where high travel demand is prevailed (Shakti Foundation 2016; TERI 2020).

Figure 2.2 shows the frequency of people visiting outside and the duration of stay at the places during pre-lockdown, lockdown, and partial lockdown from February to May 2020 in India. It shows that the social mobility behavior is dynamic, and its pattern and frequency change in response to socioeconomic security. In Fig. 2.3, the baseline is considered the median value of the corresponding days of all the five weeks, i.e., from 3 January to 6 February 2020. The visits frequency in Fig. 2.3 shows that the maximum visit corresponded to essential commodities such as groceries, workplaces, and pharmacies. The least visited places include parks and recreation spaces, and the visit to transit stations has not yet shown any sudden increase as people prefer to use private transportation.

The majority of revenue in India is collected from the transportation sector, as shown in Table 2.3. The lockdown policies of the government promoted to reduce COVID-19 transmission have severely impacted the demand–supply chain (Singh and Neog 2020). It had put the entire transportation system into a standstill mode that is crucially affecting the Micro, Small, and Medium Enterprises (MSMEs), which

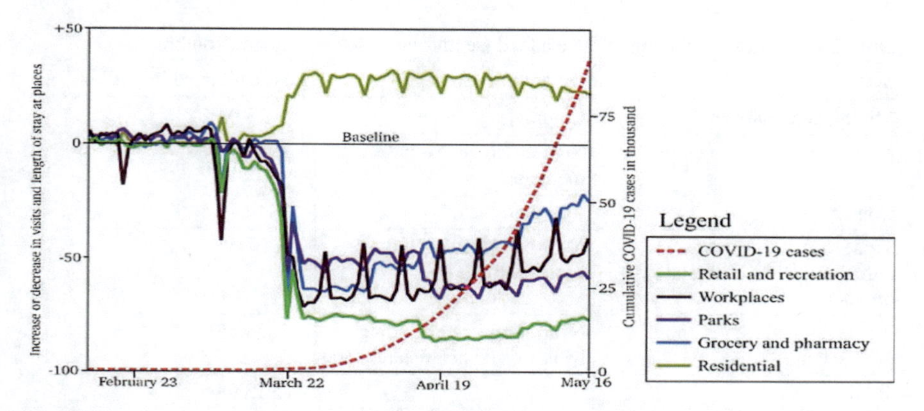

Fig. 2.3 Trends of mobility in India during coronavirus 2019 (*Source* Radhakrishnan et al. 2020)

Table 2.3 Gross revenue trends over the years in India

Years	Railways (million $)	Roadways (toll fees) (million $)	Airways (million $)
FY17	2538.64	946.76	342.44
FY18	2500.12	1177.75	407.13
FY19	2593.52	1253.76	377.29
FY20	2514.37	366.41	30.10
FY21(P)	1969.44	NA	NA

NA: Not available. (P): Provisional
Data Source Singh and Neog (2020)

are more precarious to the demand market. In 2020, exports were 2130 million USD, compared to 3.32 billion in March 2019, exhibiting a negative growth of (−) 34.57%. The exports were 21.718 billion USD in March 2020, compared to 31.02 billion in March 2019, registering a negative growth of (−) 29.98%. The decline in exports is due to the ongoing global slowdown, which got aggravated due to the current COVID-19 crisis.

During the lockdown period, in May 2020, the count of people availing the public transportation system had reduced to 40,00,000, compared to the pre-lockdown period count, which was 7,00,00,000 (Kumar and Sankar 2020). Thus, there is a drop of about 90% in mobility from the lockdown period to the pre-coronavirus situation. Also, the total volume of daily freight flow had reduced to 1 million tons from the pre-lockdown to lockdown period, i.e., the daily freight flow in the pre-lockdown period was 3 million tons, which had reduced to 2 million tons during the lockdown period. The reduced trend is also observed in the daily passenger volume in non-suburban (intercity) special migrant rail ridership. The volume during the pre-lockdown period was 10 million, which drastically reduced to 0.17 million during May 2020, i.e., the lockdown period.

Table 2.4 Impact on total gross value added (sector-wise) due to pandemic outbreak

Sectors	Disruption severity	Loss estimated in million $
Mining and quarrying	Complete	423.03
Manufacturing	Near-complete excluding medicine	2947.56
Construction	Complete	1446.49
Trade, hotel, transport, communication, and relating broadcasting	Near-complete excluding broadcasting	3302.36
Financial, real estate, and professional services	Near-complete excluding banking and healthcare	3834.55
Total		11,953.98

Data Source Mishra (2020)

Similarly, the daily passenger volume in suburban rail ridership during the pre-lockdown period was 13 million, which completely shattered during the lockdown period with zero counts. The passenger volume had reduced in road and rail transportation system, but a similar pattern was also observed in airways. The average daily domestic-cum-international passengers during the pre-lockdown period were 0.55 million, which was reduced to 0.2 million during the lockdown period.

As per the article published in Business Today (2020), the estimated loss in gross value added (GVA) is approximately 118.7 billion USD, as shown in Table 2.4. The economic sectors like trade, hospitality, and transport sectors together could register a loss of 32.7 billion USD due to the outbreak of COVID-19, which could increase the overall loss to 27.6% (Mishra 2020).

2.4 Impact of COVID-19 on the Public Transport Sector

The report of The Economic Times (2020a) stated that around 2 million people have their livelihoods as the operation of the private transportation sector have been walloped by the nation lockdown, according to the Bus and Car Operators Confederation of India (BOCI). BOCI claimed that 20 thousand operators with 1.5 million buses and maxi-cabs and 1.1 million tourist taxis directly employ one crore people. These private operators need government support in the form of waiver of taxes and interest on loans as many face closure. Furthermore, BOCI has also stated that at least out of the 10 million people, 30–40 lakh will lose their jobs shortly, whereas 2 million people have already lost their jobs.

Furthermore, the suspension of the toll collections by the National Highway Authority India (NHAI) on all the national highways and expressways had a massive impact on the overall revenue collection from the road transport sector (Livement 2021). According to CRISIL research, there is a fall of about 13% in the toll revenue during the 57 day nationwide lockdown (The Economic Times 2020b). International Credit Rating Agency (ICRA) has predicted that there would be a significant fall of about 6.5–8.0% in the toll collection and remittances in FY21.

2.4.1 Road Sector

The Indian Road transport sector carries goods worth 15.1 billion USD a year, which means the monthly business is about 1.2 billion USD. The truck fleet utilization peaked at 85% in March, which was better than the pre-Covid levels but dropped to 70%. The goods transport sector is the backbone of the country's economy. However, it is severely battered by the lockdowns promoted due to the onset of the second wave of COVID-19, especially after the devastating effect of the first pandemic wave. The fresh wave of COVID-19 cases sweeping the country and the resultant lockdown measures implemented by several states have already started hurting transporters. The trade is estimated to suffer 2.4 billion USD of revenue loss in April alone.

2.4.2 Rail Sector

India's rail network is recognized as one of the most extensive rail network systems globally, extended over 1,23,236 km, with 67,956 km, with 13,169 passenger trains and 8,479 freight trains, plying 23 million travelers and 3 million tons (MT) of freight daily from 7,349 stations. The railway network is also ideal for long-distance travel and movement of bulk commodities, apart from being an energy-efficient and economical mode of conveyance and transport. Indian Railways is the preferred carrier of automobiles in the country. The rail network contributes to a significant portion of the country's economy as industrial sectors directly depend on the rail network due to its extensive and efficient connectivity. But the outbreak of the pandemic had shaken the railway sector severely. In the pre-pandemic situation, railway mode was considered one of the ideal modes for long-distance travel. A majority of the Indian population preferred rail over other modes of transportation as it provided extensive connectivity, cheap, and comfort to the passengers. The Railways earns about 6.7 billion USD from the passenger segment and around 17.6 billion USD from freight annually. But due to the pandemic outbreak, there is a drastic reduction in the volume of passengers, i.e., from 8.44 billion in FY 2019, it has reduced to 8.1 billion in FY 20. The freight revenue till the end of August 2020 stood at 5.3 billion USD, 14.6% lower than 6.2 billion USD as of August 2019. After the declaration of the nationwide lockdown, almost all the passenger train service was canceled till the indefinite period and had put the railway sector into static mode. This situation resulted in the decline of the revenue from the passenger section for the upcoming FY. The experts have estimated that the loss that the sector has to bear can go up to 20.1 billion USD for the FY 2020–21, widening the gap by 8.5 billion

USD. Thus, leading to a whopping of 12.24 billion USD if the current gap amount is added to the earlier resource gap of over 3.8 billion USD for FY 2019–20 (Sharma 2020). In a count of FTA in March 2020 was 3,28,462 as compared to 9,78,236 in March 2019, resulting in the lowest growth rate, i.e., −66.4% in the last few decades.

2.4.3 Airway Transport Sector

Airways play a vital role as modern means of transportation, and since the last decade, the aviation industry in India has proliferated. It has emerged as one of the fastest growing industries in India. Currently, India holds ninth position in the world in the civil aviation market. However, during the lockdown period, all the operations of the flights were canceled, and only "Lifeline Udan" flights functioned by the Ministry of Civil Aviation (GoI), which aimed to support people to fight against COVID-19 by transporting medical essentials to the rural and remote locations of the country. The outbreak of COVID-19 had severely affected the Indian aviation industry, with an economic loss of around 3.2 billion USD (Manju 2020, National Aviation Report 2020). Many people have also suffered because of this pandemic situation. According to Bloomberg calculations, about 400,000 airline workers have been fired, furloughed, and lost their jobs due to coronavirus outbreak.

2.4.4 Tourism Sector

The tourism sector in India is also one of the major economic sectors which are severely impacted by the outbreak of the COVID-19. The travel and tourism industry has significant concerns with the public transportation system. There is a significant fall in the tourist count in India, as the arrivals of foreign tourists were constrained, and the volume of it had fallen to 2 million till April 2020. According to the Ministry of Tourism (Govt. of India), during 2018–2019, the total domestic tourist growth was around 25%, whereas there was a growth of 8.9% in the foreign tourist. However, from January to March 2020, a significant negative growth of 22.65% was found in the Foreign Tourist Arrival (FTA) on e-Tourist Visa. It affected the foreign exchange earnings (FEE) from tourists during 2019 and 2020. There was a sharp decline of −64.0%, i.e., 2.2–0.79 billion USD in FEEs from March 2019 to March 2020 (Data source State/Union Territory Tourism Departments, Govt. of India). Table 2.5 shows the growth percentage of FTAs on e-Tourist Visa and Foreign Exchange Earnings (FEEs) from tourism from January 2019 to March 2020.

Table 2.5 FTAs on e-tourist visa and foreign exchange earnings (FEEs) from tourism

	Duration	Amount (billion USD)	Growth rate (%)
FTAs on e-tourist visa	January–March 2019	432.47	−22.6
	January–March 2020	334.8	
	March 2019	133	−66.4
	March 2020	44.67	
Foreign exchange earnings (FEEs) from tourism	January–March 2019	7.1	−15.6
	January–March 2020	6.01	
	March 2019	2.2	−64.10
	March 2020	0.79	

Data Source Ministry of Tourism, Govt. of India (2020)

2.5 Effect of the Second Wave of COVID-19 in India

The first wave had already taken a toll on India's GDP in 2020–21. On 31 May, the Indian government released the data for GDP that during the FY 2020–21, GDP contracted by 7.3%. It is the most severe contraction from the time India got its independence. The reasons behind this trajectory are obvious—lockdown leading to the closing of business units, increasing unemployment rate, and a significant decline in domestic consumption (Economic Times). Nearly 6 months after the peak of the first wave in September 2020, the country was again under the second wave of the pandemic situation. The second wave started in the west with Maharashtra, went up north, and then to the country's south. This spread journey made a national lockdown economically suboptimal. Many experts predicted that the economic damage would not be as bad as the first wave in 2020. There were two primary reasons behind the assertion. First, India had vaccines against the virus, and second, no nationwide lockdown was imposed. However, after three months of the first sign of the second wave emerged, India started struggling to vaccinate its vast population. Strict lockdowns remain imposed to prevent virus transmission in almost all parts of the country. It resulted in the economic growth predictions were hampered drastically. The second wave has had a devastating impact on India's economy. Even the transport sector is hit hard by the second wave of COVID-19. For FY 2021–22, the Reserve Bank of India has anticipated growth of 10.5%. The rating agencies across the globe have downgraded it to 8.2% due to the impact of the second wave of COVID-19.

2.6 Post-Lockdown Scenario

The government of India has initiated unlock phase in many regions of the country. However, the fear of virus transmission persists, and the social distancing policy needs to be followed to prevent the spread of the virus. Hence, it is expected that the

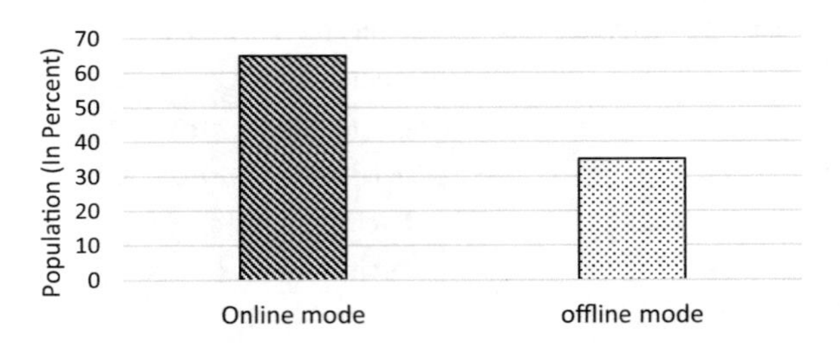

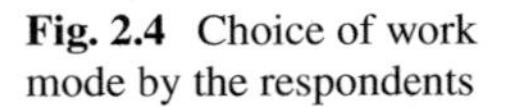
Fig. 2.4 Choice of work mode by the respondents

share of individual transport such as bikes, walking, cars will increase in response to social distancing and hygiene concerns, and the range of expectations is from zero to more than 50%, as shown in Fig. 2.4. Also, in a survey has been conducted among the private and the public transport service providers by the World Bank Group and UITP, an International Association of public transport reports that over 60% of the PTS operators believe that the service and demand may not exceed 50% of the pre-lockdown period.

Before the pandemic, the conventional wisdom was that office was critical to productivity, culture, and winning the war for talent. The idea of work from home (WFH) was not considered practicable, especially for a country like India with a huge population. Companies competed intensely for prime office space in major urban centers worldwide, and many focused-on solutions were seen to promote collaboration. But the outbreak of coronavirus has broken through cultural and technological barriers that prevented remote work in the past. During the pandemic, many people were surprised by how quickly and effectively technologies for videoconferencing and other forms of digital collaboration were adopted. According to McKinsey research, 80% of people questioned report that they enjoy working from home, 41% say that they are more productive than before, and 28% that they are as productive. According to Energy and Resources Institute (TERI) reports, the outbreak of COVID-19 had reduced the share of the Delhi metro service by 13%. In Bengaluru city, about half of the public transportation service users have shifted to a personal mode of transport, and one-fourth of the users have opted for WFH system. The remaining prefer to use carpool service, where the vehicle is shared with few regular travelers during the pandemic. Similarly, the volume of personal cars has increased in all the cities during the COVID-19 outbreak (TERI 2020).

2.6.1 Survey Analysis

A survey was conducted to observe the shifting lifestyle and preferences to depict the effect of lockdown on the public transportation system. The sample population considered for the survey was classified based on their residing locations, age, gender, level of qualification, and occupation. A digital platform survey was conducted, in which a google form was circulated through social media and emails. In total, 232 responses were received, as shown in Table 2.6. The survey results show that most of the respondents were residents from urban areas. Around 86.9% of participants belong to the age group of 15–35 years, and the maximum responses were recorded from male participants, i.e., 62%. Interestingly, 49.02% of respondents were students, and 15.32% were from the teaching profession.

Figure 2.5 shows the percentage of people preferring online and offline modes for work. In the post-lockdown period also, the preference remained the same. With the opening of several workplaces, 21% of the respondents reported that they have shifted to physical mode during post lockdown. The trend remained the same for academics. During the lockdown phase, only 6.79% of the respondents did not avail themselves of the online classes. However, there is a significant increase of 79.8% in

Table 2.6 Characteristics of the Sample data (Respondents = 232)

	Characteristics	Count (n)	Percentage
Gender	Male	143	61.65
	Female	82	35.28
	Intersex	5	2.35
	Preferred not to say	2	0.72
Age composition	<15	8	3.56
	15–35	202	86.90
	35–50	17	7.47
	>50	5	2.07
Education level	Primary	11	4.67
	Secondary/higher secondary	62	26.87
	Graduate	61	26.11
	Post-graduate	98	42.35
Occupation	Unemployed	34	14.56
	Self-employed or business	25	10.73
	Students	114	49.02
	Homemaker	20	8.54
	Teachers	35	15.32
	Others	4	1.83
Location	Rural	76	32.76
	Urban	156	67.24
Household income (annual)	<5 lakhs	25	10.80
	5–9 lakhs	156	67.20
	>9 lakhs	41	17.60
	Preferred not to say	10	4.40

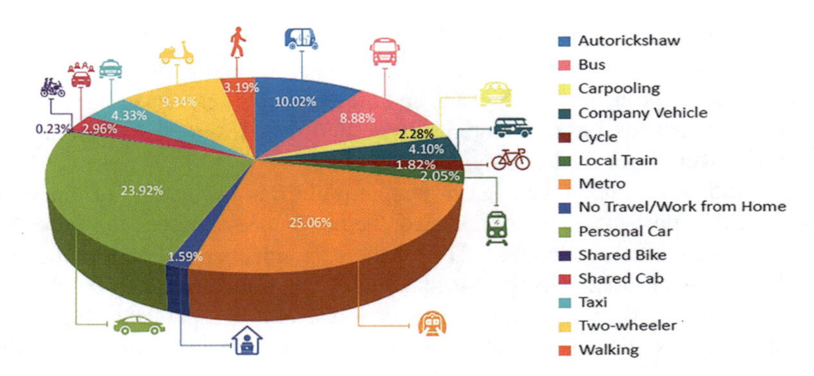

Fig. 2.5 Share of different transport modes during pre-COVID-19 used by respondents for work trips

the use of online platforms as compared to the pre-lockdown period. Around 6.05% of people stopped using the online platform for classes in the post-lockdown scenario, and the reason is that their private coaching classes had begun in a physical mode.

The result of the survey concludes that almost the entire population has adopted the virtual space. Figure 2.5 shows the share of the various mode of transportation during the pre-COVID situation. It states that the public transport such as metro, bus, and local train services was availed by around 25.06%, 8.88%, and 2.05%, respectively. The usage of intermediate public transport (IPT) modes such as taxi, auto, etc., constituted about 16% of the overall sample. The majority of the people avail their car and two-wheelers for commuting, i.e., around 23.92% and 9.34%, respectively. Respondents were questioned about their mode of transport in a post-COVID-19 situation. The questionnaire contained the option for a mode of choice during pre- and post-COVID-19 outbreaks to identify the nature of the mode shift.

To identify the nature of the mode shift, the questionnaire contained the option for a mode of choice during pre- and post-COVID-19 outbreaks. A significant decrease in public transport was observed. Most of the respondents (64%) said that they would not change their mode of transport, as they used private cars, bikes, taxis, and non-motorized transport (NMT) for their commute. Only 23% responded that their choice of mode would change due to the pandemic crisis. The remaining respondents (13%) were uncertain about their choice and selected may be an option in the questionnaire. Thus, it can be said that the public transportation mode is significantly affected as there is a vast percentage (about 36%) of people are potentially switching to a different mode choice.

Figure 2.6 shows the modal choice of initial metro users in the post-COVID situation. From Fig. 2.6, it can be observed that about 45% of the metro users have switched to other modes of transport such as the use of private cars, two-wheelers, taxis, etc. Some percent of respondents have also opted for intermediate public transport as a substitution for the metro. A similar trend was observed for the initial bus users. Figure 2.7 shows the percentage of mode choice by the initial bus users during

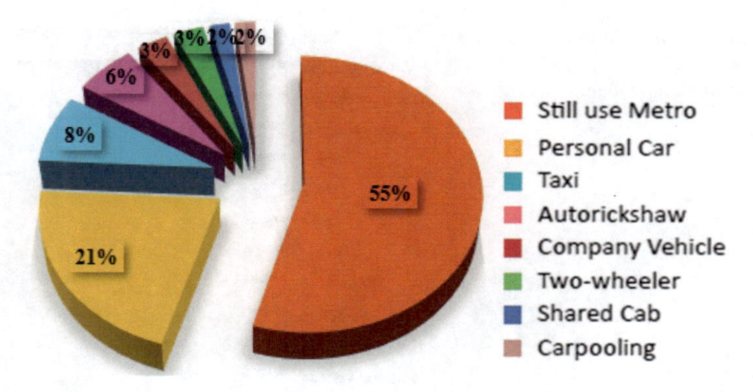

Fig. 2.6 Post-COVID-19 mode preference of respondents for work trips (initial metro users)

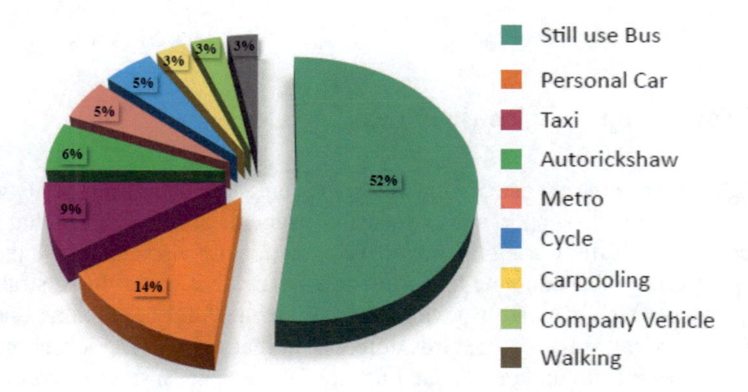

Fig. 2.7 Post-COVID-19 mode preference of respondents for work trips (initial bus users)

the post-COVID situation. From Fig. 2.7, around 52% of the bus users were willing to continue with the same mode of transport, whereas the remaining 48% were willing to shift to other modes. There is also a drastic reduction in the use of local trains as Mumbai has witnessed such a scenario, as most residents preferred to opt for other modes of transport.

This trend is not limited to work trips, but the trend remains similar for shopping and outing. Most people prefer to shop from online platforms or use a non-motorized mode or personal vehicle to access the shopping places to be in less contact with others and avoid the spread of the virus. There is a sharp decrease in buses, local trains, and metro services in the pandemic. Figure 2.8 shows the overall model share in the pre- and post-COVID-19 scenario for the considered sample. It shows that most of them prefer to switch to private cars and two-wheelers. The most considerable reduction is seen in public transportations which are vastly compensated by a significant increase in the private vehicles.

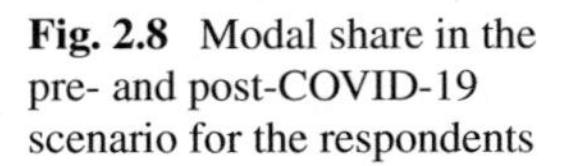

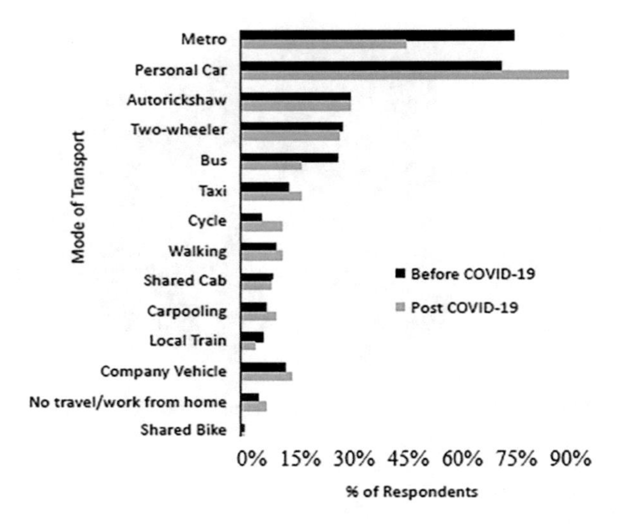

Fig. 2.8 Modal share in the pre- and post-COVID-19 scenario for the respondents

2.7 World Coping with the Crisis

In the USA, most of the agencies in the transportation sector have temporarily waived their fares. It is done to support and endorse the protocols of social distancing so that the riders are kept safer during the travel. The practice has also aided in supporting the people during this pandemic by acting as a financial cushion. In Australia, the safety protocols are enforced by embedding strict rules and regulations that vary based on the extent of the coronavirus outbreak in that region. The levels of social distancing are varied to understand the impact of the transition stage of coronavirus outbreak on public transportation capacity such as scenario 1: strict distancing (a mandatory gap of 1.5 m needs to be maintained between the people in public places and public transportation), Scenario 2, i.e., moderate distancing (no person sitting side-by-side, behind, or diagonally in public places); and Scenario 3, i.e., relaxed distancing (allows a gap of 1 m between the people facing each other). The Australian PTS strictly follows all these norms based on the situation and mode of transport (WSP 2020).

Similarly, in South Korea, a live update on the congestion level was monitored at every bus stop, and it is also integrated with a mobile application-based system. After analyzing the passenger trend, this measure also helps the passengers check out the congestion level on their mobile application and board the service. The city authority has also introduced additional buses to satisfy the demand and maintain social distancing. Similar safety protocols are being implemented to check the congestion in metro rails and manage the crowd to maintain social distancing. In Mexico City, the government plans to increase and improve its bicycle lanes to cut down the pressure on its metros. The city authority has also converted 100 km of its city street into cycle lanes to cut down the load on their public transport (Bhatt 2020).

2.7.1 Scenario in India Due to Crisis

In India, adopting social distancing protocols is a significant concern due to the mammoth volume of riders carried by the Indian transports. Though the Indian government is taking the measures by sanitizing the vehicles, seating arrangements, limiting the number of riders, e-ticketing systems are some of them. In an interview, HuffPost India asked an important question to the senior member of Urban Planning and Design from Urban Works Institute. The question was, will the country be able to recognize proper mode of transport that adequately deal the devastating demand of the people, while enforcing the social distancing. Also, how the change is going to be implemented as there is a vast gap between the operational and required number of transport vehicles (Jain 2020). There is a massive gap between the ambition, planning, and execution of the strategies. According to a survey report, during the COVID-19 pandemic, there was a massive shortage of bus services in Delhi. As per Mumbai's BEST (Brihanmumbai Electricity Supply and Transport) guidelines, there is a lack of 24 times public transport fleet compared to the existing buses for the public transport to follow the social distancing protocol (Intelligent Transport 2020; Gandhiok 2020; Sachdev 2020). The gap in this requirement would increase the risk of transmission of the virus, as maintaining social distance would be a big issue in this inadequacy.

On the other hand, the increasing number of private vehicles, city buses, e-rickshaws, and autos can be real problems. As no fixed timings are maintained by these transport, the chances of an increase in the crowd in pickup stations might be high. For short distances within the city, introducing an e-ticketing system may not be the relevant solution to increase travel costs. However, the government sector might afford to shift to the digital platform (World Bank 2020). However, the issue might arise for the private transport system as they may not agree to shift to digital due to the high initial investment (Hindustan Times 2020).

Regular sanitization might be another big issue, as to cut down the cost, some private transport sectors might use harmful chemicals that might not be listed in the government list, which might harm the rider's health. Social distancing is a vital norm enforced in all sectors by the government to avoid the transmission of the virus, which made the public opt for an alternative means of travel. Furthermore, Kolkata metro had laid a guideline which states that children and older adults are not allowed to avail of the service, so people traveling with the same were forced to use other modes of transport, leading to an increased travel cost (The Times of India 2020). However, as per fact, many Indians do not have the option of using a personal vehicle or can afford one, so there are chances that the public transportation infrastructure may ramp up (Yadav 2020). Thus, the Indian government has to ensure that people feel safe while availing the public transport service and maintaining the safety protocols of COVID-19 while traveling by continuously spreading awareness among the people.

2.8 Conclusion

The outbreak of the COVID-19 has entirely changed people's transport behavior in dramatic ways with a vast reduction in usage of public transport. It is evident to see a significant change impact on the transportation sector, as the evidence from the previous crisis says that the immediate aftermath of crisis events affects the people's behaviors towards the usage of public transportation due to the fear of the virus transmission. Therefore, to build back trust among the people, it becomes indispensable that the public transport sectors adopt the guideline imposed by the government agencies and health organizations very rigorously.

It is worth mentioning that it was impossible to imagine a modern society without any mobility and transportation a few days back. However, the outbreak of the pandemic has ultimately brought the entire transportation sector to a halt. Now the entire world has witnessed a massive shift to virtual communication from transportation base activities (The Economic Times 2021). The world needs to rethink the alternative mode of transport and try to curb the communal spread. Increasing the count of non-motorized vehicles will have a positive impact on the environment and people's health. However, post-lockdown scenarios have given a sudden rise in the number of personal vehicles, severely affecting the environment and people's health due to the increasing emission of gases. Public transport is considered an essential service that provides sustainable choice to private vehicles. Currently, there is a significant reduction in the demand for public transports. In the future, the suspension of the public transportation system will have a considerable impact on the environment, and it may increase by putting a question mark on the sustainability policy.

Instead of building a better future, we are pitching up for much chaos with no one to blame. The implementation of policies by the government policy should be more concrete and should adequately be enforced to stabilize the wobbling sector of transportation. Thankfully, for the post lockdown period, the Indian government has designed sustainable transport policies that can draw on experiences from the earlier crisis to predict likely behaviors and accordingly design the best-fit policy. Importantly, people should also need to safeguard themselves from any potential third wave of the pandemic that might occur. Thus, special care should be taken on the public transportation service, as people from low-income groups cannot avail personal and private vehicles and rely only on the public mode of transportation.

References

Bhatt A (2020) Post Covid-19 lockdown, will India's public transport systems be able to maintain social distancing? *scroll.in*, article. https://scroll.in/article/960063/post-covid-19-lockdown-will-indias-public-transport-systems-be-able-to-maintain-social-distancing

Chauhan MC (2020) COVID-19 and its effect on the rail and transport industry. Urban Transport News. https://www.urbantransportnews.com/covid-19-and-its-effects-on-rail-transport-industry/. Accessed 24 June 2021

Chen N, Zhou M, Dong X et al (2020) Epidemiological and clinical characteristics of 99 cases of 2019 novel coronavirus pneumonia in Wuhan, China: a descriptive study. Lancet 395:507–513. https://doi.org/10.1016/S0140-6736(20)30211-7

Cui J, Li F, Shi Z-L (2019) Origin and evolution of pathogenic coronaviruses. Nat Rev Microbiol 17:181–192. https://doi.org/10.1038/s41579-018-0118-9

Evan D, Over M (2020) The economic impact of COVID-19 in low and middle—income countries. Centre for Global Development. https://www.cgdev.org/blog/the-economic-impact-of-COVID-19-in-low-and-middle-income-countries. Accessed 25 Mar 2021

Foreign Tourist Inflow to India March 2021 (2021) Ministry of Tourism, Govt. of India, pdf. https://tourism.gov.in/sites/default/files/2021-06/Brief%20Note%20March%2021.pdf

Gandhiok J (2020) Social distancing: Delhi needs to triple bus fleet. http://www.timesofindia.indiatimes.com/articleshow/76719678.cms?utm_source=contentofinterest&ut_medium=text&utm_campaign=cppst. Accessed 28 May 2021

Hindustan Times (2020) Social distancing norms flouted across Bengal on first day of Unlock 1. *Hindustan Times*, article. https://www.hindustantimes.com/india-news/social-distancing-norms-flouted-across-bengal-on-first-day-of-unlock-1/story-JDC54WNBhw5ouhCfVNJTmI.html

Intelligent transport (2020) India requires 24 times more buses to allow commuters to social distance effectively. https://www.intelligenttransport.com/transportnews/100881/india-requires-24-times-more-buses-to-allow-commuters-to-social-distanceeffectively/. Accessed 26 May 2021

Jain A (2020) How can India Reopen public transport post lockdown? An expert explains. Huffpost. https://www.huffingtonpost.in/entry/public-transport-lockdown-bus-trainmasks_in_5ebbc585c5b687aa20a05128. Accessed 25 Aug 2020

Kumar A, Sankar TP (2020) Impact of Covid-19 on India's transport sector. https://www.itefoecdorg/covid-19-impact-india-transport-sector. Accessed 28 May 2021

Lai C-C, Shih T-P, Ko W-C, et al (2020) Severe acute respiratory syndrome coronavirus 2 (SARS-CoV-2) and corona virus disease-2019 (COVID-19): The epidemic and the challenges. Int J Antimicrob Agents 55(3):105924. https://doi.org/10.1016/j.ijantimicag.2020.105924

Leigh G (2020) Public transport in a pandemic. Good idea or best avoided? Forbes. https://www.forbes.com/sites/gabrielleigh/2020/03/17/public-transport-in-a-pandemic-good-ideaor-best-avoided/#2f6c2a4d3047. Accessed 20 June 2020

Livement (2021) Road developers see toll revenues pick up as highway traffic returns. https://www.livemint.com/industry/infrastructure/road-developers-see-toll-revenues-pick-up-ashighway-traffic-returns-11592306919740.html. Accessed 28 July 2020

Manju V (2020) Indian aviation industry could face Rs 24k crore revenue loss due to Covid-19 lockdown. *The Times of India* 1. https://timesofindia.indiatimes.com/business/india-business/indian-aviation-industry-could-face-rs-24k-crore-revenue-loss-due-to-covid-19-lockdown/articleshow/75595340.cms

Ministry of Statistics & Programme Implementation (MOSPI, 2019) GoI. http://www.mospi.nic.in/sites/default/files/press_release/Presss%20note%20for%20first%20advance%20estimates%202018-19.pdf

Mishra HH (2020) Coronavirus in India: COVID-19 lockdown may cost the economy Rs 8.76 lakh crore; here's how. Business Today. https://www.businesstoday.in/opinion/columns/coronavirus-in-india-covid-19-india-lockdowneconomy-cost-gdp-gva-nationwide-shutdown/story/399477.html. Accessed 28 May 2021

National Aviation Report (2020). https://www.finance.yahoo.com/news/indian-aviation-wings-remain-clipped-072942760.html

NITI Aayog (2020) Eliminating poverty: creating jobs and strengthening social programs. https://www.niti.gov.in/writereaddata/files/presentation%20for%20regional%20meetings-%20NITI%20AAYOG.pdf

Press Information Bureau (PIB) (2020) Delhi, Ministry of Finance. https://www.pib.gov.in/PressReleaseIframePage.aspx?PRID=1608345

Radhakrishnan V, Sen S, Singaravelu N (2020) Pollution levels decline in many Indian cities due to COVID-19 lockdown. The Hindu. https://www.thehindu.com/data/data-pollution-levelsdecline-in-many-indian-cities-du-to-covid-19-lockdown/article31667020.ece. Accessed 18 June 2021

Sachdev A (2020) Existing number of buses inadequate to ensure social distancing, will need 6,00,000 more: Report. CNBCTV18. https://www.cnbctv18.com/healthcare/exsiting-number-ofbuses-inadequate-to-ensure-social-distancing-will-need-600000-more-report-says6186191.htm. Accessed 26 Mar 2021

Shakti Foundation (2016) City public transportation development in India. http://www.intelligenttransport.com. Accessed 10 May 2021

Shereen MA, Khan S, Kazmi A, Bashir N, Siddique R (2020) COVID-19 infection: origin, transmission, and characteristics of human coronaviruses. J Adv Res 24:91

Singh MK, Neog Y (2020) Contagion effect of COVID-19 outbreak: Another recipe for disaster on Indian economy. J Public Affairs 20(4):e2171

Statista Research Department (2021) Gross value added from transport, communication, and related services in India between the FY 2012 and 2020, by sector. https://www.statista.com/statistics/1038649/india-gva-from-transport-communication-by-sector/

TERI (2020) Impact of COVID-19 on urban mobility in India: evidence from a perception study. The Energy and Resources Institute

The Economic Times (2020a) COVID-19 has led to 20 lakh job losses in bus, taxi sector; more on anvil: Industry Body. https://www.economictimes.indiatimes.com/jobs/covid-19-has-led-to-20-lakh-job-losses-in-bus-taxisector-more-on-anvil-industry-body/articleshow/76492675.cms?from=mdr. Accessed 1 June 2021

The Economic Times (2020b) Toll collections to see 13% drop due to 57-day coronavirus lockdown: CRISIL Research. https://www.economictimes.indiatimes.com/news/economy/finance/tollcollections-to-see-13-drop-due-to-57-day-coronavirus-lockdown-crisilresearch/articleshow/75576705.cms

The Economic Times, 24 July 2021. https://www.economictimes.indiatimes.com/topic/virtual-tours

The Times of India (2020) Kolkata: no metro on sundays; kids and elderly not allowed. https://www.timesofindia.indiatimes.com/city/kolkata/pandemic-protocol-no-metro-on-sundays-kids-elderly-not-allowed/articleshow/78025915.cms

Times Now News. https://www.timesnownews.com/auto/features/article/lockdown-4-0-recorded-more-road-accidents-than-pre-lockdown-scenario/598859. Accessed 28 May 2021

TOI (2021). https://www.timesofindia.indiatimes.com/india/covid-may-have-saved-20000-lives-on-indian-roads/articleshow/77649949.cms. Accessed 21 May 2021

UITP (2020) Management of COVID-19: guidelines for public transport operators. Int Assoc Publ Trans. https://www.uitp.org/sites/default/files/cck-focus-papersfiles/Corona%20Virus_EN.pdf

World Bank (2020) https://www.blogs.worldbank.org/transport/covid-19-brought-urban-transport-its-knees-digital-technology-will-put-it-back-its-feet. Accessed on 18 June 2021

World Health Organization (2020) Considerations for public health and social measures in the workplace in the context of COVID-19: annex to considerations in adjusting public health and social measures in the context of COVID-19 https://www.who.int/publications/i/item/considerations-for-public-health-and-social-measures-inthe-workplace-in-the-context-of-covid-19. Accessed 24 June 2020

WSP (2020) Public transportation and COVID-19: how to transition from response to recovery. https://www.wsp.com. Accessed 22 May 2021

Yadav N (2020) India needs over 600,000 buses for 25 million commuters daily to follow social distancing norms, according to a study. Bus Insider. https://www.businessinsider.in/india/news/india-needs-over-600000-buses-for-25-millioncommuters-daily-to-follow-social-distancing-norms-in-times-ofcoronavirus/articleshow/76417090.cms. Accessed 26 March 2021

Chapter 3
Life Cycle Assessment to Identify Sustainable Lime-Pozzolana Binders for Repair of Heritage Structures

Degloorkar Nikhil Kumar and Rathish Kumar Pancharathi

3.1 Introduction

With increase in population and rapid industrialization, there is a surge in infrastructure development that led to the massive requirement of cement in the entire world. It is forecasted that by the end of 2050, the cement requirement would arrive at around 9 billion tons (Aprianti 2017; DECC 2015). This huge quantity of cement manufacture, would definitely cause a harmful environmental impact. Hence, it becomes imperative to develop sustainable binders for construction. One of the solutions for sustainable binders is lime. Lime, when manufactured under similar scale and efficiency as that of cement, would cater to less energy and carbon dioxide emissions (CESA 2006).

Lime as a binder got disadvantages in terms of its slow setting times, high drying shrinkage, and low mechanical strength (Vazquez 2002). These drawbacks paved way for the invention of Portland cement, which is better than lime in all the above-mentioned aspects. However, lime has good water transmissivity and excellent durability in terms of water transport (sorptivity, water permeability, etc.,) and better frost resistance compared to that of cement (Grilo et al. 2014). It was found that usage of cement mortars in the repair of heritage structures proved to be detrimental (Veniale et al. 2003; Callebaut et al. 2001). The main reason for the deterioration was due to the incompatibility between cement mortars with ancient lime-based substrates and masonry units like brick and stone (CESA 2006, Sepulcre and Hernandez 2010). This incompatibility was caused due to the impermeable, hard, and rigid characteristics of cement-based mortars that were unnecessary in these applications. Therefore, materials that were used in past, namely lime-based mortars have to be used to address

D. N. Kumar (✉) · R. K. Pancharathi
National Institute of Technology Warangal, Warangal 506004, Telangana, India

R. K. Pancharathi
e-mail: rateeshp@nitw.ac.in

K. R. Reddy et al. (eds.), *Advances in Sustainable Materials and Resilient Infrastructure*, Springer Transactions in Civil and Environmental Engineering,
https://doi.org/10.1007/978-981-16-9744-9_3

such compatibility issues with the substrate materials. Further, to cater to the inhibitions in using lime as binder, materials like pozzolanas and chemical admixtures like air entraining agents, etc., can be used to improve its mechanical performance and durability characteristics pertaining to water permeability, frost resistance, etc.

Air lime mortars were utilized in construction till the start of twentieth century (Del Mar Barbero et al. 2014). They were part of old buildings as, joining mortar between stones or bricks, bonding material in tiles and in decorative applications like stucco, paints, etc. The ancient knowledge regarding lime usage and its mortar preparation was not transferred from one generation to another generation leading to void in the practical use of the binder (Menezes et al. 2012).

Lime as a binder is mainly categorized into two types, air lime and hydraulic lime. Air lime is basically slaked lime or hydrated lime whose mineral constituent is calcium hydroxide ($Ca(OH)_2$), whereas hydraulic lime is produced by firing and slaking siliceous limestone. The mineral phases in hydraulic lime are $Ca(OH)_2$ and dicalcium silicate (C_2S). Air lime hardens by carbonation reaction, where $Ca(OH)_2$ reacts with carbon dioxide (CO_2) in the presence of moisture to form calcium carbonate ($CaCO_3$) crystals. Hydraulic lime undergoes hardening with the combination of both hydration and carbonation reactions. C_2S component of hydraulic lime in the presence of water undergoes hydration reaction and forms calcium-silicate-hydrate gel (C-S-H). Also, due to the presence of $Ca(OH)_2$, carbonation reaction takes place and calcium carbonate crystals are formed (Li et al. 2011).

In the last few decades, many studies were carried on modern air lime mortars along with that of ancient lime mortars (Rua 1998, San Nicola 1639, Selio 1552). Primarily, studies on air lime mortars were on its proportions, carbonation phenomenon and porous structure (Peroni 1981; Cazalla et al. 2000, Lanas 2003). In addition to that, mortar preparation and application processes were also focused on assessing the possible outcomes of the study (Cavaco et al. 2003, Margalha 2011, Balksten 2005, Rosell 2014). Though, many case studies and technical information confirm usage of lime mortars as the best suitable material for conservation of ancient buildings, there is still hesitance in its usage in the rehabilitation and conservation of old heritage buildings (Papayianni 1998, Do Rosario Veiga et al. 2009). This could be prevented by increasing research on aspects such as sustainability, to study the influence of material towards environment. Due to the depriving resources throughout the world, there is an urgent need to research to encourage larger sect of construction expertise in using lime mortars.

According to (Hammond and Jones 2008) as per their research investigations noticed that sustainability of any product or service can be assessed through a universally accepted scientific method Life Cycle Assessment (LCA). Life Cycle Inventory (LCI) is the primary requisite to understand the environmental aspects of the product or service. Several countries developed authentic LCI. Further, databases of the inventory have been prepared in order to perform analysis related to energy consumption and CO_2 emissions of the product or service.

As per (Prakasan et al. 2020) mentioned India lacks LCI of building materials. LCA studies need to be conducted in India, as there is no reliable LCI data. Though some organizations have been working on developing LCI database, they are not publicly accessible.

In congruence with international standards, Indian standard codes have been developed for LCA, namely IS/ISO 14040:2006 Environmental Management-Life Cycle Assessment—Principles and framework (BIS, IS/ISO:14,040-2006a) and IS/ISO 14044:2006 Environmental Management-Life Cycle Assessment-Requirements (BIS, IS/ISO:14,044-2006b). Templates for LCI calculations have been introduced in these codes. Also, a four-phase process for understanding the LCA of any product or service have been developed. Those four phases include

(a) Goal Identification
(b) Scope Definition
(c) Life Cycle Inventory, and
(d) Life Cycle Impact Assessment and Interpretation.

LCI is the main phase of LCA study as it forms the link between the objective of study and its interpretation. In order to obtain inventory, there is need of energy factors and CO_2 emission factors. The energy factors are obtained from (US EPA 2014) and from the guidelines of (IPCC 2006). As per (Prakasan et al. 2020), it is preferable to obtain emission factors from the concerned plant of study (Cement or Steel or any other product manufacture plants) or by CHNS analysis of fuel samples. In CHNS analysis, the percentage composition of carbon, hydrogen, nitrogen, and sulfur for organic compounds were determined using an analyzer. In case of fuels, this would help in finding the CO_2 emission based on the carbon content. However, when testing is not possible CO_2 emission factors are obtained from (US EPA 2014), Cement Sustainability Initiative (CSI Protocol 2013) and also from national greenhouse gas inventory of (IPCC 2006).

From the literature, it is clear that lime-pozzolana mortars though have low mechanical strength in comparison with cement-based mortars, its durability characteristics are far better than cement-based mortars for applications such as repair of heritage structures. There is a need to understand the impact of various lime-pozzolana-based mortars on environment in terms of energy demand and CO_2 emissions. This would enhance usage of lime-pozzolana-based mortars as repair material for heritage structures. There is hence a need for extensive study in sustainable aspects of lime-pozzolana-based mortars, which forms the novel prospect of the current study. A comparative study between cement mortars and lime-pozzolana-based mortars was studied in terms of energy demand and impacts on environment like CO_2 emissions.

In the present study, two binder to sand proportions of 1:1 and 1:3 were considered for mortar preparation. Cement and lime-pozzolana-based mortars were prepared. Pozzolanas, namely fly ash and GGBS were used with percentage of replacement by weight of binder varying from 30 to 75% for preparing lime-pozzolana-based mortars. Twenty different mortar mixes were prepared and the quantities of materials required for attaining 1m^3 of mortar mix for desired workability were noted. These material quantities were used as inputs in Open LCA software for conducting

LCA study on mortars. Using various LCA impact assessment methods, energy consumption, and environmental emissions of 20 mixes were compared and analyzed.

3.2 Materials

The materials used in the current study on sustainability of mortars are hydrated lime, ordinary Portland cement, fly ash, GGBS, normal sand, and water.

3.2.1 Hydrated Lime

From visual examination, hydrated lime is greyish white in color. Further, hydrochloric acid test was conducted as per (BIS, IS:1624-2009a), wherein, there was no gel formation indicating lime as non-hydraulic lime. Chemical analysis as per (BIS, IS:712-2009a) was performed on lime and its composition obtained is shown in Table 3.1. Lime considered for the study, does not come under any class of lime mentioned in (BIS, IS:712-2009b). In India, commercial production of lime in classes as per (BIS, IS:712-2009b) was discontinued with the advent of cement in their applications.

Table 3.1 Chemical analysis of hydrated lime as per (BIS, IS:712-2009a)

S. no.	Description of oxides (in lime)	Hydrated lime Composition (%)
1	CaO and MgO, percent min (on ignited basis)	70.86
2	SiO_2, Al_2O_3, Fe_2O_3, percent min (on ignited basis)	9.66
3	Presence of insoluble residue and alkali percent, max (on ignited basis)	2.16
4	Loss on ignition (LOI)	18.80
5	Silicon dioxide (SiO_2) (on ignited basis)	8.23
6	Ferric oxide (Fe_2O_3) (on ignited basis)	0.49
7	Alumina (Al_2O_3) (on ignited basis)	0.94
8	Calcium oxide (CaO) (on ignited basis)	68.37
9	Magnesium oxide (MgO)	2.49
10	Calcium hydroxide $Ca(OH)_2$ (on dry basis)	73.34

Table 3.2 Oxide composition in (%) of OPC, GGBS, and fly ash from XRF analysis

Oxides	Oxide composition (%)		
	OPC	GGBS	Fly ash
SiO_2	19.52	37.52	61.11
Al_2O_3	3.42	18.19	27.83
Fe_2O_3	4.33	0.54	4.19
CaO	67.56	40.15	3.25
TiO_2	0.49	0.84	1.53
K_2O	0.75	0.38	1.43
SO_3	3.75	1.79	0.56
MnO	0.10	0.50	0.05
SrO	0.08	0.09	0.05

3.2.2 *Ordinary Portland Cement (OPC)*

OPC-53-grade cement adhering to (BIS, IS:12,269-2013a) was utilized in cement mortar preparation. The physical properties, viz. specific gravity, specific surface area, initial, and final setting times were observed as 3.14, 235 m^2/Kg, 50 min, and 570 min, respectively. Through X-ray fluorescence spectroscopy analysis (XRF), oxide composition of OPC is obtained and is indicated in Table 3.2.

3.2.3 *Fly Ash*

The fly ash utilized in the study was acquired from a local thermal power plant at Ramagundam, Telangana, India. From oxide composition in Table 3.1 and in reference with (BIS, IS:3812-Part-1-2013b), fly ash considered was siliceous fly ash. Also, as the Calcium Oxide (CaO) content in Fly ash is less than 18% and as per (ASTM, C618-19) it is classified as Class F fly ash.

3.2.4 *Ggbs*

Ground Granulated Blast Furnace Slag (GGBS) from a nearby JSW steel plant at Kamleswar, Nagpur was employed in the study. It can be observed from Table 3.2 that along with CaO, there are considerable amounts of SiO_2, Al_2O_3, and Fe_2O_3 in GGBS, which are though less compared to OPC, is high in comparison with fly ash. This infers the cementitious character of GGBS.

3.2.5 Sand

Sand adhering to Zone-II as per (BIS, IS:383-2016) is considered for the study.

3.2.6 Water

Potable water conforming to (BIS, IS:456-2000), with pH value not less than 6 is considered for the study.

3.2.7 Mixes

Lime-pozzolana (fly ash/GGBS) based mortars with binder to sand proportions of 1:1 and 1:3 by weight were prepared. It was observed by several researchers (Chever et al. 2010 and Lanas et al. 2004) that the binder to sand ratio of 1:3 in lime mortars showed higher carbonation, in turn helping in attaining better mechanical properties. Also, as per (Moropoulou et al. 2000 and Beck et al. 2008) 1:3 proportion mortars showed good workability characteristics with less quantity of pores. For performing comparative analysis, binder to aggregate proportion of 1:1 was also considered in the study. The pozzolanic replacement of lime (fly ash /GGBS) varied from 30 to 75%. Water required for attaining a flow of 120–150 mm is considered for the preparation of mortars. Table 3.3 shows the legend of mortar mixes considered in the study. The composition of various materials utilized in mortar mixes are showed in Table 3.4.

Material quantities obtained from Table 3.4 are considered as inputs in subsequent life cycle analysis.

3.3 Methods

3.3.1 LCA Approach and Framework

LCA emerged as an important tool in understanding the performance of various products in terms of sustainability (Del Borghi 2012). These life cycle studies were established on the concept that, for any flow of processes making a product, inputs or decisions considered in one process will have its share of impact on its subsequent processes (Janssen and Hendriks 2002).

A cradle-to-gate boundary has been considered for LCA study on the mortar mixes. According to (Manjunatha et al. 2021), in their study on LCA of concrete mixes with different binders, namely OPC, Portland Pozzolana Cement, and GGBS,

Table 3.3 Legend for the mortars

Alternative	Description of alternative
CA	Cement mortar with (1:3) mix proportion
LA	Lime mortar with (1:3) mix proportion
LF30A	Lime mortar with (1:3) mix proportion, 30% lime replaced with fly ash
LF50A	Lime mortar with (1:3) mix proportion, 50% lime replaced with fly ash
LF66A	Lime mortar with (1:3) mix proportion, 66% lime replaced with fly ash
LF75A	Lime mortar with (1:3) mix proportion, 75% lime replaced with fly ash
LG30A	Lime mortar with (1:3) mix proportion, 30% lime replaced with GGBS
LG50A	Lime mortar with (1:3) mix proportion, 50% lime replaced with GGBS
LG66A	Lime mortar with (1:3) mix proportion, 66% lime replaced with GGBS
LG75A	Lime mortar with (1:3) mix proportion, 75% lime replaced with GGBS
CB	Cement mortar with (1:1) mix proportion
LB	Lime mortar with (1:1) mix proportion
LF30B	Lime mortar with (1:1) mix proportion, 30% lime replaced with fly ash
LF50B	Lime mortar with (1:1) mix proportion, 50% lime replaced with fly ash
LF66B	Lime mortar with (1:1) mix proportion, 66% lime replaced with fly ash
LF75B	Lime mortar with (1:1) mix proportion, 75% lime replaced with fly ash
LG30B	Lime mortar with (1:1) mix proportion, 30% lime replaced with GGBS
LG50B	Lime mortar with (1:1) mix proportion, 50% lime replaced with GGBS
LG66B	Lime mortar with (1:1) mix proportion, 66% lime replaced with GGBS
LG75B	Lime mortar with (1:1) mix proportion, 75% lime replaced with GGBS

it was determined that PPC- and GGBS-based concretes exhibited less environmental impact compared to OPC-based concretes. Also, (Garces et al. 2021) in their research mentioned that geopolymer concrete since comprised of inorganic polymer and wastes like fly ash and GGBS are responsible for low greenhouse gas emissions and less environmental effects compared to OPC. These studies determine that with higher amounts supplementary cementitious materials in the binder mix of the mortars, there would be reduction in the level of environmental impacts.

The various phases of LCA study considered are outlined as follows.

3.3.1.1 Goal and Scope

To specify functional unit and system boundary. Here, CO_2 emissions and other environmental impacts are calculated using Open LCA software. However, these impacts were calculated for a cubic meter of mortar. Cubic meter is the functional unit as this is how quantity of mortar is determined.

Table 3.4 Mix proportions of materials for 1 m^3 of mortar preparation

Alternative	Binder (Kg)	Sand (Kg)	Fly ash/GGBS (Kg)	Water (Kg)
CA	486	1458	0	255
LA	442	1326	0	332
LF30A	254	1335	191	320
LF50A	224	1344	224	309
LF66A	152	1371	305	274
LF75A	115	1374	343	275
LG30A	254	1335	191	322
LG50A	224	1344	224	305
LG66A	150	1353	301	297
LG75A	115	1371	342	283
CB	876	876	0	447
LB	808	808	0	485
LF30B	472	826	354	450
LF50B	420	840	420	423
LF66B	283	848	565	404
LF75B	214	856	642	389
LG30B	470	822	352	458
LG50B	422	844	422	413
LG66B	286	855	569	391
LG75B	215	861	646	379

3.3.1.2 Inventory Analysis

Portland Cement (OPC), hydrated lime, fly ash, GGBS, sand, and water are the constituents used in the mortar mix. By definition in the Ecoinvent version 3.7 database of OpenLCA software (Green Delta 2020), the products Portland cement, hydrated lime, fly ash, GGBS, and sand inherently comprise the manufacturing energy usage till their outlet at factory or crusher site for sand. Thereafter, transportation for a distance of 100 km has been considered by light commercial vehicle on road for these materials. In OpenLCA, 0.1ton Km was used for the transportation of 1 kg of each material to the required site. Thereafter, these materials were mixed with tap water available at site manually. The quantities of materials required for making 1 m^3 of respective mortars were considered from Table 3.4.

3.3.1.3 Impact Assessment

With the help of various impact assessment methods, inbuilt in Open LCA software, various environmental impacts for each of the mortar mix design throughout their life

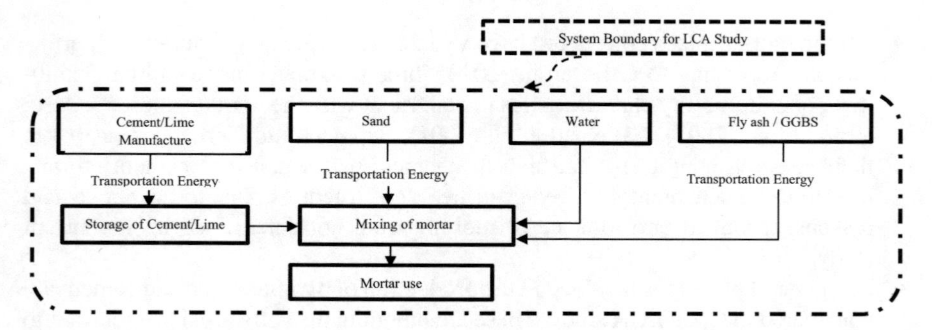

Fig. 3.1 Mortar preparation system boundary

cycle are assessed in measurable quantities. Embodied energy and CO_2 emissions are focused primarily, followed by other environmental impacts.

3.3.1.4 Interpretation

Mortar mixes will be weighed primarily in terms of embodied energy and CO_2 emissions, which is followed by other environmental impact categories. The best sustainable alternative mortar is the one with the least impact on environment.

3.3.2 System Boundary for Mortar Production

For any LCA framework, there is a need to establish a system boundary that would clearly specify the inputs and outputs of the concerned product manufacture. Here, the product is the preparation of 1m^3 mortar.

The system boundary shown in Fig. 3.1 includes manufacture of raw materials, namely cement/lime, fly ash, and GGBS. It also includes the transportation of raw materials to site where mortar mix is prepared. However, water considered to be available at site in abundant.

3.3.3 Assumptions

The following assumptions are applied in the LCA study of the mortar mixes for repair of heritage structures:

- Functional unit 1 m^3 is considered, which means that materials required for production of 1 m^3 mortar mix are used in the study. Also, the impact assessment of these mortars will be assessed per 1 m^3 of mortar.

- All the mortar mixes considered have variable mechanical and durability performance. According to Grilo et al. (2014), lime-pozzolana-based mortars exhibited better durability characteristics in congruent with the heritage mortars. Also, Venial et al. (2003), Callebaut et al. (2001), Sepulcre and Hernandez (2010) in their research mentioned that high mechanical performance of cement mortars proved to be detrimental for repair of heritage structures. Due to the above-said reasons, technical performance of mortars were not determined in the current study.
- The general processes involved in the production of hydrated lime and cement are considered in Open LCA study. These are inbuilt in the software. Data relevant to the materials and processes involved in mortar mix design within the LCA system boundary are considered.
- For supplementary cementitious materials, namely fly ash/GGBS, only emissions pertaining to their processing and transportation are considered in the study.
- Average transportation distance to the site for mortar mix design considered as 100 km for sand, hydrated lime, Portland cement, fly ash, and GGBS. This is because from the site, the distance of these materials was within the radius of 100 km. Moreover, the influence of transportation distance for materials less than 100 km and for those nearby 100 km on environment was not prominent. Hence, 100 km average transportation was considered for all the materials in the study. Also, the transportation considered is by road.

3.4 Results and Discussions of LCA Study

Three impact assessment methods: Cumulative Energy Demand, CML Baseline and Eco-indicator 99 of open LCA software are employed for analysis. These methods are generally adopted to understand the influence of manufacturing or preparation of various products on environment.

3.4.1 Cumulative Energy Demand Assessment

Cumulative Energy Demand of mortars with binder to sand ratios of 1:3 and 1:1 are indicated in Figs. 3.2 and 3.3, respectively. From Fig. 3.2, it can be observed that the highest energy demand was of cement mortar (CA) with 4807 MJ and the

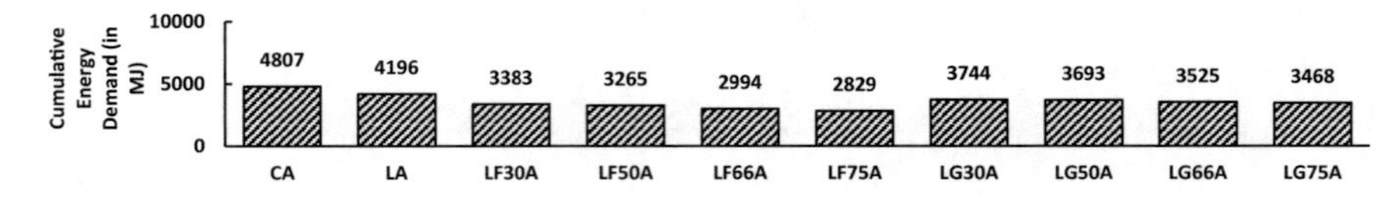

Fig. 3.2 Cumulative energy demand for 1:3 (binder: sand) mortars mixes

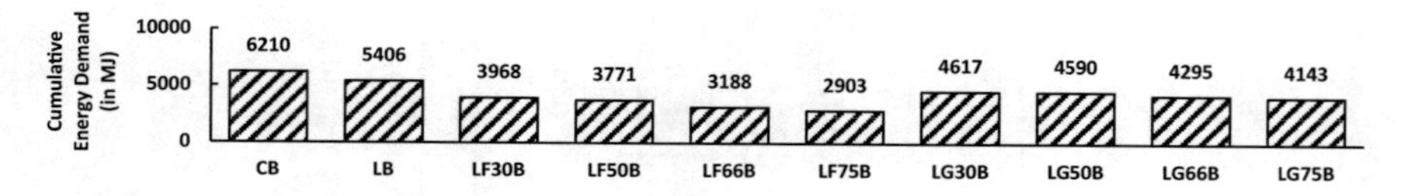

Fig. 3.3 Cumulative energy demand for 1:1 (binder: sand) mortars mixes

lowest was for lime–fly ash mortar, (LF75A) with 2829 MJ. There was a reduction of about 41% in energy demand for (LF75A) with respect to (CA). Similarly, for 1:1 proportion mortars as in Fig. 3.3, the cumulative energy demand was observed to be the highest in cement mortar (CB) with 6210 MJ and the lowest for lime–fly ash mortar (LF75B) with 2903 MJ. The reduction was 53% in (LF75B) with respect to cement mortar (CB). The main reason is attributed to the low production energy of lime, fly ash, and GGBS in comparison with cement. Moreover, energy demand of mortars with binder to sand ratios of 1:3 is less compared with 1:1. This is due to lesser binder content in 1:3 than in 1:1, whose energy consumption is the highest compared to other components, namely sand and water of mortar.

3.4.2 CML-IA (Baseline) Method Impact Assessment

CML-IA (Baseline) method was used to understand the environmental impacts of manufacture of product on environment in terms of ecotoxicity, photochemical oxidation, acidification, ozone layer depletion, eutrophication, global warming, and abiotic resources depletion. Mortars with binder to sand ratios of 1:3 and 1:1 were assessed and their impact results are shown in Tables 3.5 and 3.6, respectively. From the data, it can be observed that cement-based mortars had the highest impact over environment, whereas lime-pozzolana-based mortars, specifically lime–fly ash mortars had the lowest impact.

Mortars with binder to sand proportions of 1:3 and 1:1 had various environmental impacts. The values in Tables 3.5 and 3.6 show that lime mortars (LA and LB) showed the greatest impact in terms of ozone layer depletion. Cement mortars (CA and CB) showed the highest impact in terms of marine aquatic ecotoxicity, photochemical oxidation, fossil fuels depletion, terrestrial ecotoxicity, fresh water ecotoxicity, eutrophication, acidification, global warming (GWP100), human toxicity, and abiotic depletion resources, whereas lime–fly ash-based mortars (LF75A and LF75B) showed the least impact in the above-mentioned environmental categories.

For understanding the environmental impact of lime-pozzolana mortars with respect to cement mortars, impacts of cement mortars are taken as base of 100 and accordingly the impacts of lime-based mortars were measured and plotted in Figs. 3.4 and 3.5 for mortar of 1:3 and 1:1 proportions, respectively. Among mortars of 1:3 proportion, the least impact was observed in lime–fly ash mortar (LF75A), with a reduction of 40%, 23%, 52%, 40%, 56%, 41%, 56%, 73%, 55%, 36%, and 33% for marine aquatic ecotoxicity, ozone layer depletion, photochemical oxidation, fossil

Table 3.5 CML-IA baseline impact assessment for 1:3 mortar mixes

Impact category	Unit	Mortar mixes									
		CA	LA	LF30A	LF50A	LF66A	LF75A	LG30A	LG50A	LG66A	LG75A
Marine aquatic ecotoxicity	kg 1,4-DB eq (in 10^5)	2.424	1.529	1.468	1.465	1.466	1.452	1.692	1.731	1.801	1.849
Ozone layer depletion (ODP)	kg CFC-11 eq (in 10^-5)	2.885	3.594	2.783	2.662	2.381	2.218	2.988	2.906	2.674	2.579
Photochemical oxidation	kg C_2H_4 eq	0.138	0.108	0.083	0.079	0.071	0.066	0.088	0.086	0.078	0.075
Abiotic depletion (fossil fuels)	MJ	4123	3558	2919	2828	2619	2489	3189	3149	3012	2966
Terrestrial ecotoxicity	kg 1,4-DB eq	0.412	0.215	0.193	0.191	0.185	0.181	0.224	0.227	0.231	0.235
Fresh water aquatic ecotoxicity	kg 1,4-DB eq	123	75	73	73	73	72	83	84	88	90
Eutrophication	kg PO_4^{3-} eq	0.606	0.304	0.280	0.277	0.272	0.267	0.315	0.319	0.325	0.330
Acidification	kg SO_2 eq	3.136	1.091	0.942	0.922	0.878	0.847	1.056	1.057	1.047	1.049
Global warming (GWP100a)	kg CO_2 eq	572	540	376	351	291	259	397	375	319	295
Human toxicity	kg 1,4-DB eq	145	99	94	94	94	93	106	109	112	115
Abiotic depletion	kg Sb eq	0.0018	0.0013	0.0012	0.0012	0.0013	0.0012	0.0014	0.0015	0.0015	0.0016

Note eq means equivalent, 1,4-DB means 1,4 Dichloro Benzene, CFC-11 means chlorofluorocarbon-11 and all other compounds and elements are mentioned in their basic chemical formulae

Table 3.6 CML-IA baseline impact assessment for 1:1 mortar mixes

Impact category	Unit	Mortar mixes									
		CB	LB	LF30B	LF50B	LF66B	LF75B	LG30B	LG50B	LG66B	LG75B
Marine aquatic ecotoxicity	kg 1,4-DB eq (in 10^5)	2.970	1.505	1.406	1.407	1.369	1.355	1.813	1.911	2.050	2.123
Ozone layer depletion (ODP)	kg CFC-11 eq (in 10^-5)	3.309	4.825	3.385	3.180	2.595	2.307	3.749	3.651	3.233	3.016
Photochemical oxidation	kg C_2H_4 eq	0.199	0.150	0.106	0.100	0.083	0.074	0.116	0.113	0.099	0.092
Abiotic depletion (fossil fuels)	MJ	5219	4462	3333	3182	2725	2502	3818	3797	3557	3432
Terrestrial ecotoxicity	kg 1,4-DB eq	0.569	0.231	0.194	0.190	0.175	0.167	0.250	0.259	0.268	0.272
Fresh water aquatic ecotoxicity	kg 1,4-DB eq	152	73	69	69	68	67	87	92	98	101
Eutrophication	kg PO_4^{3-} eq	0.836	0.318	0.277	0.273	0.257	0.250	0.341	0.353	0.365	0.371
Acidification	kg SO_2 eq	4.863	1.266	1.005	0.973	0.867	0.817	1.211	1.230	1.215	1.208
Global warming (GWP100a)	kg CO_2 eq	866	835	543	500	381	322	579	548	446	393
Human toxicity	kg 1,4-DB eq	170	96	89	89	86	85	111	116	123	127
Abiotic depletion	kg Sb eq	0.0021	0.0012	0.0012	0.0012	0.0011	0.0011	0.0015	0.0016	0.0017	0.0018

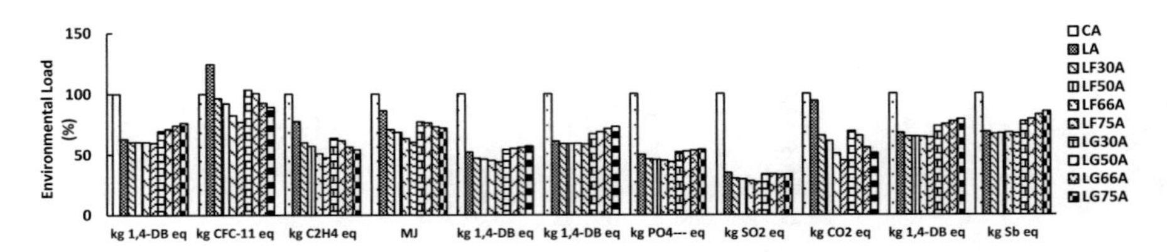

Fig. 3.4 CML-IA baseline impact assessment for 1:3 mortar mixes with cement mortar (CA) base 100

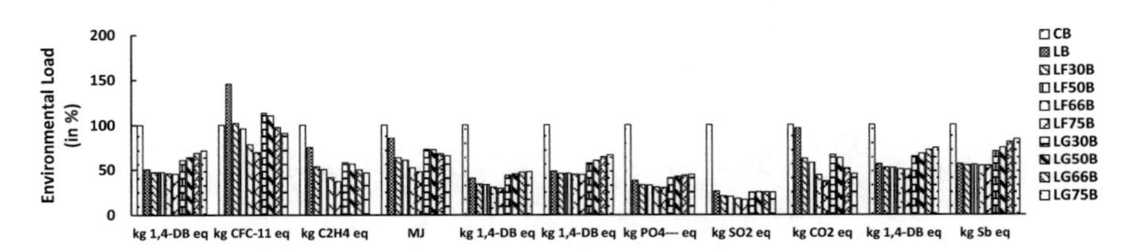

Fig. 3.5 CML-IA baseline impact assessment for 1:1 mortar mixes with cement mortar (CB) base 100

fuels depletion, terrestrial ecotoxicity, fresh water ecotoxicity, eutrophication, acidification, global warming (GWP100), human toxicity, and abiotic depletion resources, respectively, with respect of cement mortar (CA). Similarly, among mortars of 1:1 proportion the least impact was observed in lime–fly ash mortar (LF75B), with a reduction of 54%, 30%, 63%, 52%, 71%, 56%, 70%, 83%, 63%, 50% and 46% for marine aquatic ecotoxicity, ozone layer depletion, photochemical oxidation, fossil fuels depletion, terrestrial ecotoxicity, fresh water ecotoxicity, eutrophication, acidification, global warming (GWP100), human toxicity and abiotic depletion resources respectively with respect of cement mortar (CB).

The CO_2 emissions were separately plotted for mortars of 1:3 and 1:1 mix proportions with graphs as in Figs. 3.6 and 3.7, respectively. Among 1:3 proportion mortars as in Fig. 3.6, (CA) and (LF75A) were responsible for the highest and lowest CO_2 equivalent emissions with values 572 and 259 $_{kg}$ CO_2 eq., respectively. Similarly, as in Fig. 3.7, among 1:1 proportion mortars (CB) and (LF75B) were responsible for the highest and lowest CO_2 equivalent emissions with values 866 and 322 $_{kg}$ CO_2 eq., respectively. This is due to the lesser requirement and utilization of fuels and

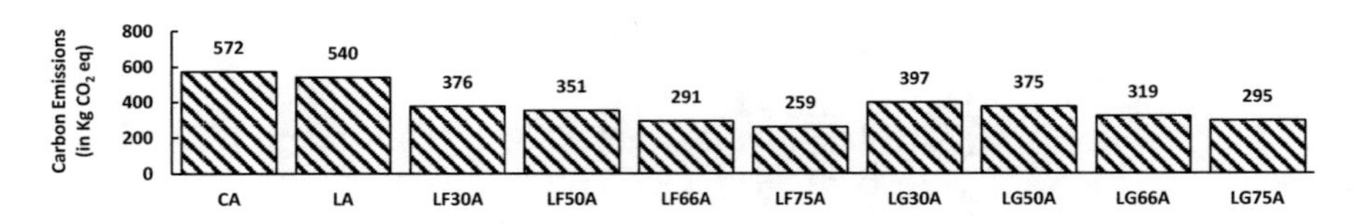

Fig. 3.6 Carbon dioxide emissions of 1:3 mortar mixes

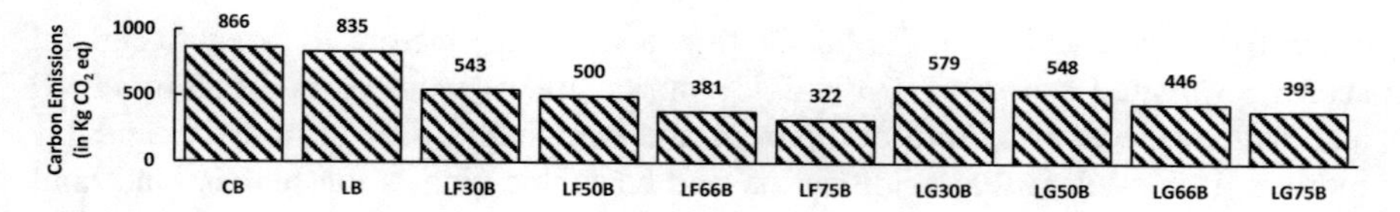

Fig. 3.7 Carbon dioxide emissions of 1:1 mortar mixes

other resources for the production of lime, fly ash, and GGBS in comparison with cement production.

Therefore, mortars (LF75A) and (LF75B) showed the least environmental impacts among all the mortars mixes in 1:3 and 1:1 proportion, respectively. However, with the help of Eco-indicator 99 impact assessment method, the highly affected impact category and the mix with the least single score in terms of environmental impact can be analyzed.

3.4.3 Eco-Indicator 99 Impact Assessment

Eco-indicator 99 impact assessment method helps in obtaining the highly affected environmental impact category. Also, with the single scores obtained from this method, the best sustainable alternative product could be attained.

Mortars with 1:3 and 1:1 mix proportions were analyzed using Eco-indicator 99 impact assessment method. Based on normalized scores graphs have been plotted as shown in Figs. 3.8 and 3.9 for 1:3 and 1:1 proportion mortars, respectively. These

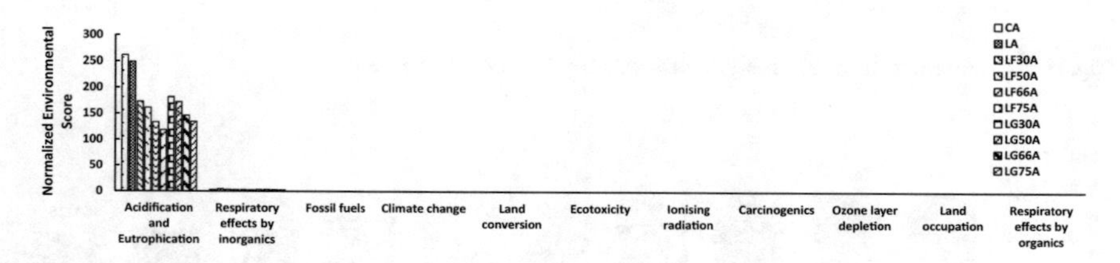

Fig. 3.8 Normalized impact assessment score for 1:3 mortar mixes

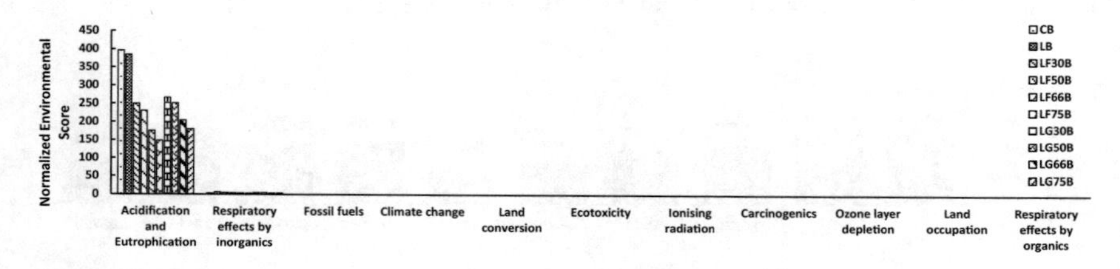

Fig. 3.9 Normalized impact assessment score for 1:1 mortar mixes

normalized scores were multiplied with the weighing factors to attain weighted scores. Weighted scores of 1:3 and 1:1 mortar mixes were plotted in Figs. 3.10 and 3.11, respectively. From normalized and weighted scores of both 1:3 and 1:1 mortars, it is evident that acidification and eutrophication is the highest and land occupation is the least affected impact categories.

The single scores of 1:3 and 1:1 proportion mortars, obtained by combining all the environmental impacts are indicated in Figs. 3.12 and 3.13, respectively. Among 1:3 proportion mortars, (CA) has the highest score of 105,868 points and (LF75A) has the least score of 48,447 points. Similarly, among 1:1 proportion mortars, (CB) and (LF75B) showed highest and lowest scores of 159,901 and 60,396 points, respectively. From these single scores, it is clear that among all the mortars, (LF75A) had

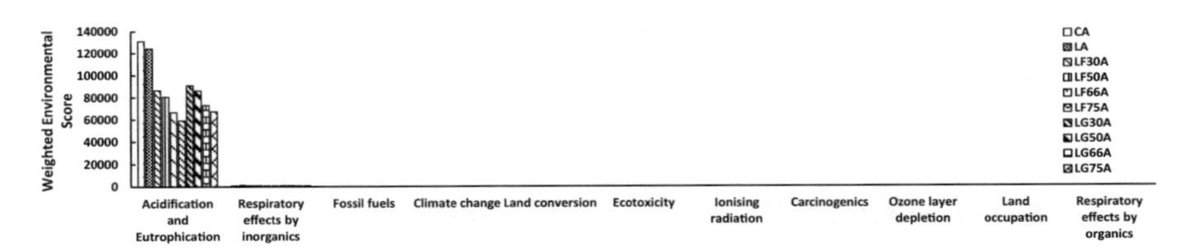

Fig. 3.10 Weighted impact assessment score for 1:3 mortar mixes

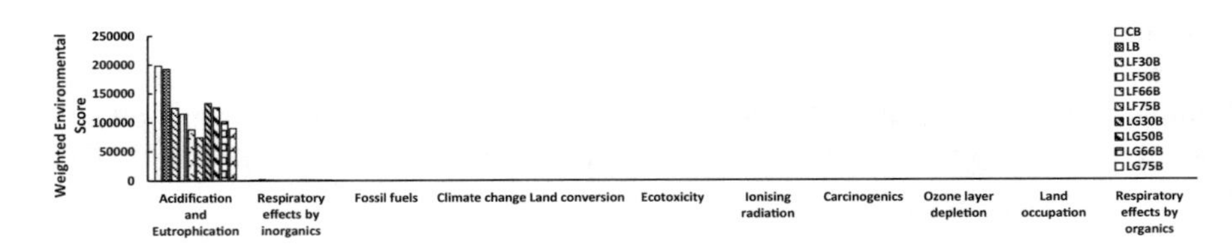

Fig. 3.11 Weighted impact assessment score for 1:1 mortar mixes

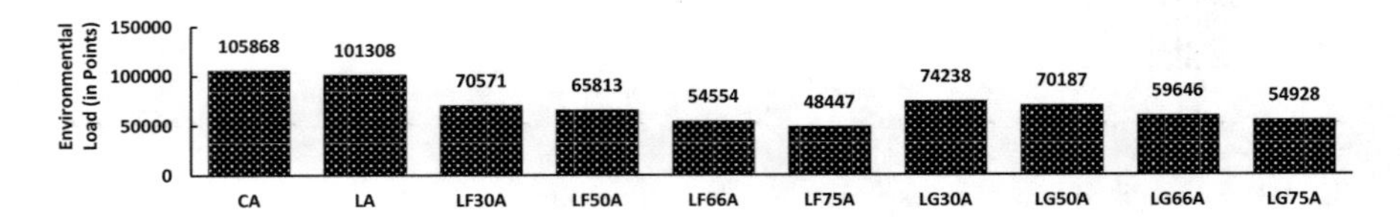

Fig. 3.12 Single impact assessment score for 1:3 mortar mixes

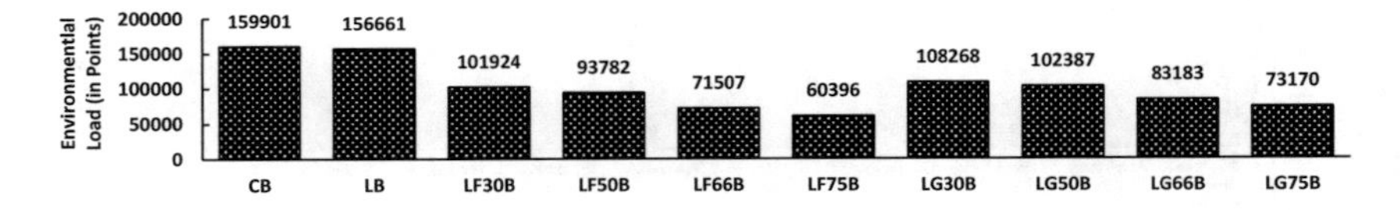

Fig. 3.13 Single impact assessment score for 1:1 mortar mixes

least impact on the environment. Therefore, it is the best mortar alternative for the repair of heritage structures in terms of sustainability.

3.5 Conclusions

The following conclusions could be derived from LCA study of cement and lime-based mortars for 1:3 and 1:1 binder to sand mix proportions:

- Cumulative energy demand for mortars (LF75A) and (LF75B) were observed to be 2829 MJ and 2903 MJ, respectively. These values were 41% and 53% less compared to their corresponding cement mortars (CA) and (CB), respectively.
- LF75A and LF75B mortars showed the least environmental impacts compared to their corresponding cement mortars (CA) and (CB). The environmental impact categories considered were ecotoxicity, ozone layer depletion, photochemical oxidation, acidification, eutrophication, global warming, and abiotic resources depletion.
- Carbon dioxide equivalent emissions of (LF75A) and (LF75B) mortars were the least with values 259 and 322 kg CO_2 eq., respectively, whereas the emissions of corresponding cement mortars (CA) and (CB) were the highest with values 572 and 866 $_{kg}$ CO_2 eq., respectively.
- From normalized and weighted environmental scores of mortars using Eco-Indicator 99 method, acidification and eutrophication attained highest score, while land occupation attained the least score. This indicates that all the mortars impacted mostly on acidification and eutrophication impact category of environment.
- LF75A and LF75B mortars obtained the lowest single environmental scores among 1:3 and 1:1 proportion mortars, respectively. Moreover, (LF75A) scored 48,447 points which is less than 60,396 points scored by (LF75B) mortar. Therefore, (LF75A) is the best alternative mortar mix for the repair of heritage structures in terms of sustainability.

References

Aprianti E (2017) A huge number of artificial waste material can be supplementary cementitious material (SCM) for concrete production–a review part II. J Clean Prod 142:4178–4194. https://doi.org/10.1016/j.jclepro.2015.12.115

ASTM C618-19 (2019) Standard specification for coal fly ash and raw or calcined natural Pozzolan for use in concrete. ASTM International, West Conshohocken, PA. http://www.astm.org

Balksten K, Klasén K (2005) The influence of craftsmanship on the inner structures of lime plasters. In Proceedings of the international RILEM workshop repair mortars for historic masonry. Rilem publications, Delft, Holland

BIS (2005) IS:456-2005, Plain and reinforced concrete -code of practice. Bureau of Indian Standards, New Delhi, India
BIS (2006) IS/ISO 14040:2006 Environmental management-life cycle assessment-principles and framework. Bureau of Indian Standards, New Delhi, India
BIS (2006) IS/ISO 14044:2006 Environmental management-life cycle assessment—requirements. Bureau of Indian Standards, New Delhi, India
BIS (2009) IS:712-2009, Specification for building limes. Bureau of Indian Standards, New Delhi, India
BIS (2009) IS:1624-2009, Methods of field-testing of building lime. Bureau of Indian Standards, New Delhi, India
BIS (2013) IS:12269-2013, Ordinary portland cement 53 grade—specification. Bureau of Indian Standards, New Delhi, India
BIS (2013) IS:3812 (Part-1):2013, Pulverized fuel ash—specification part 1 for use as Pozzolana in cement, cement mortar and concrete. Bureau of Indian Standards, New Delhi, India
BIS (2016) IS:383-2016 Coarse and fine aggregate for concrete-specification. Bureau of Indian Standards, New Delhi, India
Beck K, Al-Mukhtar M (2008) Formulation and characterization of an appropriate lime-based mortar for use with a porous limestone. Environ Geol 56(3–4):715–727. https://doi.org/10.1007/s00254-008-1299-8
Callebaut K, Elsen J, Van Balen K, Viaene W (2001) Nineteenth century hydraulic restoration mortars in the Saint Michael's Church (Leuven, Belgium): natural hydraulic lime or cement? Cem Concr Res 31(3):397–403. https://doi.org/10.1016/S0008-8846(00)00499-3
Cavaco L, Veiga MR, Gomes A (2003) Render application techniques for ancient buildings. In: International symposium on building pathology, durability and rehabilitation, vol 2. LNEC, CIB, Lisbon
Cazalla O, Rodriguez-Navarro C, Sebastian E, Cultrone G, De la Torre MJ (2000) Aging of lime putty: effects on traditional lime mortar carbonation. J Am Ceram Soc 83(5):1070–1076. https://doi.org/10.1111/j.1151-2916.2000.tb01332.x
Cement Sustainability Initiative (CSI) (2013) CSI _ Protocol V3 _ 1 _ 09December2013. Cement sustainability initiative. http://www.cement-co2-protocol.org/en/Content/Resources/Downloads/CSI_ProtocolV3_1_09December2013.xls
CESA [Internet] (2006) CO_2 emissions of various binders: St. Astier Natural Hydraulic Limes (NHL). http://www.stastier.co.uk/nhl/testres/co2emissions.htm
Chever L, Pavía S, Howard R (2010) Physical properties of magnesian lime mortars. Mater Struct 43(1):283–296. https://doi.org/10.1617/s11527-009-9488-9
DECC (2015) 2013 UK greenhouse gas emissions, final figures. In: Statistical release. National Statistics
Del Borghi A (2013) LCA and communication: environmental product declaration. Int J Life Cycle Assess 18(2):293–295
Del Mar M, Maldonado-Ramos L, Van Balen K, García A, Neila FJ (2014) Lime render layers: an overview of their properties. J Cult Herit 15(3):326–330. https://doi.org/10.1016/j.culher.2013.07.004
Garces JIT, Dollente IJ, Beltran AB, Tan RR, Promentilla MAB (2021) Life cycle assessment of self-healing geopolymer concrete. Cleaner engineering and technology, p 100147. https://doi.org/10.1016/j.clet.2021.100147
GreenDelta (2020) openLCA v1.10.3. GreenDelta Berlin. https://www.openlca.org/. Accessed 10 Jan 2021
Grilo J, Faria P, Veiga R, Silva AS, Silva V, Velosa A (2014) New natural hydraulic lime mortars–physical and microstructural properties in different curing conditions. Constr Build Mater 54:378–384. https://doi.org/10.1016/j.conbuildmat.2013.12.078
Hammond PG, Jones C (2008) Inventory of carbon & energy (ICE). University of Bath, UK. http://www.organicexplorer.co.nz/site/organicexplore/files/ICE%20Version%201.6a.pdf

Intergovernmental Panel on Climate Change (IPCC) (2006) 2006 IPCC guidelines for national greenhouse gas inventories: Stationary combustion, chap 2. https://www.ipccnggip.iges.or.jp/public/2006gl/pdf/2_Volume2/V2_2_Ch2_Stationary_Combustion.pdf

Janssen GM, Hendriks CF (2002) Sustainable use of recycled materials in building construction. Adv Build Technol 2:1399–1406. https://doi.org/10.1016/B978-008044100-9/50174-1

Lanas J, Alvarez-Galindo JI (2003) Masonry repair lime-based mortars: factors affecting the mechanical behavior. Cem Concr Res 33(11):1867–1876. https://doi.org/10.1016/S0008-8846(03)00210-2

Lanas J, Bernal JP, Bello MA, Galindo JA (2004) Mechanical properties of natural hydraulic lime-based mortars. Cem Concr Res 34(12):2191–2201. https://doi.org/10.1016/j.cemconres.2004.02.005

Li L, Zhao L, Wang J, Li Z (2011) The study of physical and mechanical properties of two traditional silicate materials in ancient Chinese architecture. Chin J Rock Mech Eng 30:2120–2127

Manjunatha M, Preethi S, Mounika HG, Niveditha KN (2021) Life cycle assessment (LCA) of concrete prepared with sustainable cement-based materials. Mater Today Proc. https://doi.org/10.1016/j.matpr.2021.01.248

Margalha G, Veiga R, Silva AS, De Brito J (2011) Traditional methods of mortar preparation: the hot lime mix method. Cement Concr Compos 33(8):796–804. https://doi.org/10.106/J.Cemconcomp.2011.05.008

Menezes M, Veiga MR, Santos AR (2012) Oral testimony of artisans as a source of knowledge for the safeguard of historical renders. In II international conference on oral tradition: orality and cultural heritage

Moropoulou A, Cakmak AS, Lohvyn N (2000) Earthquake resistant construction techniques and materials on Byzantine monuments in Kiev. Soil Dyn Earthq Eng 19(8):603–615. https://doi.org/10.1016/S0267-7261(00)00021-X

Papayianni I (1998) Criteria and methodology for manufacturing compatible repair mortars and bricks. Compat Mater Protect Eur Cult Herit PACT 56:179–190

Peroni SEA (1981) November. Lime based mortars for the repair of ancient masonry and possible substitutes. In: Mortars, cements and grouts used in the conservation of historic buildings. Symposium, Rome, 3–6 November 1981, pp 63–99

Prakasan S, Palaniappan S, Gettu R (2020) Study of energy use and CO_2 emissions in the manufacturing of clinker and cement. J Inst Eng (India) Ser A 101(1):221–232. https://doi.org/10.1007/s40030-019-00409-4

Rosell JR, Haurie L, Navarro A, Cantalapiedra IR (2014) Influence of the traditional slaking process on the lime putty characteristics. Constr Build Mater 55:423–430. https://doi.org/10.1016/j.conbuildmat.2014.01.007

Rua H (1998) The ten books of architecture by Vitruvius, Lisboa, IST

San Nicolas FL (1639) Arte y Uso de Architectura, S.I. s.f., Madrid

Selio S (1552) Tercero y Cuarto Libro de Architectura de Sebastia Serlio Bolones, Iván de Ayala, Toledo

Sepulcre A, Hernández-Olivares F (2010) Assessment of phase formation in lime-based mortars with added metakaolin, Portland cement and sepiolite, for grouting of historic masonry. Cem Concr Res 40(1):66–76. https://doi.org/10.1016/j.cemconres.2009.08.028

United States Environmental Protection Agency (US EPA) (2014) Emission factors for greenhouse gas inventories. https://www.epa.gov/sites/production/files/2015-07/documents/emissionfactors_2014.pdf

Vázquez OC (2002) Morteros de cal: aplicación en el patrimonio histórico (Doctoral dissertation, Universidad de Granada)

Veiga DRM (2009) Inglesinhos convent: compatible renders and other measures to mitigate water capillary rising problems. J Build Apprais 5(2):171–185. https://doi.org/10.1057/jba2009.28

Veniale F, Setti M, Rodriguez-Navarro C, Lodola S, Palestra W, Busetto A (2003) Thaumasite as decay product of cement mortar in brick masonry of a church near Venice. Cement Concr Compos 25(8):1123–1129. https://doi.org/10.1016/S0958-9465(03)00159-8

Chapter 4
Utilization of Recycled Industrial Solid Wastes as Building Materials in Sustainable Construction

Chinchu Cherian, Sumi Siddiqua, and Dali Naidu Arnepalli

4.1 Introduction

Waste management is a global environmental problem that significantly impacts human, animal, and ecological health; it also represents a broader challenge that affects the global economy. Due to population growth, urbanization, and shifting consumer behavior, waste generation rates are only expected to rise day by day. The continuous growth of the major businesses related to energy production, mining, manufacturing, metallurgical and civil works has tremendously increased the volume of industrial wastes, becoming a significant threat to the ecosystem. The worldwide annual waste generation is expected to increase by 70% over the next 30 years, with vast implications for the environment and health, and thus requiring urgent action for proper management and disposal (Kaza et al. 2018). Even the wealthy nations have less than 20% of their waste recycled annually and often export their garbage (viz., plastics and toxic electronic waste) to the poorer countries (Kellenberg, 2012; Abalansa et al. 2021). It is estimated that approximately 300 million tons of municipal solid waste (MSW) and 8.5 billion tons of other categories of solid waste (viz., industrial, medical, E-waste, hazardous, and agricultural waste) are produced annually in the United States of America alone (USEPA 2015). Furthermore, it is reported that

C. Cherian (✉) · S. Siddiqua
Faculty of Applied Science, School of Engineering, The University of British Columbia, 1137 Alumni Avenue, Kelowna, British Columbia V1V 1V7, Canada
e-mail: chinchu.cherian@ubc.ca

S. Siddiqua
e-mail: sumi.siddiqua@ubc.ca

D. N. Arnepalli
Department of Civil Engineering, Indian Institute of Technology Madras, Chennai, Tamilnadu 600 036, India
e-mail: arnepalli@iitm.ac.in

K. R. Reddy et al. (eds.), *Advances in Sustainable Materials and Resilient Infrastructure*, Springer Transactions in Civil and Environmental Engineering,
https://doi.org/10.1007/978-981-16-9744-9_4

about 5% of the total carbon emissions are generated from the currently inadequate solid waste treatment and disposal practices (Freitas and Magrini 2017), adding to the global warming and climate change risks. This is primarily driven by uncontrolled open dumping (~33%), incineration (~11%), and landfilling without proper gas collection systems (~10%). Experts also warn of severe economic consequences as supplies of non-renewable resources are exhausted and waste production is steadily increasing (D'Amato et al. 2012).

The circular economy model is an emerging response to the global waste challenges; it focuses on a low-carbon economy, economic growth, innovation, and new technologies for waste recycling as a climate change mitigation strategy. However, the route toward waste reduction and increased reuse of waste materials is one of the most challenging for environmental policy. Despite the relevance of the construction industry for socio-economic development, it is the main culprit generating about 40% of the total amount of waste produced globally. Even with the recent advances in reuse and recycling technologies, these nonbiodegradable wastes are mostly managed via landfill and incineration owing to stringent environmental policies and lack of standards for their efficient recycling and reuse. Moreover, decades of experience show that traditional incineration and landfill technologies emit substantial quantities of harmful greenhouse gases (GHG) with serious global climate change implications. Therefore, transforming and re-purposing the industrial residues using the latest waste-to-value technologies is the most viable solution for waste diversion from landfills and safeguarding the environment. These novel approaches for waste management view recycled waste as a potential resource or revenue stream, ensuring greater resource and energy efficiencies and productivity, and hence increasing cost savings, improving the bottom line and the economics of production.

4.1.1 Sustainable Management of Industrial Wastes—Benefits of Utilization in Construction

While the continuous exploration and depletion of natural resources are damaging the environment, on the one hand, the various toxic substances released into the atmosphere during the manufacturing process of construction materials (e.g., Portland cement) are contaminating the surroundings and influencing human health (Safiuddin et al. 2010). On the other hand, developing intelligent and sustainable industrial waste management systems is essential to building a circular economy and promoting efficient economic growth with a minimum environmental impact. The utilization of recycled industrial solid wastes as alternative raw materials in construction applications is one of such innovative efforts to compensate for the scarcity of non-renewable resources and preserve our planet and environment from further degradation (Pappu et al. 2007). Moreover, adopting efficient and eco-friendly waste recycling programs can help the parent industries significantly reduce waste disposal costs and even generate revenue from the alternative end uses.

This chapter discusses the environmental implications of various industrial solid wastes, highlights their recycling potentials and possible use in infrastructure development. Besides, the chapter also discusses the current practices of solid waste applications in actual construction (e.g., geotechnical systems, highway pavements, and construction products) and identifies future research needs. Thus, this chapter is an effort to develop the awareness and importance of industrial waste management and its utilization in a productive manner.

4.2 Industrial Solid Wastes and Environmental Implications

The term "industrial solid waste" can be defined as any solid residual matter produced by industrial activity that is rendered useless and can lead to health hazards (Millati et al. 2019). As shown in Table 4.1, modern industrial activities produce a significant amount of solid waste, including wood waste, ash, slag, sludge, scrap metal and glass, concrete and masonry, etc. Most of these wastes are non-biodegradable, and some are classified as hazardous waste due to their toxic, ignitable, corrosive, or reactive nature. The improper management of such wastes seriously impacts human health and the environment; hence, they require specialized treatment systems to reduce the toxicity levels and meet the environmental regulatory limits. For instance, fly ash (FA) and bottom ash (BA) are environmentally massive inorganic residues from municipal solid waste incinerators as well as coal and biomass power plants (Bhatt et al. 2019; Cherian and Siddiqua 2019). As per the reports, coal ash is the second largest stream of industrial waste in the United States of America at ~130 million tons of coal ash produced per year (Dwivedi and Jain 2014). The industries utilize large areas of land and incur a lot of money over ash disposal; moreover, the ash dumps contaminate the subsoil and render it unfit for agricultural activities due to its highly alkaline nature (Rastogi and Sharma 2012; Amin and Abdelsalam 2019). Further, the increased levels of fly ash suspension in the air can cause respiratory problems upon continuous inhalation.

Cement kiln dust (CKD), a waste by-product of cement manufacturing process, is produced at a rate of ~5 billion tons per year globally (54 to 200 kg of CKD per ton of produced cement clinker) (Seo et al. 2019). Slag is a waste product from the pyrometallurgical processing of various ores, and the different types of slag are blast furnace slag, ferrous and non-ferrous slag, steel slag, etc. Various metal (gold, copper, and zinc mines) and non-metal (coal mine) mining processes produce significant amounts of tailings, many folds higher than the extracted minerals themselves. It is estimated that ~14 million metric tons (15.5 million tons) of blast furnace slag is produced annually in the United States of America alone (Chesner et al. 2002). These wastes are primarily disposed of at high monetary, environmental and ecological costs. Slags and mine tailings often contain harmful heavy metals (such as Cr, Cu, Mn, Ni, Pb, and Zn), which have great potential to damage the environment and

Table 4.1 Characteristic solid wastes produced from various industrial activities

Industry	Description	Typical waste by-product
Manufacturing	• Cement • Glass • Automobile • Pulp and paper • Textile • Chemical • Wineries • Food (rice milling)	Kiln dust and fly ash Glass waste Waste rubber tires Waste wood, fibrous sludge, lime mud, dregs Waste fabric and sludge Acid slag, alkali slag, salt mud, kettle mud Spent bentonite and sludge Rice husk ash
Energy/Power	• Coal power plants • Bioenergy plants	Coal fly ash, bottom ash, slag from boilers Biomass-based fly ash and bottom ash, slag
Metallurgy	• Smelting of iron in blast furnaces • Steel/copper smelting in furnace • Aluminum extraction from bauxite • Production of elemental silicon or alloy in an electric arc furnace	Blast furnace slag Steel and copper slag Red mud Silica fume
Mining and quarrying	• Mining of coal • Metals and non-metal mining Distillation process in mine	Coal tailing Waste stones, quarry dust, phosphogypsum Tailings from gold, copper, zinc mines
Construction	• Construction and demolition activities	Concrete, bricks and blocks, tiles, aggregates, gypsum, asphalt, metals, glass, steel, timber
Waste/water treatment	• Wastewater treatment/ disposal • Solid waste incineration	Sewage sludge, sludge ash from incinerator Fly ash and bottom ash
Miscellaneous	• Wastes from recycling industries	Scrap tire, crushed glass, plastic, textile waste

contaminate food chains and drinking water (Kossoff et al. 2014). Similarly, the bauxite residue called red mud is a highly alkaline and complex composition of different non-toxic and toxic elements (Panda et al. 2017; Milačič et al. 2012; Mishra et al. 2020). Global annual production of red mud (bauxite residue) is about 150 million tonnes; Depending on the quality of the raw material processed, 1–2.5 tons of red mud is generated per ton of alumina produced (Mombelli et al. 2019). Often, we hear about major accidental failures of red mud and mine tailing dams, releasing enormous quantities of harmful waste into river catchments threatening animal and human health (Hatje et al. 2017). Research on slag, tailings, and red mud recycling focuses on their potential applications in construction materials and reprocessing for secondary metal recovery (Piatak et al. 2015; Reddy and Hanumantha Rao 2018;

Lemougna et al. 2020; Alam et al. 2021). However, there is limited understanding of the environmental behavior of such wastes and secondary products and their potential to release contaminants. Therefore, it is essential to evaluate the suitability of specific solid wastes as a potential source for environmental applications.

The different types of construction and demolition (C&D) wastes, including concrete, bricks and blocks, tiles, aggregates and rubbles, gypsum, asphalt, metals, glass, steel, timber, etc., have a high residual value and therefore could be re-purposed in several ways to minimize the disposal (Yeheyis et al. 2013). But there are many challenges for efficient C&D waste recycling, such as the lack of integrated waste management technology, unstable source, waste characteristics, and absence of explicit legislation and policies (Pariatamby 2008). Many other non-hazardous solid wastes from recycling companies, such as scrap tire, waste plastic, crushed glass, and textile waste, can be suitable raw materials for specific construction applications (Rajesh et al. 2015; Rahman et al. 2022). Several studies have proved that using such scraps as secondary resources in concrete, roads and pavements, composite building materials, etc., (Siddique and Naik 2004; Siddique et al. 2008; Reis 2009) is a sustainable waste management option and effective solution for amassing waste stockpiles.

While various wastes can technically be categorized as industrial solid waste, it is essential to determine the types of wastes that can be recycled and reused for intended purposes (Li 2009). Given that recycled industrial waste is utilized in the field of sustainable construction, the present paper reviews some major solid waste by-products that are useful to provide a potentially sustainable resource at different levels of construction.

4.2.1 Major Industrial Solid Wastes Used in Construction

With the continually and rapidly rising demand and cost of day-to-day construction materials, scientific and technological professionals are involved in extensive research to develop alternative sustainable and non-polluting building materials. Similarly, a systematic study of the recycling potential of non-hazardous and hazardous industrial wastes as primary or secondary raw materials for use in construction and their environmental implications have been conducted for several years (Vamvuka and Kakaras 2011; Cherian and Siddiqua 2021). For example, cement manufacturing is a significant contributor to the depletion of natural resources and causes carbon emissions. Hence, the construction industry must play a huge role in ensuring that the planet is conserved by incorporating various sustainable initiatives into all aspects associated with the production and utilization of cement and concrete (Adesina 2020). For example, appropriate waste materials with matching properties and free from impurities can be used as alternative cements, and aggregates in low-carbon construction (D'Alessandro et al. 2017). This practice also allows the reduction of natural resources' consumption and disposal of waste materials resulting from different production processes. Table 4.2 provides a comprehensive overview

Table 4.2 Recycling options of industrial solid wastes in construction (adapted from Safiuddin et al. 2010)

Name of solid waste	Type of waste	Reuse and recycling options
Fly ash, bottom ash, rice husk ash, sewage sludge ash	Agro-industrial	• Production of cement and concrete (blended cement, fine and coarse aggregate, geopolymer concrete) • Construction material (road and embankment fills, bituminous concrete pavement, landfill liner, and cover) • Soil remediation and stabilization • Production of supplementary cementitious material (SCM) • Production of bricks, tiles, and blocks • Production of reinforced polymer composites
Phosphogypsum, waste glass, blast furnace slag, quarry dust	Industrial/manufacturing	• Production of cement and concrete (blended cement and geopolymer concrete) • Building and construction material (aggregates for road and embankment fill) • Production of supplementary cementitious material (SCM) • Production of bricks, tiles, blocks, and ceramic products
Mine tailings (gold, copper, iron, zinc, etc.)	Mining/mineral waste	• Production of cement and concrete (blended cement and geopolymer concrete) • Construction material (fine aggregates/embankment fills) • Production of bricks, tiles, blocks • Used as surface finishing materials

(continued)

of current methods for reusing and recycling industrial solid wastes in construction applications.

The utilization of coal and biomass ash, blast furnace slag, phosphogypsum, silica fume, recycled aggregates, red mud, pulp, and paper residue, etc., to produce new construction products or as admixtures in the civil engineering applications shows some examples of the success of research in this area. The waste residues such as fly ash, rice husk ash, slag, and silica fume with pozzolanic and cementitious properties are used as supplementary cementitious materials (SCMs) to replace the conventional

Table 4.2 (continued)

Name of solid waste	Type of waste	Reuse and recycling options
Waste concrete, coarse and fine aggregates, bricks, rubble, tiles, timber, etc	Construction and demolition waste	• Production of cement and concrete (blended cement and geopolymer concrete) • Building and construction material (fine/coarse aggregate) • Production of bricks, tiles, blocks • Used as unbound granular material in pavement layers (base, sub-base, and capping layers)

cement, which has a high embodied energy and high embodied carbon (Rajamma et al. 2009; Habeeb and Mahmud 2010; Rafieizonooz et al. 2016; Karein et al. 2017; Zareei et al. 2017; Menéndez et al. 2021). The particle fineness, chemical composition (mainly the silica, alumina, and calcium content), and pH are the significant factors governing the pozzolanic properties and reactivity (Arnepalli et al. 2007). The higher pozzolanic and hydraulic characteristics lead to better strength; hence, using such waste materials can significantly improve the durability of the product and the economic and environmental benefits.

The fly ash, bottom ash, and quarry dust incorporation can substantially improve the strength and durability of concrete and earthen building materials (Hameed and Sekar 2009; Abdullah et al. 2019; Naeini et al. 2021). The relative improvement of concrete properties by the waste material depends on the composition and proportion of other components in the mixture and the environmental conditions. Concerned about the depletion of natural materials (viz., traditional soil, sand, and stone aggregates), their cost of extraction and impact on the environment, the waste materials such as fly ash, bottom ash, and waste slag are also used as admixtures for subgrade stabilization and sub-base and base aggregate fill materials for road construction (Chesner et al. 2002; Cherian and Siddiqua 2021). Other efficient reuse options for industrial wastes involve geotechnical applications, such as soil stabilization (Jayaranjan et al. 2014; Jongpradist et al. 2018); backfill for excavations, mine fill, trenches, and retaining walls (Santos et al. 2011; Mohajerani et al. 2017); landfill liners or covers (Mishra and Ravindra 2015; Brännvall and Kumpiene 2016), etc.

4.2.2 Challenges and Issues Related to Industrial Waste Recycling

In today's more environmentally conscious world, the most favorable circumstances exist for the sustainable management of industrial wastes. With strict environmental regulations and waste policies, the industries are bound to follow the most appropriate means of in-house waste management by either recycling for alternative use or disposing off safely (Rastogi and Sharma 2012; Bhatt et al. 2019). Hence, the valorization of industrial wastes is currently one of the most essential and promising technological perspectives, and a more responsible approach to the environment is to utilize waste from one industry as raw material for another industry. However, geographic proximity is considered a defining feature for implementing waste valorization approaches and generating significant environmental, economic, and social benefits (Freitas and Magrini 2017).

Extensive research and development works are carried out toward exploring suitable ways to use the different solid wastes to provide an eco-friendly solution to landfill problems, protect natural raw materials, and contribute to environmental protection. However, these technologies are relatively new and still developed and implemented at a pilot scale only. Unfortunately, there is no significant improvement in effective and large-scale utilization of wastes in the field and industrial implementations. The major constraints related to waste recycling and its use in alternative applications are the temporal/spatial properties variability, presence of harmful components, and the costs associated with the processing (viz., physical, chemical, thermal, or mechanical processes) and transportation of wastes (Pappu et al. 2007). In addition, the lack of awareness, consumer preferences, and many misunderstandings among the public have also impeded the successful waste valorization and utilization. Therefore, the policymakers must refine the regulations, guidelines, and recommended practices to utilize such recycled wastes effectively.

4.3 Potential Avenues for Future Research

The recycling and utilization potentials of different kinds of solid waste in civil engineering applications have undergone considerable development over a very long time. However, the recent research advancements encourage the researchers to undertake further study on the use of various types of fly ash and bottom ash as potential sustainable resources in construction, reducing the cost of building products and an effective method for waste recycling. Current studies indicate a greater scope for synthesizing excellent environmentally friendly building materials such as high-performing geopolymers from recycled waste materials that can withstand the test of time and minimize carbon footprint (Bajpai et al. 2020a, b; Elyamany et al. 2021). Geopolymers are mineral polymers rich in aluminosilicates with an amorphous to a semi-crystalline three-dimensional structure (Leong et al. 2018). The pozzolanic activity and hydraulic properties of residues like fly ash, slag, and red mud make it a suitable precursor for geopolymer synthesis and impart remarkable mechanical and thermal resistance to the construction product (Topcu and Canbaz 2007; Demirboga 2007). The geopolymer concrete offers a better mechanical performance compared to Portland cement concrete, with a relative improvement in the compressive strength by about 1.5–2 times (Al Bakri et al. 2013; Elyamany et al. 2021). Further, Naeini et al. (2021) reported that the partial replacement of Portland cement binder by a wood ash-based geopolymer improved the compressive strength of rammed earth material by ~2.5 times. In addition, the life cycle analysis indicates a lower carbon footprint and better endurance to climate change than other ceramic and Portland cement-based materials (Komnitsas 2011; Bajpai et al. 2020a, b).

Considering the importance of reducing global carbon emissions, much research has been invested in the carbon capture potential of recycled wastes and their geopolymer products used in construction (Jayaranjan et al. 2014; Freire et al. 2020; Taye et al. 2021). Some of the latest research features the potential of the significant amounts of solid by-products of combustion (such as boiler ash and furnace slag) to capture the atmospheric CO_2 by mineral carbonation (Uliasz-Bocheńczyk and Mokrzycki 2020). In addition, some calcium-rich fly ashes from coal and biomass plants and the C&D waste rich in calcium hydroxide are considered to have a high potential for CO_2 sequestration, forming thermodynamically stable carbonate minerals (Kaliyavaradhan and Ling 2017). Recent studies also show that the application of geopolymer mortar made from different industry by-products (such as fly ash, slag, silica fume, etc.) in the innovative digital construction technology called 3D concrete printing offers the potential to achieve a sustainable built environment (Guo et al. 2020; Dai et al. 2021). Moreover, in tandem with the unique features of 3D printing technology such as faster and precise construction, reduced labor costs, and construction waste, the use of geopolymer cement and concrete could revolutionize next generation of construction and building technology (Tay et al. 2016; Kondepudi and Subramaniam 2021).

4.3.1 Engineering and Environmental Assessment of Wastes for Suitable Applications

We can identify the potential to utilize various industrial wastes for the intended applications in construction based on detailed engineering and environmental assessment.

The physicochemical, microstructural, and mineralogical properties of waste materials analyzed using scanning electron microscopy coupled with energy-dispersive spectroscopy (SEM–EDS), X-ray diffraction (XRD), Fourier transform infrared spectroscopy (FTIR), and X-ray fluorescence (XRF) techniques are the primary criteria for the feasibility analysis as an alternative construction raw material. While the XRF analysis of industrial by-products provides the relative oxide composition, XRD technique is used for mineralogical analysis by phase identification and quantification. The presence of aluminosilicate phases and proper mineralogy allow the application of recycled waste as precursors to produce geopolymer cement and similar synthetic materials (Naeini et al. 2021). The FTIR spectrum exhibits the molecular properties of waste products and developed composites in the form of characteristic chemical bond vibrations, and the SEM–EDS analysis reveals the morphology and elemental composition (Tome et al. 2018; Cherian and Siddiqua 2021).

However, most often, the distrust of end-users in utilizing non-conventional materials limits the large-scale implementation of industrial waste recycling and reuse. One of the most significant concerns is their environmental performance. For instance, the industrial wastes may have a heterogeneous composition, and sometimes release elevated concentrations of harmful contaminants. Therefore, a comprehensive chemical and ecotoxicological characterization of waste by-products aiming at the environmental risk assessment is indispensable (Kurda et al. 2018; Rodrigues et al. 2020). The standard Toxicity Characteristic Leaching Procedure (TCLP) method is commonly used to evaluate the leaching behavior and chemical stability of products developed from recycled wastes such as the geopolymer cements and concrete (Tome et al. 2018; Diotti et al. 2021). Likewise, the batch-sorption and column tests are used to simulate the real-life scenario and determine the bioavailable heavy metal concentrations (Tigue et al. 2018; Li et al. 2019). The concentrations of leached metals are compared with the permissible limits of each contaminant stipulated by USEPA, and lower concentrations confirms the efficiency of the developed composite material to stabilize heavy metals (Cherian and Siddiqua 2021; Naeini et al. 2021). In all cases, the developed waste valorization technologies must be evaluated using various life cycle assessment (LCA) tools by benchmarking environmental impacts of the recycled waste products against the conventional material (Bajpai et al. 2020a).

4.3.2 Treatment and Processing of Wastes for Valorization

The various industrial solid wastes, including ash, slag, sludge, waste concrete, masonry, etc., may contain toxic heavy metals, contaminating the environment. Hence, they may require several necessary treatment steps and processes to reduce the toxicity levels and meet the environmental quality standards for further applications. Moreover, subjecting the waste products to specialized treatments prior to utilization in construction increases their chemical and pozzolanic reactivity (Lee et al. 2016). The different possible methods include mechanical (drying and grinding), thermal (calcination), and/or chemical (alkaline activation) treatment. Unfortunately, less attention has been paid to how the different types of industrial solid wastes are treated and processes for various applications.

Onuaguluchi and Eren (2012) describe that the tailing samples used as SCM are air dried and sieved before use to attain the required particle size distributions of cement. The wastes such as bottom ash containing coarser particles may require different levels of grinding/pulverization (in a ball mill) as a treatment method to decrease the particle size and increase specific surface area. Some studies reported that excess grinding altered the microstructure and physicochemical properties of materials (Fan et al. 1999). Besides, the residues can be ground together with some chemical activators leading to mechanical and chemical activation (Fernández-Jiménez et al. 2019).

Calcination (or heating at high temperature) is the other prominent treatment that destroys the crystalline structures and activates the amorphous minerals leading to better pozzolanic activity (Wong et al. 2004). Sometimes, different solid wastes are combined by grinding and calcined together to prepare new SCMs (Chen et al. 2017). The formulation of geopolymer products from industrial wastes involves the combined chemical and thermal treatment. Most geopolymer systems based on fly ash and slag use alkali solutions consisting of NaOH and soluble sodium silicates followed by heat curing (Wilińska et al. 2019; Darweesh 2017).

4.3.3 Environmental Implications of the Waste-Incorporated Building Products

Although, we conduct a comprehensive physicochemical and ecotoxicological analysis of the waste materials and ensure suitability for use in sustainable environmental applications. Yet, it is highly recommended to conduct further studies on the possible environmental risk aspects of the developed technology or product before its full implementation in the field and industrial-level applications. The physicochemical ecotoxicological evaluation of the technology/products (such as geopolymer cement and concrete) can complement the environmental LCA studies and contribute to achieving construction sustainability (Li et al. 2016; Carević et al. 2020). Several studies showed that the degree of mobility and bioavailability of contaminants from construction materials formulated with ecotoxic raw materials (such as fly ash, slag, and red mud) was lesser than their raw materials. This is probably due to the physicochemical reactions and cementation, stabilizing the waste and encapsulating the harmful compounds within the matrix (Cherian and Siddiqua 2021).

4.4 Conclusions

Today, the management of waste has become an essential aspect of our sustainable development and building. Even the wealthiest nations face a waste disposal crisis arising from the rapid industrial growth and amassing waste. There is a wide variety of non-hazardous industrial solid wastes that are safe for environmental applications. However, only very few applications have been found up to now. The ability to recycle such industrial wastes and transform them into functional construction materials has immense potential to reduce carbon footprints and make future construction industry much greener.

According to the specific properties, various industrial waste products (like fly ash, bottom ash, slag, red mud, and silica fume) can be used as partial or complete replacements for conventional construction materials that provide low cost, lightweight, and eco-friendly construction products. It is also proven that excellent environmentally friendly building materials such as high-performing geopolymers can be fabricated from recycled waste materials. The carbon sequestration by mineralization is yet another promising avenue for these geopolymer products complementing the circular economy concepts. However, the ongoing research must be intensified in terms of more comprehensive waste characterization, treatment/processing methods, and environmental implications. Further, the research should be broadened to cover more construction applications. Particular focus should be given to education and research in waste recycling and developing necessary waste management policies. These steps can help to cross the hurdles and realize the successful implementation of waste recycling and reuse on an industrial scale.

References

Abalansa S, El Mahrad B et al (2021) Electronic waste, an environmental problem exported to developing countries: the GOOD, the BAD and the UGLY. Sustainability 13(9):5302
Abdullah MH, Rashid ASA et al (2019) Bottom ash utilization: a review on engineering applications and environmental aspects. In: IOP Conference series: materials science and engineering (vol 527, No. 1, p 012006). IOP Publishing
Adesina A (2020) Recent advances in the concrete industry to reduce its carbon dioxide emissions. Environ Challenges 1:100004
Alam S, Jain S, Das SK (2021) Characterization and an overview of utilization and neutralization for efficient management of Bauxite residue for sustainable environment. Build Mater Sustain Ecol Environ pp 25–47
Al Bakri MMA, Kamarudin H et al (2013) Comparison of geopolymer fly ash and ordinary Portland cement to the strength of concrete
Amin M, Abdelsalam BA (2019) Efficiency of rice husk ash and fly ash as reactivity materials in sustainable concrete. Sustain Environ Res 29(1):1–10
Arnepalli DN, Das BB, Singh DN (2007) Methodology for rapid determination of pozzolanic activity of materials. J ASTM Int 4(6):1–11
Bajpai R, Choudhary K et al (2020a) Environmental impact assessment of fly ash and silica fume based geopolymer concrete. J Cleaner Prod 254:120147
Bajpai R, Shrivastava A, Singh M (2020b) Properties of fly ash geopolymer modified with red mud and silica fume: a comparative study. SN Appl Sci 2(11):1–16
Bhatt A, Priyadarshini S et al (2019) Physical, chemical, and geotechnical properties of coal fly ash: a global review. Case Stud Constr Mater 11:e00263
Brännvall E, Kumpiene J (2016) Fly ash in landfill top covers–a review. Environ Sci Process Impacts 18(1):11–21
Carević I, Štirmer N et al (2020) Leaching characteristics of wood biomass fly ash cement composites. Appl Sci 10(23):8704
Chen D, Deng M et al (2017) Mechanical properties and microstructure of blended cement containing modified quartz tailing. J Wuhan Univ Technol Mater Sci Ed 32(5):1140–1146
Cherian C, Siddiqua S (2019) Pulp and paper mill fly ash: a review. Sustainability 11(16):4394
Cherian C, Siddiqua S (2020) Formulation of a sustainable geopolymeric binder based on pulp mill fly ash for subgrade stabilization. In: Proceedings of the 73rd canadian geotechnical conference (GeoVirtual 2020). pp 8
Cherian C, Siddiqua S (2021) Engineering and environmental evaluation for utilization of recycled pulp mill fly ash as binder in sustainable road construction. J Cleaner Prod 298:126758

Chesner WH, Collins RJ et al (2002) User guidelines for waste and by-product materials in pavement construction (No. FHWA-RD-97–148, Guideline Manual). Recycled Mater Res Cent

D'Alessandro A, Fabiani C et al (2017) Innovative concretes for low-carbon constructions: a review. Int J Low-Carbon Technol 12(3):289–309

D'Amato A, Managi S, Mazzanti M (2012) Economics of waste management and disposal: decoupling, policy enforcement and spatial factors. Environ Econ Policy Stud 14(4):323–325

Dai S, Zhu H et al (2021) Stability of steel slag as fine aggregate and its application in 3D printing materials. Constr Build Mater 299:123938

Darweesh HHM (2017) Geopolymer cements from slag, fly ash and silica fume activated with sodium hydroxide and water glass. Interceram-International Ceram Rev 66(6):226–231

Demirboga R (2007) Thermal conductivity and compressive strength of concrete incorporation with mineral admixtures. Build Environ 42:2467–2471

Diotti A, Plizzari G, Sorlini S (2021) Leaching behaviour of construction and demolition wastes and recycled aggregates: statistical analysis applied to the release of contaminants. Appl Sci 11(14):6265

Dwivedi A, Jain MK (2014) Fly ash–waste management and overview: a review. Recent Res Sci Technol 6(1)

Elyamany HE, Abd Elmoaty AEM, Diab ARA (2021) Properties of slag geopolymer concrete modified with fly ash and silica fume. Can J Civ Eng (ja).

Fan Y, Yin S et al (1999) Activation of fly ash and its effects on cement properties. Cem Concr Res 29(4):467–472

Fernández-Jiménez A et al (2019) Mechanical-chemical activation of coal fly ashes: n effective way for recycling and make cementitious materials. Front Mater 6:51

Freire AL, Moura-Nickel CD et al (2020) Geopolymers produced with fly ash and rice husk ash applied to CO2 capture. J Cleaner Prod 273:122917

Freitas LA, Magrini A (2017) Waste management in industrial construction: investigating contributions from industrial ecology. Sustainability 9(7):1251

Guo X, Yang J, Xiong G (2020) Influence of supplementary cementitious materials on rheological properties of 3D printed fly ash based geopolymer. Cem Concr Compos 114:103820

Habeeb GA, Mahmud HB (2010) Study on properties of rice husk ash and its use as cement replacement material. Mater Res 13(2):185–190

Hameed MS, Sekar ASS (2009) Properties of green concrete containing quarry rock dust and marble sludge powder as fine aggregate. ARPN J Eng Appl Sci 4:83–89

Hatje V, Pedreira RM et al (2017) The environmental impacts of one of the largest tailing dam failures worldwide. Sci Rep 7(1):1–13

Jayaranjan MLD, Van Hullebusch ED, Annachhatre AP (2014) Reuse options for coal fired power plant bottom ash and fly ash. Rev Environ Sci Bio/Technology 13(4):467–486

Jongpradist P, Homtragoon W et al (2018) Efficiency of rice husk ash as cementitious material in high-strength cement-admixed clay. Adv Civ Eng

Kaliyavaradhan SK, Ling TC (2017) Potential of CO2 sequestration through construction and demolition (C&D) waste—an overview. J CO2 Utilization 20:234–242

Karein SMM, Ramezanianpour AA et al (2017) A new approach for application of silica fume in concrete: wet granulation. Constr Build Mater 157:573–581

Kaza S, Yao L et al (2018) What a waste 2.0: a global snapshot of solid waste management to 2050. World Bank Publications, Washington, DC, USA, ISBN 978-1-4648-1347-4

Kellenberg D (2012) Trading wastes. J Environ Econ Manag 64(1):68–87

Komnitsas KA (2011) Potential of geopolymer technology towards green buildings and sustainable cities. Procedia Engineering 21:1023–1032

Kondepudi K, Subramaniam KV (2021) Formulation of alkali-activated fly ash-slag binders for 3D concrete printing. Cement Concr Compos 119:103983

Kossoff D, Dubbin WE et al (2014) Mine tailings dams: characteristics, failure, environmental impacts, and remediation. Appl Geochem 51:229–245

Kurda R, Silvestre JD, de Brito J (2018) Toxicity and environmental and economic performance of fly ash and recycled concrete aggregates use in concrete: a review. Heliyon 4(4):e00611
Lee NK, An GH et al (2016) Improved reactivity of fly ash-slag geopolymer by the addition of silica fume. Adv Mater Sci Eng p12
Lemougna PN, Yliniemi J et al (2020) Utilisation of glass wool waste and mine tailings in high performance building ceramics. J Build Eng 31:101383
Leong HY, Ong DEL, Sanjayan JG et al (2018) Strength development of soil–fly ash geopolymer: assessment of soil, fly ash, alkali activators, and water. J Mater Civ Eng 30:04018171
Li J (2009) Types, amounts and effects of industrial solid wastes. Point Sources Pollut Local Eff Contr 2:204
Li YY, Zhang TT et al (2019) Mechanical properties and leaching characteristics of geopolymer-solidified/stabilized lead-contaminated soil. Adv Civil Eng
Li Y, Liu Y et al (2016) Environmental impact analysis of blast furnace slag applied to ordinary Portland cement production. J Clean Prod 120:221–230
Menéndez E, Argiz C, Sanjuán MÁ (2021) Reactivity of ground coal bottom ash to be used in Portland cement. 4(3):223–232
Milačič R, Zuliani T, Ščančar J (2012) Environmental impact of toxic elements in red mud studied by fractionation and speciation procedures. Sci Total Environ 426:359–365
Millati R, Cahyono RB et al (2019). Agricultural, industrial, municipal, and forest wastes: an overview. Sustainable Res Recovery Zero Waste Approaches 1–22
Mishra AK, Ravindra V (2015) On the utilization of fly ash and cement mixtures as a landfill liner material. Int J Geosynthetics Ground Eng 1(2):1–7
Mishra MC, Reddy NG et al (2020) A study on evaluating the usefulness and applicability of additives for neutralizing extremely Alkaline Red mud waste. In: Sustainable environmental geotechnics (pp 139–149). Springer, Cham
Mohajerani A, Lound S et al (2017) Physical, mechanical and chemical properties of biosolids and raw brown coal fly ash, and their combination for road structural fill applications. J Clean Prod 166:1–11
Naeini AA, Siddiqua S, Cherian C (2021) A novel stabilized rammed earth using pulp mill fly ash as alternative low carbon cementing material. Const Build Mater 300:124003
Mombelli D, Barella S et al (2019) Iron recovery from bauxite tailings red mud by thermal reduction with blast furnace sludge. Appl Sci 9(22):4902
Onuaguluchi O, Eren Ö (2012) Cement mixtures containing copper tailings as an additive: durability properties. Mater Res 15:1029–1036
Pappu A, Saxena M, Asolekar SR (2007) Solid wastes generation in India and their recycling potential in building materials. Build Environ 42(6):2311–2320
Panda I, Jain S et al (2017) Characterization of red mud as a structural fill and embankment material using bioremediation. Int Biodeterior Biodegradation 119: 368–376
Pariatamby A (2008) Challenges in sustainable management of construction and demolition waste. 491–492
Piatak NM, Parsons MB, Seal RR II (2015) Characteristics and environmental aspects of slag: a review. Appl Geochem 57:236–266
Rafieizonooz M, Mirza J et al (2016) Investigation of coal bottom ash and fly ash in concrete as replacement for sand and cement. Constr Build Mater 116:15–24
Rahman SS, Siddiqua S, Cherian C (2022) Sustainable applications of textile waste fiber in the construction and geotechnical industries: a retrospect. Cleaner Eng and Technol, 100420
Rajamma R, Ball RJ et al (2009) Characterisation and use of biomass fly ash in cement-based materials. J Hazard Mater 172(2–3):1049–1060
Rajesh S, Hanumantha Rao B et al (2015) Environmental geotechnology: an Indian perspective. Environ Geotech 2(6):336–348
Rastogi D, Sharma U (2012) Utilization of industrial waste in the construction industry. Int J Res Eng Technol 1(3):307–312

Reddy NG, Hanumantha Rao B (2018) Compaction and consolidation behaviour of untreated and treated waste of Indian red mud. Geotech Res 5(2):106–121

Reis JMLD (2009) Effect of textile waste on the mechanical properties of polymer concrete. Mater Res 12:63–67

Rodrigues P, Silvestre JD et al (2020) Evaluation of the ecotoxicological potential of fly ash and recycled concrete aggregates use in concrete. Appl Sci 10(1):351

Safiuddin M, Jumaat MZ et al (2010) Utilization of solid wastes in construction materials. Int J Phys Sci 5(13):1952–1963

Santos F, Li L et al (2011) Geotechnical properties of fly ash and soil mixtures for use in highway embankments. In: World of Coal Ash (WOCA) Conference. pp 9–12

Savitha K (2015) Waste generation trends and their socio-economic implications: a case study. Int J Ecol Econ Stat 36(4):104–112

Seo M, Lee SY et al (2019) Recycling of cement kiln dust as a raw material for cement. Environments 6(10):113

Siddique R, Naik TR (2004) Properties of concrete containing scrap-tire rubber–an overview. Waste Manage 24(6):563–569

Siddique R, Khatib J, Kaur I (2008) Use of recycled plastic in concrete: a review. Waste Manage 28(10):1835–1852

Tay YW, Panda B et al (2016) Processing and properties of construction materials for 3D printing. In: Materials science forum (vol 861). Trans Tech Publications Ltd, pp 177–181

Taye EA, Roether JA et al (2021) Hemp fiber reinforced red mud/fly ash geopolymer composite materials: effect of fiber content on mechanical strength. Materials 14(3):511

Tigue AAS, Malenab RAJ et al (2018) Chemical stability and leaching behavior of one-part geopolymer from soil and coal fly ash mixtures. Minerals 8(9):411

Tome S, Etoh MA et al (2018) Characterization and leachability behaviour of geopolymer cement synthesised from municipal solid waste incinerator fly ash and volcanic ash blends. Recycling 3(4):50

Topcu IB, Canbaz M (2007) Effect of different fibers on the mechanical properties of concrete containing fly ash. Constr Build Mater 21:1486–1491

Uliasz-Bocheńczyk A, Mokrzycki E (2020) The potential of FBC fly ashes to reduce CO2 emissions. Sci Rep 10(1):1–9

USEPA (2015) National overview: facts and figures on materials, wastes and recycling

Vamvuka D, Kakaras E (2011) Ash properties and environmental impact of various biomass and coal fuels and their blends. Fuel Process Technol 92(3):570–581

Wilińska I, Pacewska B, Ostrowski A (2019) Investigation of different ways of activation of fly ash–cement mixtures. J Therm Anal Calorim 138(6):4203–4213

Wong RCK, Gillott JE et al (2004) Calcined oil sands fine tailings as a supplementary cementing material for concrete. Cem Concr Res 34(7):1235–1242

Yeheyis M, Hewage K et al (2013) An overview of construction and demolition waste management in Canada: a lifecycle analysis approach to sustainability. Clean Technol Environ Policy 15(1):81–91

Zareei SA, Ameri F et al (2017) Rice husk ash as a partial replacement of cement in high strength concrete containing micro silica: evaluating durability and mechanical properties. Case Stud Constr Mater 7:73–81

Chapter 5
Development of a Machine Learning-Based Drone System for Management of Construction Sites

Kundan Meshram and Narala Gangadhara Reddy

5.1 Introduction

In order to manage construction sites, a large pool of labours, managers, site engineers, construction agencies, etc., are required. It is not easy to track the work done by these parties due to the following issues still persistent with old construction methodologies:

- Lack of stock measurement technologies for soil, cement, stone, etc., during unloading and usage.
- Lack of equipment for daily tracking of consumables.
- Incorrect evaluation of site progress on a daily basis.
- Real-time measurement of construction quality and deployment of correction processes.

In order to tackle these issues, structural researchers have devised different plans which involve work package analysis, deployment of progress monitoring rules and devices, management of different planning interfaces, progress reporting protocols, management of risks, and mitigation of issues. A flow diagram that indicates the connections between these components can be observed from Fig. 5.1, wherein all these components are seen to be connected in a circular manner, indicating continuous management of these components.

In order to deploy these architectures, a wide variety of systems have been proposed by researchers. A review of these systems is done in the next section, which allows readers to evaluate best practices for each of these steps. This is followed by the

K. Meshram (✉)
Guru Ghasidas Vishwavidyalaya (A Central University), Bilaspur 495009, India
e-mail: kundan.meshram@ggu.ac.in

N. G. Reddy
Kakatiya Institute of Technology and Science, Warangal 506015, India

K. R. Reddy et al. (eds.), *Advances in Sustainable Materials and Resilient Infrastructure*,
Springer Transactions in Civil and Environmental Engineering,
https://doi.org/10.1007/978-981-16-9744-9_5

Fig. 5.1 DJI Mavik Air 2 drone with mounted components

proposed drone-based machine learning solution for construction site management and its performance evaluation. Finally, this text concludes with some interesting observations about the proposed system and suggests ways to improve the same.

5.2 Literature Review

Construction site management involves a multitude of operations which must be done with efficient synchronization for better results. In addition to these operations, there are a large number of fields where construction management can be applied. For instance, the work in Wang et al. (2020) uses real-time monitoring and early warning model for identifying leakage in anaerobic reactors. During the construction of these reactors, different observatory parameters like methane concentration, operating temperature, operating pressure, leakage height, leakage calibre, etc., are evaluated. All these parameters are stored in real-time databases and quantitative prediction is performed. This prediction model uses knowledge-base dependent prediction, which determines if there will be a future accident due to the given parameters. In case of any accident, an emergency response guidance mechanism is activated which actuates safety measures in the system. Also, real-time condition monitoring, accident qualitative analysis and prediction, consequence quantitative calculation and emergency response guidance modules are seen to be connected. The system application is not directly related to site-based construction monitoring, but the modular deployment of this model is highly efficient. This model is the base of the proposed drone-based modular design framework, and thus is reviewed in this text. This work when combined with innovation management research (IMR) as discussed by Ellwood and Horner (2020) assists in further improving construction speed by segregating construction management into three different categories, which are technology development model, innovation acceleration and stakeholder interest. Characteristics for each of these models can be observed as follows:

- Technology development model, wherein previously repetitive testing was performed in order to generate a new model. This is replaced by simulating the model and reducing errors before deployment, while "investor-ready" propositions are directed from a future research perspective.
- Innovation acceleration previously required proof of concepts for deployment, which is replaced with generation of robust technology concepts. This can be extended with the help of stake-holder support-based technology enhancement.
- Stakeholders used to refer previous contracts, etc., but currently they are leading from the forefront to shape technology concepts. In future, investors will be involved from early stage site development

Based on this model, better technological, innovation and investor performance are expected. Adding these technologies also assists in handling large portfolios. The work in suggests addition of machine learning for time series forecasting, early warning detection and risk management to assist handling of large portfolios (Hopmere et al. 2020). Due to the use of machine learning models like Auto Regressive Integrated Moving Average (ARIMA) the system is able to detect future issues with site management with 85% accuracy. These future issues are detected with help of the following analysis:

- Detect issues at post-facto stage with very small samples for very large projects in mixed industry areas.
- Improve periodicity of data collection to collect data faster.
- Use data from one technology area to learn about nuances about other areas.

Based on these models, the overall efficiency of data processing is improved. An application of these models can be observed from Li (2019), wherein Economical Building Management System (EBMS) is discussed. This system uses multiple agents in order to improve purchasing decisions for various construction components like electricity, water, soil, etc. The same model can be used to improve purchasing decisions for other construction equipment. Due to application of EBMS, the overall cost reduction of 62% is observed for smart buildings.

The EBMS model can be extended via deep learning and machine learning models, a survey of these models as applied to construction industry is given in Akinosho et al. (2020), wherein Convolutional Neural Networks (CNNs), Autoencoders and Generative Adversarial Networks (GANs) are discussed. As per their research, around 75% of personnel in construction industries are workers, 15% are engineers, 5% are firemen, and remaining 5% are architects. They have suggested the use of these deep learning models for image classification, image captioning, activity recognition, object detection, data augmentation, and object tracking. An accuracy of 95% is obtained for each of these operations when applied to construction processes. An application of object detection for construction processes is given in Li et al. (2020), wherein a system that performs helmet detection for worker safety is described. This system uses person recognition with object recognition for detection of helmets. An accuracy of 98% is achieved using CNNs, which is very high and can be used for real-time detection purposes.

Apart from detecting these objects, it is recommended that machine learning be used for construction tracking of sites in smart cities. The work in Alahakoon et al. (2020) proposes a machine learning framework for managing construction sites and analyzing their data for improved construction performance. The work combines data from different sites to get a global perspective about them, and proposes uses of local data processing to get local insights about this data.

Due to the use of this machine learning model, overall site accidents are reduced by 8%, while the productivity is improved by 10%, when compared to traditional non-machine learning-based approaches. A similar work can be observed from Cadavid et al. (2020), wherein machine learning models are used for enhanced production planning and control on construction sites. An example application of these processes can be observed from Feng et al. (2020), wherein bidirectional LSTM (long short-term memory) units are used for identification of Building Occupancy Detection. This detection is performed with the help of Advanced Metering Infrastructure (AMI) data, wherein parameters like current consumption, voltage standards, utilized phase angle, and power consumption are extracted from Electricity Consumption and Occupancy (ECO) dataset. Based on these patterns, building occupancy is determined with 97% accuracy. The same architecture is used by the proposed drone-based solution for evaluating consumption of various on-site consumables. This concept is extended in Rashidi et al. (2016), wherein Construction Material classification is done based on different machine learning models.

Due to the use of machine learning, the following accuracy results are obtained:

- Concrete has been detected with an accuracy of 90%.
- Brick has been detected with an accuracy of 93%.
- Oriented strand board (OSB) with an accuracy 75%.

In the underlying research, the same algorithm has been used for machine learning to identify construction objects. Similar research has been done in Oprach et al. (2019), which indicates machine learning models must be used in construction industry. The concept can be extended by addition of big data systems for improved data analysis. The work in Jung et al. (2018) suggests use of analytic hierarchy process (AHP) for improving the overall accuracy of data prediction.

Due to the use of AHP, overall efficiency of construction management in terms of Order, Design¸ Purchas, Construction, and Commission is improved by 19%, which saves construction cost and reduces time needed for building the site. Due to these advantages of machine learning, site engineers often do not consider the critical factors that affect the use of these technologies on the site. A survey of these factors is given in Jia et al. (2020), wherein it can be observed that cost management, schedule management, quality management, safety management, cause degree, and center degree are the main factors which must be considered while deploying these systems. A detailed analysis of these factors is given in Supriadi and Pheng (2018), wherein Business Continuity Management (BCM) is done based on them. The BCM is applied to sites like financial services planning, education institutes, and medical institutes. It is observed that BCM can also be improved using machine learning models like Q-Learning, CNN, and LSTM.

Application of unmanned aerial vehicles (UAVs) is the center piece of this research. The work in Weng et al. (2020) showcases the use of such vehicles for structural displacement measurement. Due to this UAV-based measurement, the system is able to work in real time for any kind of environment. A similar system is deployed in the underlying work for crack and inconsistency detection on sites. The system is able to detect displacements with 95% accuracy, which assists site engineers and workers to correct the construction in real time and improve its quality. Similar application of machine learning mechanisms is adopted in Jurczenko (2020), Choudhury et al. (2021), Lee et al. (2020) wherein asset management and pattern discovery are discussed. Due to the use of these models the efficiency of asset management and pattern discovery is improved to 93%, which makes these systems applicable for real-time usage. These algorithms are extended for worker safety management using UAVs in Howard et al. (2018), wherein worker tracking, analysis and collision or accident like events are detected using machine learning with almost 90% accuracy. The concept of drones for worker safety can be extended to scheduling construction site surveillance using an integration of cloud computing with dynamic programming in Yi and Sutrisna (2021). Here, continuous scheduling-based monitoring of buildings like Innovation Complex, Quadrangle Building, Lawns, Atrium Building, Mathematical Science Building, Student Central, Recreation Center, etc., are done with 94% accuracy. Due to the use of drones, the construction site management is safer, faster, and leaner. This effect is discussed in Stroud and Weinel (2020), wherein European steel industry is discussed as a case study. Due to the use of drones, the site management is 15% safer, which indicates reduction in accidents, 25% faster due to improved tracking and control, and it is 30% leaner, due to quick decision-making on the site. It can be further optimized with the use of machine learning models as suggested in Otto et al. (2018), wherein lawn moving patterns, spiral patterns, and graphical patterns are suggested for drone movements along with CNN- and LSTM-based processing systems. An application to site safety using drones can be observed from Gheisari and Esmaeili (2016), wherein radio-frequency identification (RFID), global positioning system (GPS), ultra-sonic sensors, etc., are used for accident pre-emption and avoidance. The system is tested under working in proximity of boomed vehicles/cranes, working near unprotected edge/opening, working in blind spot of heavy equipment, conducting post-accident investigation, using boom vehicles/cranes in proximity of overhead power lines, inspecting housekeeping, inspecting proper usage of fall protection systems, working at unprotected trench, working in proximity of hazardous materials, inspecting proper usage of PPE on the site, inspecting confined space entry, inspecting requirements for ladder/scaffold, inspecting at-risk rigging operation, inspecting requirements for guarding machinery, inspecting ergonomics requirements and appropriate usage of tag out/lock out conditions. An accuracy of 94% is achieved with the help of machine learning in these systems. Similar performance is observed for managing geological construction sites as discussed in Giordan and Adams et al. 2020. Here too the performance of this system is nearly 90%, which indicates its real-time usability. The work in also confirms a combination of machine learning and artificial intelligence with drone-based systems for an improved on-site construction performance (Kim and Park 2006; Vanhoucke 2012; Yaseen et al. 2020).

5.3 Proposed Drone-Powered Machine Learning Model for Construction Site Management

The proposed drone-powered machine learning model for managing construction sites is built using two main components:

- Optimized custom drone design for monitoring and control of different site parameters.
- Cloud-based machine learning model for processing drone data and sending commands to the drone for optimized control.

The custom drone design is shown in Fig. 5.1, it consists of the following internal components:

- A high-performance low power four-direction 16 megapixel camera module to capture multi-direction imagery
- A 20,000 mAh, 3.7 V Li-polymer battery that provides a standby time of up to 35 min on a single charge.
- Following environmental sensors:
  - Temperature and humidity sensing using DH11 module.
  - pH sensor module.
  - Ultra-sonic sensor.
  - Pressure sensor BME280.
  - Light sensor TSL2591.
- DJI Mavik Air 2 interface as the base drone model.
- Buzzer module for alerting on site personnel.
- ESP8266 WiFi module for cloud connectivity.

Using these components, processing of images, environmental parameters, balancing, etc., is done on construction sites. The drone component and its usage in construction site management along with algorithmic details are given in Table 5.1.

Table 5.1 Drone components along with their use-cases

Drone component	Usage and algorithm details
DJI Mavik Air 2	This drone is selected due to its high-precision flying. The drone can be controlled from remote location, and has autobalancing feature. It is sprinkle-proof, dust-proof, and has high real-time performance. Due to these advantages, this drone is selected for monitoring construction sites. The performance of this drone is at-par with Phantom drones, but this drone has external interface capabilities

(continued)

Table 5.1 (continued)

Drone component	Usage and algorithm details
Four-direction camera modules	Each camera is able to capture images from each direction of the site, the following sub-modules are implemented with this camera: • Crack detection module is implemented using neural networks mentioned in Weng et al. (2020). This module is able to analyze any kind of cracks or dis-orientations on site walls. The module is able to detect this with 97% accuracy due to use of CNN • Soil measurement module is implemented using LSTM-based CNN via the architecture used in Alahakoon et al. (2020), this architecture is able to produce an accuracy of 95% in measuring quantity of soil and stone unloaded on sites. Due to this, the overall cost of construction comes down by 3.5% • Worker safety module is able to measure any irregular worker movements, and alert workers about their own safety. This module uses the same architecture of Weng et al. (2020) for reducing worker accidents to 1% • Worker efficiency tracking module also uses CNN, via which classes like idle worker, running worker, working worker, and sleeping worker are identified. These classes assist in alerting owners and contractors about worker positions, and allow them to take corrective actions as and when needed. This module improves worker efficiency by 15% when compared to a non-efficiency tracking scenario • Site quality monitoring module uses BiLSTM-based classification module from Feng et al. (2020) for better classification accuracy. It is able to detect width of walls, height of rooms, height of floors, tile fitting, window size, and other parameters with 94% accuracy. Due to this, the overall construction quality improves by 18% and reduces re-construction and modification costs by 4% • Site progress monitoring module uses continuous and periodic surveillance using the system defined in Yi and Sutrisna (2021), which allows continuous site updates and progress monitoring. This module is able to reduce the overall surveillance cost by 70%, thereby reducing site visits and reducing site construction cost by 1.5%

(continued)

Table 5.1 (continued)

Drone component	Usage and algorithm details
Temperature and humidity sensing using DH11 module	This module is used for sensing temperature and humidity levels of various site regions. In case the values go beyond a certain threshold, then the Arduino controller alerts cloud device to take corrective actions
pH sensor module	This sensor is used to test pH of the water used on the site for curing purposes. This module ensures that non-acidic water is used to cure the site, thereby reducing corrosion and other side effects
Ultra-sonic sensor	During flying the drone is alerted in case of any nearby obstacles using this module
Pressure sensor BME280	Pressure on different site regions is kept under determined thresholds using this module
Light sensor TSL2591	While construction, if there is lack of light in given site areas, then this module alerts and provides a temporal light report for better understanding of light conditions in the given regions
Buzzer	A high-efficiency buzzer is used to alert all workers and working personnel on site

All this information is given to the cloud, wherein processing of this data is done. The following CNN architecture showcased in Fig. 5.2 is used for processing different images for crack detection, worker detection, and quality monitoring. Using these modules, the overall efficiency of the system is improved. The high-powered battery allows for continuous and periodic monitoring of the site. Using cloud interface, the following data communication is performed:

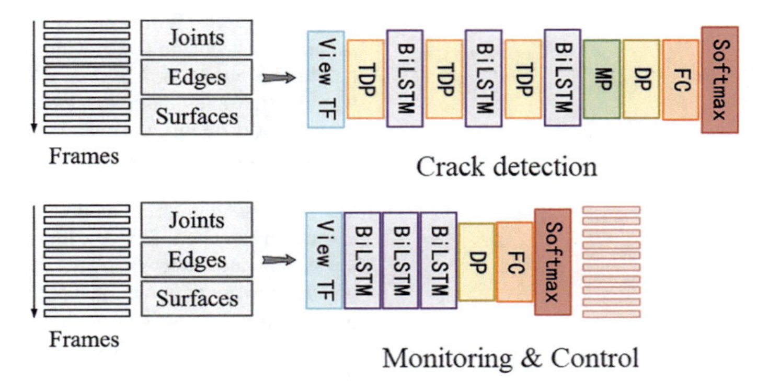

Fig. 5.2 CNN models for crack detection, monitoring and control

- Drone sends image and parametric data to the cloud.
- Cloud processes this data using the CNN architecture as described in Fig. 5.2.
- Using this processing, cloud generates the following control signals:
  - If crack detected, then send CRACK = 1, else CRACK = 0.
  - If soil & stone measurement is proper, then send SS = 1, else SS = Difference in soil and stone amount from what is prescribed.
  - If worker safety is ok, then send WS = 1, else send WS = 0 and blow alarm on the site.
  - If worker is in working mode, then send WORKER = 1, else send WORKER = 0, and alert workers on the site.
  - If all parameters are normal, then send PARAMETER = 1, else send PARAMETER = < List of Abnormal Parameters > and alert workers on site.

Based on this architecture, the drone is deployed, and tested on five different sites. The next section describes results obtained from these sites.

5.4 Performance Analysis

In order to evaluate real-time performance of the proposed drone-powered system, it was tested on five different sites. Each site was chosen such that there are two flats from the same builder being constructed in the same scheme. Many builders were approached for this research, but Lodha construction and Omkar builders agreed to use this system on their sites. The results as observed from Table 5.2, were obtained when the system was used different builder sites. Each of the construction sites was carefully examined before deploying the system, and complete drone operation control was given to the builder's executives for continuous monitoring.

It can be observed that the overall site cost reduces by almost 12%, which can either be used to improve builder profits or can be used to benefit the customer. In both ways, the proposed system assists in improving site's financial efficiency. Moreover, the site construction quality is also improved by almost 10% when compared to a non-drone-based solution.

5.5 Conclusion and Future Scope

From the evaluation done for the drone-based system on different sites, it is observed that the proposed solution improves customer feedback by 10%, which indicates superior construction quality. The system also reduces construction costs by almost 12%, which is a major impact factor of this design, this cost reduction assists in lowering site prices, thereby giving better value to the customer. Number of cracks are also reduced by over 80%, which is an indication of better construction quality.

Table 5.2 Performance evaluation of the drone-based system on real-time sites

Construction site	Parameter observed	Without drone system	With drone system	Improvement (%)
Lodha Palava Dombivli	Completion time	2 Years	1 Year, 8 Months	16
Lodha Palava Dombivli	Final site cost	1.9 Cr	1.7 Cr	10.5
Lodha Palava Dombivli	Number of accidents	85	14	83
Lodha Palava Dombivli	Average rating by users	4.2	4.9	14
Lodha Golden Dream Khoni	Cracks at final site	14	2	85
Lodha Golden Dream Khoni	Completion time	1 Years, 6 Months	1 Year, 4 Months	12
Lodha Golden Dream Khoni	Final site cost	1.1 Cr	1.05 Cr	4.5
Lodha Golden Dream Khoni	Number of accidents	61	15	75
Lodha Golden Dream Khoni	Average rating by users	4.3	4.85	10
Lodha Golden Dream Khoni	Cracks at final site	10	3	70
Lodha Vista Lower Parel	Completion time	2 Years, 1 Month	1 Year, 7 Months	24
Lodha Vista Lower Parel	Final site cost	1.95 Cr	1.6 Cr	17.9
Lodha Vista Lower Parel	Number of accidents	102	26	74
Lodha Vista Lower Parel	Average rating by users	4.1	4.95	17
Lodha Vista Lower Parel	Cracks at final site	31	5	83
Omkar Vive, BKC	Completion time	1 Years, 3 Months	1 Year	20
Omkar Vive, BKC	Final site cost	1.1 Cr	0.98 Cr	10.9
Omkar Vive, BKC	Number of accidents	148	16	90
Omkar Vive, BKC	Average rating by users	3.9	4.8	20.8
Omkar Vive, BKC	Cracks at final site	16	3	81
Omkar Signet, Malad East	Completion time	2 Years, 3 Months	1 Year, 10 Months	18.5

(continued)

Table 5.2 (continued)

Construction site	Parameter observed	Without drone system	With drone system	Improvement (%)
Omkar Signet, Malad East	Final site cost	1.7 Cr	1.5 Cr lePara>	11.7
Omkar Signet, Malad East	Number of accidents	79	8	89.8
Omkar Signet, Malad East	Average rating by users	4.5	4.9	8
Omkar Signet, Malad East	Cracks at final site	18	3	83.3

The system also ensures highly safe environment by reducing number of accidents by almost 80%, thereby making the site a safer place for workers. Due to all these factors, the final site completion time is reduced by 10%, which allows builders to follow Government led rules for building construction. In future, this work can be extended by adding big data and improved reporting for obtaining an even better performance in terms of site quality, worker efficiency, and site costs perspective.

References

Akinosho TD, Oyedele LO et al (2020) Deep learning in the construction industry: A review of present status and future innovations. J Build Eng 101827. https://doi.org/10.1016/j.jobe.2020.101827

Alahakoon D, Nawaratne R et al (2020) Self-building artificial intelligence and machine learning to empower big data analytics in smart cities. Inf Syst Frontiers 1–20. https://doi.org/10.1007/s10796-020-10056-x

Cadavid JP, Lamouri S et al (2020) Machine learning applied in production planning and control: a state-of-the-art in the era of industry 4.0. J Intell Manuf 1–28. https://doi.org/10.1007/s10845-019-01531-7

Choudhury P, Allen RT, Endres MG (2021) Machine learning for pattern discovery in management research. Strateg Manag J 42(1):30–57. https://doi.org/10.1002/smj.3215

Ellwood P, Horner S (2020) In search of lost time: the temporal construction of innovation management. R&D Management 50(3):364–379. https://doi.org/10.1111/radm.12405

Feng C, Mehmani A, Zhang J (2020) Deep learning-based real-time building occupancy detection using ami data. IEEE Trans Smart Grid 11(5):4490–4501. https://doi.org/10.1109/TSG.2020.2982351

Gheisari M, Esmaeili B (2016) Unmanned aerial systems (UAS) for construction safety applications. In: Construction Research Congress 2016:2642–2650. https://doi.org/10.1061/9780784479827.263

Giordan D, Adams MS et al (2020) The use of unmanned aerial vehicles (UAVs) for engineering geology applications. Bull Eng Geol Env 79(7):3437–3481. https://doi.org/10.1007/s10064-020-01766-2

Hopmere M, Crawford L, Harré MS (2020) Proactively monitoring large project portfolios. Project Manage J 51(6):656–69. https://doi.org/10.1177%2F8756972820933446

Howard J, Murashov V, Branche CM (2018) Unmanned aerial vehicles in construction and worker safety. Am J Ind Med 61(1):3–10. https://doi.org/10.1002/ajim.22782

Jia M, Xu Y, He P, Zhao L (2020) Identifying critical factors that affect the application of information technology in construction management: a case study of China. Frontiers Eng Manage 1–6. https://doi.org/10.1007/s42524-020-0122-4

Jung M, Ko S, Chi S (2018) A progress measurement framework for large-scale urban construction projects. KSCE J Civ Eng 22(7):2188–2194. https://doi.org/10.1007/s12205-017-0245-2

Jurczenko E (2020) Machine learning for asset management: new developments and financial applications. John Wiley & Sons

Kim D, Park HS (2006) Innovative construction management method: assessment of lean construction implementation. KSCE J Civ Eng 10(6):381–388. https://doi.org/10.1007/BF02823976

Lee D, Huang HY, Lee WS, Liu Y (2020) Artificial intelligence implementation framework development for building energy saving. Int J Energy Res 44(14):11908–11929. https://doi.org/10.1002/er.5839

Li W (2019) Application of economical building management system for singapore commercial building. IEEE Trans Industr Electron 67(5):4235–4243. https://doi.org/10.1109/TIE.2019.2922946

Li Y, Wei H et al (2020) Deep learning-based safety helmet detection in engineering management based on convolutional neural networks. Adv Civ Eng 1–10. https://doi.org/10.1155/2020/9703560

Oprach S, Bolduan T et al (2019) Building the future of the construction industry through artificial intelligence and platform thinking. Digitale Welt 3(4):40–44. https://doi.org/10.1007/s42354-019-0211-x

Otto A, Agatz N et al (2018) Optimization approaches for civil applications of unmanned aerial vehicles (UAVs) or aerial drones: a survey. Networks 72(4):411–458. https://doi.org/10.1002/net.21818

Rashidi A, Sigari MH, Maghiar M, Citrin D (2016) An analogy between various machine-learning techniques for detecting construction materials in digital images. KSCE J Civ Eng 20(4):1178–1188. https://doi.org/10.1007/s12205-015-0726-0

Supriadi LS, Pheng LS (2018) Business continuity management in construction. Springer Singapore. https://doi.org/10.1007/978-981-10-5487-7

Stroud D, Weinel M (2020) A safer, faster, leaner workplace? Technical-maintenance worker perspectives on digital drone technology 'effects' in the European steel industry. N Technol Work Employ 35(3):297–313. https://doi.org/10.1111/ntwe.12174

Vanhoucke M (2012) Project management with dynamic scheduling: baseline scheduling, risk analysis and project control. Springer, Berlin Heidelberg. https://doi.org/10.1007/978-3-642-40438-2

Wang F, Deng F, Wang Y (2020) Construction method and application of real-time monitoring and early-warning model for anaerobic reactor leakage. Process Saf Prog 39(4):1–11. https://doi.org/10.1002/prs.12144

Weng Y, Shan J et al (2020) Homography- based structural displacement measurement for large structures using unmanned aerial vehicles. Computer-Aided Civ Infrastruct Eng 1–15. https://doi.org/10.1111/mice.12645

Yaseen ZM, Ali ZH, Salih SQ, Al-Ansari N (2020) Prediction of risk delay in construction projects using a hybrid artificial intelligence model. Sustainability 12(4):1–14. https://doi.org/10.3390/su12041514

Yi W, Sutrisna M (2021) Drone scheduling for construction site surveillance. Computer-Aided Civ Infrastruct Eng 36(1):3–13. https://doi.org/10.1111/mice.12593

Chapter 6
Influence of Binder Chemical Properties on the Elastic Properties of Asphalt Mixes Containing RAP Material

Ramya Sri Mullapudi, Venkata Joga Rao Bulusu, and Sudhakar Reddy Kusam

6.1 Introduction

The Indian road network consists mostly of flexible pavements that contain bituminous structural and surface layers. Rehabilitation and reconstruction of the existing pavements generates humongous quantity of Reclaimed asphalt pavement or simply RAP material. The RAP material usage is gaining importance in pavement construction due to the scarcity of resources (both aggregates and binders). The properties of the bituminous mixtures prepared using RAP material will be generally different from those of the mixes prepared using virgin binders. The incorporation of RAP material into the asphalt mixtures increases the stiffness as well as the brittleness of the mixtures (Daniel and Lachance 2005; Izaks et al. 2015; Noferini 2017; Yan et al. 2017). The properties (modulus and time lag/phase angle) of the RAP mixtures need to be evaluated to further assess the performance characteristics of RAP mixes with the help of these mechanical properties. Phase angle of the binder and time lag of the bituminous mixtures are important parameters that indicate the elastic behavior of the binders and mixtures. In the past, time lag of bituminous mixtures was demonstrated to have good relation with the rutting characteristics of mixtures (Radhakrishnan et al. 2017).

The chemical makeup of the binders affects the mechanical and rheological properties of the binders as well as mixtures. Various mechanical properties of the binders have been sufficiently correlated with the chemical characteristics of the binders by different researchers in the past (Griffin et al. 1959; Simpson et al. 1961; Oyekunle

R. S. Mullapudi (✉)
Civil Engineering Department, Indian Institute of Technology, Hyderabad 502285, India
e-mail: ramyamullapudi@ce.iith.ac.in

V. J. R. Bulusu · S. R. Kusam
Department of Civil Engineering, IIT Kharagpur, Kharagpur 721302, West Bengal, India
e-mail: ksreddy@civil.iitkgp.ernet.in

K. R. Reddy et al. (eds.), *Advances in Sustainable Materials and Resilient Infrastructure*, Springer Transactions in Civil and Environmental Engineering,
https://doi.org/10.1007/978-981-16-9744-9_6

2006, 2007; Hofko et al. 2015, 2016; Mullapudi and Sudhakar Reddy 2020). The identification of the molecules and their quantification can be done from the spectra obtained from Fourier transform infrared spectroscopy. Ketones and sulfoxides form during the aging process and in turn impart stiffness and viscosity to the binders. Aromatics and aliphatics are among the generic fractions of asphalt binders and the properties of the binders change with change in their amount. Aromatics and aliphatics are the fractions that impart viscous properties to the binder.

Few of the research studies explored the correlation of mechanical properties of the binders with their elemental composition and functional groups (Petersen, 1983; Martin et al., 1990; Oyekunle, 2006; Jung, 2006; Firozifaar et al., 2013; Wei et al., 2014; Mullapudi et al. 2018). Mullapudi et al. (2019a) presented the relation between the mechanical properties and the chemical makeup of the RAP binder blends (functional groups). Past literature has not focused on examining relation between the chemical composition of the RAP binders and RAP mixes mechanical properties, except for the study that reported the relation between chemical makeup of RAP binder blends and resilient modulus of the RAP mixes (Mullapudi et al. 2019b). The present study was conducted to explore the relation b/w the chemical composition of the binder and the elastic properties of the RAP binders and mixtures.

6.2 Objective of the Work

The objective of the current study is to examine the relationship between the chemical composition of binders and the elastic properties (phase angle) of binders and mixtures (time lag) containing RAP material.

6.3 Scope of the Present Work

The scope of the present study is as follows:

- Aggregates, binder, and RAP material used were characterized to determine their physical properties. The target gradation used for preparing the RAP mixtures is bituminous concrete (BC-1) mid-point gradation having the maximum nominal aggregate size of 26.5 mm given by MoRTH (2013). The RAP contents used were 0, 15, 25, 35, and 45% by weight of the total mixture. Mix design for the mixtures with different RAP contents was done according to Marshall mix design methodology to find the design binder contents.
- RAP-virgin binder blends were prepared by blending the RAP binder carefully extracted from the RAP material and VG30 viscosity grade (base) virgin bitumen in different proportions as per the RAP binder to total binder ratio obtained from the mix design exercise. Rheological studies and Fourier transform infrared spectroscopy studies were conducted on the prepared RAP-virgin binder blends.

- Stiffness modulus test was conducted on the samples of RAP mixes prepared at their optimum binder contents [at three distinct temperatures (15 to 35 °C at an interval of 10 °C) and four frequencies (0.5 to 2.0 Hz at an interval of 0.5 Hz)]. Time lag which is the time gap between the times of incidence of peak stress and peak strain in the stiffness modulus test was evaluated from the test data.
- The relation between phase angle, time lag, and chemical composition have been examined and the correlations are presented in the paper.

6.4 Materials and Methods Considered

6.4.1 Materials

The target gradation used to prepare the mixes was the mid-point line of the bituminous concrete (BC-1) gradation as per MoRTH (2013). The details of the gradation are given in Table 6.1. The properties of the viscosity grade binder (VG30) are presented in Table 6.2. The RAP material that was used for the preparation of mixtures was obtained from the Kulpi plant which was located on national highway 117. The RAP material was characterized to determine the binder content, binder quality, and the gradation of aggregates in the RAP. Details of the gradation of the aggregates in the RAP material are given in Table 6.1. The adopted gradation of the RAP aggregate under study was determined by using the aggregates obtained from the centrifuge extractor by extracting the bitumen available in the RAP material. Once the particle size distribution of the extracted RAP aggregates is known, the virgin aggregates (individual fraction) required are added to meet the target gradation (BC-I). The binder content present in the RAP material is found to be 3.96%. The binder properties of the RAP binder extracted and recovered from the RAP material are presented in Table 6.2.

6.4.2 Mix Design of RAP Mixes

The standard Marshall method of mix design with 75 blows compaction on either face of the cylindrical specimen was adopted for determining the design binder contents of the five asphalt mixtures (with RAP proportion of 0, 15, 25, 35, 45%) considered. 4 inch diameter Marshall samples were prepared with varying total (RAP + Virgin) binder content in the mix (4.5, 5.0, 5.5, and 6.0%). The mixing temperature and compaction temperature for preparing the mixes using VG30 virgin binder without any RAP content were identified corresponding to viscosity levels (Brookfield viscometer was employed) of 0.17 ± 0.02 Pa.s for mixing and 0.28 ± 0.03 Pa.s for compaction (MS-2, 2015) respectively (ASTM D2196, 2015). Virgin aggregates were heated separately to mixing temperatures that are higher than the normal mixing temperatures by 0.5 °C for every 1% RAP content addition. The virgin binder was

Table 6.1 BC-1 mid-point and RAP aggregates gradation

Sieve sizes, mm		26.5	19	13.2	9.5	4.75	2.36	1.18	0.6	0.3	0.15	0.075
% passing	Aggregate extracted from RAP material	100	99	96	91	60	39	23	16	11	7	5
	BC-1	100	95	69	62	45	36	27	21	15	9	5

Table 6.2 RAP and virgin binder properties

Binder	Penetration at 25 °C, dmm	Softening point, °C	Viscosity at 60 °C, P	Kinematic viscosity at 135 °C, cSt
RAP	15	82	346,000	–
VG30	61	52	3516	522

Table 6.3 Characteristics of RAP mixes at optimum binder contents

Mix Id	Total (design) binder content, %	Virgin binder content, %	Stability, kN	Flow, mm	Voids in mineral aggregate (VMA), %	Voids filled with asphalt (VFA), %
V3-0D	4.7	4.7	13.1	2.9	15.8	74.4
V3-15D	4.7	4.1	15.3	2.6	15.7	73.4
V3-25D	5.0	4.0	15.8	2.4	16.0	75.6
V3-35D	5.2	3.8	16.5	2.3	16.3	75.4
V3-45D	5.5	3.7	18.3	2.2	16.3	76.1

heated to its mixing temperature and the RAP material was only heated to 110 °C (MS-2, 2015). The binder content that gives the required 4% air void content in the compacted mix was taken as the design binder content for each of the mixes (Mullapudi et al. 2020a; Mullapudi et al. 2020b). The properties of the mixes at their corresponding design binder contents are given in Table 6.3. V3-XD represents VG30 virgin binder, percent RAP content in the mix prepared at respective design binder content.

RAP-virgin binder blends were carefully prepared by using extracted RAP binder and VG30 binder. Both the binders were heated individually to reach the mixing temperature of VG30 binder and then required amount of each of the binders were transferred to another container and mixed thoroughly (hand mixing) to obtain the blends.

6.4.3 Oscillation Test on RAP-Virgin Binder Blends

Oscillation test was carried out for the RAP-virgin binder blends using dynamic shear rheometer at four different temperatures (58, 64, 70, and 76 °C), and at a frequency of 10 rad/s at a constant stress of 100 Pa. The purpose was to understand the viscoelastic behavior of the binder blends. In the procedure, a sample of bitumen film is placed between two (upper and lower) parallel plates with the lower plate being fixed and the upper plate used to apply torque by rotation. As bituminous binders are viscoelastic in nature the stress and strain will have some time lag (as illustrated in Fig. 6.1). The

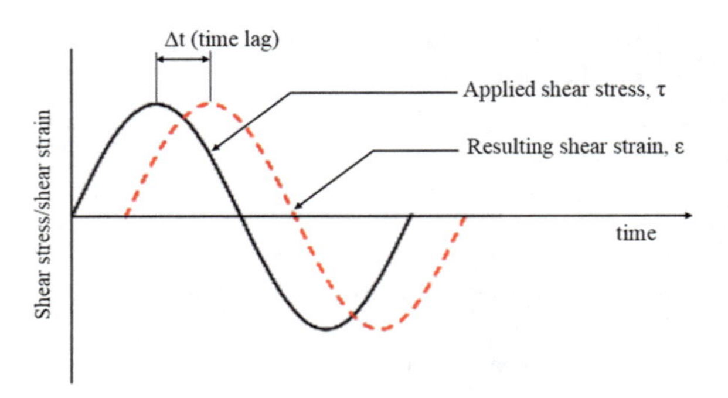

Fig. 6.1 Variation in shear stress and shear strain in oscillation test

magnitude of the time lag will depend on the binder, temperature, and frequency of loading.

6.4.4 *Fourier Transform Infrared Spectroscopy on Binder Blends*

Infrared spectroscopy is the most frequently used method to quantify the amount of functional groups in a binder. FTIR Nexus 870 was used Fourier transform infrared (FTIR) spectroscopy testing. Pellets were prepared using spectroscopic grade KBr, which was further used to prepare the binder samples. The samples for the spectroscopy were made by dropping a drop of 10 wt% (solution of 0.5 g binder in 5 ml CS_2) on the prepared KBr pellet and air drying them for 10 min. The binder sample is positioned so that it interferes the path of the IR light. The detector measures the amount of light absorbed by the binder sample for the wavelengths of infrared light ranging from 400 to 4000 cm^{-1}. The spectra obtained for VG30 binder are shown in Fig. 6.2. The Ketones, sulfoxides, aliphatics, and aromatics are commonly used to represent/explain the strength or aging characteristics of binders. Equations 6.1–6.4 are used to estimate the indices pertinent to ketones (ICO), sulfoxides (ISO),

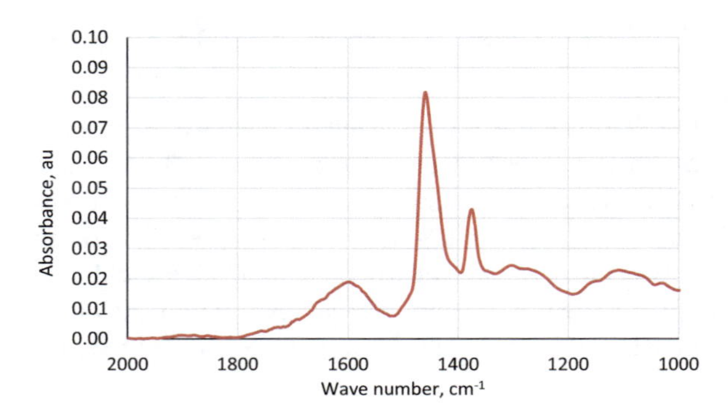

Fig. 6.2 Infrared spectra of VG30 binder

Aliphatics (ICH), and aromatics (ICC) (Mullapudi et al. 2019b).

$$\text{ICO} = \frac{Areaaround1700cm^{-1}}{\sum Areaaround1460cm^{-1}andAreaaround1375^{-1}} \quad (6.1)$$

$$\text{ISO} = \frac{Areaaround1030cm^{-1}}{\sum Areaaround1460cm^{-1}andAreaaround1375^{-1}} \quad (6.2)$$

$$\text{ICC} = \frac{Areaaround1600cm^{-1}}{\sum Areaaround1460cm^{-1}, 1375^{-1}, 1700cm^{-1}, 1030cm^{-1}, 1600cm^{-1}} \quad (6.3)$$

$$\text{ICH} = \frac{Areaaround1460cm^{-1}andAreaaround1375^{-1}}{\sum Areaaround1460cm^{-1}, 1375^{-1}, 1700cm^{-1}, 1030cm^{-1}, 1600cm^{-1}} \quad (6.4)$$

6.4.5 Stiffness Modulus Test

Bituminous/asphalt mixture is a viscoelastic material that has both viscous and elastic properties. Due to the viscous nature of the binder used in the asphalt mixtures there will be a time difference between the applied stress pulse and the obtained strain/deformation response of the mixture. The time lag, which is the time difference between the peaks of stress and strain is an important parameter to explain the performance of mixtures. Marshall specimens of different mixes were tested for the stiffness modulus as per EN 12,697–26 (2012). The ratio of loading time to rest period was maintained as 1:9. The test was conducted essentially at three different temperatures (15, 25, and 35 °C) and at four different frequencies (0.5, 1.0, 1.5, and 2.0 Hz). Figure 6.3 indicates the variation of stress and deformation with time. The figure also illustrates the concept of time lag.

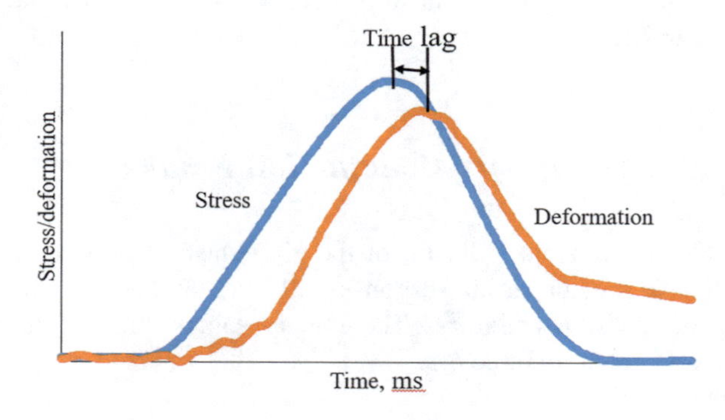

Fig. 6.3 Plot showing time lag between stress and deformation pulse

Table 6.4 Phase angles of binders

Temperature, °C	V3-0D	V3-15D	V3-25D	V3-35D	V3-45D
58	86.3	84.0	83.7	81.9	80.3
64	87.4	85.9	85.7	84.3	83.0
70	87.9	87.3	87.2	86.2	85.2
76	88.1	88.3	88.4	87.6	86.8

6.5 Results and Discussion

6.5.1 Oscillation Test Results

The phase angles for the binder blends obtained from oscillation tests at different temperatures are given in Table 6.4. From the data it is visible that the phase angle value reduced with incorporation of higher RAP content in the binder blend. This could be due to the increase in the aged binder content in the blend due to which binder becomes stiff and indicates the improvement in the elastic nature of the binder with RAP binder proportion.

6.5.2 FTIR Test Results

During the process of aging, ketones and sulfoxides are formed by interaction of carbon and sulfur atoms with oxygen, respectively. Aliphatics turn into aromatics (higher molecular weight molecules) during the aging process. The chemical indices related to ketones, sulfoxides, aliphatics, and aromatics calculated using Eqs. 6.1–6.4 are given in Fig. 6.4.

It is observed that the ketones, sulfoxides, and the aromatics increased and the aliphatics reduced with the increase in the proportion of RAP in the binder blend. This phenomenon is also due to the increase in the aged binder proportion in the binder blend which in turn increased the heavier molecular groups present in the binder.

6.5.3 Stiffness Modulus Test Results

Time lag is an indicator of the viscoelastic behavior of the mix. The mixtures that have greater elastic response will have smaller time lag values when compared to more viscous mixtures. The time lag values obtained for the RAP mixes are tabulated and given in Table 6.5.

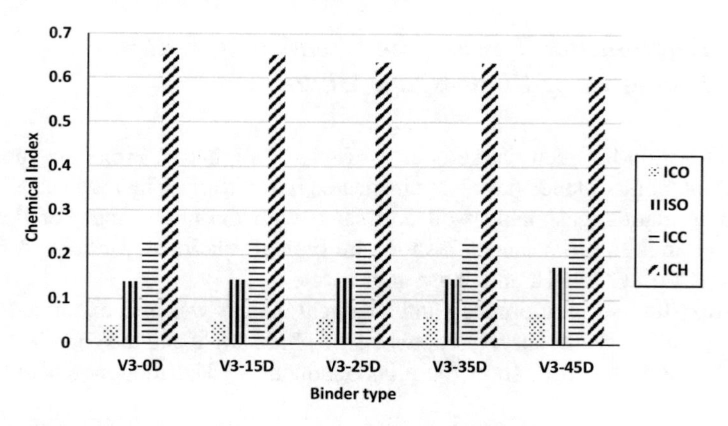

Fig. 6.4 Chemical indices for various RAP-virgin binder blends

Table 6.5 Time lag values of various mixes at different frequencies and temperatures

Temperature	Frequency	Time lag, ms				
		V3-0D	V3-15D	V3-25D	V3-35D	V3-45D
15 °C	2 Hz	4.17	4.08	3.64	3.68	3.18
	1.5 Hz	5.73	6.77	5.46	5.04	3.35
	1 Hz	7.96	8.19	6.56	4.93	3.96
	0.5 Hz	16.80	16.53	15.90	12.13	8.10
25 °C	2 Hz	7.07	7.00	6.12	6.28	4.00
	1.5 Hz	9.29	9.97	8.31	8.27	6.62
	1 Hz	13.44	12.85	11.07	10.83	9.73
	0.5 Hz	24.33	22.67	19.93	20.00	18.00
35 °C	2 Hz	10.22	9.95	9.64	8.72	9.57
	1.5 Hz	12.02	11.28	10.72	10.84	11.68
	1 Hz	19.31	15.07	14.16	13.28	13.36
	0.5 Hz	25.6	22.30	23.20	22.00	24.87

It can be seen that the time lag of the bituminous mixes generally reduced with increase in the proportion of the RAP content, reduction in the temperature, and increase in the testing frequency. This shows that inclusion of RAP increases the elastic nature of the mixes.

6.5.4 Relationship Between the Chemical and Elastic Properties of Binders and Mixtures

The relationship between the ketones, sulfoxides, aliphatics, aromatics, and phase angle of the binder blends have been inspected in the study. The ketones, aliphatics, and the aromatics reasonably well correlated with the phase angle of the binder blends. Figures 6.5, 6.6, and 6.7 show the typical relationship obtained between ketones (ICO), ICC, ICH, and phase angle.

Relationship between time lag and chemical indices was also examined as there was a good relationship observed between the phase angle and the chemical indices. Figures 6.8, 6.9, and 6.10 show the plots obtained at 1 Hz frequency between time

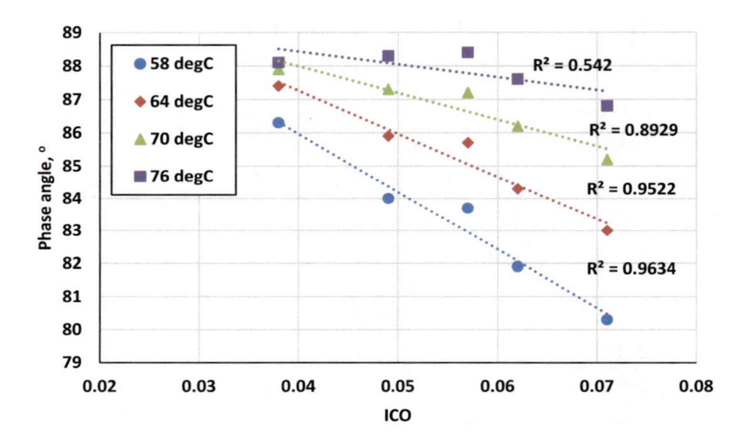

Fig. 6.5 Relationship between ICO and phase angle

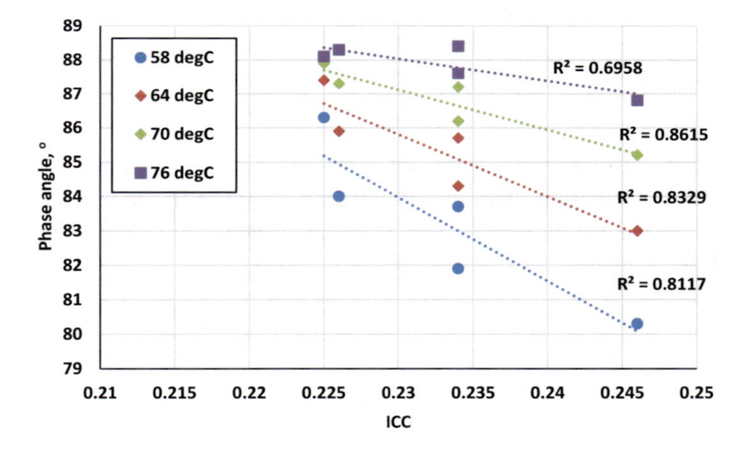

Fig. 6.6 Relationship between ICC and phase angle

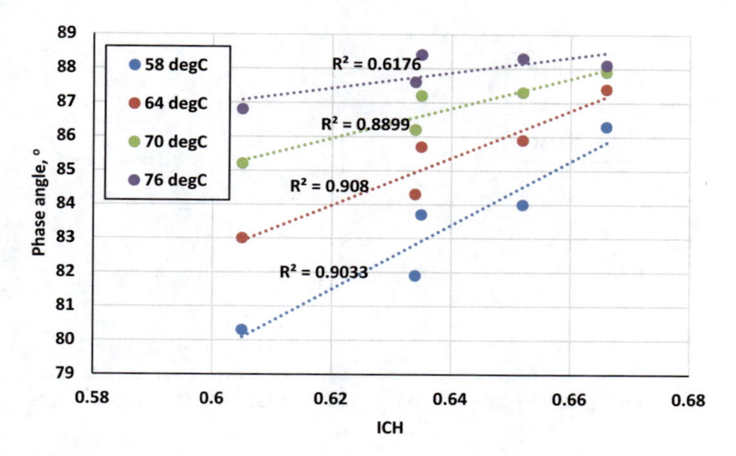

Fig. 6.7 Relationship between ICC and phase angle

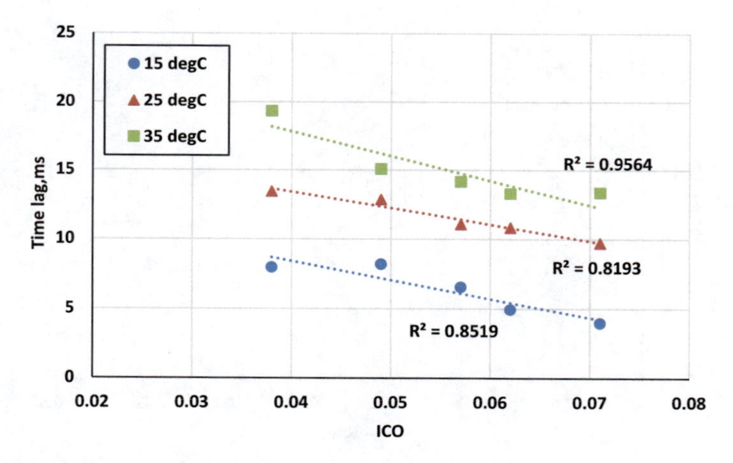

Fig. 6.8 Relationship between ICO and time lag

lag and ICO, ICC, and ICH, respectively. Inclusion of higher RAP binder content in the blend increased the oxides and aromatics contents in the binder and reduced the amount of saturates suggesting that there is an increase in binder viscosity and stiffness. From Figs. 6.8 and 6.9, it is evident that ICO and ICC have an inverse relationship with time lag values. From Fig. 6.10, it can be noted that increase in saturates increased the time lag of the mixture. The correlation between ISO and elastic properties of binders and mixtures was observed to be poor. Thermal instability of sulfoxides (instability of the sulfoxides when subjected to higher temperatures) might be the reason behind the poor relationship of ISO with time lag and phase angle.

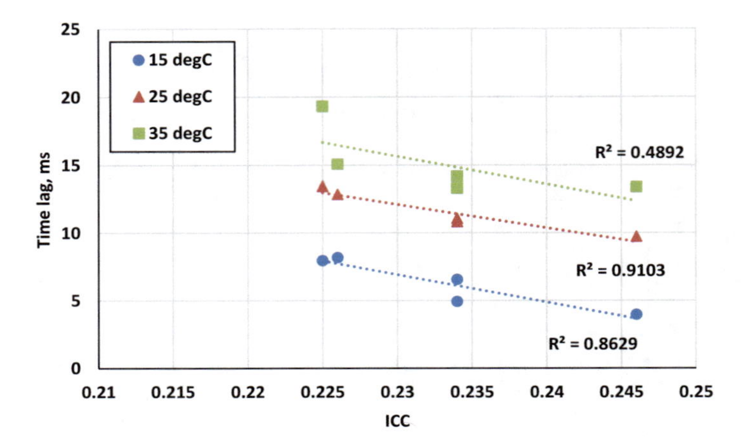

Fig. 6.9 Relationship between ICC and time lag

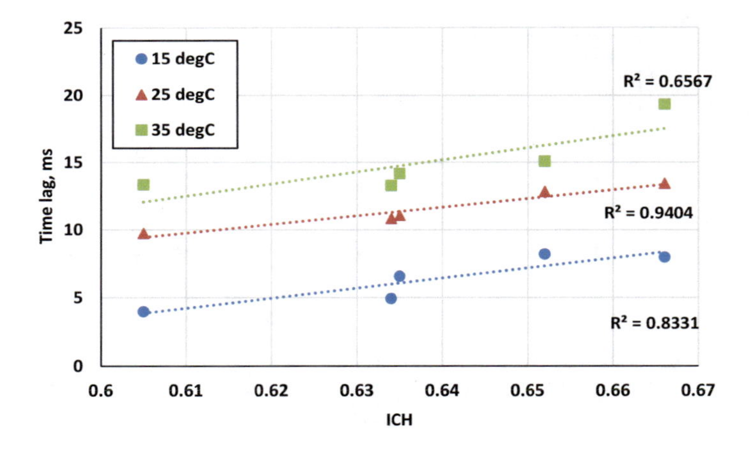

Fig. 6.10 Relationship between ICH and time lag

6.6 Conclusions

The current study is a step advancement to increase the level of understanding about the relation between the chemical composition of binders and the elastic properties of binders and mixtures. In the current study, the elastic property of the binders and mixtures was evaluated in terms of phase angle and time lag, respectively. The chemical characteristics of the binders were characterized through FTIR indices. The relationship between the chemical indices and phase angle and time lag has been examined. The following are the key conclusions obtained based on the results:

- Ketones, sulfoxides, and aromatics increased with inclusion of higher RAP binder proportion in the binder blend. This is due to the increase in the proportion of the aged binder in the blend. Aliphatics, which are the lowest molecular weight compounds which convert to the next higher molecular weight compounds during the aging process, decreased with increase in the RAP content
- The phase angle of the binder blends and the time lag of the asphalt mixtures reduced with incorporation of higher RAP content. This is due to the improvement in the elastic nature of the binder with increase in RAP binder proportion.
- There is a good correlation between ketones, aromatics, aliphatics, and the elastic properties of binders (phase angle) as well as mixes (time lag). Sulfoxides have poor correlation with time lag and phase angle due to their instability at higher temperatures.

References

ASTM D2196 (2015) Standard test methods for rheological properties of non-Newtonian materials by rotational viscometer. ASTM International: West Conshohocken

Daniel J, Lachance A (2005) Mechanistic and volumetric properties of asphalt mixtures with recycled asphalt pavement. Transp Res Rec J Transp Res Board 1929:28–36. https://doi.org/10.1177/0361198105192900104

EN 12697–26 (2012) Bituminous mixtures. Test methods for hot mix asphalt, Part 26: Stiffness. Brussels

Firoozifar SH, Foroutan S, Foroutan S (2011) The effect of asphaltene on thermal properties of bitumen. Chem Eng Res Des 89(10):2044–2048. https://doi.org/10.1016/j.cherd.2011.01.025

Griffin RL, Simpson WC, Miles TK (1959) Influence of composition of paving asphalt on viscosity, viscosity-temperature susceptibility, and durability. J Chem Eng Data 4(4):349–354. https://doi.org/10.1021/je60004a019

Hofko B, Eberhardsteiner L, Füssl J, Grothe H, Handle F, Hospodka M, Scarpas A (2016) Impact of maltene and asphaltene fraction on mechanical behavior and microstructure of bitumen. Mater Struct 49(3):829–841. https://doi.org/10.1617/s11527-015-0541-6

Hofko B, Handle F, Eberhardsteiner L, Hospodka M, Blab R, Füssl J, Grothe H (2015) Alternative approach toward the aging of asphalt binder. Transp Res Rec J Transp Res Board 2505(2505):24–31. https://doi.org/10.3141/2505-04

Izaks R, Haritonovs V, Klasa I, Zaumanis M (2015) Hot mix asphalt with high RAP content. Procedia Eng 114:676–684

Jung SH (2006) The effects of asphalt binder oxidation on hot mix asphalt concrete mixture rheology and fatigue performance, Doctoral dissertation, Texas A&M University

Martin KL, Davison RR, Glover CJ, Bullin JA (1990) Asphalt aging in Texas roads and test sections. Transp Res Rec 1269:9–19

MoRTH (2013) Specifications for road and bridge works. Ministry of Road Transport and Highways, Indian Roads Congress, New Delhi

MS-2 (7th Ed.). (2015) Mix design methods for asphalt concrete and other hot-mix types (No. 2). Asphalt Institute

Mullapudi RS, Sudhakar Reddy K (2018) An investigation on the relationship between FTIR indices and surface free energy of RAP binders. Road Mater Pavement Des 1–15. https://doi.org/10.1080/14680629.2018.1552889

Mullapudi RS, Sudhakar Reddy K (2020) Relationship between rheological properties of RAP binders and cohesive surface free energy. J Mater Civ Eng 32(6):04020137. https://doi.org/10.1061/(ASCE)MT.1943-5533.0003199

Mullapudi RS, Aparna Noojilla SL, Reddy KS (2020a) Fatigue and healing characteristics of RAP mixtures. J Mater Civ Eng 32(12):04020390

Mullapudi RS, Noojilla SLA, Kusam SR (2020b) Effect of initial damage on healing characteristics of bituminous mixtures containing reclaimed asphalt material (RAP). Constr Build Mater 262:120808

Mullapudi RS, Deepika KG, Reddy KS (2019a) Relationship between chemistry and mechanical properties of RAP binder blends. J Mater Civ Eng 31(7):04019124. https://doi.org/10.1061/(ASCE)MT.1943-5533.0002769

Mullapudi RS, Karanam GD, Kusam SR (2019b) Influence of chemical characteristics of RAP binders on the mechanical properties of binders and mixes. Int J Pavement Res Technol 12(6):632–637. https://doi.org/10.1007/s42947-019-0075-3

Noferini L (2017) Investigation on performances of asphalt mixtures made with reclaimed asphalt pavement: effects of interaction between virgin and rap bitumen. Int J Pavement Res Technol https://doi.org/10.1016/j.ijprt.2017.03.011

Oyekunle LO (2006) Certain relationships between chemical composition and properties of petroleum asphalts from different origin. Oil & Gas Science and Technology-Revue De l'IFP 61(3):433–441. https://doi.org/10.2516/ogst:2006043a

Oyekunle LO (2007) Influence of chemical composition on the physical characteristics of paving asphalts. Petrol Sci Tech 25(11):1401–1414. https://doi.org/10.1080/10916460500528854

Petersen JC (1993) Asphalt oxidation–a overview including a new model for oxidation proposing that physicochemical factors dominate the oxidation kinetics. Fuel Sci Tech Int 11(1):57–87. https://doi.org/10.1080/08843759308916058

Radhakrishnan V, Dudipala RR, Maity A, Sudhakar Reddy K (2017) Evaluation of rutting potential of asphalts using resilient modulus test parameters. Road Mater Pavement Des 20(1):20–35. https://doi.org/10.1080/14680629.2017.1374994

Simpson WC, Griffin RL, Miles TK (1961) Relationship of asphalt properties to chemical constitution. J Chem Eng Data 6(3):426–429. https://doi.org/10.1021/je00103a029

Wei J, Dong F, Li Y, Zhang Y (2014) Relationship analysis between surface free energy and chemical composition of asphalt binder. Cons Build Mat 71:116–123. https://doi.org/10.1016/j.conbuildmat.2014.08.024

Yan Y, Roque R, Cocconcelli C, Bekoe M, Lopp G (2017) Evaluation of cracking performance for polymer-modified asphalt mixtures with high RAP content. Road Mater Pavement Des 18(sup1):450–470. https://doi.org/10.1080/14680629.2016.1266774

Chapter 7
Principles and Prospects of Using Lignosulphonate as a Sustainable Expansive Soil Ameliorator: From Basics to Innovations

Nauman Ijaz, Zia ur Rehman, and Zain Ijaz

7.1 Introduction

The momentous increase in industrialization has not only escalated the ever-increasing demand for raw materials but has also increased the volume of undesirable waste by-products. Thus, industrialization not only results in the depletion of non-renewable natural resources but also poses threats to the ecosystem through residues (Habert et al. 2010; Harris 1999). Therefore, efforts should be lined to manage the productive disposal of such industrial waste by-products by converting them into utilizable raw materials which can be sustainably used in various fields. Similar to other industries, the Paper/wood industry is also facing waste (i.e., lignosulphonate (LS)) disposal-related problems. One common practice in the pulping industry to get rid of LS is through incineration, but this generates anthropogenic carbon dioxide (CO_2) while simultaneously producing large quantities of residues such as fly ash and bottom ash that also become a formidable problem on account of disposal. The scale of this waste management problem can be gauged by the fact that the global production of LS is almost 50 Mt (Gandini and Belgacem 2008; Alzigha et al. 2018a, b). In recent years, many researchers have attempted to effectively utilize LS, in the field of geotechnical engineering as an additive material (Puppala and Hanchanloet 1999a;

N. Ijaz · Z. Ijaz
Key Laboratory of Geotechnical and Underground Engineering of Ministry of Education, College of Civil Engineering, Tongji University, Shanghai 200092, China
e-mail: nauman_ijaz99@tongji.edu.cn

Z. Ijaz
e-mail: zain@tongji.edu.cn

Z. Rehman (✉)
Department of Civil Engineering, University of Engineering and Technology, Taxila 47080, Pakistan
e-mail: ziaur.rehman@uettaxila.edu.pk

K. R. Reddy et al. (eds.), *Advances in Sustainable Materials and Resilient Infrastructure*, Springer Transactions in Civil and Environmental Engineering,
https://doi.org/10.1007/978-981-16-9744-9_7

Tingle and Santoni 2003; Indratna et al. 2008; Koohpeyma et al. 2013; Tasalloti et al. 2015; Canakci et al. 2015; Vakili et al. 2018a, b).

The problematic soil, namely, expansive or swelling soil is considered to be a threat to the stability of the civil engineering structures owing to its adverse swell-shrinkage behavior (Dif and Bluemel 1991; Basma et al. 1996; Mitchell and Soga 2005; ur Rehman et al. 2021; ur Rehman et al. 2018). The presence of hydrophilic smectite clay minerals group is held accountable for such adverse characteristics (Moore Reynolds 1989; Bergaya and Lagaly 2006). Expansive soils are regarded as unsuitable for construction, and guidelines suggest that they should be replaced or their properties should be altered to meet the specific engineering properties for safe and durable construction (Nelson and Miller 1997). Among various available techniques, to deal with such problematic soils, chemical stabilization is one of the effective ways (Al-Rawas and Goosen 2006; Puppala and Pedarla 2017). Many studies on chemical soil stabilization have been conducted in the past, to improve physical, mechanical, and hydraulic properties. However, the main focus of these studies was traditional stabilizers i.e., cement, lime (LM), gypsum, etc. (Elert et al. 2018; Naseem et al. 2019). Regardless of the effectiveness, the extensive use of traditional stabilizers is criticized on account of sustainability, natural resource consumptions, and environmental protection (Mujtaba et al. 2020; Du et al. 2020). These shortcomings deemed the researchers to find new viable and sustainable ways to meet the ever-increasing demand for stabilizers for expansive soils in the construction industry (Chenarboni et al. 2021; Khalid et al. 2019; Reddy et al. 2015; ur Rehamn and Khalid 2021; Wu et al. 2020a). Therefore, in recent years many researchers have attempted to use paper/wood industrial waste, i.e., LS as a soil stabilizer for sustainable development in the perspective of environmental protection and resilience of the structure (Athukorala et al. 2013; Bolander 1999; Ijaz et al. 2020a, b; Indraratna et al. 2010, 2013; Landlin 2021; Li et al. 2019; Madurwar et al. 2013; Monje et al. 2021; Ravishankar and Reddy 2017; Qian et al. 2006; Tingle et al. 2007; Santoni et al. 2002; Sharmila et al. 2021; Vinod et al. 2010; Wu et al. 2020b) This strategy presents a productive solution for the disposal of LS as well as an alternate sustainable raw material for the construction.

This chapter manifests (1) comprehensive insight into the efficacy of LS as an expansive soil stabilizer accounting for short and long-term stability (i.e., under wetting-drying condition), (2) review on the technological advances in the geotechnical application of LS, (3) elaboration of the pertinent features of stabilization mechanism by which LS addition improves the adverse characteristics of expansive soils, (4) research progress that has laid the basis of implementation of LS as a soil stabilizer in real-scale projects.

7.2 Formation, Structure, and Properties of LS

LS is a by-product derived from the paper/wood industry during the pulping reaction. The two major constituents of wood are "lignin" and "cellulose". The pulping of woods primarily involves the separation of cellulose and the dissolution of lignin. The process involves the digestion of wood chips at a high temperature of 125–145 °C in an aqueous solution of Sulfur (IV) oxide and salt of sulfurous acid such as calcium bisulfite (Pearl 1967). The lignin present in the wood will react with liquor (i.e., comprise of sulfur (IV) oxide and salt of sulfurous acid) in the digestion tank and results in the formation of soluble LS. Further, LS is isolated from the residual cellulosic pulp by employing filtration and washing techniques, for example, solvent extraction, ion-exchange chromatography, dialysis method, precipitation by calcium hydroxide, salting out method, or by precipitation using quaternary ammonium salts (Theng 2012). Figure 7.1 shows the diagrammatical illustration of LS formation.

The LS is an organic compound having a benzene ring as the basic building unit. The LS is considered water-soluble and exhibits acidic nature due to the presence of strongly disassociated sulfonic acid groups present in their molecules (Pearl 1967). However, the pH value of different commercially available LS falls under a wide range between 2 and 7 and can form complexes with different cations such as Ca, Mg, and Na enabling it to compete for adsorption sites in clays (Alzigha et al. 2018b). Due to the vague concept of LS structure, many researchers have proposed structural formulae that put forward the chemical structure of LS (Alzigha et al. 2018a). Figure 7.2 presents the proposed chemical structure of LS.

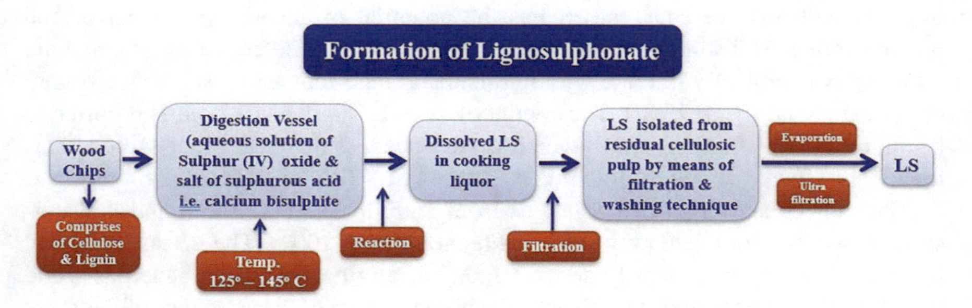

Fig. 7.1 Diagrammatical illustration of the formation of LS

Fig. 7.2 **a** Binding unit of LS molecule given by Sulfite Pulp Manufacturers' Research League 1963 (Alzigha et al. 2015); **b** proposed chemical structure of LS given by Vinod et al. 2010) Proposed chemical structure of LS given by (Alzigha et al. 2018a)

Table 7.1 Chemical and physical properties of LS (Alzigha et al. 2015, 2016)

Chemical composition		Physical properties	
Composition	Mass (%)	Property	Value
SiO_2	0.126	Lignosulphonate (%)	55 (min)
CaO	2.16	Ash (%)	12 (max)
Al_2O_3	0.02	Dry maters	95 (min)
SO_3	72.6	Appearance	Light yellow or light brown
Fe_2O_3	0.573		
MgO	18.4	pH	4–7
Na_2O	1.12	Moisture content (%)	7% (max)
K_2O	2.31	Bulk density (kg/m^3)	205

The LS surface comprises both negative potential (i.e., hydrophilic, hydroxyl, sulfite phenolic, and carboxylic groups) and positive potential (i.e., hydrophobic aromatic structure). Other studies show that the net potential on the LS surface is negative (Alzigha et al. 2018a; Theng 2012). Because of the presence of both negative and positive charges, LS has the potential to adsorb both negative and positive ions in soil solution. Also, the LS comprises intramolecular hydrogen bond which may chemically interact with multivalent metal ions and result in a covalent coordinate bond (Pearl 1967). The conductivity of LS-derived carbons is reported to be approximately 0.005 S cm^{-1}, which is 10 times less than any other carbon. This manifests a lower degree of graphitic structure in LS (Wu et al. 2020a).

The molecular weight of LS ranges from 4600 to 398,000 g/mol and is water-soluble over the complete range of pH (Fredheim et al. 2002). The LS structure and its molecular weight primarily depend upon the origin of extraction such as wood (i.e., Fir, Pine), hardwood (i.e., Beech, Eucalyptus, Aspen), grass, paper mill, etc. In comparison to the traditional stabilizer, LS has the major advantage of being non-corrosive and nontoxic. Another major advantage is that it is not listed as biodegradable hydrocarbons like bacteria, fungi, and yeast. Table 7.1 presents the chemical and physical properties of LS documented in the literature (Noorzad and Ta'negonbadi 2018; Puppala and Hanchanloet 1999b).

7.3 Efficacy of LS as an Expansive Soil Stabilizer

Being the second most widely available natural polymer numerous researchers have attempted to analyze its potential use in various fields. In view of this, different studies were carried out in recent years on the efficacy of LS as an expansive soil stabilizer. Various geotechnical properties such as soil consistency, swell-shrinkage behavior, deformation, strength, and hydraulic parameters were the focus of the study

(Alzigha et al. 2015, 2016; Ijaz et al. 2020c; Noorzad and Ta'negonbadi 2017, 2018). Under this section, the efficacy of LS on various geotechnical properties has been discussed in detail.

7.3.1 Soil Plasticity and Grain Size Distribution

The LS treatment has a positive impact in reducing the soil plasticity index (*PI*) of expansive soil. Numerous researchers have evaluated this property by varying the LS content at a wide range (Indraratna et al. 2012; Fernandez et al. 2020; Singh et al. 2021; Sabitha and Evangeline 2021; Giahi et al. 2021; Prasad and Reshmarani 2020). In a study conducted by Alazigha et al. (2015), 2% of LS has substantially reduced the liquid limit (LL) by 16.5%, with some increase in the plastic limit (~10%), exhibiting an overall reduction in the *PI* value of 37%. Similar kinds of results were also reported by Ijaz et al. (2020a), in which 2% LS was found to be optimum with an overall reduction in the *PI* value of about 30%. A study conducted by Landline et al. (2021), reported a reduction of 20% in *PI* value. In addition, the impact of curing time has also been found to affect the *PI* value of the soil. As per Ijaz et al. (2020b) for LS stabilization of expansive soil seven days is found to be the optimum curing period for accurate determination of the *PI* value of treated soil. The reduction in *PI* values is attributed to the interaction of LS with soil particles which induces aggregation and reduction in double diffused layer owing to the hydrophobic nature of LS, which is further verified by grain size distribution (Giahi et al. 2021; Ijaz et al. 2020b). Some other studies have reported adverse effects on *PI* values at higher LS content (Ijaz et al. 2020a, b; Orlandi et al. 2019). According to Wu et al. (2020b), at an optimum value of 4% LS content, an increased *PI* value was observed highlighting the increase in water holding capacity of the expansive soil; however, the strength parameter was found to be appreciably enhanced. Some studies also highlight the adverse effect of wetting-drying cycles on the efficacy of LS and propose an LS-based novel admixture which results in a tremendous reduction in the *PI* value up to 41% and reduce the LM consumption by near to one-half (Ijaz et al. 2020a, c). It is pertinent to mention here that for durability the chemical interaction of added admixture with smectite clay minerals is of prime importance.

7.3.2 Volumetric Change Behavior

Numerous researchers have investigated the effect of LS treatment in mitigating the swell-shrinkage behavior of expansive soil (Alzigha et al. 2015, 2016, 2018a; b; Ijaz et al. 2020a, b). In an investigation conducted on the efficacy of LS on swelling behavior of highly expansive soil, the swelling extent and pressure were observed to reduce by 23% and 20% respectively at optimum LS content of 2% (Alzigha et al. 2015, 2016, 2018a; b). Similar kinds of results were also documented by Ijaz et al.

(2020a, b), wherein a low dosage of LS (i.e., 2%) was found to be optimum and reverse effects (i.e., increase in swelling behavior) were observed at higher dosages. However, some other studies exhibit good efficacy in reducing the swelling percentage at a higher dosage of LS (Landlin et al. 2021). For example, in a study conducted by Fernandez et al. (2020), the swelling behavior of highly expansive soil reduced to the moderate and low degree of expansion at 3 and 5% LS addition. However, non-linear behavior was observed in the case of swelling pressure; at 3 and 5% LS addition the swelling pressure was reduced by 60 and 70%. It is important to note that LS treatment of expansive soils at higher dosages exhibits non-linear behavior considering the volumetric change response of the soil. Some other studies account for the effect of climatic variations, i.e., wetting-drying cycles to evaluate the long-term effectivity of LS treatment. A study conducted by Ijaz et al. (2020b) subjected the LS-treated expansive soil to partial shrinkage and wetting cycles, the results showed a reduction in swelling percentage and swelling pressure and attainment of equilibrium at the third wetting-drying cycle. A similar kind of behavior was also observed in the case of untreated expansive soil when subjected to partial shrinkage. In addition, the desiccation-induced cracking of expansive soils was found to be moderately improved by LS treatment. Keeping in view, the limitations of LS as a lone expansive soil stabilizer, Ijaz et al. (2020b) proposed an LS-based composite admixture to account for the long-term climatic variations. The results showed that under wetting-drying cycles, the volumetric change behavior was completely mitigated with the newly proposed LS-based composite admixture (integrating LS with LM). Figure 7.3 shows the variation in swelling behavior of untreated and LS-based treated soils. In various studies 2% LS was found to be an optimum percentage of LS after 7 days of curing taking into account the *PI* value and the swelling potential (Ijaz et al. 2020a, c; Alzigha et al. 2016). Moreover, it can be also inferred that swelling potential inhibits completely and the dosage of optimum LS reduces to 0.875% by integrating it with 2.63% lime. It can be inferred that among LS-based additives, the only optimized

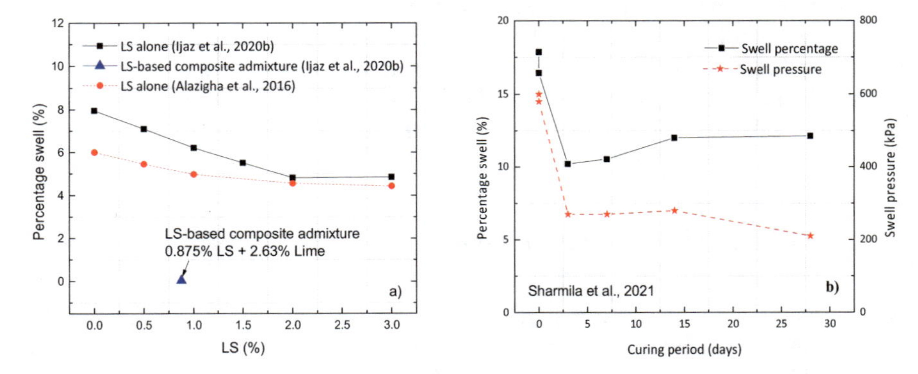

Fig. 7.3 Swelling potential; **a** swell percentage with respect to LS content (Ijaz et al. 2020b; Sharmila et al. 2021); **b** swell percentage and swell pressure of LS-treated expansive soil with respect to curing period (Sharmila et al. 2021)

LS-based composite admixture proposed by Ijaz et al. (2020b) completely mitigates the swelling potential of the expansive soil.

7.3.3 Compressibility and Permeability Parameters

Compressibility and permeability are important parameters of expansive soils as it controls the settlement and strength characteristics of the soil with respect to suction. Expansive soils normally exhibit high compressibility and low permeability (Alzigha et al. 2018a; Ijaz et al. 2020a). Different researchers have carried out extensive laboratory testing to evaluate the efficacy of LS on these parameters (Alzigha et al. 2018b; Ijaz et al. 2020a). Fernandez et al. (2020) conducted a study by treating the highly expansive soil with 3 and 5% LS admixture. A non-linear behavior was observed with LS treatment, the coefficient of compression (C_c) increased by 30% at 3% LS content while the opposite trend was observed at 5% LS content, i.e., C_c value reduced by 30% compared to untreated soil. In addition, the coefficient of consolidation (C_v) and permeability (k_w), increased with an increase in LS content. Another study conducted by Alzigha et al. (2018b), evaluated the efficacy of LS by analyzing various consolidation parameters, (i.e., coefficient of consolidation (C_v), coefficient of volume change (m_v), and corresponding soil permeability (k_w) for a confining pressure range between 50 and 3500 kPa under saturated conditions. At optimum LS content (i.e., 2%), the treated soil exhibited relatively higher C_v values at low confining pressure (50–300 kPa) highlighting rapid initial settlement followed by relatively constant C_v values with an increase in consolidation pressure from 300 to 2000 kPa. Such behavior of LS-treated expansive soil is similar to silty soils (Alzigha et al. 2018b). Further, the LS treatment improves the reduction in m_v value against the consolidation pressure associated with the resistance offered by soil particles due to the presence of LS. Besides, LS treatment induces a negligible increase in soil permeability.

A study conducted by Ijaz et al. (2020b), also evaluated the effect of LS treatment on C_c, coefficient of recompression (C_r), and vertical effective yield stress (σ_y) under saturated and wetting-drying conditions. At optimum LS content (i.e., 2%), the treated soil showed improvement in C_c and σ_y values with some reduction in C_r values. Whereas wetting-drying cycles induced adverse effects, highlighting the inability of LS under long-term durability. In view of it, Ijaz et al. (2020b) proposed a novel LS-based composite admixture by integrating it with LM. The results showed tremendous improvement in C_c, C_r, and σ_y taking into account the long-term durability subjected to wetting–drying cycles. Besides, falling head permeability test results showed little variation with LS addition and results are in-lined with Alzigha et al. (2018b) under normal laboratory conditions, however, under wetting-drying cycles, an increase in permeability was observed (Fig. 7.4) associated with desiccation-induced cracking. In addition, the LS-based proposed novel admixture provides better resistance to desiccation-induced cracking and curtails the permeability to lower levels under successive wetting-drying cycles.

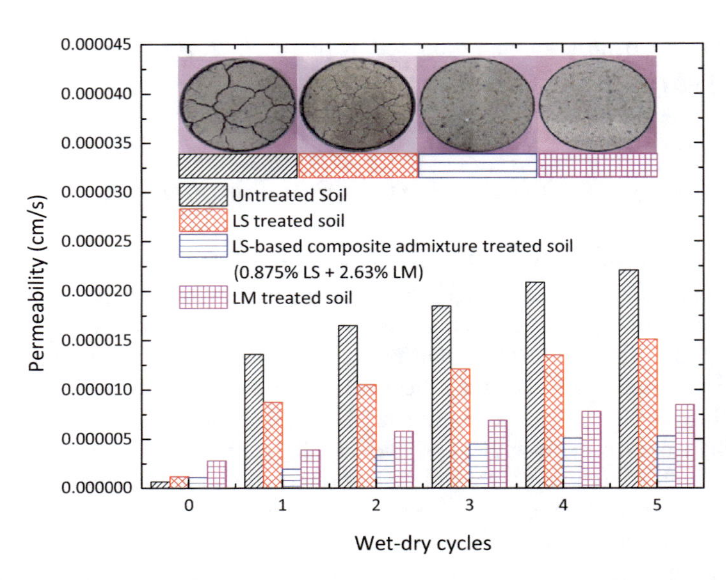

Fig. 7.4 Comparative graph of the variation in permeability of untreated and treated expansive soil with different admixtures taking into account the wetting-drying cycles (Ijaz et al. 2020b)

7.3.4 Strength Characteristics

The strength (i.e., shear and compressive) is an important property of soil that governs the stability of the structure. The expansive soils normally exhibit low strength (Alzigha et al. 2018b; Ijaz et al. 2020a, b). Many researchers have incorporated various doses of LS to analyze the effect on strength of expansive soil (Alzigha et al. 2018b; Ijaz et al. 2020a; Sarker et al. 2021). Indraratna et al. (2012) reported that LS addition in expansive soils induced stiffness and increase the unconfined compressive strength (UCS) values along with the curing period. Another study conducted on high expansive soil also showed an increase in UCS value of about 7.5% at optimum LS content of 2% (Alzigha et al. 2018b; Ijaz et al. 2020a). Ijaz et al. (2020b) also observed a marginal improvement in UCS value of soil with LS addition at a curing period of 28 days. Ta'negonbadi and Norzad (2018) worked on low expansive clays, the results exhibited an increase in UCS and cohesion (*c*) values with slight variation in the angle of internal friction. Furthermore, the effect of wetting–drying cycles was also incorporated which showed a reduction in *c*-value from 7 to 1 kPa at the fourth wetting-drying cycle. Such adverse effects were also reported by Ijaz et al. (2020b), under successive wetting-drying conditions the *c*-value tends to reduce to the minimum. Such reduction in shear strength is attributed to desiccation-induced cracking which results in breaking of the bond between soil particles which in turn results in the reduction of *c*-value. Literature reveals that LS as a lone expansive soil stabilizer insufficiently improves the mechanical properties of expansive soils due to absentia of pozzolanic reaction that forms the cementing gel to contribute to the mechanical characteristics of the soil. Thus, LS must be coupled with calcareous materials that can initiate a pozzolanic reaction with soil minerals. In view of this, Ijaz et al. (2020b) analyzed the efficacy of the proposed LS-based composite admixture (LS-drying cycles.

7.3.5 *Stabilization Mechanism*

Numerous researchers attempted to evaluate the effect of LS treatment on the micro-morphological properties of the expansive soil (Alzigha et al. 2018a; Ijaz et al. 2018b; Ta'negonbadi and Noorzad 2018). Various techniques were employed in the literature including X-ray diffraction (XRD), Scanning Electron Microscope (SEM), Fourier Transform Infrared (FTIR), Nuclear Magnetic Resonance (NMR), Computed Tomography (CT-Scan), Zeta potential test (ZPT); Cation Exchange Capacity (CEC), and Specific Surface Area (SSA) (Alzigha et al. 2018a; Ijaz et al. 2018b; Wu et al. 2020a). The smectite group of clay minerals was found to be the major reason for adverse swelling characteristics (Mitchell and Soga 2005). The XRD results showed that upon LS treatment the smectite peaks tended to decrease with some shift in the 2-theta angle which highlights the reduction in peak surface area and the average size of crystallite (Alzigha et al. 2018a; Ijaz et al. 2018b). The addition of LS induces a waterproofing effect and inhibits the water penetration into the inner layered structure of clay minerals owing to its amorphic nature but it remains chemically inert (Ijaz et al. 2018a, b). Moreover, Alzigha et al. (2018a) found an initial increase in d-spacing of smectite minerals attributed to ion-exchange and corresponding intrusion of LS which intercalates in the inner layered structure within crystalline realms of the mineral lattices, resulted in initial expansion. Besides different pertinent literature show that the efficacy of LS treatment is predominant in the case of expansive clay minerals (i.e., smectite) compared to non-expansive clay minerals (i.e., kaolinite) with no formation of new cementing material (Alzigha et al. 2018a; Ijaz et al. 2018b; Wu et al. 2020b). The X computed tomography results showed that LS reduced porosity and improved the pore size distribution (Alzigha et al. 2018a; Wu et al. 2020a). The ZPT showed an increase in zeta potential value with an increase in LS addition showing the dispersion stability and adsorption of the LS acid group creating a shielding effect (Wu et al. 2020b). The FTIR study of LS-treated expansive soil revealed that the intercalated admixture was then absorbed via hydrogen-bonding and/or bonded directly with the dehydrated cations (Alzigha et al. 2018a; Wu et al. 2020a). It was also observed by the researchers that with LS addition the cation exchange between calcium ions (Ca^{+2}) present in the admixture and soil cations (Na^{+}, Ti^{+}, Mg^{+2}) occurred which resulted in flocculation, which was further confirmed by SEM micrographs (Alzigha et al. 2015, 2018a; b; Ijaz et al. 2018a, b). The study conducted by Ijaz et al. (2018a, b) showed that the proposed LS-based composite admixture was three times more effective in reducing the montmorillonite peak intensities, compared to LS-alone. The better results attributed to the chemical reaction of LM with clay minerals coupled with amorphicity of LS admixture. Besides, the peak intensities of non-expansive soil minerals such as quartz, albite, etc., tended to increase with LS-based composite admixture treatment along with the curing period promoting the crystalline structure.

7.4 Conclusions and Recommendations

This article reviews the impending efficiency of LS as an expansive soil stabilizer on important geotechnical and micro-morphological properties based on short and long-term effects. Based on the pertinent literature following conclusions and recommendations are drawn:

- The LS exhibits a positive potential in mitigating the adverse swell-shrinkage characteristics. However, the efficacy of LS is based on the degree of expansion of the subjected expansive soil. The studies show that LS as a lone stabilizer is only effective in the case of expansive soil with a low degree of expansion.
- LS can appreciably reduce the swelling potential and soil plasticity, however, LS treatment in case of highly expansive soil demands some chemical triggering mechanism that could effectively intercalate the inner layer structure of the smectite minerals along with the waterproofing effect of LS.
- LS treatment reduces the compressibility of the soil owing to the aggregation of the soil particles and has an insignificant effect on the soil permeability. However, under wetting-drying conditions, the desiccation-induced cracking tends to increase and affect the long-term durability of the soil. Therefore, LS should be integrated with some other potential stabilizers to improve its long terms durability, i.e., LM, cement, etc.
- The strength parameters with LS addition show limited improvement. In addition, such improvement tends to diminish accounting for the long-term wetting-drying cycles. Therefore, for a particular structure, special attention should be given subjected to its long-term exposure to the environment. Moreover, depending on the nature of the project it is advisable that the LS-based admixture coupled with some other potential admixture such as LMand cement should be developed which address its shortcoming.
- The stabilization mechanism shows that LS treatment reduces the smectite peaks, reduces the porosity and improves the pore size distribution, and induces aggregation of soil particles along with waterproofing effect which provides sufficient reasoning for improvement in various geotechnical properties. For high expansive soils, it is advisable to integrate LS with some other potential traditional admixtures.
- Furthermore, it is recommended to test and analyze LS and LS-based composite admixtures as a stabilizer for additional problematic soils such as loess, soft, and collapsible soils in future studies to promote the geotechnical application of LS in the construction industry.

References

Alazigha DP, Indraratna B, Vinod JS, Ezeajugh LE (2016) The swelling behaviour of lignosulfonate-treated expansive soil. Proc Inst Civ Eng Ground Improv 169(3):182–193. https://doi.org/10.1680/jgrim.15.00002

Alazigha DP, Vinod JS, Indraratna B, Heitor A (2018a) Potential use of lignosulfonate for expansive soil stabilization. Environ Geotech 6(7):480–488. https://doi.org/10.1680/jenge.17.00051

Alazigha DP, Indraratna B, Vinod JS, Heitor A (2018b) Mechanisms of stabilization of expansive soil with lignosulfonate admixture. Transp Geotech 14:81–92. https://doi.org/10.1016/j.trgeo.2017.11.001

Alazigha DP (2015) The efficacy of lignosulfonate in controlling the swell potential of expansive soil and its stabilization mechanisms. UOW, PhD, Dissertation

Al-Rawas AA, Goosen MF (2006) Expansive soils: recent advances in characterization and treatment. Taylor and Francis

Athukorala R, Indraratna B, Vinod JS (2013) Modeling the internal erosion behavior of lignosulfonate treated soil. In: Geo-congress 2013: stability and performance of slopes and embankments III, March 2013, pp 1865–1874. https://doi.org/10.1061/9780784412787.188

Basma AA, Al-Homoud AS, Malkawi AIH, Al-Bashabsheh MA (1996) Swelling-shrinkage behavior of natural expansive clays. Appl Clay Sci 11(2–4):211–227. https://doi.org/10.1016/S0169-1317(96)00009-9

Bergaya F, Lagaly G (2006) General introduction: clays, clay minerals, and clay science. Dev Clay Sci 1:1–18. https://doi.org/10.1016/S1572-4352(05)01001-9

Bolander P (1999) Laboratory testing of nontraditional additives for stabilization of roads and trail surfaces. Transp Res Rec 1652(1):24–31. https://doi.org/10.3141/1652-38

Canakci H, Aziz A, Celik F (2015) Soil stabilization of clay with lignin, rice husk powder and ash. Geomech Eng 8(1):67–79. https://doi.org/10.12989/gae.2015.8.1.067

Chenarboni HA, Lajevardi SH, MolaAbasi H, Zeighami E (2021) The effect of zeolite and cement stabilization on the mechanical behavior of expansive soils. Constr Build Mater 272:121630. https://doi.org/10.1016/j.conbuildmat.2020.121630

Dif AE, Bluemel WF (1991) Expansive soils under cyclic drying and wetting. Geotech Test J 14(1):96–102. https://doi.org/10.1520/GTJ10196J

Du J, Zhou A, Lin X, Bu Y, Kodikara J (2020) Revealing expansion mechanism of cement-stabilized expansive soil with different interlayer cations through molecular dynamics simulations. Am J Phys Chem 124(27):14672–14684. https://doi.org/10.1021/acs.jpcc.0c03376

Elert K, Azañón JM, Nieto F (2018) Smectite formation upon lime stabilization of expansive marls. Appl Clay Sci 158:29–36. https://doi.org/10.1016/j.clay.2018.03.014

Fernandez MT, Orlandi SG, Codevilla M, Piqué TM, Manzanal D (2020) Performance of calcium lignosulfonate as a stabiliser of highly expansive clay. Transp Geotech 27:100469. https://doi.org/10.1016/j.trgeo.2020.100469

Fredheim GE, Braaten SM, Christensen BE (2002) Molecular weight determination of lignosulfonates by size-exclusion chromatography and multi-angle laser light scattering. J Chromatogr A 942(1–2):191–199. https://doi.org/10.1016/S0021-9673(01)01377-2

Gandini A, Belgacem MN (2008) Lignins as components of macromolecular materials. In: Monomers, polymers and composites from renewable resources, pp 243–271. https://doi.org/10.1016/B978-0-08-045316-3.00011-9

Giahi A, Jiryaei Sharahi M, Mohammadnezhad B (2021) Evaluation of a by-product and enviromental-friendly chemical additives for clay soils with different mixing and curing methods. Amirkabir J Civ Eng 53(2):659–674. https://doi.org/10.22060/CEEJ.2020.16461.6238

Habert G, Bouzidi Y, Chen C, Jullien A (2010) Development of a depletion indicator for natural resources used in concrete. Resour Conserv Recycl 54(6):364–376. https://doi.org/10.1016/j.resconrec.2009.09.002

Harris DJ (1999) A quantitative approach to the assessment of the environmental impact of building materials. Build Environ 34(6):751–758. https://doi.org/10.1016/S0360-1323(98)00058-4

Ijaz N, Dai F, Meng L, ur Rehman Z, Zhang H (2020a) Integrating lignosulphonate and hydrated lime for the amelioration of expansive soil: a sustainable waste solution. J Clean Prod 254(119985). https://doi.org/10.1016/j.jclepro.2020.119985

Ijaz N, Dai F, ur Rehman Z (2020b) Paper and wood industry waste as a sustainable solution for environmental vulnerabilities of expansive soil: a novel approach. J Environ Manage 262(110285). https://doi.org/10.1016/j.jenvman.2020.110285

Ijaz N, Dai F, ur Rehman Z, Ijaz Z, Zahid M (2020c) Laboratory evaluation of curing period for stabilized expansive soil by a new paper/timber industry waste based cementing material. IOP Conf Ser Earth Environ Sci 442(1):012008. https://doi.org/10.1088/1755-1315/442/1/012008

Indraratna B, Muttuvel T, Khabbaz H, Armstrong R (2008) Predicting the erosion rate of chemically treated soil using a process simulation apparatus for internal crack erosion. J Geotech Geoenviron Eng 134(6):837–844. https://doi.org/10.1061/(ASCE)1090-0241(2008)134:6(837)

Indraratna B, Mahamud MAA, Vinod JS, Wijeyakulasuriya V (2010) Stabilization of an erodible soil using a chemical admixtures. Proc Inst Civ Eng Ground Improv 163(1):45–54. https://doi.org/10.1680/grim.2010.163.1.43

Indraratna B, Mahamud MAA, Vinod JS (2012) Chemical and mineralogical behaviour of lignosulfonate treated soils. In: GeoCongress 2012: State of the art and practice in geotechnical engineering, pp 1146–1155. https://doi.org/10.1061/9780784412121.118

Indraratna B, Athukorala R, Vinod J (2013) Estimating the rate of erosion of a silty sand treated with lignosulfonate. J Geotech Geoenviron Eng 139(5):701–714. https://doi.org/10.1061/(ASCE)GT.1943-5606.0000766

Khalid U, ur Rehman Z, Liao C, Farooq K, Mujtaba H (2019) Compressibility of compacted clays mixed with a wide range of bentonite for engineered barriers. Arab J Sci Eng 44(5):5027–5042. https://doi.org/10.1007/s13369-018-03693-7

Koohpeyma HR, Vakili AH, Moayedi H, Panjsetooni A, Nazir R (2013) Investigating the effect of lignosulfonate on erosion rate of the embankments constructed with clayey sand. Sci World J 2013(587462). https://doi.org/10.1155/2013/587462

Landlin G, Soundarya MK, Bhuvaneshwari S (2021) Behaviour of lignosulphonate amended expansive soil. In: Sustainable practices and innovations in civil engineering. Springer, Singapore, pp 151–162. https://doi.org/10.1007/978-981-15-5101-7_15

Li GY, Hou X, Mu YH, Ma W, Wang F, Zhou Y, Mao YC (2019) Engineering properties of loess stabilized by a type of eco-material, calcium lignosulfonate. Arab J Sci Eng 12(22):1–10. https://doi.org/10.1007/s12517-019-4876-0

Madurwar MV, Ralegaonkar RV, Mandavgane SA (2013) Application of agro-waste for sustainable construction materials: a review. Constr Build Mater 38:872–878. https://doi.org/10.1016/j.conbuildmat.2012.09.011

Mitchell JK, Soga K (2005) Fundamentals of soil behavior. John Wiley & Sons, New York

Monje V, Nobel P, Junicke H, Kjellberg K, Gernaey KV, Flores-Alsina X (2021) Assessment of alkaline stabilization processes in industrial waste streams using a model-based approach. J Environ Manage 293:1128s06. https://doi.org/10.1016/j.jenvman.2021.112806

Moore DM, Reynolds Jr RC (1989) X-ray diffraction and the identification and analysis of clay minerals. Oxford University Press (OUP)

Mujtaba H, Khalid U, Farooq K, Elahi M, Rehman Z, Shahzad HM (2020) Sustainable utilization of powdered glass to improve the mechanical behavior of fat clay. KSCE J Civ Eng 24 (12):3628–3639. https://doi.org/10.1007/s12205-020-0159-2

Naseem A, Mumtaz W, De Backer H (2019) Stabilization of expansive soil using tire rubber powder and cement kiln dust. Soil Mech Found Eng 56(1):54–58. https://doi.org/10.1007/s11204-019-09569-8

Nelson J, Miller DJ (1997) Expansive soils: problems and practice in foundation and pavement engineering. John Wiley and Sons, New York

Noorzad R, Ta'negonbadi B (2018) Mechanical properties of expansive clay stabilized with lignosulphonate. Q J Eng Geol Hydrogeol 51(04):483–492. https://doi.org/10.1144/qjegh2017-050

Orlandi S, Manzanal D, Miranda E, Robinson M (2019) Use of lignin as stabilizer in expansive soils. In: Geotechnical engineering in the XXI century: lessons learned and future challenges, vol 1, issue 1. IOS Press, pp 2291–2298
Pearl IA (1967) The chemistry of lignin. Publisher, Marcel Dekker INC
Prasad CRV, Reshmarani B (2020) Characterization of expansive soils treated with lignosulfonate. Int J Geo-Eng 11(1):1–10. https://doi.org/10.1186/s40703-020-00124-1
Puppala AJ, Hanchanloet S (1999) Evaluation of a new chemical (SA-44/LS-40) treatment method on strength and resilient properties of a cohesive soil. In: 78th Annual meeting of the transportation research board, Washington, DC, vol 1, issue 1, pp 1–8
Puppala AJ, Hanchanloet S (1999) Evaluation of a new chemical treatment method on strength and resilient properties of a cohesive soil. Transp Res Board 1(990389):36–42
Puppala AJ, Pedarla A (2017) Innovative ground improvement techniques for expansive soils. Innov Infrastruct Solut 2(1):1–15. https://doi.org/10.1007/s41062-017-0079-2
Qian G, Cao Y, Chui P, Tay J (2006) Utilization of MSWI fly ash for stabilization/solidification of industrial waste sludge. J Hazard Mater 129(1–3):274–281. https://doi.org/10.1016/j.jhazmat.2005.09.003
Ravishankar AU, Reddy MKJC (2017) Experimental investigation of lateritic soil treated with calcium lignosulfonate. In: Indian geotechnical conference, vol 1, issue 1, pp 1–7
Reddy NG, Tahasildar J, Rao BH (2015) Evaluating the influence of additives on swelling characteristics of expansive soils. Int J Geosynth Ground Eng 1(1):1–7. https://doi.org/10.1007/s40891-015-0010-x
Sabitha BS, Evangeline YS (2021) Stabilisation of Kuttanad soil using calcium and sodium lignin compounds. In: Proceedings of the Indian geotechnical conference 2019, vol 1, issue 1, pp 249–258. https://doi.org/10.1007/978-981-33-6444-8_22
Santoni RL, Tingle JS, Webster SL (2002) Stabilization of silty sand with nontraditional additives. Transp Res Rec 1787(1):61–70. https://doi.org/10.3141/1787-07
Sarker D, Shahrear Apu O, Kumar N, Wang JX, Lynam JG (2021) Application of sustainable lignin stabilized expansive soils in highway subgrade. In: IFCEE 2021, vol 1, issue 1, pp 336–348. https://doi.org/10.1061/9780784483435.033
Sharmila B, Bhuvaneshwari S, Landlin G (2021) Application of lignosulphonate-a sustainable approach towards strength improvement and swell management of expansive soils. Bull Eng Geol Environ 1(1):1–19. https://doi.org/10.1007/s10064-021-02323-1
Singh SP, Palsule PS, Anand G (2021) Strength properties of expansive soil treated with sodium lignosulfonate. In: Problematic soils and geoenvironmental concerns, vol 1, issue 1, pp 665–679. https://doi.org/10.1007/978-981-15-6237-2_55
Ta'negonbadi B, Noorzad R (2017) Stabilization of clayey soil using lignosulfonate. Transp Geotech 12(1):45–55. https://doi.org/10.1016/j.trgeo.2017.08.004
Ta'negonbadi B, Noorzad R (2018) Physical and geotechnical long-term properties of lignosulfonate-stabilized clay: an experimental investigation. Transp Geotech 17(1):41–50. https://doi.org/10.1016/j.trgeo.2018.09.001
Tasalloti SM, Indraratna B, Rujikiatkamjorn C, Heitor A, Chiaro G (2015) A laboratory study on the shear behavior of mixtures of coal wash and steel furnace slag as potential structural fill. Geotech Test J 38(4):361–372. https://doi.org/10.1520/GTJ20140047
Theng BKG (2012) Formation and properties of clay-polymer complexes. Elsevier Scientific Publishing Company, Amsterdam, The Netherlands
Tingle JS, Santoni RL (2003) Stabilization of clay soils with nontraditional additives. Transp Res Rec 1819(1):72–84. https://doi.org/10.3141/1819b-10
Tingle JS, Newman JK, Larson SL, Weiss CA, Rushing JF (2007) Stabilization mechanisms of nontraditional additives. Transp Res Rec 1989(1):59–67. https://doi.org/10.3141/1989-49
ur Rehman Z, Khalid U (2021) Reuse of COVID-19 face mask for the amelioration of mechanical properties of fat clay: a novel solution to an emerging waste problem. Sci Total Environ 794(1):148746. https://doi.org/10.1016/j.scitotenv.2021.148746

ur Rehman Z, Farooq K, Mujtaba H, Khalid U (2021) Unified evaluation of consolidation parameters for low to high plastic range of cohesive soils. Mehran Univ Res J Eng Technol 40(1):93–103. https://doi.org/10.22581/muet1982.2101.09

ur Rehman Z, Khalid U, Farooq K, Mujtaba H (2018) On yield stress of compacted clays. Int J Geo-Eng 9(1):1–16. https://doi.org/10.1186/s40703-018-0090-2

Vakili AH, Ghasemi J, bin Selamat MR, Salimi M, Farhadi MS (2018a) Internal erosional behaviour of dispersive clay stabilized with lignosulfonate and reinforced with polypropylene fiber. Constr Build Mater 193(1):405–415. https://doi.org/10.1016/j.conbuildmat.2018.10.213

Vakili AH, Kaedi M, Mokhberi M, bin Selamat MR, Salimi M (2018b) Treatment of highly dispersive clay by lignosulfonate addition and electroosmosis application. Appl Clay Sci 152(1):1–8. https://doi.org/10.1016/j.clay.2017.11.039

Vinod JS, Indraratna B, Mahamud MA (2010) Stabilisation of an erodible soil using a chemical admixture. Proc Inst Civ Eng Ground Improv 163(1):43–51. https://doi.org/10.1680/grim.2010.163.1.43

Wu D, She W, Wei L, Zuo W, Hu X, Hong J, Miao C (2020) Stabilization mechanism of calcium lignosulphonate used in expansion sensitive soil. J Wuhan Univ Technol Mater Sci 35(5):847–855. https://doi.org/10.1007/s11595-020-2329-y

Wu X, Jiang J, Wang C, Liu J, Pu Y, Ragauskas A, Yang B (2020) Lignin-derived electrochemical energy materials and systems. Biofuels Bioprod Bioref 14(3):650–672. https://doi.org/10.1002/bbb.2083

Chapter 8
Use of Photocatalyst in Self-Cleaning Constructions Material: A Review

Naveen Thakur, S. B. Singh, and Anshuman

8.1 Introduction

Global climate change due to industrialization and urbanization has badly affected our environment. Although industries are essential for our economic growth and development, their effects on ecosystem are adverse. Byproducts of the industries have significantly affected our environment including agriculture and water resources. It also has negative impact on air quality consequently human health. Further, rapidly developing concrete industry to fulfill human needs also impacted our environment (Nath and Zain 2016; Ziming and Shen 2020). Extensive use of cement and sand has hampered the sustainable development and has degraded the air and water quality (Ziming and shen 2020). Portland cement which is an important ingredient of concrete is the major contributor of pollution. Production of cement generates large amount of CO_2 along with other hazardous gases such as NOx and SO_3 that leads to enhanced greenhouse effect and acid rain (Shen and Lyu 2020). However, by selecting eco-friendly cement material in construction environment, pollution can be reduced to some extent (Zhong and Haghinghat 2015). Recently, a new construction material with self-cleaning ability has been developed (Zhong and Haghinghat 2015; Sikkema and Alleman 2014; Ohama and Gemert 2011). Self-cleaning mixture can retain its color for longer period of time than existing construction matrix, so need

N. Thakur · S. B. Singh (✉) · Anshuman
Department of Civil Engineering, Birla Institute of Technology and Science, Pilani 333031, Rajasthan, India
e-mail: sbsingh@pilani.bits-pilani.ac.in

N. Thakur
e-mail: naveen.thakur@secs.ac.in

Anshuman
e-mail: anshu@pilani.bits-pilani.ac.in

K. R. Reddy et al. (eds.), *Advances in Sustainable Materials and Resilient Infrastructure*, Springer Transactions in Civil and Environmental Engineering,
https://doi.org/10.1007/978-981-16-9744-9_8

not be substituted frequently and can also minimize the risk of air pollution (Sikkema and Alleman 2014; Ohama and Gemert 2011).

8.1.1 How Self-Cleaning Material was Invented and It's History

Self-cleaning property of cement was described by Italian chemist Luigi Cassar in his pioneer work. He incorporated TiO_2 with cement, which gets oxidized with sunlight and breaks down pollutants (Husken and Hunger 2009; Cassar 2004). A building material with long-term stability and ability to retain a bright color in toxic environment was prepared by a method called "photocatalysis", which keeps away dirt with the help of sun's energy. Surprisingly, the air around the tested concrete depicted 80% reduction in nitrous oxide. Reduction in other toxic compounds such as lead, carbon monoxide, and sulfur dioxide were also observed. Some of the important historical events about the research and development of photocatalytic compound are shown in Fig. 8.1 (Sikkema and Alleman 2014; Hamidi and Aslani 2019). In early 2000, in a church Dives in Misericordia, in Italy, photocatalytic material was analyzed using Rhodamine B discoloration test. Lately, first generation of photocatalytic material was prepared with start of PICADA project (Fig. 8.1). Thus, a thin layer of cement coating with photocatalytic compound enabled the cement to clean concrete and also clean the air in the city (Husken and Hunger 2009; Cassar 2004).

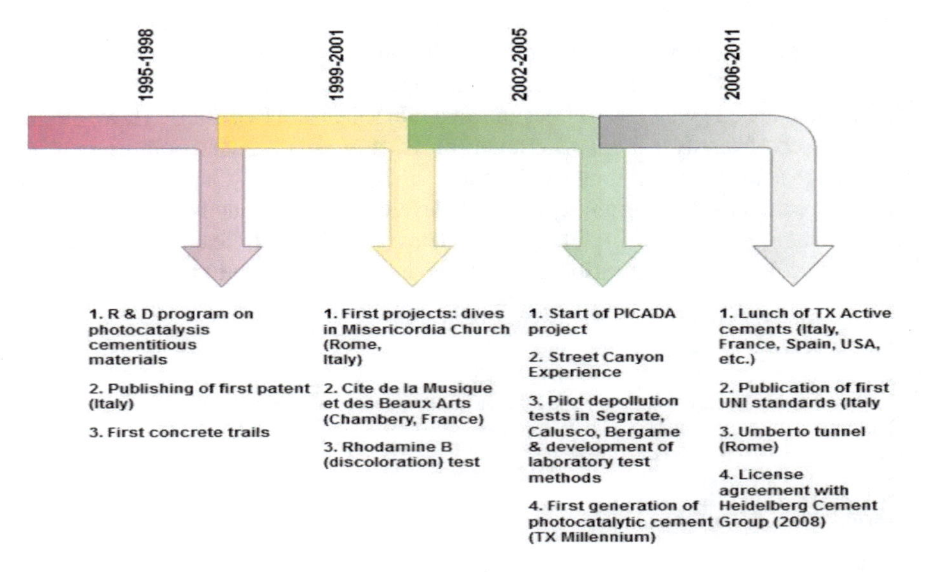

Fig. 8.1 History of photocatalysis cementitious materials (Bondioli et al. 2009; Cassar 2004)

8.1.1.1 Self-Cleaning Concrete

Concrete and cementitious building material are generally used in the construction of offices, residential buildings, public places and area of aesthetic values, etc. (Nath and Zain 2016). Dust produced by construction and demolition of buildings causes the pollution and makes building surfaces looks dull and dirty (Nath and Zain 2016; Shen and Lyu 2020). These situations have forced the researchers and industries to design an eco-friendly alternative material known as self-cleaning concrete for the maintenance of healthy environment. Self-cleaning concrete is also known as "Green concrete" as it is capable of retaining the visual properties of buildings for longer duration even in highly polluted environment (Shen and Lyu 2020). Self-cleaning concrete keeps itself dirt free and maintains its brightness in polluted urban areas (Shen and Lyu 2020). Some of the important aspects of self-cleaning materials are shown in (Fig. 8.2).

Geo-polymer concrete, supplementary cementitious materials, and alkali-activated materials are some of the environment friendly concrete with detoxification abilities. (Nath and Zain 2016; Fujishima and Rao 2000; Paola and Marci 2012). Geo-polymer cement is able to tolerate high temperature, high salt concentration, and acidic environment. This material not only decreases the CO_2 emissions but is also economical and durable (Nath and Zain 2016; Shen and Lyu 2020). Other alternative cementitious materials such as fly ash, furnace slag, silica fume, limestone dust, rice husk ash, palm oil fuel ash, cement kiln dust, and metakaolin instead of Portland cement have also been used to reduce CO_2 emission (Hipolito and Martinez 2019; Morsy and Alsayed 2012).

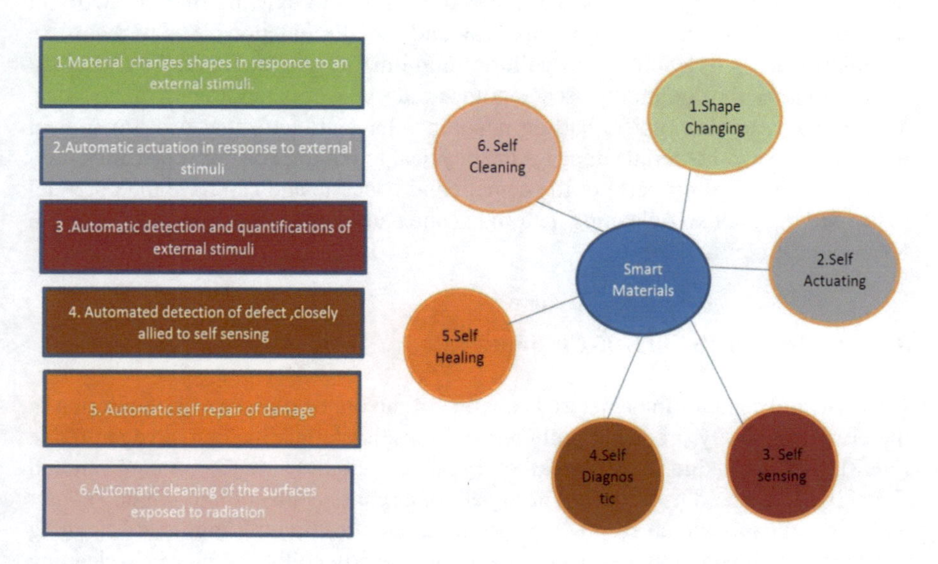

Fig. 8.2 Smart functions added to the structural materials and their capabilities (Fujishima and Rao 2000)

Furthermore, addition of "photocatalyst" also provides self-cleaning effect to the building material (Nath and Zain 2016; Fujishima and Rao 2000; Paola and Marci 2012). Photocatalytic compound is applied to the surface of concrete as a thin layer which adds the property of air purification to the concrete (Nath and Zain 2016; Shen and Lyu 2020). Photocatalyst removes pollutants like hydrocarbons, sulfur dioxide, carbon monoxide, carbon dioxide, and nitrogen oxides. It converts organic pollutants into carbon dioxide and water by photodegradation reaction (Nath and Zain 2016; Shen and Lyu 2020). TiO_2 is widely used photocatalyst material, it can be used on wide range of surface like mortars, paints, and tiles (Diamanti and Ormellese 2008; Bondioli and Ferrari 2009). But, TiO_2 has certain limitations as its activation due to solar radiation is limited therefore it is modified with other transition metals or non-metallic anionic species (Pinho and Rojas 2015; Luna and Juan 2018). Further, Portland cement is also modified by adding Titanium Oxynitride (TiO_2-xNy) to enhance its activation under visible light spectrum (Pinho and Rojas 2015; Luna and Juan 2018). Recently, an attempt has been made to replace traditional TiO_2 with bismuth due to its high photocatalytic activity and cost efficiency (Pinho and Rojas 2015; Luna and Juan 2018). The tendency of the photocatalytic self-cleaning coating activities is: $Bi_2O_2CO_3$ (49%) > BiOI (30%) > $BiVO_4$ (15%) > $BiPO_4$ (14%) > Bi_2O_3 (5%) (Hipolito and Martinez 2019). $Bi_2O_2CO_3$ exhibited the lowest crystallite size (27 nm) among the studied compound (Hipolito and Martinez 2019). Importantly, all the titanium and bismuth-based photocatalyst exhibit self-cleaning ability of photodegradation of organic compounds. Moreover, photocatalytic compounds are low cost and they won't affect the mechanical property of the concrete.

In this view, in future, nanotechnology can play a crucial role in the construction of functional buildings. Nanosized materials addition to existing material will not only improve basic properties but also can add specific functionalities to them like antimicrobial, self-cleaning, and pollution minimizing properties (Zaho and Zhou 2020). Some of the commonly used nanomaterials are carbon nanotubes, nano silica, and graphene oxide. Typical carbon nanotubes have strength that of normal steel tubes. All the nanomaterials improve the physical (strength, elasticity, ductility, etc.) as well as chemical property of the cement and concrete and can have the potential to be used as better self-cleaning material (Zaho and Zhou 2020,Chuh and Pan 2014; Kumar and Kolay 2012).

8.1.1.2 Importance of Self-Cleaning

Concrete and cementitious materials are one of the major sources of air pollution in the city. Commonly produced pollutants by these materials include nitrogen oxide (NO2), Sulfur dioxide (SO2), volatile organic compounds (VOCs), etc. (Nath and Zain 2016; Shen and Lyu 2020). These pollutants cause deposition of organic matter and contaminants which result in external damage to the buildings. Self-cleaning concrete can be potential approach for keeping the city pollution free. Self-cleaning concrete helps to minimize the air pollutants, reduce maintenance cost, and extend the life of buildings (Nath and Zain 2016; Zhong and Haghinghat 2015; Sikkema and Alleman 2014). In addition, self-cleaning materials reflect light and reduce the heat builds up on buildings, and keep the city cool.

Furthermore, various photocatalytic materials improve air quality and keep city clean and beautiful. Self-cleaning is greatly beneficial for cementitious materials as it maintains its mechanical strength and functions. It has been reported that photocatalyst-TiO_2 nanoparticles contribute to increased tensile and flexibility of cement (Hamidi and Aslani 2019; Paola and Lopez 2012; Witkowski and Hubert 2019). TiO_2 in concrete gets activated in the presence of light radiation and mostly remains unutilized. Studies have investigated the exploitation of TiO_2 as surface coating of concrete as a protective coating for cement hydration products. TiO_2 is used as photocatalyst due to its low cost, resistance to corrosion, low toxicity, and it is activated by solar radiation (Quiroga and Viles 2018; Folli and Pade 2012). Recently bismuth has been proposed as alternative to TiO_2 with much higher photocatalytic activity than TiO_2(Hipolito and Martinez 2019). The hydroxyl radicals and superoxide anions produced in photocatalytic reactions can react with pollutant molecules (SO_2, NO^2, VOCs, etc.) and remove them. Photocatalytic materials use atmospheric O_2 as an oxidant agent. Apart from this, outer coating with TiO_2 maintains the visual aspects and brightness of buildings due to photocatalytic action. Use of natural wollastonite powder as the binder also reduces the CO_2 emission and global warming (He et al. 2019). Results of some of the important studies on self-cleaning material are presented in Table 8.1. Thus, self-cleaning coatings reduce aesthetic damage and associated deterioration of building materials.

8.2 Photocatalytic Cementitious Materials

Photocatalytic structure materials have been considered as a good alternative to existing environmental polluting construction materials. Long-term maintenance of aesthetic properties of white cement is the additional benefit of photocatalysis cement-based structures (Nath and Zain 2016, Shen and Lyu 2020). Mixing of Photocatalytic material such as TiO_2 in construction material does not alter the final characteristics of cementitious products. The development of construction materials mainly depends on two factors:

(1) Surface quality and visual appearance.
(2) Structural stability.

Thus, materials should be selected carefully to keep a balance between mixture constituents and rheological properties of resulting mixture (Paola and Marci 2012). Materials selection and processing are the most critical parameters which affect properties and functionality of cementitious matrix. (Quiroga and Viles 2018; Awadalla and Arafa 2011; Teoh and Scott 2012). Factors influencing the performance of TiO_2-based photocatalytic construction material and major steps used for dispersion of TiO_2 in construction material are shown in Fig. 8.3.

Table 8.1 Literature review on different existing research techniques of self-cleaning

Sr. no.	References	Description
1	Awadalla (2011)	Efficiency of Self-cleaning photocatalyst Titanium dioxide (TiO_2) on porosity of different types of waste materials was studied. Photodegrading of CO_2 was related to porosity cement
2	Dalawai (2020)	Discussed the advantages of SCT (Self-cleaning technology) durability and different methods of fabrication of photocatalytic surfaces and potential utility in different sectors
3	Folli and Aandrea (2012)	Self-cleaning photocatalyst, Titanium dioxide TiO_2 is implemented into cement to test its depolluting properties by using Rhodamine B (RhB) dye
4	Garcia (2018)	Investigated glass reinforced concrete (GRC) based fabrication and characterization of self-cleaning and pollutant degradation. The GRC panel fabrication maintains the TiO_2 on the surface of material. But NO_x degradation decreased with aged treatment
5	He et al. (2019)	Self-cleaning cement-based composite materials (SSCCM) were prepared by mixing material with hydrophobic and luminescent properties. Afterward, luminous properties and hydrophobicity were examined
6	Hamidi and Aslani (2019)	Reported TiO_2-based photocatalysis cement material, applications, and future prospects and challenges in photocatalytic material technology. TiO_2 in combination with cement-based material decrease the concentration of pollutant NO_x
7	Hipolito (2019)	In this experiment alternative building materials such as fly ash, sodium carbonate as an alkaline activator were coated with bismuth-based photocatalysts. $Bi_2O_2CO_3$ showed very high photocatalytic efficiency
8	Li and Liu (2016)	A novel type of photocatalytic construction materials (TiO_2-EMR cement) was formulated by sol–gel dip-coating method. TiO_2-EMR cement showed high self-cleaning and mechanical strength
9	Li and Zhang (2019)	Common cement is replaced with supplementary cementitious materials (SCMs)-diatomite and limestone. It is "green concrete" with 50% lower global warming potential
10	Papanikolaou and Arena (2019)	Investigated the role of graphene nanoplatelets (GNPs) as functional fillers matrix for self-cleaning. GNPs are environmentally friendly and lesser damaging in terms of global warming as it produces 248 times lesser CO_2 as compared to normal Portland cement
11	Quiroga (2018)	Evaluated the efficiency of self-cleaning coating of TiO_2 in degrading artificial stain (rhodamine B) after long time incubation. Effect of photocatalytic activity on durability of cement was also analyzed

(continued)

Table 8.1 (continued)

Sr. no.	References	Description
12	Shen (2015)	Self-cleaning concrete was prepared by using main hydration product of cement Calcium Silicate Hydrates (C-S-H) with photocatalyst-TiO_2 nano particles. Due to properties of these compounds, contaminants were washed away by rain
13	Szymanowski (2019)	Studied the different testing methods and the factors for characterization of the binding between surface coating and construction material
14	Vaidevi and kala (2020)	Various testing methods to evaluate stability and resistance against segregation of self-compacting concrete are shown
15	Witkowski (2019)	Measured the air detoxification efficiency of photocatalytic concrete after 7 years in a moderate climate in Zielona Gora (Poland). They showed that TiO_2 was present in form of agglomerates with diamond and has nitric oxide (NO) removal capability
16	Yang (2019)	Photocatalytic cementitious material was prepared by adding g-C_3N_4 nanosheets (CNNs) into cement. It enhanced the NO_x abatement and self-cleaning performance

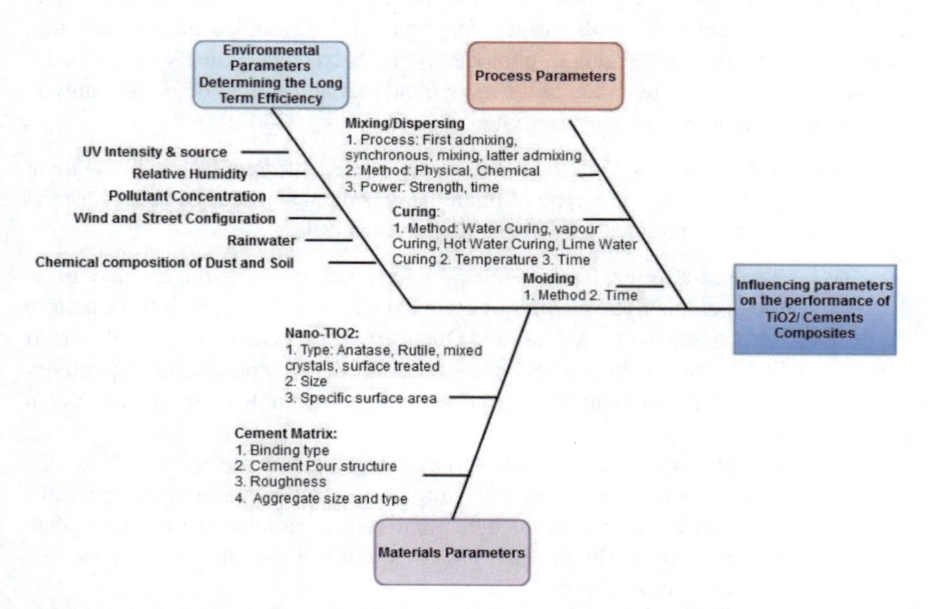

Fig. 8.3 Factors affecting the performance of photocatalytic cement-based materials (Tsang and Cheng 1997; Vittoriadiamanti and Pedeferri 2013)

8.3 Process Parameters

Processing mainly includes mixing/dispersion, molding, and curing. Amalgamation procedure of TiO_2 nanoparticles in cement material is one of the vital steps in production of construction mix. Mixing method affects consistency and the properties of the final products (Addamo and Augugliaro 2008). Nano-size TiO_2 particles owing to high surface energy can easily agglomerate and it's hard to break these aggregates (Addamo and Augugliaro 2008; Palmisano and Augugliaro 2011). Homogeneous distribution of nano-TiO_2 in the construction matrix is complicated step. Therefore, this is a challenging task for researchers to develop an optimum method for dispersion of TiO_2 particles in construction material (Addamo and Augugliaro 2008, Tsai and Cheng 2007).

8.3.1 Environmental Parameters

The important features of photocatalytic construction material are their self-cleaning property and long-term photocatalytic ability for the degradation of air pollutants. Various studies have indicated the same properties of photocatalytic construction material with a reduction in the photocatalytic efficacy of the TiO_2-based cement composites with time. The photocatalytic ability of TiO_2-based materials may decrease significantly for both surface coatings and TiO_2 added in the main mix after a 4-month duration. Aging of photocatalytic concrete also affects the air pollutants removal due to carbonation of the matrix and partial inactivation of the catalytic sites on the Titanium oxide surface (Guerrini and Plassais 2007).

a. **Cement Parameters**: Various studies have reported different factors like cement matrix pore organization, type of binder, and cement surface coarseness for the performance of photocatalytic construction materials.
    (i) **Type of Binder**: The chemical characteristics of the binder also affect the photocatalytic activity of composite material (Cassar 2004; Cucitore and Cassar 2011a; Murata and Obata 1999; Vittoriadimanti and Pedeferri 2013). The ordinary Portland cement shows lower photocatalytic activity them white cement because of metallic compounds (Hamidi and Aslani 2019).
    (ii) **Roughness**: Few researchers have investigated the effect of surface roughness on the performance of photocatalytic construction materials. The surface areas for medium and rough samples were, larger than sample with fine area. However, NOx removal was directly proportional to surface topography.
    (iii) **Cement Pore Structure**: Highly porous structure does not lead to higher photo-activity (Hamidi and Aslani 2019). However, pores $> 1\ \mu m$ (pores

of air) and < 0.05 μm exhibited decreased degradation of NOx and organic dyes.

8.3.1.1 Techniques to Evaluate Photocatalytic Efficiency in Cementitious Materials

Qualitative and Quantitative parameters of Photocatalysis technology need to be assessed for better performance and functionality of cementitious materials. Various assessment tests have been designed to measure the detoxification effect of different photocatalytic materials, however no standard testing method has been approved. Similarly, there is no optimum method available for the evaluation of self-cleaning ability.

a. Efficiency on the basis of environment toxicants: Different test methods are used for photocatalytic cement on the basis of category of environment toxicants (e.g., NOx, organics, etc.), formulation of construction matrix and physio-chemical properties, etc. (Quiroga and Viles 2018; Folli and Pade 2012).

   1. NOx Tests: NOx test methods include four main types as given below:
      (i) NOx flow-through test: In this, air purification efficiency of the photocatalytic material is assessed on the basis of relative concentration of NOx in the water and sample.
      (ii) Dynamic method: It is used for inorganic materials to reduce the NOx concentration.
      (iii) Static method: Like dynamic method, it is also used for inorganic materials to reduce the NOx concentration.
      (iv) Photocatalytic Innovative Coating: The photo-conversion of NOx is assessed with time on the basis of chamber coated with photocatalytic materials on wall. The efficiency of the photocatalytic method for NOx removal is proportional to contact time), high temperature, and less relative humidity ().
   2. (BTEX) Tests: BTEX (Benzene, Toluene, Ethylbenzene, and Xylene) test series is a quantitative test method to assess the photocatalytic ability for degradation of hydrocarbon-based molecules. This test measures the degradation of hydrocarbon molecule in air and at the topmost layer of the cement-based materials by using stirred flow reactor. Stirred flow reactor maintains the homogeneous concentration of reactant deposited on the surface of the material (Vittoriadimanti and Pedeferri 2013).
   3. Colorimetric Tests: Colorimetric tests are used to measure the dye degradation and self-cleaning abilities of photocatalytic construction materials. Degradation of rhodamine B, on the TiO_2 (photocatalyst) in the cement matrix confirmed the photocatalytic activity (TiO_2-sensitized photoreaction). However, this test is invalid for spongy, coarse, and colored materials because in the case of spongy material, homogeneous distribution of the dye is not possible and even in red-color materials, the red color

of the dye cannot be transformed. One study has advised the use of a terephthalic acid-based fluorescence probe to quantify the rate of hydroxyl radical production, and ultimately photocatalytic activity. The main advantages of this test are that it is rapid, highly sensitive, and can be used for colored materials. Incorrect hypothesis of existing test methods related to the removal of air pollutants may result in incorrect measurement of photocatalytic efficiency of a construction material. Most of these existing standard test methods require costly equipment and time (Guerrini and Plassais 2007; Cassar 2004; Zhong and Haghighat 2015).

8.4 Various Techniques for Self-Repairing

Broadly five parameters are used to measure the efficiency of self-repairing. These parameters are shelf life, perverseness, quality, reliability, and versatility which have been discussed with different techniques in Table 8.2. However, reliability data is not available for any of the discussed methods.

Table 8.2 Various parameters to test efficiency of self-repairing

<table>
<tr><th>S. no.</th><th>Techniques</th><th>Shelf life</th><th>Pervasiveness</th><th>Quality</th><th>Versatility</th></tr>
<tr><td>1</td><td>Chemical encapsulation</td><td>Depends upon encapsulated repairing chemical used</td><td rowspan="2">Encapsulation helps the homogenous distribution throughout the matrix</td><td>Good mechanical strength, self-cleaning not evaluated</td><td>Depends on environment conditions</td></tr>
<tr><td>2</td><td>Microorganism encapsulation</td><td>Depends upon spores used as they provides stability</td><td>Good self-repairing ability, low mechanical strength</td><td rowspan="2">Poor, it requires continuous moisture exposure</td></tr>
<tr><td>3</td><td>Admixtures</td><td>Depends upon water content and additives activity</td><td>Agents mixed and distributed evenly</td><td>Complete self-cleaning, Mechanical strength not analyzed</td></tr>
<tr><td>4</td><td>Glass tubing</td><td>Long and affected by encapsulating agent</td><td>Tubes are placed only at cracks</td><td rowspan="2">Complete self-cleaning, Good mechanical strength</td><td rowspan="2">Good, it is independent of environment</td></tr>
<tr><td>5</td><td>Intrinsic cleaning</td><td>Long</td><td>Good due to homogenous distribution</td></tr>
</table>

8.4.1 *Taxonomy of Self-Cleaning Techniques*

In this section different techniques of self-cleaning have been discussed. These are as follows:

a. **Autogenous/natural cleaning**: In traditional concrete material, approximately 20–30% cement is deficient in water. After formation of cracks in cement, ingress water interacts with cement particles and hydrations start again to repair the cracks. Self-repairing of cracks is termed as autogenous cleaning.

b. **Autonomic cleaning:** In this process repairing of cracks is performed with help of repairing agent to concrete at normal temperature (Cucitore and Cangiano 2011b). This can be either biological or chemical based.

   i .**Chemical self-cleaning**: In this process chemical self-repairing of concrete, chemical molecules like glue are added to concrete. Mainly hollow pipette and encapsulation are used for preparation of this mix. Active and passive modes are used in self-repairing concrete. In active, mode chemical agent is added externally (Virginie and Junker 2011).

   ii **Biological self-cleaning**: Biologically based self-repaired concrete is an eco-friendly process which utilizes micro-organisms in development of self-repairing concrete (Papanikolaou and Arena 2019; Jiaqi and Zhang 2019). Micro-organisms consist of bacteria, virus, and fungi. Mainly bacteria are used for this process. The main advantage of biological-based method is ease of growth of micro-organisms. Micro-organisms can be added in concrete in form of broth, as spores, in immobilized or encapsulated form. Mainly spore form and encapsulation method are preferred because of the harsh environment in concrete. However, encapsulation method is quite complicated and costly.

   iii **Engineered self-cleaning**: Although different self-repairing methods of cracks in concrete have been investigated, still a convenient method is not designed. Every process has its own pros and cons. Bacteria-based method is efficient but it's hard to grow bacteria in harsh environment (Wang and Dewanckele 2014). Vascular method may have pre-maturation of repairing agents even before the appearance of crack. Out of available approaches, concrete technology has shown promising result. Different areas and environmental conditions also affect self-repairing techniques (Table 8.3). Bacteria-based and autogenous methods require water for repairing process to occur, which makes this optimum for water exposed structures. However, the glue-based agents are not optimum for under water structures, because water presence can interfere with release of glue-based agents. In case of underground structures, any repairing method can be used but if water level is high then use of adhesive agent should be avoided.

Table 8.3 Self-cleaning techniques versus structural environment where symbol √ represents highly preferred and symbol × represents rarely preferred

	Self-cleaning techniques	Under water	Under ground	Open air	Indoor elements
1	Adhesive agent-based	×	Water availability not required	√	√
2	Bacterial self-cleaning	√	√	Preferred with Water availability	×
3	Autogenous self-cleaning	√	√	Preferred with Water availability	×
4	Self-cleaning due to admixtures	√	√	Preferred with Water availability	×

8.5 Cost Analysis

Yang and colleagues (Alfani 2013) investigated the photocatalytic efficiency of TiO_2 maintained on mortar surfaces vs TiO_2 scattered as in mortar in 2019. Considering photonic efficiency as just an indicator, the impacts of environmental factors like NO concentration as well as fluid velocity, UV light intensity, with relative humidity upon photocatalytic performance also were studied in this work. In photocatalytic mortars, the research resulted in much greater utilization efficiency (approximately 150 times higher) over TiO_2. TiO_2's advantages in photocatalytic concrete technology were further proven by its efficiency and affordable cost. Chen et al. (Yang and Hakki 2019) created compound photocatalysts by coating recycled clay with brick sands, as well as recycled glass with nano- TiO_2. They also conducted research on the photocatalytic mortar created using such photocatalysts· rheological behavior, the mechanical performance, including NOx elimination. The usage of nano- TiO_2 in combination with recycled clay with brick sand, as well as recycled glass improves rheological behavior. Photocatalysis was already discovered to be improved by the use of composite photocatalysts. Furthermore, due to mixing, NOx elimination was claimed to be boosted by 18.8%, whereas cost is lowered by 80%.

8.6 Challenges and Future Prospects

Formulation of photocatalytic construction matrix with improved function is vital for environment safety and socio-economic growth. Efficient exploitation of sunlight with the help of a suitable light stimulating photocatalyst could help in the achievement of desired goals related to photocatalysis-based construction materials. Photocatalytic material degrades pollutants and cleans the surface as well as air. Mixing of TiO_2 in construction material may decrease its band gap and can affect the efficacy of photocatalytic activity (Alfani 2013). TiO_2 has been used as photocatalytic material due to its non-toxic nature, cheap availability, less corrosive property, etc. It absorbs ultraviolet rays and oxidizes most organic and some inorganic pollutants.

Additionally, to expand the potential use of photocatalysis structure materials, their efficiency must be increased. Increased surface area may prove to be a significant approach to enhance efficiency of structure material. However, for stable and long-term increase in efficacy of material, different parameters should be considered and optimized like (1) electron-hole recombination, (2) number of active sites on the surface, (3) dispersion control of TiO_2 for optimum pore size and to maximize photocatalytic activity for both organic dyes and gases, (4) use of efficient photocatalysts, and (5) fine pore arrangement.

Furthermore, byproducts released in photocatalytic reaction must be assessed carefully and their effect on health and ecosystem must be investigated. A deep study about the effect of photocatalyst addition on cement mix structure and stability is also required. Energy consumption is an important factor that will decide the applicability of photocatalytic material (Yang and Hakki 2019). Effectiveness of TiO_2 addition to other eco-friendly material like magnesium phosphate cement can be investigated as an alternative in the construction sector (Chen and Kou 2020). TiO_2 included cement has many characteristics like sustainability, self-cleaning photo-induced hydrophilicity and elimination of the urban heat from urban building, etc.

8.7 Conclusions

In the present review, main emphasis was on self-cleaning process based on photocatalytic method. Photocatalyst keeps the environmental air clean. This approach had been followed by various researchers to develop various photocatalytic material with different photodegradation abilities to keep the indoor and outdoor air pollution free. Photocatalyst acts as an oxidizing agent and decomposes various organic and inorganic pollutants by photodegradation. Thus, photocatalyst does not allow the pollutants to accumulate on the surface of buildings. In this review an attempt has been made to analyze various characteristics, effect of different parameters, and various tests to assess the performance of construction material. Based on this review, following research gaps are identified:

- TiO_2 is used as photocatalyst in most of the composite mixture. It should be replaced with more effective photocatalyst such as bismuth, graphene, nano platelets g-C3N4 nano sheets having better self-cleaning properties.
- Alternative cementitious material should be made with fly ash and other alkaline activators. Nevertheless, to better apply cementitious material containing highly efficient photocatalyst, research should be done in-depth in future.
- At present, research on self-cleaning construction material is mostly based on laboratory tests and theoretical analysis. It should be available at field level and at field level these materials should be available at low cost and with better efficiency

References

Addamo M, Augugliaro V, Bellardita M, Di Paola A, Loddo V, Palmisano G, Palmisano L, Yurdakal S (2008) Environmentally friendly photocatalytic oxidation of aromatic alcohol to aldehyde in aqueous suspension of brookite TiO_2. Catal Lett 126:58–62

Alfani R (2013) Coatings based on hydraulic binders with an optimal rheology and high photocatalytic activity. U.S. Patent No 8,377,579

Awadalla A, Zain MFM, Kadhum AMH, Abdalla Z (2011) Titanium dioxide as photocatalyses to create self-cleaning concrete and improve indoor air quality. Int J Phys Sci 6(29):6767–6774

Bondioli F, Taurino R, Ferrari AM (2009) Functionalization of ceramic tile surface by sol–gel technique. J Colloid Interf Sci 334:195–201

Carmona-Quiroga PM, Martínez-Ramírez S, Viles HA (2018) Efficiency and durability of a self-cleaning coating on concrete and stones under both natural and artificial ageing trials. Appl Surf Sci 433:312–320

Cassar L (2004) Photocatalysis of cementitious materials: clean buildings and clean air. MRS Bull 2004:1–4

Cassar L (2004) Photocatalysis of cementitious materials. Clean buildings and clean air. Mrs Bull 29:328–331

Chen X, Kou S, Sun Poon C (2020) Rheological behaviour, mechanical performance, and NOx removal of photocatalytic mortar with combined clay brick sands-based and recycled glass-based nano-TiO_2 composite photocatalysts. Constr Build Mater 240:117698

Chuah SZP, Sanjayan JG, Wang CM, Duan WH (2014) Nano reinforced cement and concrete composites and new perspective from graphene oxide. Constr Build Mater 73:113–124

Cucitore R, Cangiano S, Cassar L (2011a) High durability photocatalytic paving for reducing urban polluting agents. U.S. Patent No. 8,039,100, 18 October 2011

Cucitore R, Cangiano S, Cassar L (2011b) High durability photocatalytic paving for reducing urban polluting agents. U.S. Patent No. 8,039,100, 18 October 2011

Dalawai SP, Mohamed ASA, Latthe SS, Xing R, Sutar RS, Nagappan S, Ha C-S, Sadasivuni KK, Liu S (2020) Recent advances in durability of superhydrophobic self-cleaning technology: a critical review. Progr Org Coat 138(105381):1–15

Dewanckele WJ, Cnudde V, Van Vlierberghe S, Verstraete W De Belie N (2014) X-ray computed tomography proof of bacterial-based self-cleaning in concrete. Cement Concr Compos 53:289–304

Diamanti MV, Ormellese M, PiaPedeferri M (2008) Characterization of photocatalytic and super hydrophilic properties of mortars containing titanium dioxide. Cem Concr Res 38:1349–1353

Folli A, Pade C, Hansen TB, De Marco T, Macphee DE (2012) TiO_2 photocatalysis in cementitious systems: insights into self- cleaning and depollution chemistry. Cem Concr Res 42:539–548

Fujishima A, Rao TN, Tryk DA (2000) Titanium dioxide photocatalysis. J Photochem Photobiol C Photochem Rev 1:1–21

García LD, Pastor JM, Peña J (2018) Self-cleaning and depolluting glass reinforced concrete panels: fabrication, optimization and durability evaluation. Constr Build Mater 162:9–19

Guerrini L, PlassaisA, Pepe C, Cassar L (2007) Use of Photocatalysis of cementitious materials. In: Proceedings of the International RILEM symposium on photocatalysis, environment and construction materials, Florence, Italy, 8 October 2007, pp 131–145

Hamidi F, Aslani F (2019) TiO_2-based photocatalytic cementitious composites: materials, properties, influential parameters, and assessment techniques. Nanomaterials 9(10):1444

He B, Gao Y, Qu L, Duan K, Zhou W, Pei G (2019) Characteristics analysis of self-luminescent cement-based composite materials with self-cleaning effect. J Clean Prod 225:1169–1183

Hipólito E, Torres-Martínez LM, Cantú-Castro LVF (2019) Self-cleaning coatings based on fly ash and bismuth-photocatalysts: Bi_2O_3, $Bi_2O_2CO_3$, BiOI, $BiVO_4$, $BiPO_4$. Constr Build Mater 220:206–213

Husken G, Hunger M, Brouwers H, Baglioni P, Cassar L (2009) Experimental study of photocatalytic concrete products for air purification. Build Environ 44(12):2463–2474

Junker VHM (2011) Self-cleaning of cracks in bacterial concrete. In: 2nd International life symposium on service life design for infrastructures, 2011, pp 825–831. ISBN: 978-2-35158-097-4

Kolay KSP, Malla S, Mishra S (2012) Effect of multiwalled carbon nanotubes on mechanical strength of cement paste. Mater Civ Eng 24(1):84–91

Li Q, Liu Q, Peng B, Chai L, Liu H (2016) Self-cleaning performance of TiO_2-coating cement materials prepared based on solidification/stabilization of electrolytic manganese residue. Constr Build Mater 106:236–242

Li J, Zhang W, Li C, Monteiro PJM (2019) Green concrete containing diatomaceous earth and limestone: workability, mechanical properties, and life-cycle assessment. J Clean Prod 223:662–679

Luna M, Delgado JJ, Almoraima Gil ML, Mosquera MJ, TiO_2-SiO_2 coatings with a low content of AuNPs for producing self-cleaning building materials. Nanomaterials 8:177–203

Morsy MSSH Alsayed, Salloum YA (2012) Development of eco-friendly binder using metakaolin-fly ash–lime-anhydrous gypsum. Constr Build Mater 35:772–777

Murata Y, Obata H, Tawara H, Murata K (1999) NOx-cleaning paving block. U.S. Patent No 5,861,205

Nath RK, Zain M, Jamil M (2016) An environment-friendly solution for indoor air purification by using renewable photocatalysts in concrete: a review. Renew Sustain Energy Rev 62:1184–1194

Ohama YY, Van Gemert D (2011) Application of titanium dioxide photocatalysis to construction materials: state-of-the-art report of the RILEM technical committee 194-TDP; Springer Science & Business Media, Dordrecht, The Netherlands

Paola Di A, García-López E, Marcì G, Palmisano L (2012) A survey of photocatalytic materials for environmental remediation. J Hazard Mater 211:3–29

Papanikolaou I, Arena N, Al-Tabbaa A (2019) Graphene nanoplatelet reinforced concrete for self-sensing structures—a lifecycle assessment perspective. J Clean Prod 240:118202

Palmisano L, Augugliaro V, Bellardita M, Paola Di A, García-López E, Loddo V, Marcì G, Palmisano G, Yurdakal S (2011) Titania photocatalysts for selective oxidations in water. Chemsuschem 4:1431–1438

Pinho L, Rojas M, Mosquera MJ (2015) Ag-SiO_2-TiO_2 nanocomposite coatings with enhanced photoactivity for self-cleaning application on building materials. Appl Catal B 178:144–154

Shen W, Zhang C, Li Q, Zhang W, Cao L, Ye J (2015) Preparation of TiO_2 nanoparticle modified photocatalytic self-cleaning concrete. J Clean Prod. https://doi.org/10.1016/j.jclepro.2014.10.014,pp1-19

Sikkema JK, Alleman JE, Cackler T, Taylor PC, Bai B, Ong S-K, Gopalakrishnan K (2014) Photocatalytic pavements. Climate change, energy, sustainability and pavements. Springer-Verlag, New York, NY, pp 275–307. ISBN 978-3-662-44718-5 2014

Szymanowski J (2019) Evaluation of the adhesion between overlays and substrates in concrete floors: literature survey recent nondestructive and semi-destructive testing method and research gap. Buildings 203(9):1–23

Teoh WY, Scott JA, Amal R (2012) Progress in heterogeneous photocatalysis: from classical radical chemistryto engineering nanomaterials and solar reactors. J Phys Chem Lett 3:629–639

Tsai S-J, Cheng S (1997) Eect of TiO_2 crystalline structure in photocatalytic degradation of phenolic contaminants. Catal Today 33:227–237

Vaidevi C, Kala FT, Kalaiyarrasi ARR (2020) Mechanical and durability properties of self-compacting concrete with marble fine aggregate. Mater Today Proc 22:829–835

Vittoria M, Pedeferri M (2013) Concrete, mortar and plaster using titanium dioxide nanoparticles: applications in pollution control, self-cleaning and photo sterilization. In: Nanotechnology in eco-ecient construction. Amsterdam, The Netherlands, Elsevier, pp 299–326

Vittoriadiamanti M, Pedeferri M (2013) Concrete, mortar and plaster using titanium dioxide nanoparticles: applications in pollution control, self-cleaning and photo sterilization. In Nanotechnology in eco-ecient construction. Elsevier, Amsterdam, The Netherlands, pp 299–326

Witkowski H, Jackiewicz-Rek W, Chilmon K, Jarosławski J, Tryfon-Bojarska A, Gąsiński A (2019) Air purification performance of photocatalytic concrete paving blocks after seven years of service. Appl Sci 9(9):1735

Yang Y, Ji T, Su W, Yang B, Zhang Y, Yang Z (2019) Photocatalytic NOx abatement and self-cleaning performance of cementitious composites with g-C3N4 nanosheets under visible light. Constr Build Mater 225:120–131

Zhao ZT Qi, Zhou W, Hui D, Xiao C, Qi J, Zheng Z, Zhao Z (2020) A review on the properties, reinforcing effects, and commercialization of nanomaterials for cement-based materials. Nanotechnol Rev 9:1

Zhong L, Haghighat F (2015) Photocatalytic air cleaners and materials technologies—abilities and limitations. Build Environ 91:191–203

Zimimng H, Shen A, Lyu Z, Yue L, Wu H, Wang W (2020) Effect of wollastonite microfibers as cement replacement on the properties of cementitious composites: a review. Constr Build Mater 261:1–13

Chapter 9
Investigations on Chemical, Mechanical, and Long-Term Characteristics of Alkali-Activated Concrete

Arkamitra Kar, Kruthi Kiran Ramagiri, Sriman Pankaj Boindala, Indrajit Ray, Udaya B. Halabe, and Avinash Unnikrishnan

9.1 Introduction

Rapid global development of infrastructure has resulted in the consumption of 4.1 billion tons (BTs) of limestone-based portland cement (PC), with an anticipated increase to 4.83 BTs by 2030 (Garside 2021). This accounts for 5–8% of the total global CO_2 emission, as the 1 kg of PC leads to nearly 1 kg of CO_2 production. India alone is estimated to produce 0.298 BTs of PC per annum (PA), whereas the USA produces 0.09 BTPA with a steady increase anticipated through the decade (BEEINDIA 2021). It is also reported that India produces an estimated 0.225 BTPA

A. Kar (✉) · K. K. Ramagiri
Department of Civil Engineering, BITS-Pilani, Hyderabad Campus, Hyderabad 500078, Telangana, India
e-mail: arkamitra.kar@hyderabad.bits-pilani.ac.in

K. K. Ramagiri
e-mail: p20170008@hyderabad.bits-pilani.ac.in

S. P. Boindala
Civil and Environmental Engineering, Technion—Israel Institute of Technology, Haifa, Israel

I. Ray
Department of Civil and Environmental Engineering, University of West Indies, St. Augustine, Trinidad and Tobago
e-mail: Indrajit.Ray@sta.uwi.edu

U. B. Halabe
Department of Civil and Environmental Engineering, West Virginia University, Morgantown, WV 26506, USA
e-mail: udaya.halabe@mail.wvu.edu

A. Unnikrishnan
Department of Civil and Environmental Engineering, Portland State University, Portland, OR 97201, USA
e-mail: uavinash@pdx.edu

K. R. Reddy et al. (eds.), *Advances in Sustainable Materials and Resilient Infrastructure*, Springer Transactions in Civil and Environmental Engineering,
https://doi.org/10.1007/978-981-16-9744-9_9

of fly ash, with around 30% being discarded as waste and less than 1% in concrete production (Yousuf et al. 2020). In the USA, out of the 0.04 BTPA of fly ash produced, around 55% is reused with a goal of 90% utilization by 2030 (Ghazali et al. 2019). In Europe, almost 80% of the fly ash produced as waste is utilized in construction activities to reduce PC usage (Euorstat 2020). Fly ash is popularly used as a mineral admixture in cementitious systems up to nearly 70% by volume of PC in high-volume applications. More recently, another technique to utilize fly ash is the implementation of alkali-activated concrete (AAC). The corresponding alkali-activated binder (AAB) is developed through the reaction of industrial waste like fly ash as a precursor and an alkaline activator composed of sodium silicate (SS) and sodium hydroxide (SH). The chemical composition of the resulting sodium aluminosilicate matrix is different from the calcium silicate hydrate matrix in traditional hydrated PC paste. The hardened characteristics of AAC aided with thermal curing at 60–80°C are comparable to those of PC concrete (Provis 2018). Presently, AAC is utilized in the form of precast members such as the Toowoomba Wellcamp Airport in Brisbane, Australia, but the implementation of ambient curing can encourage its widespread practical usage (Glasby et al. 2015). To achieve a sustainable practice, it is important to prevent energy consumption due to thermal curing and simultaneously compensate for the slow strength gain due to the lower reaction rate of fly ash. The inclusion of slag to partially replace fly ash is implemented in this regard (Ramagiri et al. 2021).

It is reported that the mechanical strength values of AAC are comparable or sometimes even higher than that of PC concrete (Provis 2018; Ramagiri et al. 2021). However, there is limited information reported on the long-term characteristics of AAC. Increasing global development of infrastructure with greater risk of fire exposure mandates the need for highly fire-resistant concrete. It is reported that thermally cured fly ash-based AAC exposed to severely high temperatures displays higher residual strength and greater resistance to spalling compared to PC concrete. The reduction of mechanical strength of PC concrete upon exposure to 400–600 °C is attributed to the decomposition of its calcium silicate hydrate (C-S-H) matrix and calcium hydroxide ($Ca(OH)_2$) with the loss of chemically bound water, followed by the complete disintegration of the matrix beyond 600 °C. The performance of AAC is comparatively enhanced due to the formation of a pseudo-viscous phase, resulting from the sintering of the fly ash (Martin et al. 2015). The response of AAC to elevated temperatures depends on its precursors and the subsequent formation of a sodium aluminosilicate hydrate (N-A-S-H) polymeric matrix, which is starkly different from the C-S-H in PC (Kar 2013). The associated porosity in hardened PC concrete and AAC are consequently different, thereby directly governing the resistance to elevated temperatures. The selected precursor-activator and curing temperature combination governs the initial mechanical strength of AAC, and subsequently, its residual strength. It is reported that the initial strength and residual strength of fly ash-based AAC are inversely proportional when the exposure temperature is increased to 600 °C. This happens due to further polymerization of the unreacted precursor residue at elevated temperatures, leading to the greater formation of the N-A-S-H polymer matrix. However, in the case of fly ash-slag blended precursor, the N-A-S-H matrix coexists with a calcium aluminosilicate hydrated (C-A-S-H) matrix formed by the alkali-activation of slag. Hence, the residual strength of blended AAC decreases on exposure to elevated temperatures due to the disintegration of this C-A-S-H, similar to that of the C-S-H in PC-based systems (Ramagiri and Kar 2020).

A critical parameter for the widespread application of reinforced AAC is its bond strength, especially at elevated temperatures caused by fire incidents. The bond strength of concrete is expressed as idealized uniform stress developed at the concrete-steel interface. The bond strength of thermal-cured AAC is governed by its mix proportions, duration of curing, effective cover, size and type of rebar, and development length (Ramagiri et al. 2021; Adak et al. 2017; Castel and Foster 2015; Sarker 2011). Observations from pull-out tests conducted on AAC and PC concrete reveal superior bond strength results in the former, owing to enhanced resistance to tensile stress and a more compact interfacial transition zone (ITZ) (Sarker 2011). In the case of rebars having smaller diameters, enhanced bonding of the rebar and concrete facilitates breaking of the bar over slippage. In contrast, larger bar diameter fails due to concrete splitting (Castel and Foster 2015; Fernandez-Jimenez 2006). The bond strength of AAC reinforced with mild steel or deformed bars is improved in the presence of thermal curing or nanosilica addition (Adak et al. 2017; Castel and Foster 2015). However, limited literature is reported on the characteristics of ambient-cured AAC and its bond behavior at elevated temperatures. Hence, this study focuses on assessing the strength characteristics of ambient-cured AAC through compressive and bond strength tests after exposure to elevated temperatures; and to determine an optimal precursor combination based on these results. Mineralogical, chemical, and morphological analyses are correlated with these findings after validation using existing data (Ramagiri and Kar 2019). The following section presents the materials used for this study and the detailed methodology adopted for the relevant tests conducted.

9.2 Materials and Methods

9.2.1 Materials

Class F fly ash (specific surface 490 m^2/kg; sp. gr. 2.06; loss on ignition, LOI 3%; pozzolanic activity index, PAI 96.46%) conforming to ASTM C618-19 (2019) is supplied by the National Thermal Power Corporation (NTPC) plant in Ramagundam, India. The slag used for this study, conforming to Grade 100 (specific surface 580 m^2/kg; sp. gr. 2.71; LOI 1.41%; PAI 114.46%) of the ASTM C989/C989M-18a (2018), is procured from JSW Cement Ltd. The relevant oxide compositions and particle size information are provided in Table 9.1.

SH used in this study to prepare the alkaline activator conforms to rayon grade with 99% purity. SS solution is composed of 55.1% water, 29.5% SiO_2, and 14.7% Na_2O by weight (to maintain initial silicate modulus of 2:1). Both these components are obtained from HYCHEM laboratories. The liquid blend of crushed SH pellets

Table 9.1 Specification of raw materials

Component (%)	Fly ash	Slag
CaO	1.78	40.64
SiO_2	60.13	35.15
Al_2O_3	28.37	19.60
Fe_2O_3	5.10	0.53
SO_3	0.11	1.89
K_2O	2.16	0.40
TiO_2	1.42	0.92
d_{50} (μm)	51.90	13.93

and SS solution generates heat. So, it is allowed to cool down for 24 h before concrete preparation to alleviate this heat.

Fine aggregate in the form of river sand and coarse aggregate in the form of crushed granite are used to comply with Zone II of IS 383 (2016). The respective water absorption values are 0.5% and 0.1%; the corresponding specific gravity values are 2.65 and 2.72. High-range water-reducing admixture (HRWRA), conforming to Type F of ASTM C494-13 (Specification for Chemical Admixtures for Concrete) is used. For the testing of bond strength, deformed steel bars having 0.2% proof stress of 500 MPa and 12 mm diameter are used.

9.2.2 *Mix Proportions and Specimen Preparation*

Fly ash: slag ratios in the AAC mixes are varied as 100% fly ash (FS 0), 70% fly ash + 30% slag (FS 30), 60% fly ash + 40% slag (FS 40), and 50% fly ash + 50% slag (FS 50) respectively. The silicate modulus, Ms, for the activator is maintained at 1.4 based on a previous study (Kar et al. 2013). Coarse aggregate quantity is maintained at 1209 kg/m^3 and fine aggregate at 651 kg/m^3 for each mix, respectively. A slump height of 75–100 mm is maintained for each mix by maintaining water to total solids (w/s) ratio of 0.3. The extra water required to maintain w/s is calculated after taking into account the water present in the SS solution (Table 9.2).

During mixing, a uniform blend of the dry precursors with the coarse aggregates and fine aggregates is ensured. Then, the preblended activator is added to the dry mix. After mixing for two to three minutes, previously calculated additional water is added, followed by the required HRWRA to achieve the desired workability. FS 0 specimens, being solely based on the lower activation of fly ash under ambient conditions, are kept in sealed condition and demolded after a week to allow for full hardening and to prevent any damage to the specimens. The other blended AAC mixes are demolded after 48 h and moist-cured for a week. Then, they are stored at ambient laboratory conditions (31 ± 2 °C average temperature and 70% average

Table 9.2 AAC mix proportions in (kg/m^3)

Mix components	FS 0	FS 30	FS 40	FS 50
Fly ash	400	280	240	200
Slag	0	120	160	200
SH	10.57	10.57	10.57	10.57
SS solution	129.43	129.43	129.43	129.43
Extra water	77.38	77.38	77.38	77.38
HRWRA (l/m^3)	0	3.14	3.57	4

relative humidity) prior to testing. For this study, all specimens are ambient-cured, and triplicates are tested at the age of 28 days, as elaborated below.

9.2.3 Experimental Methods

The time–temperature curves presented in ASTM E119-20 (2020) are used to select exposure temperatures of 31 ± 2 °C (ambient), 538 °C, 760 °C, and 892 °C for the AAC specimens. Using an AIMIL muffle furnace, the specimens are exposed to these environments 2 h prior to the mechanical strength tests to attain thermal equilibrium with subsequent gradual cooling to room temperature (Park et al. 2016). Compressive strength test is conducted as per ASTM C39/C39M-21 (2021) on cylindrical AAC specimens of 150 mm diameter and height 300 mm. Capacity of the compression test setup is 2000 kN.

The mineralogical, chemical, and microstructural analyses are conducted on powdered or hardened paste specimens to avoid interference due to the presence of aggregates in mortar and concrete. X-ray diffraction (XRD) analyses are performed using a RIGAKU Ultima X-ray Diffractometer. Rate of scanning equal to 1°/min with intervals of 0.02° is adopted over a 2θ range of 5–90°. The CuK$_\alpha$ X-rays are operated at 40 kV and 30 mA. Scanning electron microscopy (SEM) using an FEI Apreo setup is used for microstructural analyses. Oven-dried samples are sputtered with a gold–palladium layer of 10 nm thickness using LEICA EM ACE200 setup before the SEM–EDS analyses.

For the bond strength determination, the pull-out tests on different AAC mixes are conducted according to the IS 2770-1 (1967). The test specimens have size 100 mm × 100 mm × 100 mm containing an embedded centrally placed rebar. The test setup is shown in Fig. 9.1. The specimens are exerted to displacements of 0.01 mm/sec till they reach failure. The average bond stress (MPa), σ_{bond}, corresponding to the specimen failure load (N), P, is determined using the following equation:

$$\sigma_{bond} = \frac{P}{\pi d L},$$

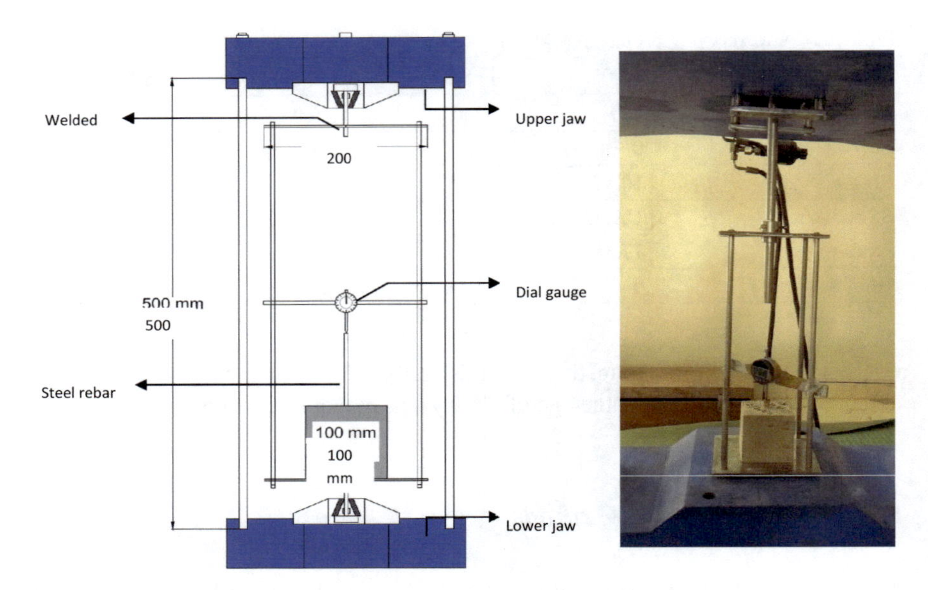

Fig. 9.1 Pull-out test apparatus

where d and L represent rebar diameter and the length of specimen parallel to the rebar, respectively (mm)In this case, $L = 100$ mm that is equal to the length of the rebar embedded inside the concrete specimen. For this study, the rebar diameter used is 12 mm. The detailed compressive and bond strength values are presented in Figs. 9.2 and 9.6, respectively.

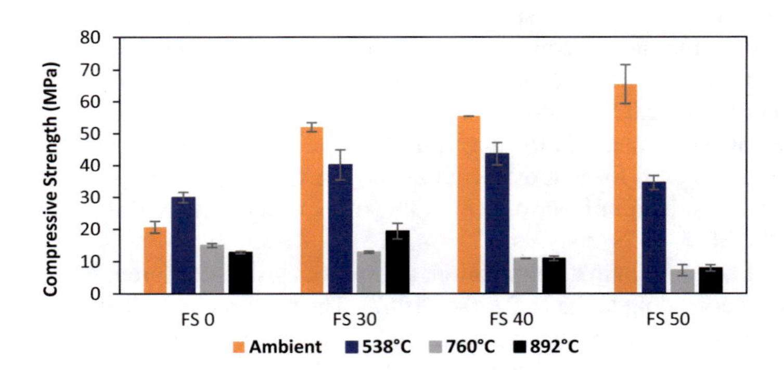

Fig. 9.2 Compressive strength results

9.3 Results

The compressive strength values for the selected exposure conditions 28 days are presented in Fig. 9.2. Ambient-cured AAC exhibits enhanced strength with increasing slag content, as reported in the literature (Kar et al. 2013). The development of C-A-S-H yields a denser microstructure and consequent pore refinement. At elevated temperatures, C-A-S-H disintegrates, leading to increased strength loss as slag proportion increases. Ambient curing results in lower fly ash reaction rates and hence lower strength development in FS 0 mixes. However, additional geopolymerization at temperatures of about 500 °C leads to an increase in strength.

Up to a temperature of 760 °C, FS 30 and FS 40 show a reduction in strength, but exposure to 892 °C leads to a relative increase in the strength. The formation of akermanite ($2CaO{\cdot}MgO{\cdot}2SiO_2$) and gehlenite ($2CaO{\cdot}Al_2O_3{\cdot}SiO_2$) crystalline phases at this elevated temperature causes an increase in the compactness of the microstructure, leading to this observation. These findings are corroborated by the diffractograms presented in Fig. 9.3 (1: analcime, 2: akermanite, 3: albite, 4: anorthite, 5: Calcite, 6: C-A-S-H, 7: gehlenite, 8: Hydrotalcite, 9: mullite, 10: nepheline, 11: quartz). FS 0 mixes exhibit a relatively lesser loss in strength at elevated temperatures due to greater pore content in the microstructure. Fly ash comprises hollow spherical particles at the molecular level, known as cenospheres. The sintering of these cenospheres at elevated temperatures leads to the formation of a highly dispersed network of pores, as observed from the micrographs presented in Fig. 9.4. This network serves as an escape route for high pore pressure, which can otherwise cause failure due to spalling or disintegration of the concrete.

Visual inspection of the specimens exposed to elevated temperatures reveals changes in superficial color due to dehydration and corresponding bond failure modes (Fig. 9.5).

Figure 9.6 presents the bond strength test values for the different AAC mixes, with the trends resembling those of compressive strength values.

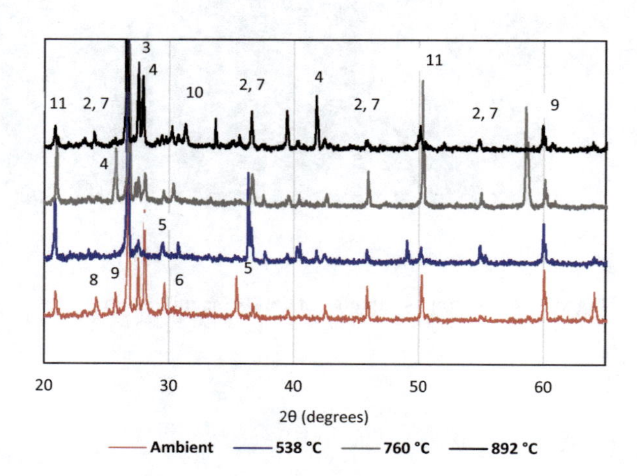

Fig. 9.3 Diffractogram for FS 50

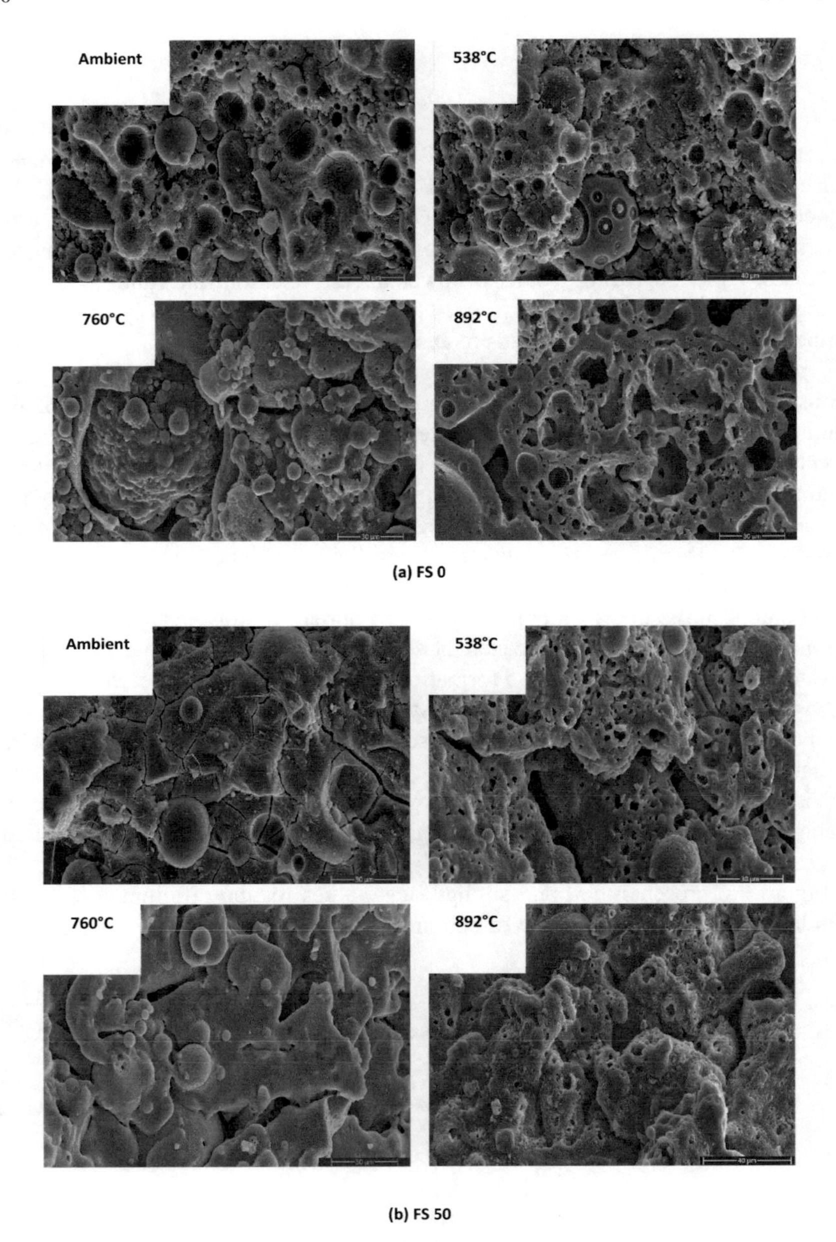

Fig. 9.4 Micrographs captured at 2500x magnification

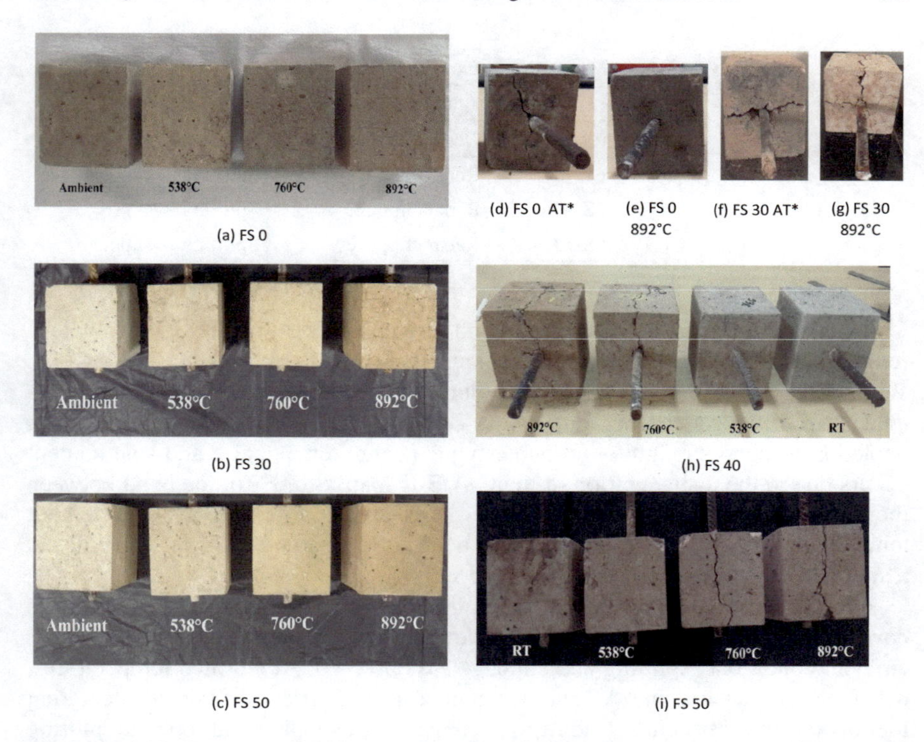

Fig. 9.5 Surface deterioration and bond failure patterns due to elevated temperature. *AT = ambient temperature

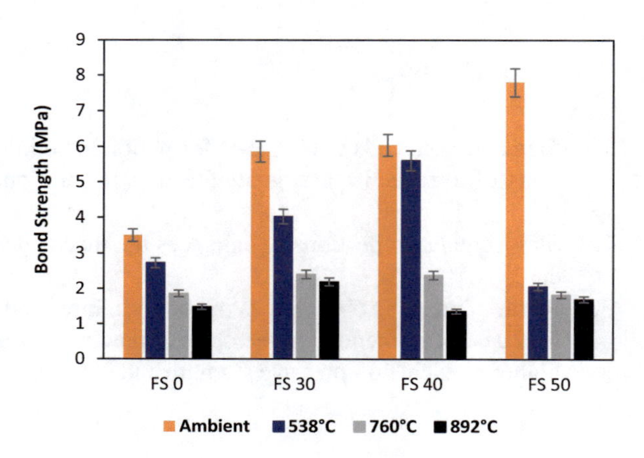

Fig. 9.6 Bond strength variation with exposure temperature

Table 9.3 Summary of observed bond failure

Mix ID	Exposure temperature	
	Ambient	Elevated
FS 0	Concrete splitting	Splitting + slippage
FS 30	Rebar slippage	Splitting + slippage
FS 40	Rebar slippage	Splitting + slippage
FS 50	Rebar slippage	Splitting + slippage

FS 50 exhibits a 122.8% higher bond strength value than FS 0, owing to a refined pore structure and a relatively homogenous and compact AAC-rebar ITZ. With increasing slag content, greater volumes of reaction products are formed, as evident from the coexistence of the C-A-S-H and N-A-S-H matrices. This greater depletion in bond strength with increasing exposure temperature and slag content occurs due to the disintegration of the C-A-S-H matrix. In FS 0, the bond between the rebars and the polymeric N-A-S-H is less prone to decomposition at higher temperatures, resulting in higher residual bond strength after exposure. The pull-out test observations presented in Fig. 9.6 are summarized in Table 9.3.

Bond failure mode depends on the splitting tensile strength of AAC, concrete cover, geometry of the rebar, and the friction between steel and concrete. Under ambient conditions, splitting and slippage are observed. At elevated temperatures, differential expansions in AAC and steel reduce inherent friction while disintegrating the concrete microstructure, leading to combined rebar slippage and concrete splitting failure.

9.4 Conclusion

1. Slag addition leads to C-A-S-H formation in addition to N-A-S-H. The formations of akermanite and gehlenite impart additional residual strength at 892 °C.
2. Slag content proportionally enhances the mechanical performance of ambient-cured AAC.
3. Rebar slippage is observed as the bond failure mode upon slag addition.
4. FS 30 is recommended as the optimal precursor combination as it shows 145% higher residual compressive strength and 29% higher bond strength than FS 50.

References

Adak D, Sarkar M, Mandal S (2017), Structural performance of nano-silica modified fly ash based geopolymer concrete. Constr Build Mater 135:430–439. https://doi.org/10.1016/j.conbuildmat.2016.12.111

ASTM C989/C989M-18a (2018) Standard specification for slag cement for use in concrete and mortars. ASTM International, West Conshohocken, PA, 2018, vol 04.02. http://www.astm.org. https://doi.org/10.1520/C0989_C0989M-18A

ASTM E119-20 (2020) Standard test methods for fire tests of building construction and materials. ASTM International, West Conshohocken, PA, 2020, vol 04.07. http://www.astm.org. https://doi.org/10.1520/E0119-20

ASTM C39/C39M-21 (2021) Standard test method for compressive strength of cylindrical concrete specimens. ASTM International, West Conshohocken, PA, 2019, vol 04.02. http://www.astm.org. https://doi.org/10.1520/C0039_C0039M-21

ASTM C618-19 (2019) Standard specification for coal fly ash and raw or calcined natural Pozzolan for use in concrete. ASTM International, West Conshohocken, PA, 2019, vol 04.02. http://www.astm.org. https://doi.org/10.1520/C0618-19

BEEINDIA (2021) Cement, bureau of energy efficiency, Government of India, Ministry of Power. https://www.beeindia.gov.in/node/166. Accessed 04 July 2021

Castel A, Foster SJ (2015) Bond strength between blended slag and Class F fly ash geopolymer concrete with steel reinforcement. Cem Concr Res 72:48–53. https://doi.org/10.1016/j.cemconres.2015.02.016

Eurostat (2020) Treatment of waste by waste category, hazardousness and waste management operations. Eurostat. http://www.data.europa.eu/data/datasets/8bxb7vunmkpy3c2mnoelw?locale=en. Accessed 04 July 2021

Fernandez-Jimenez AM, Palomo A, Lopez- C (2006) Engineering properties of alkali-activated fly ash concrete. ACI Mater J 103(2):106–112

Garside M (2021) Major countries in worldwide cement production 2010–2020. Statista. https://www.statista.com/statistics/267364/world-cement-production-by-country/. Accessed 04 July 2021

Ghazali N, Muthusamy K, Wan Ahmad S (2019) Utilization of fly ash in construction. IOP Conf Ser Mater Sci Eng 601:012023. https://doi.org/10.1088/1757-899X/601/1/012023

Glasby T, Day J, Genrich R, Aldred J (2015) EFC geopolymer concrete aircraft pavements at Brisbane West Wellcamp Airport. In: Concrete 2015 Conference, Melbourne Australia, 2015, pp 1–9. https://www.wagner.com.au/media/1512/bwwa-efc-pavements_2015.pdf. Accessed 04 July 2021

IS 2770-1 (1967) Methods of testing bond in reinforced concrete, Part 1: Pull-out test. Bureau of Indian Standards, New Delhi, India

IS 383 (2016) Coarse and fine aggregate for concrete—specification. Bureau of Indian Standards, New Delhi, India

Kar A, Halabe UB, Ray I, Unnikrishnan A (2013) Nondestructive characterizations of alkali activated fly ash and/or slag concrete. Eur Sci J 9(24):52–74. https://doi.org/10.19044/esj.2013.v9n24p%25p

Kar A (2013) Characterizations of concretes with alkali-activated binder and correlating their properties from micro-to specimen level. West Virginia University—Graduate Theses, Dissertations, and Problem Reports, 165. https://www.researchrepository.wvu.edu/etd/165

Martin A, Pastor JY, Palomo A, Jiménez AF (2015) Mechanical behaviour at high temperature of alkali activated aluminosilicates (geopolymers). Constr Build Mater 93:1188–1196. https://doi.org/10.1016/j.conbuildmat.2015.04.044

Park SM, Jang JG, Lee NK, Lee HK (2016) Physicochemical properties of binder gel in alkali-activated fly ash/slag exposed to high temperatures. Cem Concr Res 89:72–79. https://doi.org/10.1016/j.cemconres.2016.08.004

Provis JL (2018) Alkali-activated materials. Cem Concr Res 114:40–48. https://doi.org/10.1016/j.cemconres.2017.02.009

Ramagiri KK, Kar A (2019) Effect of precursor combination and elevated temperatures on the microstructure of alkali-activated binder. Indian Concr J 93(10):34–43. http://www.icjonline.com/explore_journals/2019/10

Ramagiri KK, Kar A (2020) Effect of high-temperature on the microstructure of alkali-activated binder. Mater Today Proc 28(2):1123–1129. https://doi.org/10.1016/j.matpr.2020.01.093

Ramagiri KK, Chauhan D, Gupta S, Kar A, Adak D, Mukherjee A (2021) High-temperature performance of ambient-cured alkali-activated binder concrete. Innov Infra Sol 6(2):1–11. https://doi.org/10.1007/s41062-020-00448-y

Sarker PK (2011) Bond strength of reinforcing steel embedded in fly ash-based geopolymer concrete. Mater Struct 44(5):1021–1030. https://doi.org/10.1617/s11527-010-9683-8

Yousuf A, Manzoor SO, Youssouf M, Malik ZA, Khawaja KS (2020) Fly ash: production and utilization in India-an overview. J Mater Environ Sci 11(6):911–921. http://www.jmaterenvironsci.com/Document/vol11/vol11_N6/JMES-2020-1182-Yousuf.pdf. Accessed 04 July 2021

Chapter 10
Ash Utilization Strategy in India—A Way Forward

P. N. Ojha, Brijesh Singh, Puneet Kaura, and Rajiv Satyakam

10.1 Introduction

10.1.1 Ash Generation from Thermal Power Plants

The increase in Indian population over the past decades, the electrical energy obtained from coal-based thermal power plants as of now is at an all-time high due to which there is a serious enhancement in the generation of ash thereby creating environmental concerns for its disposal. The ash generated from thermal power plant are mainly fly ash and of total generated ash; about 20–25% is bottom ash. The slurry part disposed off in pond or dykes which is mixture of fly ash and bottom ash is termed as pond ash. Due to low pozzolanic property of pond ash, it is not suitable as part replacement of cement. Apart from being a management challenge for plant operators, such vast quantities of ash itself poses several public health challenges for communities living around the ash disposal sites. Just in terms of land use the conventional disposal of flyash in form of slurry currently occupies nearly 40000 hectares of land and require about 1040 mn m^3 of water annually (Soundaram et al. 2020; Anon 2019). The Ministry of Environment, Forest, and Climate Change vide notification dated 14th September 1999 and subsequent amendments in year 2003, 2009, and 2016 has emphasized 100% fly ash utilization to lower the impact on environment (MOEF 1999). As per the current estimates of Central Electricity Authority (CEA) (CEA Report 2018–2019), Indian power plants generated 217.04 million metric tonnes of ash in the year 2018–2019 due to coal or lignite combustion of about 667.43

P. N. Ojha (✉) · B. Singh · P. Kaura
National Council for Cement & Building Materials, Ballabgarh, Haryana 121004, India

R. Satyakam
Netra, National Thermal Power Corporation Limited, Noida 201307, India

K. R. Reddy et al. (eds.), *Advances in Sustainable Materials and Resilient Infrastructure*, Springer Transactions in Civil and Environmental Engineering,
https://doi.org/10.1007/978-981-16-9744-9_10

Table 10.1 Fly ash consumption in India as per Central Electricity Authority (CEA) (AS 3582.1)

Year	Parameter				
	Thermal power Stations (Nos.)	Coal consumption (Million Tons)	Ash production (Million Tons)	Ash utilization (Million Tons)	% Utilization
2014–2015	145.00	549.72	184.14	102.54	55.69
2015–2016	151.00	536.64	176.74	107.77	60.97
2016–2017	155.00	536.40	169.25	107.10	63.28
2017–2018	167.00	624.88	196.44	131.87	67.13
2018–2019	195.00	667.43	217.04	168.40	77.59

million tonne. The data published by the CEA indicates 78% ash utilization in year 2018–2019 (Table 10.1).

10.1.2 Trends in Ash Utilization in India

As per document by Soundaram et al. (2020), titled "An Ashen Legacy: India's thermal power ash mismanagement", states like Uttar Pradesh and Chhattisgarh generates the highest ash quantity compared to other states like West Bengal, Maharashtra, Andhra Pradesh, Madhya Pradesh and Odisha. All these states have large quantity of unutilized ash accumulated over a last decade as the power generation plants are coal based. The actual ash utilization in different major sectors in India is given in Fig. 10.1. In the report submitted to National Green Tribunal (NGT), total unused quantity of ash from coal power sector is about 1650 million tonnes. Out of

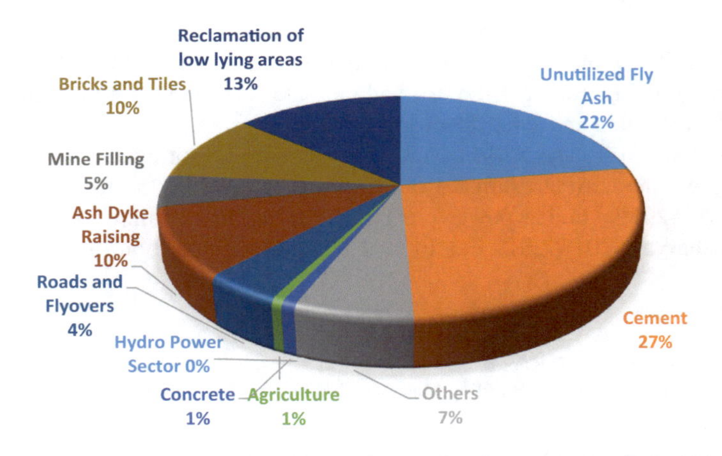

Fig. 10.1 Sector-wise ash utilization during the year 2018–2019 (CEA annual report 2018–2019)

186 power plants, only 103 power plants have been able to achieve full ash utilization whereas remaining power plants were not able to achieve full ash utilization as per CEA 2018–2019 annual report. Ash utilization rate has increased from 55.69% in 2014–2015 to 77.59% as per CEA 2018–2019 annual report. But unresolved area of concern is the utilization of residual or unutilized ash accumulated over the past years (Ujjwal and Kandpal 2000; Yadav and Fulekar 2018). To promote sustainable infrastructure and circular economy, National Institution of Transforming India (NITI) Aayog from last three years has taken up task to revisit existing notifications and formulate policies for effective implementation of 100% ash utilization.

10.1.3 Usage of Ash in Construction Industry and Other Allied Sectors

For the past one and half decade government of India is taking measures to ensure 100 percent utilization of ash but still the same is not achieved due to lack of stringent policies, incentives, technical as well as general awareness among construction fraternity. Ash being pozzolanic material has a huge potential for its utilization in different areas pertaining to construction sector such as (a) manufacturing of cement, (b) back fill material, (c) construction material for roads, bridges, dams, embankments, etc. (d) factory made products like paver block, tiles, fly ash bricks, kerb stone, etc. (e) fly ash-based ceramic products, (f) ingredient for distemper.

10.1.4 Challenges in Ash Utilization in India

The major challenges associated with ash utilization in India are (a) lack of consumers in construction sector due to technical and general awareness (b) lack of business opportunity (c) logistics related problem in remote location (d) low demand in states with high ash generation (e) lack of coordination among different stakeholders like power plants, state government, central government, research organizations and civil society at large. The research-based technologies and practices are need of an hour to maximize ash utilization in India to 100% wherein still a lot of focused and application-based research is missing. Utilization of fly ash-based products shall be made compulsory in construction activities after the sufficient indigenous research data has been evaluated by competent panel of experts and standards/specifications are made by Bureau of Indian Standard or Indian Road Congress for application in construction sector. The way forward to achieve 100% ash utilization to promote durable and environment friendly infrastructure and circular economy in India is strict implementation of policies, promotion of technologies and practices based upon Technology Readiness Level (TRL).

10.2 Classification of Ash

Quality of Indian coal is of low grade wherein ash content is about 30–45% in comparison to imported coals with low ash content of about 10–15%. Depending upon the technologies available for collection including mechanical to electrical precipitators, etc. approximately 80% of the ash from the flue gases is obtained in the form of fly ash and the remaining is collected in the form of bottom ash or pond ash or mound ash. Fly ash is broadly classified on basis of their chemical composition and physical characteristics. The physical and chemical characteristics of fly ash depend upon the quality of coal and coal burning technique (Mukherjee et al. 2008).

10.2.1 Classification on the Basis of Chemical Composition

Fly ashes are mainly rich in silica (amorphous and crystalline), alumina, and iron oxide and also consist of other oxides such as sodium oxide, potassium oxide, magnesium oxide, and sulfate. The total amount of SiO_2, Al_2O_3, and Fe_2O_3 and active lime content (CaO) decides the type of class. Fly ash is therefore classified as siliceous fly ash (class F), if total amount of SiO_2, Al_2O_3, and Fe_2O_3 is more than 70 % and active lime content (CaO) is less than 10%. Calcareous fly ash (class C) constitutes of at least 50% of SiO_2, Al_2O_3, and Fe_2O_3 and active lime content (CaO) is more than 10%. Siliceous fly ash (class F) or low lime fly ash has pozzolanic nature whereas calcareous fly ash (class C) or high lime fly ash has both pozzolanic and hydraulic nature (Bhatta et al. 2019; Bouaissi et al. 2020; Wesche 2005). In case of class F fly ash, the silica content is almost double of alumina content whereas in class of class C fly ash, the content of these two oxides is almost comparable to each other. The amount of iron oxide in class F fly ash is significantly higher than in class C fly ash (Chatterjee 2010). Even pH and calcium/sulfur ratio affects the nature of fly ash (Bhatta et al. 2019).

10.2.2 Classification on the Basis of Physical Characteristics

The performance of cement, mortar, and concrete in terms of rheology, volume stability, strength development, and durability are significantly impacted by the physical behavior of fly ash (Bhatta et al. 2019). The parameters like fineness, particle size, lime reactivity, residue on 45-micron sieve govern the performance criteria. The strength activity index with Portland cement is considered only as an indication of reactivity and should not be used as a measure to predict the compressive strength of concrete containing the fly ash.

10.2.3 Standards and Specification on Fly Ash

Development of standards is always being an ongoing process that largely comprises establishment of a consensus among different interests (Malhotra and Mehta 1996). Table 10.2 represents a compilation of the designations of the relevant national standards and specifications from the different countries of the world. A summary of the main technical features, i.e., chemical and physical requirements of some of the standards as mentioned in Table 10.3 is given in Tables 10.3A and 10.3B.

ASTM C 618 defines fly ash types based upon their chemical composition; Class C and class F. In terms of physical requirements, class C and class F have same limits. Class N fly ash as per ASTM C 618 represents raw or calcined natural pozzolans such as diatomaceous earths, opaline cherts, and shales; tuffs and volcanic ashes complying to the requirements of Table 10.3. IS 3812 cater to the requirements of fly ash for two specific uses: Part 1 covers the use of fly ash as a pozzolana in cement and concrete whereas Part 2 covers the use of fly ash as an admixture for concrete. IS 3812 Part1 classifies the fly ash into two classes based upon their chemical composition, i.e., siliceous fly ash (with reactive lime content less than 10%) and calcareous fly ash (with reactive lime content more than 10%). The European Union Standards (EN 450-1) classify fly ashes based on the Loss of ignition and particle fineness as shown in Table 10.3. The difference in fineness of fly ash from a given lot leads to variations in the water demand and mechanical properties of the resultant concrete (Sear 2001).

Table 10.2 Some of the National standard for fly ash to be use in cement and concrete

Sl	Country	Designation of standard	
1	India	IS 3812 Part1	Pulverized fuel ash—Specification: Part 1; For use as pozzolana in cement, cement mortar and concrete
		IS 3812 Part 2	Pulverized Fuel Ash—Specification Part 2; For Use as Admixture in Cement Mortar and Concrete
2	USA	ASTM C 618	Standard Specification for Coal Fly Ash and Raw or Calcined Natural Pozzolan for Use in Concrete
3	Canada	CSA A3000	Cementitious materials compendium
4	Australia	AS 3582.1	Supplementary cementitious materials—Part 1: Fly ash
5	European Union	EN 450–1	Fly ash for concrete—Part 1: Definition, specifications and conformity criteria
6	Japan	JIS A 6201	Fly Ash for use in Concrete

Table 10.3A Chemical requirements of fly ashes in different countries

Country	USA			India		European		
Chemical Composition (Mass %)	Class			Siliceous	Calcareous	Category		
	N	F	C			A	B	C
Silicon Oxide (SiO_2), min	–	–	–	35.0	25.0	–	–	–
$SiO_2 + Al_2O_3 + Fe_2O_3$, min	70.0	70.0	50.0	70.0	50.0	70.0	70.0	50.0
Sulfur Trioxide (SO_3), max	4.0	5.0	5.0	3.0	3.0	–	–	–
Free Calcium Oxide (CaO), max	–	–	–	–	–	1.50	1.50	1.50
Reactive Calcium Oxide (CaO)	–	–	–	10.0 max	10.0 min	10.0 max	10.0 max	10.0 max
Magnesium Oxide (MgO), max	–	–	–	5.0	5.0	4.0	4.0	4.0
Phosphate (P_2O_5), max	–	–	–			5.0	5.0	5.0
Sodium Oxide Equivalent ($Na_2O + 0.658K_2O$), max	–	–	–	1.50	1.50	–	–	–
Moisture Content, max	3.0	3.0	3.0	2.0	2.0	3.0	3.0	3.0
Loss on Ignition, max	10.0	6.0	6.0	5.0	5.0	5.0	7.0	9.0
Reactive silica, max	–	–	–	20.0	20.0	25.0	25.0	25.0
Total chlorides, max	–	–	–	0.05	0.05	0.10	0.10	0.10

10.3 Studies Done at Nccbm for Ash Utilization—A Way Forward

To pursue full ash utilization in India, a strict implementation of policies, promotion of technologies, and practices based upon Technology Readiness Level (TRL) is a need of the hour. National Council for Cement & Building Materials (NCCBM) in association with NETRA (NTPC Energy Technology Research Alliance) and other government bodies has carried out research for ash utilization in construction which is discussed hereunder.

10.3.1 Bottom Ash as Replacement of Fine Aggregate

Study has been carried out at NCCBM to replace natural and crushed sand with bottom ash at various percentages for making concrete and study its effect on mechanical and durability properties of concrete. Bottom ash was collected from Vindhyachal thermal power plant of India. Experimental studies were conducted on w/c ratio

Table 10.3B Physical requirements of fly ashes in different countries

Physical parameters	Class			Siliceous	Calcareous	Category	
	N	F	C				
						N	S
Fineness (m^2/kg)	–	–	–	320.0 min	320.0 min	–	–
Amount retained on a 45 μm sieve (%)	34 max	34 max	34 max	34 max	34 max	40.0 max	12.0 max
Strength Activity Index Ratio to Control @ 7 days	75 min	75 min	75 min	–	–	–	–
Strength Activity Index Ratio to Control @ 28 days	75 min	75 min	75 min	80 min	80 min	75 min	75 min
Strength Activity Index Ratio to Control @ 90 days	–	–	–	–	–	85 min	85 min
Water Requirement (% of Control)	115 max	105 max	105 max	–	–	–	95 max
Soundness by Autoclave Expansion (%)	0.8 max	0.8 max	0.8 max	0.8 max	0.8 max	–	–
Soundness by le chatelier's (mm)	–	–	–	–	–	10 mm	10 mm
Lime reactivity	–	–	–	4.5 min	4.5 min	–	–
Particle Density (kg/m^3)	–	–	–	–	–	± 200 kg/m^3 from declared value	
Initial setting time (max)	–	–	–	–	–	2.0 times the setting time of test cement	

of 0.65 and 0.40. Thirty-Two concrete mixes were prepared using bottom ash as a substitution of natural and crushed sand at various percentages and tested for various mechanical and durability parameters such as compressive and tensile strength, rapid chloride penetrability test, water permeability test, carbonation depth, and chloride depth. Based on fresh concrete properties and strength development studies, 50% replacement of bottom ash is technically feasible. Durability properties of concrete mixes at 50% replacement of natural and crushed sand with bottom ash are comparable with the control mixes. Based on the findings by the authors (Ojha et al. 2020a), the Indian Standard IS: 383-2016 is proposed for revision which will enhance the use of bottom ash as fine aggregate in construction. Research indicated that specific gravity of bottom ash is lower in comparison to conventional fine aggregate and fineness modulus is also on lower side. Therefore, there is an urgent need for development of methodology and concrete mix design guidelines when bottom ash replacement is to be done in different percentage of conventional fine aggregate.

10.3.2 Flyash Concrete—A Cost Effective Solution

Research has been conducted by NCCBM on fly ash concrete as well as on the use of PPC in M40 grade concrete and above for utilization in prestressed concrete and cement concrete roads. Studies indicated that except for the slow gain in strength at early ages, the mechanical properties and durability of the fly ash concrete were superior to the normal concrete (Arora and Singh 2016). Based on the durability and corrosion studies, it can be concluded that PPC is more durable than OPC in coastal environment/aggressive environment owing to high resistance to chloride ion penetration in PPC. Results of the accelerated carbonation test indicate that carbonation induced corrosion is more in PPC made concrete as compared to OPC and OPC is more durable as compared to PPC in areas other than coastal environment. This increase in carbonation in case of PPC made concrete may enhance the rate of corrosion and precautions that needs to be taken are related to late strength development, enhanced curing regime, and additional cover to cater carbonation induced corrosion (Arora and Singh 2017). The further study is required to investigate high performance fly ash concrete using multifunctional admixtures (plasticizer and accelerator) for pavements and dams in India.

10.3.3 Fly Ash and Slag-Based Geopolymer Concrete

Comprehensive studies were carried out at NCCBM to optimize ratios of SiO_2/Al_2O_3, SiO_2/Na_2O for the production of geopolymer concrete. Findings of research show that with minor modifications in test, formulation and usage methodology, plain alkali activated precast products can be used as a substitute to normal concrete products (Yadav et al. 2020). The study highlighted that cost can be brought to comparable

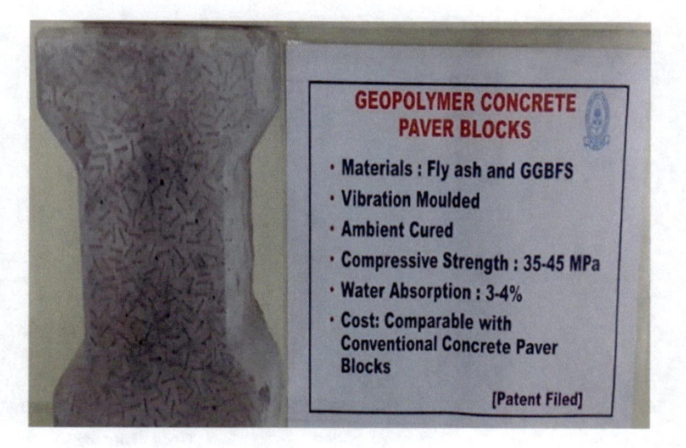

Fig. 10.2 Flyash based geopolymer concrete paver block

Fig. 10.3 Flyash based geopolymer concrete block road stretch

to conventional concrete by resorting to commercial grade of chemicals and maximizing usage of relatively cheaper materials like fly ash. A trial stretch was cast with plain alkali activated precast paver blocks, as shown in Figs. 10.2 and 10.3. Findings of the study contributed in the development of Indian standard on precast geopolymer concrete products. Since flyash-based geopolymer concrete is a low calcium system, there is a need to develop admixtures which can improve early strength development even at ambient conditions. Further studies are also needed to develop design parameters for geopolymer concrete for its structural applications.

10.3.4 Sintered Fly Ash Light Weight Aggregate for Structural Concrete

Research was conducted to evaluate the potential of sintered fly ash coarse aggregate in masonry, structural, and light weight blocks (both hollow and solid). Based on the detailed study, it can be concluded that both the fractions of lightweight aggregate

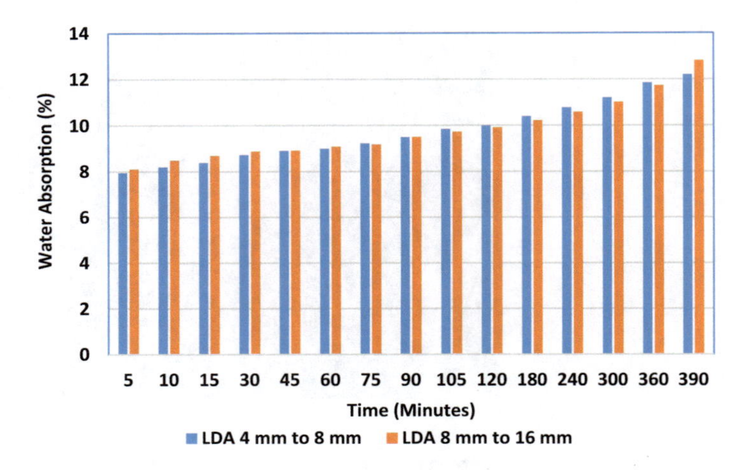

Fig. 10.4 Water absorption at different interval for fraction 4–8 mm LDA and 8–16 mm LDA (Soundaram et al. 2020)

can be used in concrete masonry units, production of hollow and solid lightweight concrete blocks. Results of abrasion, crushing, and impact values indicate that lightweight aggregate shall not be used for concrete to be used in wearing surfaces. Study on mechanical and durability properties SLC indicate that concrete made with lightweight coarse aggregate can be used as structural concrete. However, various structural design codal provisions need to be established for structural lightweight concrete since parameters such as flexural strength and modulus of elasticity values are lower than that of normal weight concrete for same compressive strength. Codal provision such as limit on cement content, free water-cement ratio, and concrete cover to the reinforcement shall be further strict than that of normal weight aggregate concrete to ensure similar level of durability in the same exposure conditions. The findings of the study have resulted in the development of new Bureau of Indian standard IS: 9142 (Part-2)-2018 on sintered flyash lightweight coarse aggregate in structural applications (Ojha et al. 2021). Further studies are needed to enhance the physical properties of sintered flyash aggregate and to explore its application in reinforced and prestressed concrete (Fig. 10.4).

10.3.5 Controlled Low Strength Material for Backfilling

Detailed study was conducted on fresh, hardened, and water penetration property of Controlled Low Strength Material (CLSM) at NCCBM (Ojha et al. 2020b). Some of the mix design trails conducted for CLSM are mentioned in Table 10.4. Test results corresponding to compressive strength and density are given in Table 10.5. Study indicated that due to the presence of fine materials and fillers in place of coarse aggregate there was no issues in fresh behavior of CLSM. Findings highlighted that

Table 10.4 Trials conducted for CLSM

Sl. no	Water content (Kg/m^3)	Cement content (Kg/m^3)	Fly ash content (Kg/m^3)	Fine aggregate content (Kg/m^3)
Mix-1	250	30	460	1364
Mix-2	250	50	460	1347
Mix-3	250	70	460	1330
Mix-4	290	30	800	839
Mix-5	304	50	800	786
Mix-6	355	50	1000	405
Mix-7	375	30	1200	122

Table 10.5 CLSM test results

Sl. No	Slump of concrete (mm)	7-Day cube comp. strength (N/mm^2)	28-Day cube comp. strength (N/mm^2)	CLSM density (Kg/m^3)
Mix-1	180	0.80	1.45	2122
Mix-2	170	1.20	2.99	2126
Mix-3	170	1.40	3.62	2132
Mix-4	165	0.85	1.77	1976
Mix-5	195	1.82	3.68	1954
Mix-6	185	1.65	3.09	1832
Mix-7	180	0.66	1.47	1736

coefficient of permeability of CLSM (1×10^{-8} cm/s) is lower than clay (1×10^{-7} cm/s). The fly ash used in the study of CLSM conforms to IS: 3812 (Part 2). As the fly ash content used in the study is as high as 1200 kg/m^3, it has a great potential to be used in CLSM as landfill and backfill material.

10.3.6 Pond Ash—Clay Fired Bricks

Prashant and Dwivedi (2013) carried out the comparison of bricks with varying percentage of pond ash substitution with clay bricks. Findings of the study indicated that substitution of pond ash with clay bricks beyond 20% was not technically feasible as it lead to issues related to dimension exceeding tolerance limit, significant increase in water absorption, and decrease in compressive strength. In field of bricks and blocks, a new standard on fly ash cement bricks (IS 16720) has been already formulated. Hence, there is huge potential for application of pond ash clay fired bricks. Application of pond ash in non-structural elements like kerb stone, paver blocks needs to be explored.

10.4 Conclusion and Need of Further Research

With the advancement in the techniques and various government policies, ash generated from thermal power plants has acquired a status of value-added materials. The use of ash in terms of bottom ash as a fine aggregate, blended cement or fly ash concrete, low calcium or high calcium geopolymer concrete, sintered fly ash light weight aggregate for structural concrete, controlled low strength material for backfilling, and pond ash for brick production has shown promising performance. Currently IS: 6491-1972 method for sampling fly ash is about 50 years old. During this period there has been significant changes with respect to technology of coal firing system, conveyance, and collection system of ash. Present standard does not include sampling method for bottom ash. In view of this there is an urgent need for revision of Indian standard pertaining to method of sampling fly ash. Further research is needed in different areas like (a) methodology and concrete mix design guidelines for usage of bottom ash as fine aggregate (b) development of multifunctional admixtures (plasticizer and accelerator) for production high performance fly ash concrete for pavements and dams (c) development of admixtures which can improve early strength development even at ambient conditions in low calcium-based geopolymer concrete (d) development of design parameters for low calcium geopolymer concrete for its structural applications (e) development of technology or methodology for improvement in the physical properties of sintered fly ash aggregate and to explore its application in reinforced and prestressed concrete.

References

ACI 232.2R, Use of fly ash in concrete

AS 3582.1, Supplementary cementitious materials—Part 1: fly ash

Anon (2019) Executive summary on the power sector. Central Electricity Authority. Available at the http://cea.nic.in/reports/monthly/executivesummary/2019/exe_summary-10.pdf. Accessed 21 Sept 2020

Arora VV, Singh B (2017) Use of fly ash concrete and recycled concrete aggregates-a cost effective solution and durable option for construction of concrete roads. Indian Concr J 91(1)

Arora VV, Singh B (2016) Durability studies on prestressed concrete made with portland pozzolana cement. Indian Concr J 90(8)

Bhatta A, Priyadarshinia S, Mohanakrishnana AA, Abria A, Sattlera M, Techapaphawitc S (2019) Physical, chemical, and geotechnical properties of coal fly ash: a global review. Constr Mater. https://doi.org/10.1016/j.cscm.2019.e00263

Bouaissi A, Li LY, Abdullah MMAB, Ahmad R, Razak RA, Yahya Z (2020) Fly ash as a cementitious material for concrete. Zero Energy Build-New Approaches Technol. http://dx.doi.org/https://doi.org/10.5772/intechopen.90466

CEA: Central Electricity Authority Report (2018–2019) Available online at: http://cea.nic.in/reports/others/thermal/tcd/flyash_201819.pdf. Accessed 05 Apr 2020

Chatterjee KA (2010) Indian Fly ashes, their characteristics, and potential for mechano-chemical activation for enhanced usability. In: Second international conference on sustainable construction materials and technologies

CSA A3000, Cementitious materials compendium

EN 450–1, Fly ash for concrete—Part 1: definition, specifications and conformity criteria
JIS A 6201, Fly Ash for use in Concrete
Wesche K (2005) Fly ash in concrete. Chapman & Hall, Properties and performance, London, UK
Malhotra VM, Mehta PK (1996) Pozzolanic and cementitious materials. CRC Press
MOEF (1999) Gazette notification for ministry of environment and forests, vol 563. Ministry of Environment and Forests, New Delhi
Mukherjee AB, Zevenhoven R, Bhattacharya P, Sajwan KS, Kikuchi R (2008) Mercury flow via coal and coal utilization by-products: a global perspective. Resour Conserv Recycl 52(4):571–559
Ojha PN, Singh B, Behera AK (2021) Sintered fly ash light weight aggregate—its properties and performance in structural concrete. Indian Concr J
Ojha PN, Kumar S, Singh B, Mohapatra BN (2020a) Pervious concrete, plastic concrete and CLSM-a special application concrete. J Build Mater Struct
Ojha PN, Trivedi A, Singh B, Singh A (2020b) Evaluation of mechanical and durability properties of concrete made with indian bottom ash as replacement of fine aggregate. Asian Concr Fed J
Prashant GS, Dwivedi AK (2013) Technical properties of pond ash—clay fired bricks—an experimental study. Am J Eng Res 02(09):110–117
Sear LKA (2001) The properties and use of coal fly ash. Thomas Telford Ltd.
Soundaram R, Arora S, Trivedi V (2020) Coal-based power norms—where do we stand today? Centre for science and environment. Delhi
Ujjwal B, Kandpal TC (2000) Potential of fly ash utilisation in India. Energy 27(1):151–166
Yadav KV, Fulekar MH (2018) The current scenario of thermal power plants and fly ash: production and utilization with a focus in India. Int J Adv Eng Res Dev 05(04)
Yadav L, Trivedi A, Arora VV, Mohapatra BN (2020) Case study on field trials of developed geopolymer (Slag and Fly Ash Based) precast concrete paver blocks. Indian Concr J

Chapter 11
Sustainable Pavements for Low-Impact Developments in Urban Localities

B. R. Anupam, Anush K. Chandrappa, and Umesh Chandra Sahoo

11.1 Introduction

Low-impact development (LID) is a design and implementation approach to balance the infrastructure needs of an urban setup and the environment (Dietz and Clausen 2008). With the incessant urbanization especially in the emerging countries like India, China, Bangladesh, etc. the infrastructure demand has gradually converted the natural soil into an impervious fabric primarily constituted by the pavements (Ligtenberg 2017). The natural soil, which had the ability to in-filter the stormwater, slowdown the runoff, reduce the runoff quantity, and keep the surroundings cool has been buried beneath the impervious surface such as pavements (Paul and Meyer 2001). Due to this, two significant changes in the urban setups have been reported and investigated for several decades. The first significant change is in the domain of hydrological characteristics of the urban setup, where the increased imperviousness leads to decrease in infiltration, reduction in groundwater recharge, and increase in the surface runoff thus leading to frequent flash floods (Meena et al. 2017). Figure 11.1 shows the schematic of hydrograph corresponding to pre- and post-urbanization.

In the pre-urbanization era, it is seen that the runoff quantity (Q1) is low, and time of concentration (t1) is longer contrary to post-urbanization era. This observation has been well documented in various cities of developed and developing nations in past research studies (Miller et al. 2014; Li et al. 2018; Wang et al. 2020). An et al. (2018) calculated that in the case of urban setup, the flooding can occur every 0.20–1.8 years

B. R. Anupam · A. K. Chandrappa (✉) · U. C. Sahoo
School of Infrastructure, Indian Institute of Technology, Bhubaneswar, India
e-mail: akc@iitbbs.ac.in

B. R. Anupam
e-mail: abr10@iitbbs.ac.in

U. C. Sahoo
e-mail: ucsahoo@iitbbs.ac.in

K. R. Reddy et al. (eds.), *Advances in Sustainable Materials and Resilient Infrastructure*,
Springer Transactions in Civil and Environmental Engineering,
https://doi.org/10.1007/978-981-16-9744-9_11

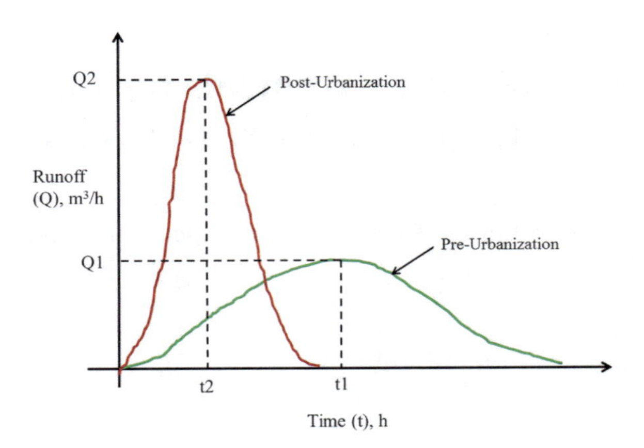

Fig. 11.1 Typical schematic of hydrograph under different scenarios

with a larger impact. The conventional stormwater management practices have been found to be insufficient to cater the increased runoff due to urbanization. Although the urban floods are location specific, these will result in huge financial loses owing to high density of population, high valued assets, disturbance in day-to-day activities, damaged infrastructure, and other associated losses (Svetlana et al. 2015).

On the other hand, the second significant change is in the domain of thermal characteristics of urban setup. For a given urban setup, the pavement occupies almost 30–40% of the urban built-up area, which is directly exposed to the sun (Cheela et al. 2021). The building materials used in the pavement construction tend to store more heat unlike the natural soil due to high volumetric heat capacity, lower latent heat capacity, reduced thermal conductivity and diffusivity, which increases the thermal mass in urban setup (Chandrappa and Biligiri 2015). Due to this, the temperature of the pavement surface increases up to almost 30–35 °C higher than the ambient air temperature during a typical summer season. However, both the air and pavement do not reach their respective peak temperatures at the same time owing to the differences in thermal properties, which is generally referred to as hysteresis effect (Song et al. 2017). The hysteresis lag results in high nighttime minimum temperature, which has a significant impact on electricity consumption for cooling. The high temperature and hysteresis lag effect associated with pavements increase the ambient temperature of the urban setup leading to a phenomenon referred to as *urban heat island (UHI) effect*. The UHI effect has been under the study since several decades, where an average difference of 3–6 °C is found between urban and its surrounding rural environment. The UHI effect poses several health hazards for urban dwellers such as heat strokes, dehydration and nausea, migraine, heat cramps, lack of productivity, and heat-related mortality (Cheela et al. 2021). A detailed review on effect of UHI on health of urban dwellers is well documented by Heaviside et al. (2017).

The above two significant changes keep escalating and continue to affect the quality of life of urban dwellers due to continuous modification in the natural ground due to urbanization. The conventional practice of urban infrastructure development can be termed as "high impact development (HID)" due to the significant changes it brings with it. With the escalating problems associated with urbanization such as reduced groundwater, improper stormwater management, urban flooding and increased ambient temperature, the researchers are now focusing on "Low-impact development (LID)". The environmental protection agency (EPA) of the USA refers to the LID as *engineered practices and systems of infrastructure development, which use or mimic the natural processes of stormwater infiltration, evapotranspiration, ground water recharge and use of stormwater in order to protect the water quality.* In other words, using LID, the development and infrastructure needs of the urban dwellers can be achieved without adversely impacting the environment. The LID methods focus on land development strategies and principles that function in tandem with nature to create a liveable urban setup. Some of the most commonly referred LID principles include bio-retention facilities, rain gardens, vegetated rooftops, permeable pavements, and cool pavements. In this chapter, the cool pavements and pervious concrete pavements are comprehensively described as strategies for LID in developed and emerging countries.

11.2 Cool Pavements

The phenomenon of higher temperature experienced in urban setup than that of its surrounding rural area is termed as UHI effect. The conversion of naturally vegetated area to artificial structures, which store more thermal energy, is the main reason leading to UHI effect. The UHI effect contributes to air pollution, disruption in local wind patterns, loss of human thermal comfort, and degradation of water quality (Akbari et al. 2001). As the pavement surface temperature has a substantial impact on the ambient temperature (Lin et al. 2007), a higher pavement temperature adds to the UHI effect (Akbari et al. 2001). According to Lin et al. (2007), the pavement can be up to 15–30 °C warmer than natural vegetation in summertime. Therefore, technologies to cool down the pavement temperature, termed as *cool pavements*, are being extensively investigated.

"Cool pavements include a range of established and emerging technologies that communities are exploring as part of their heat island reduction efforts. The term currently refers to paving materials that reflect more solar energy, enhance water evaporation, or have been otherwise modified to remain cooler than conventional pavements" (US-EPA). Reflective pavements, evaporative pavements, and heat storage modified pavements are the three main forms of cool pavements (Fig. 11.2). Reflective and evaporative pavements are the widely used technologies and hence can be called as traditional cool pavements. Modifying heat storage characteristics to achieve cool pavements has not been widely investigated and is still in the research phase (Anupam et al. 2021b).

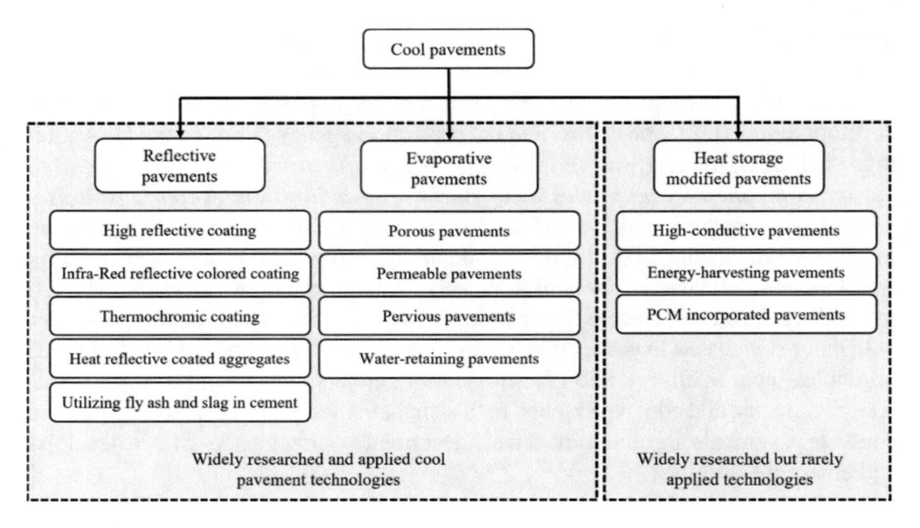

Fig. 11.2 Different types of cool pavements

11.2.1 Reflective Pavements

11.2.1.1 Conventional Reflective Pavements

Reflective pavements are typically lighter in color and have a high level of solar reflection. Due to the greater solar reflectance, the pavement surface can reflect a large portion of the incident solar energy. As a result, these pavements absorb less heat energy and remain cooler than normal pavements (Balan et al. 2021). Application of a high reflective coating (Xie et al. 2020), Infra-Red reflective colored coating (Xie et al. 2019), thermochromic coating on the pavement surface (Chen et al. 2021), and incorporation of heat reflective coated aggregates (Ma et al. 2002) or lighter colored aggregates (Anupam et al. 2021a) in pavement construction can be adopted to construct reflective pavements. Applying a light-colored (pink) coating to the pavement surface has been reported to decrease the temperature by 10 °C in the field (Zheng et al. 2015).

11.2.1.2 Limitations of Conventional Reflective Pavements

The ultraviolet radiations reflected from these pavements may cause damage to human tissues. Glaring issues are also another major disadvantage of reflective pavements. The reflected radiation in the visible area has visual effects on the corona of the human eye, impairing the drivers' vision. However, glare-related problems can be reduced to a certain extent by using near-infrared coatings (Santamouris

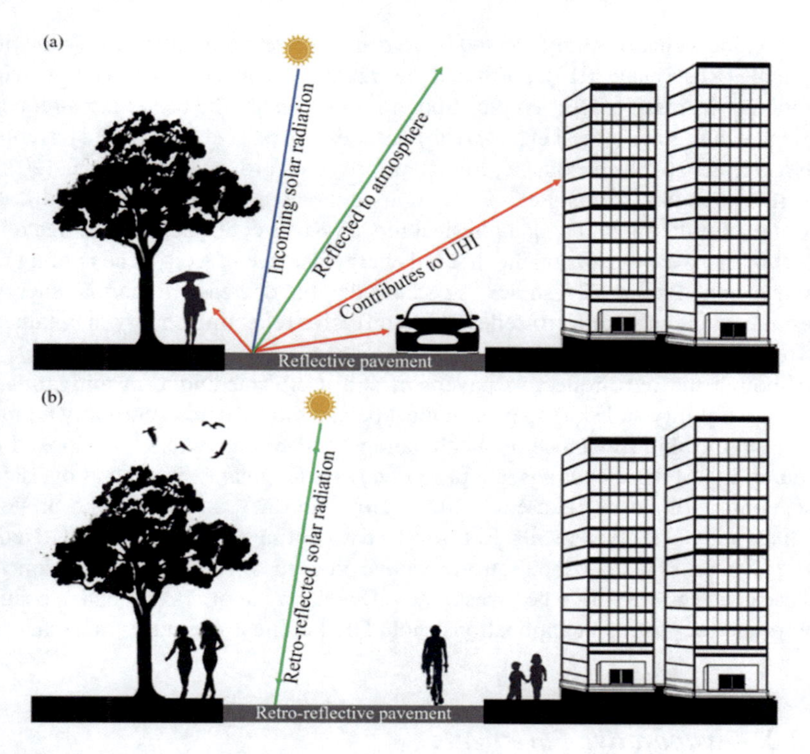

Fig. 11.3 Reflectivity of **a** Traditional Reflective Pavement and **b** Retro-reflective Pavement

et al. 2011). After being polluted by the environment, the durability of these coatings drops dramatically. The limited durability of these coatings is also restricting its wide application. Further, such pavements are only effective in UHI mitigation, if the incoming radiations are reflected to the sky, else the reflected radiations will have a negative impact on the thermal comfort of the people, cars, and surrounding structures (Anupam et al. 2021b), as shown in Fig. 11.3a. Therefore, it may be inferred that reflective pavements remain cool than regular pavements, however, they do not necessarily help with UHI alleviation.

11.2.1.3 Retro-Reflective Pavements

To redirect the solar energy out of the urban environment, Rossi et al. (2014) recommended the use of retro-reflective coatings for buildings and paving applications. As illustrated in Fig. 11.3b, a retro-reflective pavement can reflect incoming solar radiation in the same direction of incidence, thus reducing their negative impact on the thermal comfort of the commuters, automobiles, and surrounding structures. In this, the incoming radiations are reflected by retro-reflective materials at a very low incident angle (Rossi et al. 2014). Further, there was also a considerable reduction in inter radiation transmission between the structures. As a result, the use of retro-reflective coatings in buildings (Rossi et al. 2015b), facades (Rossi et al. 2015), and pavements (Rossi et al. 2016) has become a research priority. According to Rossi et al. (2015b), retro-reflective pavements are 3–7 °C cooler compared to standard white diffusive coverings. Furthermore, using a physical model, Rossi et al. (2016) investigated the impact of these pavements on the thermal energy balance of a city. The cooling ratio was determined to be 0.37, indicating a considerable decrease in thermal energy as measured as the ratio of retro-reflective to diffusive reflecting energy maintained in the study area.

Although the retro-reflective pavement is a useful technique in mitigating the UHI, its durability and skid resistance must be assessed. There is currently no information on the skid resistance of such coatings. Some researchers have looked into the durability of these coatings and proposed ways to enhance it, such as overlaying the coating with glass (Yuan et al. 2016). However, these are only appropriate for buildings and not for pavements. Retro-reflective coatings are also intended to reduce glaring concerns owing to their retro-reflective properties. The glare-related concerns of these coatings are still to be investigated. Development of retroreflective coatings appropriate for pavement applications should be the focus of future research.

11.2.2 Evaporative Pavements

In evaporative pavements, water is held inside the pavements for evaporative cooling. The latent heat collected by the pavement during the liquid-to-vapor phase transition of water is used to cool it. Therefore, such pavements are often recommended for rainy and humid climates, where water supply is never an issue. Evaporative pavements can be divided into four categories: porous pavements, permeable pavements, pervious pavements, and water-retaining pavements.

Porous pavements are a modified version of interlocking block pavements (ICBP). Porous pavements get their name based on the fact that the pores are not linked. Blocks are often laid out in a cellular grid pattern that occupies 20–50% of the pavement surface. In case of the ICBP, solid conventional concrete blocks are used, which are usually impervious except for the permeability through the jointing sand. However, in the modified version, these blocks are made in such a way that each one has a circular hole to improve permeability, which allows the grass to grow and/or can be filled with gravel/sand. Both of them improve the water holding capacity and therefore boost the cooling action since the water stored increases the latent heat capacity.

Pervious pavements are a type of porous pavement that could infiltrate rainwater, reduce runoff, and improve groundwater recharge. With permeability in the range of 97–0.097 mm/s, they can hold more water than porous pavements (Roseen et al.

2009). This water, on the other hand, will drain through the reservoir layer and finally reach the subgrade.

Permeable pavements enable rainwater to flow around the surface rather than through it. An open-graded friction course is put atop a dense asphaltic concrete to minimize hydroplaning is a good example. This water is retained for a short period in the underlying pavement layers and is evaporated through evaporation channels, reducing the pavement temperature.

Water-retaining pavements are intended to keep water at the top layers of the pavement. They are usually made up of a pervious top layer and an impervious bottom layer. Grouting the pores with pervious fillers with high capillarity can achieve this. Fillers like blast furnace slag and pervious mortar can be used for this. Water is kept near the pavement surface by the capillary action of fillers, which allows effective evaporative cooling.

11.2.2.1 Limitations of Conventional Evaporative Pavements

The hardest part in evaporative pavement construction is attaining good mechanical strength and durability by maintaining an appropriate air void content. Poor durability of evaporative pavements is one of the major drawbacks, as they are susceptible to raveling and water damage due to higher air voids content (Alvarez et al. 2011). According to Will and Liv (2012), permeable concrete pavements appropriate for collector streets and residential streets can be planned for 20–30 years with mechanical strength comparable to traditional pavements. Nevertheless, in order to encourage their widespread use, it is important to increase their lifespan without sacrificing their infiltration characteristics.

Furthermore, evaporative pavements have a lower solar reflectance than traditional pavements, which increase the amount of solar radiation absorbed (Haselbach et al. 2011). Also, when the pavement is dry, the reduced thermal mass of evaporative pavements results in higher pavement temperatures in summertime (Zhang et al. 2015). Evaporative cooling is only efficient when water is present at the pavement surface (0–25 mm) (Nemirovsky et al. 2013). Furthermore, evaporative pavements elevate the humidity nearer to the pavement surface and reduce pedestrian sweat cooling (Erell et al. 2012). Evaporative pavements may further enhance the risk of contaminants flowing into groundwater (Vairagade 2016).

11.2.2.2 Evaporation Enhancing Permeable Pavements

Traditional permeable pavements are generally efficient in reducing the pavement temperature for 48 h after rain. As a result, water must be sprayed over permeable pavements, which may not be practicable or economical. As a result, Liu et al. (2018) created evaporation-enhancing pavements, illustrated in Fig. 11.4, that include capillary columns to elevate water collected by the liner to the pavement surface. Since these pavements are intended to manage rainwater in areas where the groundwater

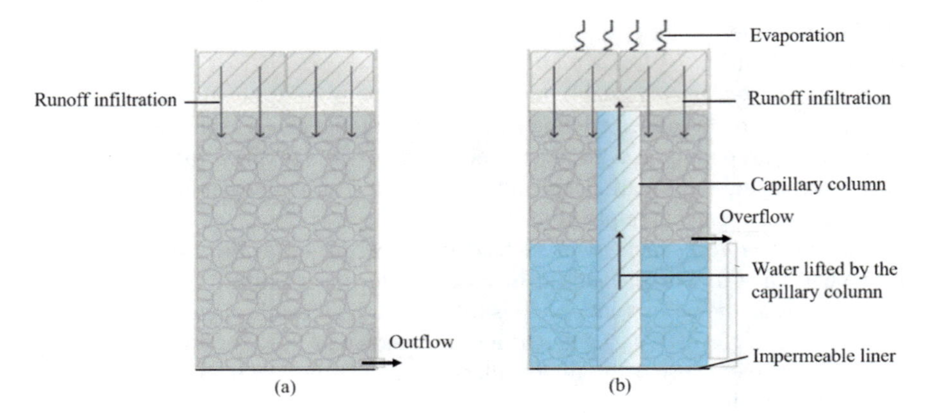

Fig. 11.4 Cross-sectional view of **a** Traditional Permeable Pavements and **b** Evaporation Enhancing Permeable Pavements (Liu et al. 2018)

table is high, the liner helps to prevent contaminants from entering the groundwater. While a traditional permeable pavement consists of permeable paver blocks, laying course, geotextile, and aggregates, the evaporation-enhancing pavements have capillary columns inside the aggregates. Capillary action elevates the stored water in the bottom layers to flow upward via these capillary columns. As a result, water is always accessible for evaporative cooling at the top region of the pavement. The findings demonstrate that evaporation-enhancing permeable pavements can reduce pavement temperature for several days after rain. These pavements have also been shown to be up to 9.4 °C cooler than traditional permeable pavements.

Further, Liu et al. (2020) looked at the cooling potential and feasibility of these pavements under field conditions in another research. At Tongji University in Shanghai, China, evaporation-enhancing permeable pavement sections reported 15.3, 15.8, and 14.4 °C lower temperatures than that of impermeable concrete pavement, coarse concrete paver, and fine concrete paver, respectively. When these pavements are compared to traditional concrete pavements, the stormwater volume is reduced by 90.6%, whereas the reductions for coarse and fine permeable concrete pavers are 40.2 and 41.9%, respectively.

11.2.2.3 Drainable Water Retaining Pavements

Water retaining pavements are typically used to collect rainwater for evaporative cooling. Despite the fact that such pavements are efficient in temperature reduction, heavy rainfall causes overflow and urban flooding. Furthermore, the water-retaining medium takes up space, thereby reducing the amount of water available for evaporative cooling. To address these constraints, Qin et al. (2018) designed an advanced type of water retaining pavement targeted to minimize stormwater runoff and enhance water storage capacity. Instead of using the water-holding medium, the structure is sealed with impermeable concrete on the sides and bottom, as illustrated in Fig. 11.5a. As a result, the pore structure of this pavement could hold a greater amount of water. The overflow from pervious concrete is similar to that of typical dense concrete. Drainable water-retaining pavements, on the other hand, are able to hold 9.5 L/m^2 of water and postpone the overflow by about 10 min. Even though the solar reflectance of drainable water holding pavements is 0.05–0.10 and 0.10–0.15 less than conventional pervious and dense concrete pavements, a temperature drop of 2–10 °C is achievable. This means that, despite high absorbed solar radiation, evaporative cooling of drainable water-retaining pavements is sufficient to reduce the pavement temperature.

However, the impermeable bottom of drainable water retaining pavements leads to frequent flooding. To address this, Bao et al. (2019) proposed certain modifications to drainable water retaining pavements, as illustrated in Fig. 11.5b. To drain the surplus water, drain pipes made of an impermeable wall and filled with permeable concrete are installed. The drain pipes are installed before the application of impermeable mortar, and their height was up to 10 mm below the pavement surface. This modified drainable water-retaining pavement is successful in draining surplus water while maintaining evaporative cooling. In comparison to ordinary concrete pavements, the modified drainable water retaining pavements decrease the pavement temperature by 13 °C during the day and 3 °C at night. Furthermore, the structural performance of the pavement is supposed to improve as a result of the integrated

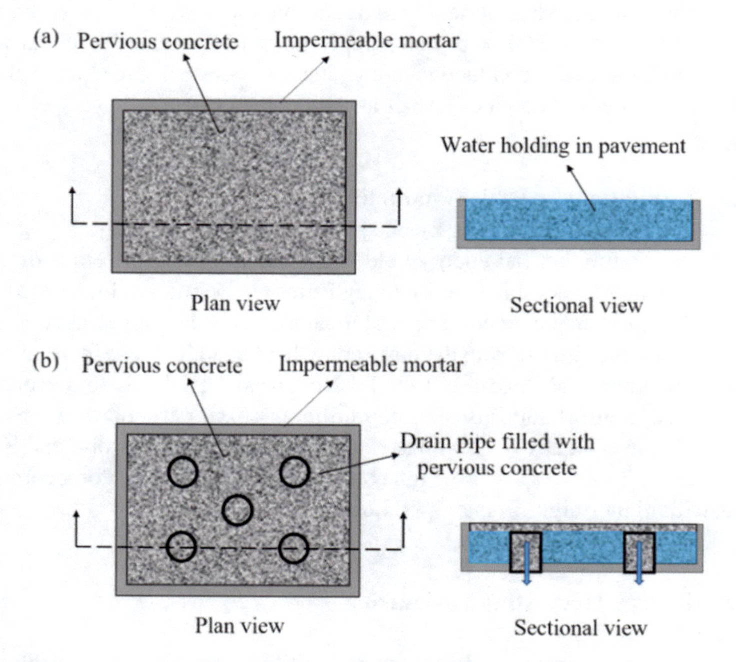

Fig. 11.5 **a** Drainable Water Retaining Pavements, **b** Modified Drainable Water Retaining Pavements

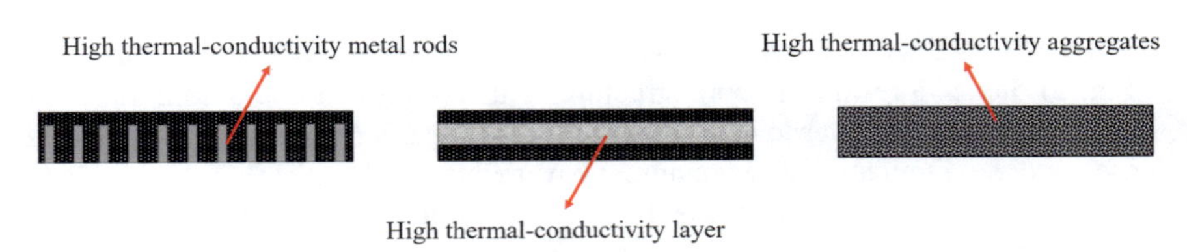

Fig. 11.6 High conductive pavements

drainpipes and mortar sealing. However, no research has been done to support this claim.

11.2.3 Heat Storage Modified Pavements

11.2.3.1 High Conductive Pavements

The heat transmission from the pavement surface to the subgrade can be accelerated by improving the thermal conductivity of the pavement (Gavin et al. 2007). Due to the existence of moisture and micropores, the subgrade soil has a larger latent heat capacity, which can dilute the transferred heat and lower the pavement temperature. Imparting high conductive materials such as carbon fibers, steel fibers, carbon black, and graphite (Wu et al. 2005), or using high thermally conductive metal rods, or integrating high thermally conductive aggregates such as steel slag (Jiao et al. 2020), can enhance the thermal conductivity of asphalt (Fig. 11.6).

11.2.3.2 Limitations of High Conductive Pavements

During the night, the thermal energy held in the top pavement layers is discharged into the environment, resulting in a rise in nighttime temperature (Guntor et al. 2014). Even though the rise in pavement temperature at night is relatively small considering the advantages it provides during the day, it may lead to an increase in urban temperature at night, known as "nocturnal UHI". Nocturnal UHI leads to a reduction in human thermal comfort and raises air conditioning costs, particularly in residential structures (Zhang et al. 2017), as well as potentially cause health problems (Hancock and Vasmatzidis 2003). As a result, high conductive pavements are not recommended for areas with higher nighttime temperatures.

11.2.3.3 Energy Harvesting Pavements

The method of transforming and conserving unutilized or wasted renewable energy, such as solar, wind, and tidal energy, into a usable form, is known as energy

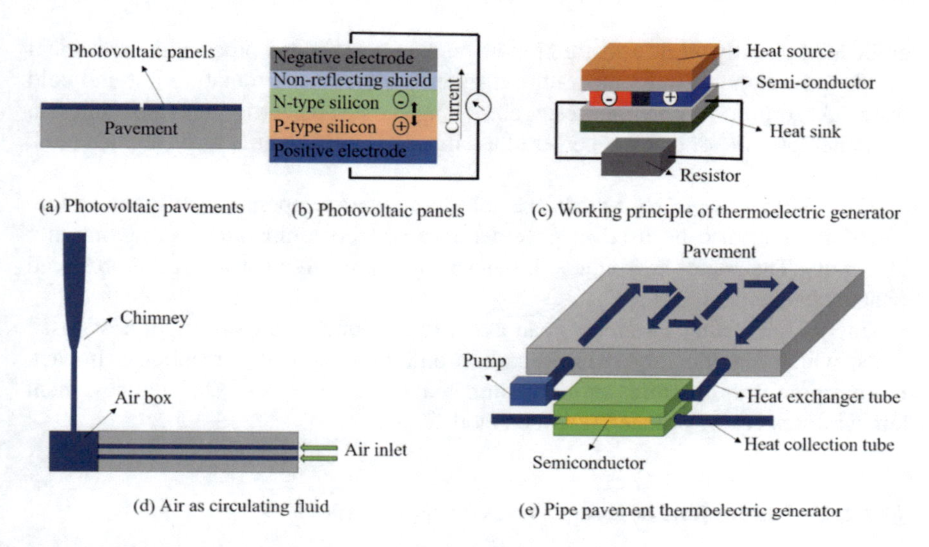

Fig. 11.7 Types of energy harvesting pavement technologies

harvesting. Heat energy harvesting from pavements has acquired scientific relevance since the pavements are continually open to solar energy for a longer period. Energy harvesting in pavements can be accomplished using any of the methods discussed below (Fig. 11.7).

Liquid circulation may be used to harvest energy and keep pavement cool. Cooling fluids can be pumped through buried pipes inside these pavements, generally water from local rivers or the sea. These pavements are known as hydronic pavements when water is utilized as the circulating fluid (Balbay and Esen 2010). The thermal energy may be stored in a heat sink, which can be used to melt snow and heat the buildings during winter (Bobes-Jesus et al. 2013). A thermoelectric generator can transform the thermal energy collected by the circulating liquid into electrical energy (Wang et al. 2018a).

In the embedded pipes, *air circulation* can also be used for heat exchange. At lower elevations, cold air is permitted to enter through pipes and flow through the pavement, which absorbs the thermal energy by convention, and is removed by a chimney (Chiarelli et al. 2015). The airflow is driven by the temperature difference between the pavement structure and the surrounding air. For air circulation, stainless steel, copper, and concrete pipes are commonly used. These pavements are up to 6 °C cooler than traditional pavements (Chiarelli et al. 2017).

The incident sunlight on the pavement can be converted into electrical energy *using photovoltaic cells* made up of P-type and N-type semiconductors. According to previous research, incorporating photovoltaic cells into pavements can lower pavement temperature by 5 °C (Efthymiou et al. 2016).

Thermoelectric generators can turn the temperature differential between the upper and lower pavement layers into electricity. Thermoelectric generators can be embedded directly while construction or linked to embedded pipes in the pavement to take advantage of the temperature differential between circulating hot and cold water. According to previous research, a single thermoelectric generator unit can generate 250 mW of electric power at a temperature differential of 5 °C (Hsu et al. 2011).

Pyroelectric materials, which are able to convert temperature differences into electricity, can also be used in pavements to convert temperature variations into electricity. The rate of heating/cooling determines the pyroelectric current produced (Bhattacharjee et al. 2011).

Energy harvesting pavements can assist to minimize the consumption of fossil fuels, which significantly reduces carbon emissions and the greenhouse impact. Furthermore, the structural and environmental consequences of elevated pavement temperatures can be reduced (Dawson et al. 2014).

11.2.3.4 Limitations of Energy Harvesting Pavements

The principal job of pavements is to carry vehicular loads, and every pavement layer plays an important part in this. Integration of energy harvesting components, such as embedded pipes, may have a negative impact on pavement strength (Dawson et al. 2014). Energy harvesting pavements, according to studies, are not strong enough to sustain severe traffic loads unless other procedures are used to increase their strength. Furthermore, while creating the upper layers, care should be taken to preserve the energy harvesting components, and this may slow down the construction process. Traditional maintenance procedures like milling, overlaying, trenching, etc. can reduce the structural and thermal efficiency of an energy harvesting system (Dawson et al. 2014). Furthermore, because of the difficulties in compaction, voids around the implanted pipes degrade the thermal efficiency of the liquid circulating thermoelectric generator (Dehdezi 2012). Based on the cross slope of the pavement, more energy may be required to maintain the fluid flow in the pipes under site conditions (Pan et al. 2015).

11.2.3.5 PCM Incorporated Cool Pavements

PCMs are capable to store heat energy as latent heat without significantly increasing their temperature. During the day, the PCM undergoes a solid–liquid phase change due to the heat energy absorbed by the pavements. Hence, the pavements remain cool as the heat energy is utilized for phase transition by the PCM. When the pavement temperature decreases during the night, the PCM undergoes liquid–solid phase transition by discharging the stored thermal energy. As a result, the PCM in pavements manages both high and low temperature extremities. Direct inclusion of PCM in asphalt pavement has a negative effect on the physical and rheological properties of the bitumen. PCM interacts with cement hydration in concrete pavements and decreases their mechanical strength. As a result, before integrating PCMs

Table 11.1 Studies PCM incorporated cool pavements

PCM used	Encapsulation	Temperature reduction (°C)	
		Lab	Field
Paraffin (Chen et al. 2010) (T_m = 40–50 °C, ΔH = 150 J/g)	Expanded graphite as carrier	–	2.0
Unsaturated organic acid (Ma et al. 2014) (T_m = 22.1 °C, ΔH = 46.97 J/g)	Polypropylene as carrier	1.5	0.7
Ceresin (Ryms et al. 2015) (T_m = 42–78 °C)	Expanded clay aggregates as carrier	8.5	7.0
Polyethylene glycol (Jin et al. 2017) (T_m = 52–55 °C, ΔH = 100–115 J/g)	Expanded perlite as carrier	7.0	4.3
Eicosane (Refaa et al. 2018) (T_m = 36.5 °C, ΔH = 247.3 J/g)	Mineral fillers as carrier	–	2.7
Paraffin Wax (Dehdezi et al. 2013) (T_m = 26 °C, ΔH = 160 J/g)	Acrylic outer shell	4.0	3.5
Organic mixture 42 (Anupam et al. 2021c) (T_m = 44 °C, ΔH = 199 J/g)	Expanded clay aggregates as carrier	–	2.8

into pavements, they must be encapsulated. PCM can be encapsulated in a metallic shell (core–shell encapsulation) or impregnated in a porous media like lightweight aggregates (shape stabilization) (Anupam et al. 2020).

The most desirable property of a PCM for pavement applications is a high latent heat capacity with a narrow range of phase change temperature. According to studies, the latent heat should be as higher as possible, and the phase change temperature should be selected carefully based on climatic conditions. Because of their excellent melting point and relatively low cost, paraffin compounds are commonly recommended for use in pavements (Regin et al. 2008). Table 11.1 summarizes the findings of key studies on PCM incorporated pavements.

11.2.3.6 Limitations of PCM Incorporated Cool Pavements

Although the PCM is encapsulated, high traffic loads may cause the encapsulation to fail. The desired properties of the pavements will be affected by the PCM leakage (Ryms et al. 2015). Furthermore, the PCM and the encapsulating material must withstand the higher temperatures during asphalt mixing, placement, and compaction. Moreover, most of the organic PCMs have poor thermal conductivity, and hence limits the effective heat transfer and adversely affects PCM melting (Anupam et al. 2020).

11.2.3.7 Proposed Modifications in PCM Incorporated Cool Pavements

Suitable additives to improve the thermal conductivity of the pavement such as high conductivity fines, and the use of high conductive nanoparticles and nanofibers containing PCM, can be a useful method for further improving the cooling potential of PCM incorporated pavements (Sebti et al. 2013).

11.2.4 Future Research Prospects

Given the complicated working circumstances for the investigation of the durability and design, a complete durability assessment technique for heat reflective coatings is important. Moreover, it is necessary to create durable retro-reflective coatings with improved skid resistance. The adherence of the pavement surface to the coatings requires a great deal of care and research.

Further, the latent heat capacity of porous pavements can be enhanced by lowering the aggregate size, resulting in smaller pores. A smaller pore holds moisture for longer periods of time, increasing the latent heat capacity. Further, no research has been done on the influence of drainpipe features on the thermal performance of drainable water retaining pavements. A smaller number of drain pipes could be utilized, the height of the drain pipes could be raised, a thicker pavement could be built, and the porosity of the concrete could be enhanced to increase the water-retaining capacity. This feature, as well as the improvement of thermal performance and runoff reduction, should be the focus in the future studies in this area.

Similar to the importance of choosing the proper PCM, it is important to choose the right encapsulation method for pavement applications. The mechanical properties of both rigid and flexible pavements will be harmed if PCM leaks. To date, no ideal encapsulation technique with zero PCM leakage has been developed that can be used in pavement applications. Methodologies for incorporating PCMs into asphalt and concrete mixes should be devised in such a way that the material integrity is preserved. Further, energy harvesting pavements and PCM incorporated pavements have a relatively high initial investment and need complicated construction techniques. To justify the large initial expenditure, researchers should work on enhancing the efficiency of these technologies.

To compare cool pavement technologies with traditional pavements, a life cycle study of these technologies is necessary. To calculate the total cost–benefit ratio, the thermal advantages should be translated to monetary profits. The development of techniques to quantify the effects of pavements in UHI is required to analyze the impact of cool pavements on UHI mitigation. In addition, detailed construction techniques for these cool pavements must be established to expedite their implementation.

11.3 Pervious Concrete

Pervious concrete is a class of sustainable concrete, which is characterized by the interconnected pore structure with porosity in the range of 15–35%. The materials used in pervious concrete are similar to that used in conventional concrete with only the gradation consisting of limited to no-fine aggregate and hence it is sometimes referred to as no fines concrete. In addition to this, there are several synonyms for pervious concrete such as eco-concrete, permeable concrete, limited fines concrete, etc.

A typical pore structure of pervious concrete is shown in 2D and 3D images in Fig. 9. The pores in the pervious concrete may be either interconnected or isolated. The interconnected pores assist in percolation of water while the isolated will just store the water. Depending on the interconnectivity, the porosity in pervious concrete is classified as total porosity and effective porosity. The total porosity takes into account both the interconnected porosity and isolated porosity, while the effective porosity takes only interconnected porosity into consideration. As seen in the 3D image of Fig. 11.8, the blue color pores represent the isolated pores, while the green-colored pores represent interconnected pores. The total and effective porosity can be determined using different methods such as ASTM C1754-12 and X-ray tomography (XRT) (ASTM C1754-12, Cosic et al. 2015; Chandrappa and Biligiri 2017; Tan et al. 2020).

The interconnectivity of pores allows the percolation of stormwater reducing the runoff quantity and recharging the groundwater (Chandrappa and Biligiri 2016b). Further, the moisture present inside the pores increases the latent heat capacity of the pervious concrete (Li et al. 2014; Meena et al. 2017; Wang et al. 2018b). Due to its porous structure, with an ability to reduce runoff and UHI effects, it is considered as

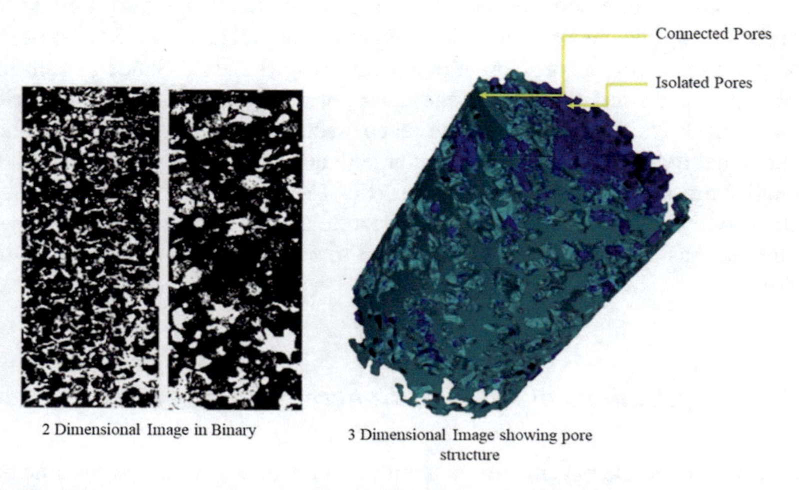

Fig. 11.8 Pore structure of pervious concrete (Chandrappa and Biligiri 2018c)

one of the LID methods by Environmental Protection Agency (EPA), USA and the sponge city initiative by Chinese government (Hu et al. 2018).

11.3.1 Materials and Properties of Pervious Concrete

The materials used in the pervious concrete are similar to that used in the conventional concrete with only difference being the type of gradation adopted. In order to achieve interconnected pore structure, the aggregate gradation for pervious concrete is mainly composed of single-sized coarse aggregates in the size range of 4.75 to 19.0 mm or blend of coarse aggregates at different proportions with no or limited fine aggregates (Deo and Neithalath 2011; Chandrappa and Biligiri 2016a). However, a few studies have employed fine aggregate up to certain extent and also single-sized fine aggregates to enhance the strength characteristics (Bonicelli et al. 2015; Zhong and Wille 2016; Maguesvari and Sundararajan 2017). The aggregate can be bound by using cement and supplementary cementitious materials (SCMs) such as silica fume, fly ash, ground granulated blast furnace slag, metakaolin, etc., which will not only improve the performance but also addresses the environmental problem. However, careful assessment of the cement paste with SCMs should be performed before producing pervious concrete. Higher dosages of SCM will increase the harshness of the mix leading to lack of workability and low density. (Saboo et al. 2019; Adil et al. 2020).

The properties of pervious concrete significantly differ from that of a conventional concrete due to higher porosity. The properties most commonly studied in pervious concrete include porosity, density, permeability, compressive strength, splitting tensile strength, and flexural strength. The permeability is one of the important functional requirements of pervious concrete, which varies over a broad range depending on aggregate size, gradation, cement-to-aggregate ratio, water-to-cement ratio, and compaction effort (Chandrappa and Biligiri 2016a; Sahdeo et al. 2021). The compressive strength of first-generation pervious concrete generally varies in the range of 3–25 MPa. However, the recent second-generation pervious concrete mixtures that focus on high strength are able to achieve a strength of up to 80 MPa by modifying the cement paste characteristics (Yang and Jiang 2003; Shen et al. 2020). The flexural strength of pervious concrete is determined using beams, where the strength has been found to vary from 0.8 to 5.0 MPa (Chandrappa and Biligiri 2018b).

11.3.2 Applications of Pervious Concrete in Pavements

The pervious concrete pavements are mainly designed to serve in low volume traffic such as parking lots, walkways, cycle tracks, pavement shoulders, etc. where the vehicle loads are significantly lower compared to highway traffic. The parking lots are considered to be ideal as they consist of large paved area, which can generate a considerable quantity of runoff if it is impervious.

11.3.3 Mix Design of Pervious Concrete

The mix design of pervious concrete can be performed using different principles as there are no universally accepted methods. Some of the most commonly used mix design methods include ACI 522 R-10 and national ready mix concrete association (NRMCA) method (Obla 2010). The mixture variables include aggregate size, aggregate gradation, aggregate-to-cement ratio, and water-to-cement ratio. The aggregate-to-cement ratio plays a very critical role in pervious concrete and generally a ratio from 5:1 to 3:1 has shown a balance between strength and permeability. The water-to-cement ratio is less significant than aggregate-to-cement and can be varied from 0.25 to 0.33. A higher w/c ratio will lead to choking of the paste at the bottom rendering pervious concrete to become impervious.

According to NRMCA approach, the dry aggregate air voids (ASTM C29) is occupied by design air voids and paste volume. The paste volume is mainly composed of cementitious materials, water and admixture. The paste volume can be determined using Eq. 11.1.

$$\text{Paste volume} = \text{Dry aggregate air voids} + \text{compaction index} - \text{design air voids} \tag{11.1}$$

The compaction index (CI) is a parameter to account for compaction effort. In general, higher compaction effort indicates lower CI and vice versa. This is because the lower compaction effort requires more paste volume to achieve similar density corresponding to high compaction effort. The CI is generally adopted to be 4–5% for standard proctor hammer compaction (ASTM C1688-14), which is commonly used to fabricate pervious concrete cylinders. A typical mix design of pervious concrete is shown below:

Assuming,

- Specific gravity of cement = 3.150; Specific gravity of aggregate = 2.750; Water-to-cement (w/c) = 0.30; Dry aggregate air voids = 38%; Design air voids = 20%; CI = 4%

The proportions are determined as follows.

- Paste volume = 38 + 4 − 20 = 22%
- As the paste volume is composed of cement volume and water volume:
- Paste volume = Cement volume + water volume
- Paste volume = Cement content/(Specific gravity of cement * 1000) + water content / (specific gravity of water * 1000)
- Cement content = 1000 * [Paste volume/(0.315 + 0.30)]

- Cement content = 357.72 kg/m^3
- Water content = cement content * w/c = 107.31 kg/m^3
- Total volume = aggregate volume + paste volume + design air voids
- Aggregate volume = 1 − 0.22 − 0.20 = 0.58
- Aggregate proportion = 0.58 * 2.75 * 1000 = 1595 kg/m^3
- Aggregate to cement ratio (a/c) = 4.45

In addition to this methodology, there are several other studies, which have proposed design equations considering the effect of gradation, size, w/c and a/c ratio (Chandrappa and Biligiri, 2018d).

11.3.4 Pervious Concrete Pavement Design

The pervious concrete pavement is mainly provided to function as low-impact development strategy to reduce runoff, recharge groundwater, and mitigate UHI effects. The typical pervious concrete pavement cross-section is shown in Fig. 11.9. The top layer is made up of pervious concrete slab, which is resting on a reservoir layer. The reservoir layer mainly consists of single-sized coarse aggregates of size above 25 mm, which functions as a temporary storage for the stormwater before slowly infiltration into the subgrade. The subgrade soil plays a very important role in controlling the exfiltration of the stormwater from the pavement system. The national ready mix concrete association (NRMCA) recommends that the infiltration rate of subgrade soil should be greater than 12 mm/hr.

The design of pervious concrete pavements consists of two parts: *Structural design and Hydrological design.* The structural design is not evolved owing to low traffic conditions and hence pervious concrete slab thickness is generally adopted to be 150–200 mm. However, the hydrological design plays an important part as the stormwater should percolate and stored temporarily before it completely drains into subgrade. Therefore, the thickness of the reservoir layer should be sufficient for this purpose, which depends on the local rainfall conditions and subgrade infiltration. The thickness of the reservoir layer is determined as a function of detention time and subgrade infiltration rate. The detention time basically indicates the time duration for the

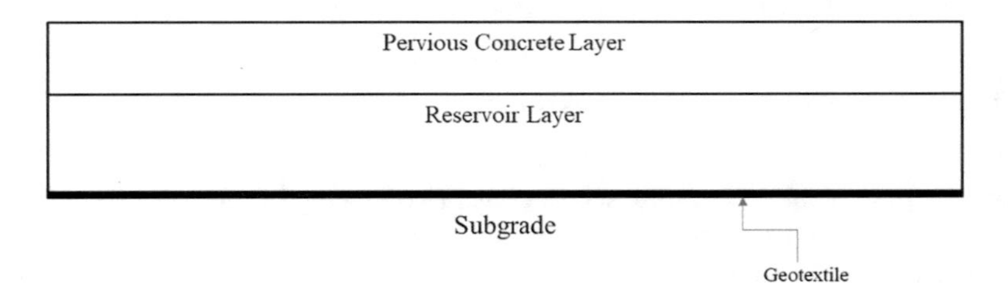

Fig. 11.9 Cross-Section of Pervious Concrete Pavement

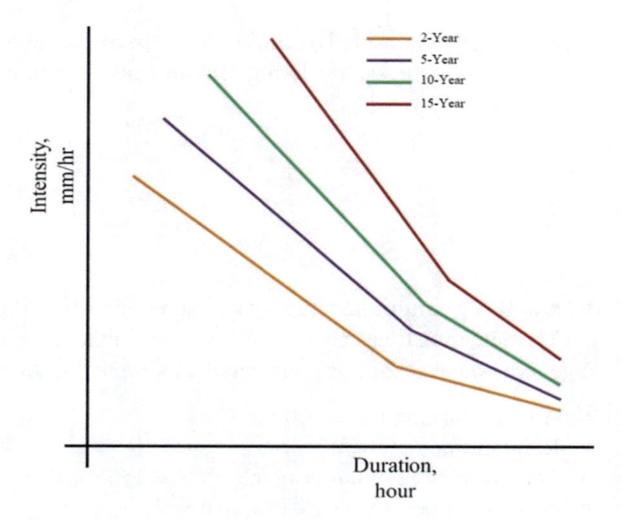

Fig. 11.10 Schematic of IDF Curve

reservoir layer to regain the initial porosity after the rainfall event. In order to design the thickness of the reservoir layer, the intensity–duration–frequency (IDF) curves corresponding to rainfall of different return periods for a given location should be available to select the design storm. A typical schematic of IDF curve is shown in Fig. 11.10. Generally, a 2-year return period rainfall is considered for the design of reservoir layer to make the design more economical.

The hydrological design procedure begins with first determining the runoff quantity (Q), which can be calculated using Eq. 11.2.

$$Q = A * I \tag{11.2}$$

where, Q = runoff quantity, m^3; A = area of catchment, m^2; I = rainfall intensity, mm.

The area of catchment can be the area of parking lot or the surface area of pavement exposed to rainfall. The intensity can be determined from IDF curves for a given return period and duration. The maximum storage capacity (S) of the pervious concrete pavement system can be determined using Eq. 11.3.

$$S = A[P * H + r * h] \tag{11.3}$$

where, A = surface area of pervious concrete pavement system, m^2; P = porosity of pervious concrete slab; H = thickness of pervious concrete layer, m; r = porosity of reservoir layer; h = thickness of reservoir layer, m.

The thickness of reservoir layer can be determined by rearranging Eq. 11.2 as shown in Eq. 11.4. In this Equation, the porosity of pervious concrete slab is not considered to make the design more conservative. Further as the thickness of reservoir layer is largely controlled by infiltration capacity of subgrade, Eq. 11.5 can be used as a safety design check considering detention time and infiltration capacity of subgrade.

$$h = \frac{S}{A * r} \tag{11.4}$$

$$h = \frac{IFC * t}{r} \tag{11.5}$$

where, IFC = infiltration capacity of subgrade soil, m/hr; t = detention time, h.

The detention time can be considered as per the requirement of the designer, where a general value of 24 h is considered. A simple design example is shown below:

- Area of parking lot = 200 m^2
- Rainfall intensity = 60 mm
- Thickness of pervious concrete slab = 150 mm
- Porosity of pervious concrete slab = 20%
- Porosity of reservoir layer = 35%
- Infiltration capacity of subgrade = 16 mm / hr
- Q = 200 * 60 / 1000 = 6 m^3

The pervious concrete pavement system should in-filter 12 m^3 of stormwater with a detention period of 4 h. Using Eq. 11.4, thickness of reservoir layer is calculated to be 171 mm. The obtained design thickness is used to calculate the detention time for a given infiltration capacity of subgrade, which is calculated to be 3.74 h. Since the calculated detention time is less than assumed detention time of 4 h, the thickness of reservoir layer is found to be sufficient for the above-given conditions.

11.3.5 Limitations and Way Forward

Albeit pervious concrete pavement is gaining importance in the domain of LID strategy, the long-term performance data for these pavements are still very limited. Further, due to its porous structure, pervious concrete tends to undergo clogging thus resulting in loss of permeability and hence the functional performance. However, the clogging in pervious concrete is mostly confined to the top 25 mm of the layer as the clogging particles cannot easily get carried deeper due to tortuous structure of the pores (Kayhanian et al. 2012). Further, routine maintenance such as sweeping and periodic maintenance such as vacuum/pressure washing can be used to restore the permeability (Chopra et al. 2010). On the other hand, the open texture of pervious concrete tends to undergo raveling or loss of aggregate over time due to movement of vehicles. This has been found significant at the entry and exit points of the parking lots due to acceleration and deceleration movement. In addition, the implementation of pervious concrete pavements is very limited in emerging countries due to lack of expertise and construction records unlike developed countries where separate pavers

Fig. 11.11 Two-layered pervious concrete

are manufactured to pave pervious concrete pavements (Chandrappa et al. 2018a). In order to improve the performance and implementation of pervious concrete, several studies have come up with different alternatives of pervious concrete such as interlocking pervious concrete paver block, porous concrete with vertical pores, etc. Further, the authors have come up with an innovation of developing two-layered pervious concrete as shown in Fig. 11.11, which is first-of-its-kind.

The two-layered pervious concrete consists of a 50 mm top layer made up of aggregate of size between 6.75 and 4.75 mm, while the bottom layer is 150 mm made of aggregate size between 13.2 and 9.5 mm. The aggregate sizes are chosen such that the interface between two layers should not get choked. The finer top layer results in reduced clogging, reduce abrasion and easy cleaning, while the bottom layer stays free from clogging and raveling during its service life.

11.4 Summary

The low-impact development (LID) strategies are need of the hour in urban setups considering the increase in weather-related calamities. The increase in temperature, runoff quantity, and increased flash flood frequency is already having serious consequences on quality of life of urban dwellers. The weather-related calamities will not only result in frequent loss of properties but also cause long-term health impacts, which will affect the day-to-day activities of the urban dwellers. However, these LID strategies may provide a balanced approach towards development and environment. These strategies being slowly practiced in several developed countries will also be beneficial for the developing countries, which should implement these strategies before facing the consequences. In this chapter, two different LID strategies that can be implemented in urban setups have been discussed comprehensively. The aspects related to design, performance, merits and demerits, and future research scope related to cool pavements and pervious concrete have been provided for the benefit of the readers.

References

Adil G, Kevern JT, Mann D (2020) Influence of silica fume on mechanical and durability of pervious concrete. Constr Build Mater 247. https://doi.org/10.1016/j.conbuildmat.2020.118453

Akbari H, Pomerantz M, Taha H (2001) Cool surfaces and shade trees to reduce energy use and improve air quality in urban areas. Sol Energy 70(3):295–310

Alvarez AE, Martin AE, Estakhri C (2011) A review of mix design and evaluation research for permeable friction course mixtures. Constr Build Mater 25(3):1159–1166

An CL, Billa L, Azari M (2018) Anthropocene climate and landscape change that increases flood disasters. Int J Hydrol 2(2)

Anupam BR, Anjali Balan L, Sharma S (2021a) Thermal and mechanical performance of cement concrete pavements containing PVC-glass mix. Road Mater Pavement Des 290:123238. Elsevier Ltd. https://doi.org/10.1080/14680629.2020.1868328

Anupam BR, Sahoo UC, Rath P (2020) Phase change materials for pavement applications: a review. Constr Build Mater 247:118553. Elsevier Ltd. https://doi.org/10.1016/j.conbuildmat.2020.118553

Anupam BR, Sahoo UC, Rath P (2021c) Thermal and mechanical performance of phase change material incorporated concrete pavements. Road Mater Pavement Des 1–18. Taylor & Francis. https://doi.org/10.1080/14680629.2021.1884590

Anupam BR, Sahoo UC, Chandrappa AK, Rath P (2021b) Emerging technologies in cool pavements: a review. Constr Build Mater 299:123892. Elsevier Ltd. https://doi.org/10.1016/j.conbuildmat.2021.123892

ASTM C1688/C1688M-14a (2014) Standard test method for density and void content of freshly mixed pervious concrete. ASTM International, West Conshohocken, PA. www.astm.org, https://doi.org/10.1520/C1688_C1688M-14A

Balan LA, Anupam BR, Sharma S (2021) Thermal and mechanical performance of cool concrete pavements containing waste glass. Constr Build Mater 290:123238. https://doi.org/10.1016/j.conbuildmat.2021.123238

Balbay A, Esen M (2010) Experimental investigation of using ground source heat pump system for snow melting on pavements and bridge decks. Sci Res Essays 5(24):3955–3966

Bao T, Liu Z (Leo), Zhang X, He Y (2019) A drainable water-retaining paver block for runoff reduction and evaporation cooling. J Clean Prod, 228:418–424. Elsevier Ltd.

Bhattacharjee S, Batra AK, Cain J (2011) Energy harvesting from pavements using pyroelectric single crystal and nano-composite based smart materials. In: T and DI congress 2011: integrated transportation and development for a better tomorrow—proceedings of the 1st congress of the transportation and development institute of ASCE, pp 741–750

Bobes-Jesus V, Pascual-Muñoz P, Castro-Fresno D, Rodriguez-Hernandez J (2013) Asphalt solar collectors: a literature review. Appl Energy 102:962–970. Elsevier Ltd.

Bonicello A, Giustozzi F, Crispino M (2015) Experimental study on the effects of fine sand addition on differentially compacted pervious concrete. Constr Build Mater 91:102–110. https://doi.org/10.1016/j.conbuildmat.2015.05.012

Chandrappa AK, Maurya R, Biligiri KP, Rao J, Nath S(2018a) Laboratory Investigations and field implementation of pervious concrete paving mixtures. Adv Civil Eng Mater 7:447–462. https://doi.org/10.1520/ACEM20180039

Chandrappa AK, Biligiri KP (2018b) Investigation on flexural strength and stiffness of pervious concrete for pavement applications. Adv Civil Eng Mater 7(2):223–242. https://doi.org/10.1520/ACEM20170015

Chandrappa AK, Biligiri KP (2018c) Pore structure characterization of pervious concrete using X-Ray microcomputed tomography. J Mater Civil Eng ASCE 30. https://doi.org/10.1061/(ASCE)MT.1943-5533.0002285

Chandrappa AK, Biligiri KP (2015) Development of pavement-surface temperature predictive models: parametric approach. J Mater Civil Eng ASCE 28. https://doi.org/10.1061/(ASCE)MT.1943-5533.0001415

Chandrappa AK, Biligiri KP (2016a) Comprehensive investigation of permeability characteristics of pervious concrete: a hydrodynamic approach. Constr Build Mater 123:627–637. https://doi.org/10.1016/j.conbuildmat.2016.07.035

Chandrappa AK, Biligiri KP (2016b) Pervious concrete as a sustainable pavement material–research findings and future prospects: a state-of-the-art review. Constr Build Mater 111:262–274. https://doi.org/10.1016/j.conbuildmat.2016.02.054

Chandrappa AK, Biligiri KP (2017) Relationships between structural, functional, and X-ray micro-computed tomography parameters of pervious concrete for pavement applications 2629. https://doi.org/10.3141/2629-08

Chandrappa AK, Biligiri KP (2018d) Methodology to develop pervious concrete mixtures for target properties emphasizing the selection of mixture variables. J Transp Eng Part B: Pavements 144. https://doi.org/10.1061/JPEODX.0000061

Cheela VRS, John M, Biswas W, Sarker P (2021) Combating urban heat island effect—a review of reflective pavements and tree shading strategies. Build MDPI 11. https://doi.org/10.3390/buildings11030093

Chen M, Xu G, Wu S, Zheng S (2010) High-temperature hazards and prevention measurements for asphalt pavement. In: 2010 international conference on mechanic automation and control engineering, MACE2010. IEEE, pp. 1341–1344

Chen Z, Zhang H, Duan H, Wu C, Zhang S (2021) Long-term photo oxidation aging investigation of temperature-regulating bitumen based on thermochromic principle. Fuel 286:119403

Chiarelli A, Dawson AR, García A (2015) Parametric analysis of energy harvesting pavements operated by air convection. Appl Energy 154:951–958. Elsevier Ltd.

Chiarelli A, Dawson AR, García A (2017) Pavement temperature mitigation by the means of geothermally and solar heated air. Geothermics, CNR-Istituto Di Geoscienze e Georisorse 68:9–19

Chopra M, Kakuturu S, Ballock C, Spence J (2010) Effect of rejuvenation methods on the infiltration rates of pervious concrete pavements. J Hydrol Eng 15:426–433. https://doi.org/10.1061/(ASCE)HE.1943-5584.0000117

Cosic K, Korat L, Ducman V, Netinger I (2015) Influence of aggregate type and size on properties of pervious concrete. Constr Build Mater 78:69–76. https://doi.org/10.1016/j.conbuildmat.2014.12.073

Dawson A, Mallick R, Garc A, Dehdezi PK (2014) Climate change, energy, sustainability and pavements

Dehdezi PK (2012) Enhancing pavements for thermal applications 97–100

Dehdezi PK, Hall MR, Dawson AR, Casey SP (2013) Thermal, mechanical and microstructural analysis of concrete containing microencapsulated phase change materials. Int J Pavement Eng 14(5):449–462

Deo O, Neithalath N (2011) Compressive response of pervious concretes proportioned for desired porosities. Constr Build Mater 25:4181–4189. https://doi.org/10.1016/j.conbuildmat.2011.04.055

Dietz ME, Clausen JC (2008) Stormwater runoff and export changes with development in a traditional and low impact subdivision. J Environ Manage 87:560–566

Efthymiou C, Santamouris M, Kolokotsa D, Koras A. (2016) Development and testing of photovoltaic pavement for heat island mitigation. Solar Energy 130:148–160. Elsevier Ltd.

Erell E, Pearlmutter D, Boneh D (2012) Effect of high-albedo materials on pedestrian thermal comfort in urban canyons. In: ICUC8—8th international conference on urban climates 8–11

Gavin GJ, E., P. P., E., K. K., and S., G. J. (2007) Impact of pavement thermophysical properties on surface temperatures. J Mater Civ Eng, Am Soc Civ Eng 19(8):683–690

Guntor NAA, Fadhil M, Ponraj M, Iwao K (2014) Thermal performance of developed coating material as cool pavement material for tropical regions. J Mater Civ Eng 26(4):755–760

Hancock PA, Vasmatzidis I (2003) Effects of heat stress on cognitive performance: the current state of knowledge. Int J Hyperthermia 19(3):355–372. Taylor & Francis

Haselbach L, Boyer M, Kevern JT, Schaefer VR (2011) Cyclic heat island impacts on traditional versus pervious concrete pavement systems. Transp Res Rec 2240(1):107–115

Heaviside C, Macintyre H, Vardoulakis S (2017) The urban heat island: implications for health in a changing environment. Curr Health Environ Rep 4:296–305. https://doi.org/10.1007/s40572-017-0150-3

Hsu CT, Huang GY, Chu HS, Yu B, Yao DJ (2011) Experiments and simulations on low-temperature waste heat harvesting system by thermoelectric power generators. Appl Energy 88(4):1291–1297. Elsevier Ltd.

Hu M, Zhang X, Siu YL, Li Y, Tanaka K, Yang H, Xu Y (2018) Flood mitigation by permeable pavements in Chinese Sponge city construction. Water 10. https://doi.org/10.3390/w10020172

Jiao W, Sha A, Liu Z, Jiang W, Hu L (2020) Utilization of steel slags to produce thermal conductive asphalt concretes for snow melting pavements. J Clean Prod 261:121197. Elsevier Ltd.

Jin J, Lin F, Liu R, Xiao T, Zheng J, Qian G, Liu H, Wen P (2017) Preparation and thermal properties of mineral-supported polyethylene glycol as form-stable composite phase change materials (CPCMs) used in asphalt pavements. Sci Rep 7(1):1–10. Springer US

Kayhanian M, Anderson D, Harvey JT, Jones D, Muhunthan B (2012) Permeability measurement and scan imaging to assess clogging of pervious concrete pavements in parking lots. J Environ Manag 95:114–123. https://doi.org/10.1016/j.jenvman.2011.09.021

Li C, Liu M, Hu Y, Shi T, Zong M, Walter MT (2018) Assessing the impact of urbanization on direct runoff using improved composite CN method in a large urban area. Int J Environ Res Public Health, MDPI 15. https://doi.org/10.3390/ijerph15040775

Li H, Harvey J, Ge Z (2014) Experimental investigation on evaporation rate for enhancing evaporative cooling effect of permeable pavement materials. Constr Build Mater 65:367–375. https://doi.org/10.1016/j.conbuildmat.2014.05.004

Ligtenberg J (2017) Runoff changes due to urbanization: a review. Umeå University

Lin TP, Ho YF, Huang YS (2007) Seasonal effect of pavement on outdoor thermal environments in subtropical Taiwan. Build Environ 42(12):4124–4131

Liu Y, Li T, Peng H (2018) A new structure of permeable pavement for mitigating urban heat island. Sci Total Environ 634:1119–1125. Elsevier B.V.

Liu Y, Li T, Yu L (2020) Urban heat island mitigation and hydrology performance of innovative permeable pavement: A pilot-scale study. J Clean Prod 244:118938. Elsevier Ltd.

Ma B, Si W, Ren J, Wang HN, Liu FW, Li J (2014) Exploration of road temperature-adjustment material in asphalt mixture. Road Mater Pavement Des 15(3):659–673

Ma Y, Zhang X, Zhu B, Wu K (2002) Research on reversible effects and mechanism between the energy-absorbing and energy-reflecting states of chameleon-type building coatings. Sol Energy 72(6):511–520

Maguesvari U, Sundararajan T (2017) Influence of fly ash and fine aggregates on the characteristics of pervious concrete. Int J Appl Eng Res 12(8):1598–1609

Meena K, Chandrappa AK, Biligiri KP (2017) comprehensive laboratory testing and evaluation of the evaporative cooling effect of pavement materials. J Test Eval 5:1650–1661. https://doi.org/10.1520/JTE20150462

Miller JD, Kim H, Kjeldsen TR, Packman J, Grebby S, Dearden R (2014) Assessing the impact of urbanization on storm runoff in a peri-urbancatchment using historical change in impervious cover. J Hydrol 515:59–70. https://doi.org/10.1016/j.jhydrol.2014.04.011

Nemirovsky EM, Welker AL, Lee R (2013) Quantifying evaporation from pervious concrete systems: methodology and hydrologic perspective. J Irrig Drain Eng 139(4):271–277

Obla KH (2010) Pervious concrete—an overview. Indian Concr J 9–18

Pan P, Wu S, Xiao Y, Liu G (2015) A review on hydronic asphalt pavement for energy harvesting and snow melting. Renew Sustain Energy Rev 48:624–634. Elsevier

Paul MJ, Meyer JL (2001) Streams in urban landscape. Annu Rev Ecol Syst 32:333–365. https://doi.org/10.1146/annurev.ecolsys.32.081501.114040

Qin Y, He Y, Hiller JE, Mei G (2018) A new water-retaining paver block for reducing runoff and cooling pavement. J Cleaner Prod 199:948–956. Elsevier B.V.

Refaa Z, Kakar MR, Stamatiou A, Worlitschek J, Partl MN, Bueno M (2018) Numerical study on the effect of phase change materials on heat transfer in asphalt concrete. Int J Therm Sci 133(March):140–150

Regin AF, Solanki SC, Saini JS (2008) Heat transfer characteristics of thermal energy storage system using PCM capsules: a review. Renew Sustain Energy Rev 12(9):2438–2458

Roseen RM, Ballestero TP, Houle JJ, Avellaneda P, Briggs J, Fowler G, Wildey R (2009) Seasonal performance variations for storm-water management systems in cold climate conditions. J Environ Eng 135(3):128–137

Rossi F, Castellani B, Presciutti A, Morini E, Anderini E, Filipponi M, Nicolini A (2016) Experimental evaluation of urban heat island mitigation potential of retro-reflective pavement in urban canyons. Energy Build 126:340–352. Elsevier B.V.

Rossi F, Castellani B, Presciutti A, Morini E, Filipponi M, Nicolini A, Santamouris M (2015a) Retroreflective façades for urban heat island mitigation: Experimental investigation and energy evaluations. Appl Energy 145:8–20. Elsevier Ltd.

Rossi F, Morini E, Castellani B, Nicolini A, Bonamente E, Anderini E, Cotana F (2015b) Beneficial effects of retroreflective materials in urban canyons: Results from seasonal monitoring campaign. J Phys: Conf Ser 655(1)

Rossi F, Pisello AL, Nicolini A, Filipponi M, Palombo M (2014) Analysis of retro-reflective surfaces for urban heat island mitigation: a new analytical model. Appl Energy 114:621–631. Elsevier Ltd.

Ryms M, Lewandowski WM, Radziemska EK, Denda H, Wcislo P (2015) The use of lightweight aggregate saturated with PCM as a temperature stabilizing material for road surfaces. Constr Build Mater 81:313–324

Sahdeo SK, Chandrappa AK, Biligiri KP (2021) Effect of compaction type and compaction efforts on structural and functional properties of pervious concrete. Transp Dev Econ 7. https://doi.org/10.1007/s40890-021-00129-0

Sahoo N, Shivhare S, Kori KK, Chandrappa AK (2019) Effect of fly ash and metakaolin on pervious concrete properties. Constr Build Mater 223:322–328. https://doi.org/10.1016/j.conbuildmat.2019.06.185

Santamouris M, Synnefa A, Karlessi T (2011) Using advanced cool materials in the urban built environment to mitigate heat islands and improve thermal comfort conditions. Solar Energy 85(12):3085–3102. Elsevier Ltd.

Sebti SS, Mastiani M, Mirzaei H, Dadvand A, Kashani S, Hosseini SA (2013) Numerical study of the melting of nano-enhanced phase change material in a square cavity. J Zhejiang Univ: Sci A 14(5):307–316

Shen P, Zheng H, Liu S, Lu J, Poon CS (2020) Development of high-strength pervious concrete incorporated with high percentages of waste glass. Cement Concr Compos 114. https://doi.org/10.1016/j.cemconcomp.2020.103790

Song J, Wang Z, Myint SW, Wang C (2017) The hysteresis effect on surface-air temperature relationship and its implications to urban planning: an examination in Phoenix, Arizona, USA. Landscape Urban Plan 167:198–211. Elsevier. https://doi.org/10.1016/j.landurbplan.2017.06.024

Svetlana D, Radovan D, Jan D (2015) The economic impact of floods and their importance in different regions of the world with emphasis on Europe. Procedia Econ Finance 34:649–655. Elsevier. https://doi.org/10.1016/S2212-5671(15)01681-0

Tan Y, Zhu Y, Xiao H (2020) Evaluation of the hydraulic, physical, and mechanical properties of pervious concrete using iron tailings as coarse aggregates. Appl Sci MDPI 10. https://doi.org/10.3390/app10082691

United States Environmental Protection Agency (US-EPA), Using Cool Pavements to Reduce Heat Islands. Using Cool Pavements to Reduce Heat Islands | US EPA

Vairagade V (2016) Quality, testing and engineering applications of pervious concrete -a state of art. IOSR J Mech Civil Eng (IOSR—JMCE) 13:84–90

Wang H, Jasim A, Chen X (2018) Energy harvesting technologies in roadway and bridge for different applications—a comprehensive review. Appl Energy 212:1083–1094. Elsevier

Wang J, Hu C, Ma B, Mu X (2020) Rapid Urbanization Impact on the Hydrological Processes in Zhengzhou, China. Water, MDPI 12. https://www.mdpi.com/2073-4441/12/7/1870

Wang J, Meng Q, Tan K, Zhang L, Zhang Y (2018) Experimental investigation on the influence of evaporative cooling of permeable pavements on outdoor thermal environment. Build Environ Elsevier 140:184–193. https://doi.org/10.1016/j.buildenv.2018.05.033

Wei K, Li W, Li J, Wang Y, Zhang L (2017) Study on a design method for hybrid ground heat exchangers of ground-coupled heat pump system. Int J Refrig 76:394–405

Will G, Liv H (2012) Investigation into the structural performance of pervious concrete. J Transp Eng Am Soc Civil Eng 138(1):98–104

Wu S, Mo L, Shui Z, Chen Z (2005) Investigation of the conductivity of asphalt concrete containing conductive fillers. Carbon 43(7):1358–1363

Xie N, Li H, Zhang H, Zhang X, Jia M (2020) Effects of accelerated weathering on the optical characteristics of reflective coatings for cool pavement. Solar Energy Mater Solar Cells 215:110698

Xie N, Li H, Zhao W, Zhang C, Yang B, Zhang H, Zhang Y (2019) Optical and durability performance of near-infrared reflective coatings for cool pavement: laboratorial investigation. Build Environ 163:106334. Elsevier

Yang J, Jiang G (2003) Experimental study on properties of pervious concrete pavement materials. Cem Concr Res 33:381–386

Yuan J, Emura K, Sakai H, Farnham C, Lu S (2016) Optical analysis of glass bead retro-reflective materials for urban heat island mitigation. Solar Energy 132:203–213. Elsevier Ltd.

Zhang R, Jiang G, Liang J (2015) The Albedo of Pervious Cement Concrete Linearly Decreases with Porosity. In: Aggelis DG (eds) Advances in materials science and engineering. Hindawi Publishing Corporation, p 746592

Zhang Y, Murray AT, Turner BL (2017) Optimizing green space locations to reduce daytime and nighttime urban heat island effects in Phoenix, Arizona. Landsc Urban Plan 165:162–171

Zheng M, Han L, Wang F, Mi H, Li Y, He L (2015) Comparison and analysis on heat reflective coating for asphalt pavement based on cooling effect and anti-skid performance. Constr Build Mater 93:1197–1205. Elsevier Ltd.

Zhong R, Wille K (2016) Compression response of normal and high strength pervious concrete. Constr Build Mater 109:177–187. https://doi.org/10.1016/j.conbuildmat.2016.01.051

Chapter 12
Review on Biopolymer Stabilization—A Natural Alternative for Erosion Control

S. Anandha Kumar, G. Kannan, M. Vishweswaran, and Evangelin Ramani Sujatha

12.1 Introduction

Surface erosion of soil is one of the major issues to be tackled in the field of soil engineering. The primary external agents causing soil erosion are wind and water (Almajed et al. 2020; Tran et al. 2019). Wind erosion involves the process of removing topsoil from the surface of the earth and causes severe environmental issues like dust emissions in the bare lands, construction sites as well as agricultural lands in arid and semi-arid regions (Ham et al. 2018; Swain et al. 2018). The primary factors that govern the susceptibility of soil to wind erosion include the particle size, soil morphology, collapsibility of macro-pores, loose structural arrangement and the magnitude and velocity of the wind. The loose accumulations of soil like loess that are prone to wind erosion are also strongly collapsible. The collapsible nature of the soil can cause severe distress like foundation failure or the collapse of infrastructure facilities like tunnels that are constructed on such soil. The high affinity of these collapsible soils to water can cause geologic hazards like mudslides and landslides. The individual soil particle depending primarily on its size tends to get separated from the soil continuum when the velocity of wind reaches a certain threshold velocity and thereafter is lifted, transported and deposited by the wind in another area. It causes severe health risks like asthma and other lung diseases (Kavazanjian et al. 2009).

S. A. Kumar
Department of Civil Engineering, Aditya Engineering College (Autonomous), Surampalem 533437, Andhra Pradesh, India

G. Kannan · M. Vishweswaran
School of Civil Engineering, SASTRA Deemed University, Thanjavur 613401, Tamil Nadu, India

E. R. Sujatha (✉)
Centre for Advanced Research On Environment, School of Civil Engineering, SASTRA Deemed University, Thanjavur 613401, Tamil Nadu, India

K. R. Reddy et al. (eds.), *Advances in Sustainable Materials and Resilient Infrastructure*, Springer Transactions in Civil and Environmental Engineering,
https://doi.org/10.1007/978-981-16-9744-9_12

Various methods adopted in general to combat wind erosion are watering the soil frequently, providing soil covers and use of chemical stabilizers.

Erosion or scouring induced by moving water like a river tends to hamper the stability of the nearshore or off-shore structures. Precipitation of high intensity and subsequent runoff is also another cause of surface erosion. Less favorable geotechnical characteristics of soil like the low relative density, loose accumulation and less/absence of clay content lead not only to erosion along the waterfronts but also can cause major geological disasters like slope failures and foundation failures (Dejong et al. 2013; Wade et al. 2021). The ill-effects of erosion strongly underline the need for countermeasures to control surface erosion. Several methods are commonly used to mitigate erosion like reinforcing soil and stabilizing the soil with suitable chemical additives (Dejong et al. 2013). Bio-geotechnical stabilization like growing a green cover over the susceptible zone is also in practice. Of all the adopted techniques, stabilization of soil with additives is often adopted in many scenarios owing to the ease of application and advantageous economics. Conventionally stabilizers like lime, cement, fly ash, and reinforcements like geosynthetics, geomembranes and geotextiles were used for improving the resistance of the soil to erosion (Horpibulsuk et al. 2010; Khan et al. 2018; Mishra and Ravindra 2015; Saygili and Dayan 2019).

Nowadays, researchers focus on the biological methods for enhancing the engineering performance of the soil in line with the need for sustainable development. Biological materials like biochar and biopolymers have shown greater efficiency in improving the physical and engineering properties of the soil. Several studies demonstrate that biopolymers offer a promising option to stabilize soil and prevent erosion (Anandha Kumar and Sujatha 2021; Chang et al. 2016b; Choi et al. 2020). Biopolymers like xanthan gum, gellan gum, sodium alginate, guar gum, chitosan, pre-gelatinized starch and casein, etc. are commonly used to stabilize soil (Bitar 2020; Chang et al. 2016a, 2016b; Fatehi et al. 2019; Hataf and Jamali 2018; Kumar and Sujatha 2020). Biopolymers are natural, eco-friendly, economic and simple to extract (Anandha Kumar and Sujatha 2021; Chang et al. 2016b; Reddy et al. 2020; Sujatha et al. 2020).

Also, biopolymers are less susceptible to degradation and, therefore, can offer a longer service period (Dehghan et al. 2018). Table 12.1 presents an overview of biopolymers used for improving the geotechnical properties of the soil. The addition of biopolymers influences the particle size of the soil and changes the soil morphology through the mechanism of inter-particle bonding, aggregation and bio-clogging. Several studies show that the biopolymer addition enhances the durability of the soil (Kumar et al. 2021; Muguda et al. 2020) and improves the adhesion in the soil matrix. These traits indicate that biopolymers can mitigate soil erosion effectively. The soil properties that influence the erosion are the particle size, soil morphology, cohesion, strength of the soil before and after treatment. The results of these studies and their findings are summarized and discussed in detail in this study. Also, several experimental investigations are used to assess the effect of biopolymers in controlling the susceptibility to erosion. The efficacy of these laboratory investigations, their merits and limitations are also discussed. The transfer of this technology to the field in terms of implementation is discussed through available case studies

Table 12.1 Biopolymers used to improve the geotechnical properties

Biopolymer	Chemical formula	Charge	Source	Geotechnical application	References
Xanthan gum	$C_{36}H_{58}O_{29}P_2$	Anion	Xanthomonas Campestris (Bacteria)	Strengthens soil Reduces the hydraulic conductivity	Anandha Kumar and Sujatha (2021), Cabalar et al. (2017), Chang et al. (2018), Dehghan et al. (2018)
Guar gum	$C_{10}H_{14}N_5Na_2O_{12}P_3$	Neutral	Cyamopsis Tetragonoloba (Plant based)	Enhances the mechanical properties Decreases the compressibility Contaminant removal from the pollutant soil	Bouazza et al. (2009), Dehghan et al. (2018), Kumar et al. (2021), Reddy et al. (2020)
Gellan gum	$C_{24}H_{37}O_{20}$	Anion	Spingomonas elodea (Microbes)	Enhances the soil strength	Bitar (2020), Chang et al. (2016a), Chang and Cho (2018a)
Agar gum	$C_{14}H_{24}O_9$	Neutral	seaweed (algae)	Increases the shear strength Most susceptible to temperature	Smitha et al. (2019), Smitha and Sachan (2016)
Betaglucan	$C_{18}H_{32}O_{16}$	Anion	Saccharomyces Cerevisiae (the cell wall of yeast)	Provides high strength and less hydraulic conductivity	Anandha Kumar and Sujatha (2021), Chang and Cho (2014, 2012), Soldo et al. (2020)
Chitosan	$C_{18}H_{35}N_3O_{13}$	Cation	Crustacean shells	Reduces the hydraulic conductivity of sandy soils	Aguilar et al. (2016), Hataf et al. (2018), Khachatoorian et al. (2003)
Starch	$C_{27}H_{48}O_{20}$	Cation	Plant derived	Adhesives for drilling fluid	Khatami and Kelly (2013), Smitha and Rangaswamy (2019)

(continued)

Table 12.1 (continued)

Biopolymer	Chemical formula	Charge	Source	Geotechnical application	References
Sodium Alginate	$C_5H_7O_4COONa$	anion	Cell walls of marine brown algae	Enhances the strength Reduces the erosion rate Pavement application	Fatehi et al. (2019), Tran et al. (2017), Zhao et al. (2020)

to underline the advantage of using biopolymers to stabilize the soil susceptible to erosion.

12.2 Particle Size of Untreated and Biopolymer Admixed Soils

Particle size significantly impacts the erodible nature of the soil. The soil loss due to erosion is quantified by the Universal Soil Loss Equation (USLE) and the Revised Universal Soil Loss Equation (RUSLE). An important variable of this USLE and RUSLE is K-factor, which depends on soil texture and soil particle size (Zhao et al. 2018). Biopolymer amendment causes variation in the particle size of the soil. Further, from a geotechnical engineering perspective, the nature and behavior of the soil vary with the particle size. For instance, clayey soil with particle size less than 2 microns was predominantly cohesive, whereas fine sand with particle size ranging between 75 and 425 microns has friction between particles. Particle size analyzer (PSA) is conventionally used to study the variation in the particle size using the principle of diffraction of light by the particle under study. The PSA produces diffraction pattern rings, and the distance between the rings determines the size of the particles.

Larson et al. (2013) demonstrated the usage of 0.2 and 0.5% R. tropici bacteria-derived biopolymer as an alternative to petroleum-based synthetic polymers to control soil erosion for the US Department of Defense (Larson et al. 2013). Their objective was to establish an effective, easy and long-lasting methodology to maintain slope angle with long-term performance. Particle size analysis revealed that biopolymer amendment increased the quantity of soil with particle size greater than 300 microns by 22%. This increase is justified by the binder effect of biopolymer in treated soil and thus emphasized the capability of biopolymer in reducing sediment loss during surface runoff. The behavior of kaolinite and bentonite with xanthan gum was investigated and observed a better improvement in the strength and consolidation characteristics (Latifi et al. 2016a). Latifi et al. (2016a) conducted PSA tests on 1% xanthan gum treated bentonite, 1.5% xanthan gum treated kaolinite in addition to untreated soil after 7, 28 and 90 days of curing. A significant reduction in the

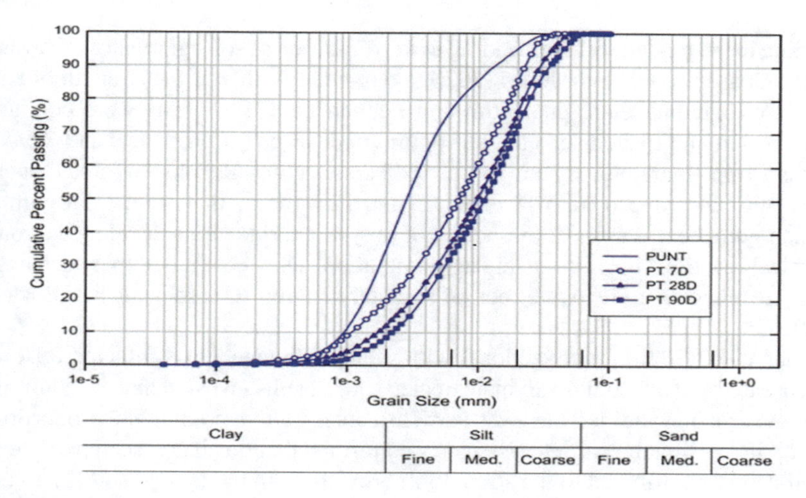

Fig. 12.1 Grain size distribution plots of xanthan gum treated and untreated peat (Latifi et al. 2016b)

proportion of clay-sized particles from 35 to 12% in treated bentonite after 28 days of curing was observed. The corresponding increase of 23% is reflected in the increase of silt-sized particles by 23%. After 90 days of curing, an marginal reduction of the clay-sized particle was observed by just 3.5%. Similarly, the clay proportion in kaolinite reduced from 31 to 4% after 28 days of curing. The loss of 27% is reflected in the increase of silt-sized particles. Grain size distribution revealed a general shift of grain size curves to the right due to larger clusters and cementitious bond formation. This bond and cluster formation stands as the reason for strength and stiffness improvement in the treated soil.

Latifi et al. (2016b) stabilized organic peat with xanthan gum biopolymer for dosages ranging between 0.5% and 2.5%. Figure 12.1 shows the grain size distribution plots of untreated and xanthan gum-treated peat. The inclusion of xanthan gum created agglomeration among particles, and the same has been interpreted with the results from the PSA test. Comparison of the grain size distribution curves of untreated and 2% xanthan gum-treated soil reveals that the additive caused a reduction in the clay-sized particles in the soil sample from 37 to 11% after 28 days of curing. Consequently, the silt-sized particles in the soil increased from 63 to 89%. Further increase in curing period from 28 to 90 days did not show much change in particle size of the treated soil. This reduction in particle size variation revealed that no appreciable reactions happened in the treated soil after 28 days of curing (Latifi et al. 2016b).

Viswanath (2019) conducted an extensive experimental analysis on xanthan and guar gum usage as an alternative to cement stabilizers in earthen construction materials. PSA results revealed that there was a general increase in the coarser fraction and reduction in a finer fraction, which could be due to particle aggregation. Xanthan gum being an anionic biopolymer created a strong ionic bonding in addition to hydrogen bonding in the treated soil thus leading to the formation of soil aggregates (Viswanath 2019). Ghasemzadeh and Modiri (2020) attempted to enhance the strength of kaolinite using xanthan gum, guar gum and Persian gum. PSA tests were conducted on untreated and treated soils at their optimum dosages (i.e., 1.5, 1, 2 and 2.5% for xanthan gum, guar gum, dry and wet mixed Persian gum). Results revealed that clay-sized particles reduced from 18 to 16, 15 and 12% for xanthan gum, guar gum and Persian gum-treated soils. Introduction of polysaccharide gums developed stronger chemical reactions in soil, causing agglomeration. They concluded that such aggregation resulted in a positive effect on soil stabilization (Ghasemzadeh and Modiri 2020).

Zhao et al. (2020) improved loess with sodium alginate and conducted PSA tests on untreated and 3% sodium alginate treated soil. Results revealed that the addition of sodium alginate solidified the loess particle, causing a reduction in the proportion of clay by increasing the particle size. They confirmed that this increase in particle size significantly contributed to the strength improvement in the treated soil (Zhao et al. 2020). Barani and Barfar (2021) conducted fracture studies on xanthan gum-treated clay. A small shift of grain size distribution curve towards the right is observed in 1.5% xanthan gum-treated clay. This shift implies a reduction in the proportion of clay-sized and silt-sized particles in the treated soil, confirming the particle aggregation that happened with biopolymer treatment (Barani and Barfar 2021).

Based on the survey of literature as mentioned above, it is evident that the inclusion of gel-based biopolymers leads to an increase in the grain size of treated soil by the action of soil aggregation. Cementitious bonding between the treated soil and the gum fills the pores, leading to increased particle size (Ghasemzadeh and Modiri, 2020; Latifi et al. 2016a, 2016b). This particle aggregation decreases the specific surface area of the soil after the biopolymer treatment. This increased particle size plays a major role in strength enhancement with time (Latifi et al. 2016b).

12.3 Soil Morphology

The interaction between the soil and the biopolymer greatly relies on the intrinsic chemical structure of the materials. Scanning electron microscopy (SEM) is a technique commonly used to interpret the reaction mechanism of materials at a microstructural level by producing magnified images of the specimen. SEM works in such a way that a thin beam of electrons incident on the specimen gets scattered back and is sensed by a detector to produce the final enlarged images of the microscopic structure. Chang et al. (2016a) attempted to enhance the strength of sand using gellan gum. SEM micrographs revealed that gellan gum coagulates to form films between sand particle pores, which later dehydrate, harden and densify to give improved strength and elastic modulus. Chang and Cho (2018a) studied the shear strength variation and behavior of gellan gum-treated sand, clay and clay-sand combinations (Chang and Cho 2018b). Investigation of SEM patterns revealed that no chemical interactions happened between gellan gum and the sand particles.

However, the clay particles formed hydrogen bonding with the gellan gum. Being anionic in nature, gellan gum monomers attract to the edges of kaolinite, leading to stacking of particles and showed a predominant strength improvement mechanism compared to sand-gellan gum interactions (Chang et al. 2016a).

Smitha and Sachan (2016) amended sand with 0.5–3.0% of agar biopolymer and investigated their shear strength characteristics. SEM images of all the investigated dosages after three days of curing were studied. Micrographs revealed that fewer aggregated particles were found on 0.5% agar treated soil and the aggregation effect increased with an increase in dosage of biopolymer. The agar gel being hydrophilic, absorbed water and formed a cementitious gel between the soil particles which filled the pores, increased the contact area between the particles and thereby improved the shear strength of the soil. No chemical reactions were observed between the sand and agar biopolymer despite the formation of gel (Smitha and Sachan 2016). Fatehi et al. (2018) analyzed the strength improvement on the sand with protein-based casein and sodium caseinate biopolymers obtained from milk. The polymer chains of casein adhere to the soil particles through van der Waal bonds and electrostatic interactions. Further, the sand grains were coated with sodium caseinate and increased their surface of contacts. Introduction of sodium in casein phosphate formed polar groups through hydrolysis of amino acid chains. Hence a stronger adhesion has been formed between the charged sodium caseinate and sand particles (Fatehi et al. 2018).

Fatehi et al. (2019) attempted to strengthen dune sand with sodium alginate biopolymer. The inclusion of sodium alginate reduced the pore space and improved the particle contact area. Availability of carboxylic charge density enhances the ionic interaction and formation of hydrogen bonds between sand and polymer chains (Fatehi et al. 2019). Further sodium alginate had other advantages like enhanced chain mobility and physical entanglement of chains. Similarly, Zhao et al. (2020) enhanced loess using sodium alginate and observed that sodium alginate gel reduced porosity, improved surface contact, formed a thin film around the particles and promoted aggregation in soil samples (Zhao et al. 2020). Ghasemzadeh and Modiri (2020) studied the effect of anionic xanthan gum and Persian gum, and non-ionic guar gum. Investigation of SEM micrographs from their studies reveals that all three biopolymers acted as cementitious agents that bridged the soil particles with the adhesive gel. But the interparticle interaction of all the three gums was different. Among the three, a dense and uniform microstructure has been observed in the Persian gum amended soil. The presence of divalent calcium ions acted as a medium of gel formation in Persian gum amended soil (Ghasemzadeh and Modiri 2020).

Anandha Kumar and Sujatha (2021) investigated the hydro-mechanical behavior of xanthan gum, beta-glucan and guar gum treated clayey sand and observed a reduction in permeability and improvement in the strength of the treated soil. A detailed investigation on the mechanism of strength improvement has been carried out through SEM images. Figure 12.2 represents the SEM images of 2% biopolymer treated samples at various curing periods. The images reveal that the anionic xanthan gum and beta-glucan exhibited better binding and aggregation through greater surface absorption. Also, it is evident from Fig. 12.2 that the gel plugging mechanism is common in the treated soil despite the ionic nature of the treated soil. At initial

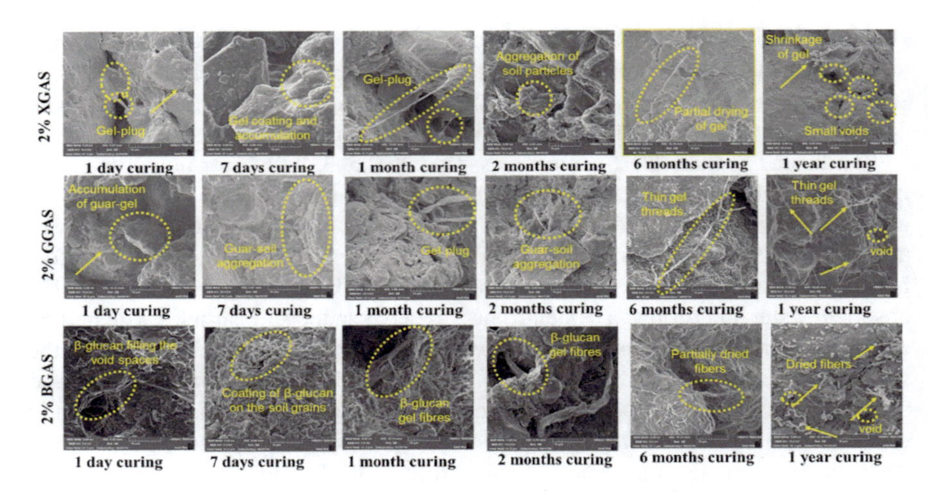

Fig. 12.2 SEM images of xanthan gum, guar gum and beta-glucan treated soil at various curing periods (Anandha Kumar and Sujatha 2021)

curing periods up to 120 days, swelling of hydrogels happened by imbibing the water. However, at a higher curing period of 1 year, shrinkage of gel plugs was observed owing to dehydration, thus creating voids in the soil matrix. This dehydration and void formation also enlighten the impact of aging on such biopolymer-treated soil (Anandha Kumar and Sujatha 2021).

The above-discussed results indicate that SEM images played a key role in identifying behavioral changes in biopolymer-treated soil. In case of sand treated with biopolymer, a thin gel film formed as a coating over the grain, leading to reduced porosity. This thin film in turn improves the surface contact between the particles. At one point, the aggregation of particles started taking place leading to enhanced shear strength properties in soil. When it comes to clayey soil treated with biopolymers like xanthan gum, sodium alginate, gellan gum etc., the formation of hydrogen bonding leads to enhanced strength. Similarly, casein formed van der Waal bond and electrostatic interactions with the soil. Further, hydrophilic biopolymers tend to absorb water leading to swelling of hydrogels, but with time, dehydration causes shrinkage in gel threads. This would also serve as the reason for the degradation in the behavior of biopolymer amended soil after a prolonged period.

12.4 Strength of the Biopolymer Treated Soils

Numerous studies have reported the phenomenal increase in strength of the biopolymer treated soils (Ayeldeen et al. 2016; Chang et al. 2015a; Gopika and Mohandas 2019; Lee et al. 2017). Caballero et al. (2016) demonstrated the ability of guar gum on slope stabilization of clay soils. The addition of 0.5% guar gum increased the shear strength parameters of the soil and the factor of safety of the slope improved by 300% (Caballero et al. 2016). Khatami and Kelly (2013) used a combination of agar and modified starch and enhanced the shear strength of sand (Khatami and Kelly 2013). Cohesion was imparted to the cohesionless sand resulting in maximum unconfined compressive strength of 487 kPa. Chen et al. (2016) accentuated the application of xanthan gum biopolymer on mine tailings and found that the cohesion, friction and strength of the mine tailing improved with 0.5% addition of biopolymer (Chen et al. 2016). Hataf et al. (2018) demonstrated the application of chitosan biopolymer in clay soil stabilization and improvement in mechanical properties was observed (Hataf et al. 2018). At optimum and moist conditions, the interparticle interaction in the chitosan-clay matrix facilitated improvement in the shear strength of the soil. Soldo et al. (2020) investigated the effect of five different biopolymers i.e., xanthan gum, beta 1,3/1,6 glucan, guar gum, chitosan, and alginate in strength improvement of silty sand by a series of unconfined compression test, triaxial test and direct shear test (Soldo et al. 2020). Xanthan gum, guar gum and β-1,3/1,6 glucan exhibited superior performance with higher strength gain as compared to the other two biopolymers. Lee et al. (2019) applied xanthan gum on soft soils and observed an improved shear strength and ductile resistance of the treated soil (Lee et al. 2019). Arab et al. (2019) revealed the capability of sodium alginate with clayey and silty soils. The addition of sodium alginate in concentrations of 2% for clayey and 4% for silty soils resulted in significant improvement in strength until 28 days (Arab et al. 2019).

12.5 Laboratory Assessment of Erosion

A pressure plate extractor was used to determine water retention of poorly graded sand (Tran et al. 2019). The stabilized soil was compacted into a ring of circumference 50 mm and height 15 mm before wetting. Volumetric moisture of the soil was determined after the application of pneumatic air pressure and the increase in volumetric moisture content at varying suction pressures was validated by field study. Natural coconut fiber mats were laid on a 45° slope to counteract problems associated with erosion. A solution comprising xanthan gum and starch was then sprayed accompanied by vegetative seeding. It was observed that the treated soil retained a higher quantity of water than the untreated soil which in turn enhanced the vegetative growth and water holding capacity of the soil. The application was monitored for 307 days. An increase in water retention was observed at the constant temperature for the treated than untreated soils, and this increase was validated by pressure plate extractor laboratory test (Tran et al. 2019).

The erosion resistance of Korean residual soil was analyzed through a series of heavy rain simulations (Chang et al. 2015b). The rainfall simulation consisted of 15 cycles each lasting 10 min. The results revealed that β-glucan treated soil underwent no erosion while the untreated soil experienced 60% erosion. Erosion of untreated soil accelerated after the fourth cycle. It was observed that the surface of the β-glucan treated soil did not allow the intrusion of sprayed water and thus the simulated precipitation flowed down with no loss of the residual soil. To assess the vegetative characteristics of the treated soil, a steel tray was used for compaction and placement of untreated and treated soils. An increase in vegetation for about 300%–400% in the β-glucan treated soil was attributed to enhanced stimulation of seed germination.

Orts et al. (2001) developed a mini-furrow for 6.3 mm × 6.3 mm by thrusting it longitudinally on the compacted gravelly clay loam soil. Discharge was set to 7 mL/min, and water and the quantity of sediments were evaluated by determination of turbidity. The laboratory mini-furrow test was conducted to determine the efficacy of chitosan against erosion during irrigation. It was noted that 20 ppm of chitosan biopolymer caused 8.7% sediment erosion as compared to 100% erosion by the untreated soil (Orts et al. 2001). Ayeldeen et al. (2018) used four different types of biopolymers, i.e., xanthan gum, guar gum, carrageenan and modified starches to analyze their effectiveness in reducing silty soil erosion. Loss in weight of the specimens was measured after they were subjected to wind simulation corresponding to field conditions. An electric fan motor was adopted to assign the velocity of air at 41.7 m/s, which would be in contact with the specimens from above at 25° inclination for 600 s. In addition, uniformly distributed sand was subjected to air circulation in order to simulate field conditions. The entire setup was assembled inside an air chamber, and the resistance against erosion was measured by weighing the treated soil before and after the test. Guar gum exhibited the least erosion and carrageenan recorded the highest erosion among the four biopolymers (Ayeldeen et al. 2018).

The efficacy of biopolymers against water erosion was documented by the researcher (Chang et al. 2015b). Precipitation was simulated by a sprinkling apparatus against the inorganic silty loam stabilized by β-glucan and xanthan gum. Severe precipitation for 5 s was applied by simulated rainfall intensity of 800 cm/h by placing the trays at 20° inclination and the fall of water at 30 cm from the top of the soil. The test was carried out nine times after every 2 days interval. Results revealed that the increase in rainfall intensity for increased duration led to increased erosion for the treated soil. The study emphasized the need for a detailed study on the mechanism of soil interaction with biopolymers for practical applications (Chang et al. 2015b).

A laboratory wind tunnel of 4500 × 600 × 600 mm was fabricated using a transparent polyvinyl chloride sheet for easy observation. At one end of the tunnel, a fan was installed while the other end consisted of air filters to collect the dust particles. Mine tailing samples were subjected to a wind velocity of 17.6 m/s for 600 s, and a change in mass of the samples was observed. The results revealed that increasing the concentration of xanthan gum and guar gum led to a substantial decrease in reduction in weight (Chang et al. 2015b). Also, a penetration test was performed using a penetrometer to assess the surface strength of the mine tailings. The biopolymer treated mine tailings exhibited superior surface strength as compared to untreated samples and as the penetration force increased under different penetration depths. The third test is the alternate wetting and drying test, was performed for the biopolymer-treated mine tailings, and the increase in water content indicates interparticle bonding in the treated mine tailings and thus reduction of dust generation

(Chang et al. 2015b). Even though the technical processes of wind erosion and water erosion are different, the engineering properties of the soil play an important role in resistance against erosion. The selection of the type of erosion test depends on the expected erosion of the particular site such as rainfall intensity, wind velocity, etc. (Chang et al. 2015b).

Joga and Varaprasad (2020) adopted the crumb test for the assessment of dispersive soils by visual observation after submersion. The test is outlined in ASTM D 6572 and revealed that xanthan gum-treated soil demonstrated reduced dispersivity at 1% concentration. A double hydrometer test was used to analyze the dispersive attributes of the fine-grained soil to become suspended. The obtained percentage of dispersion could be used to categorize the soil as dispersive, intermediate dispersive and non-dispersive. A pinhole erosion test was used to identify the extent of erosion by allowing water to pass through the soil samples at different piezometric heads. The results revealed that xanthan gum-treated soil enhanced the erosion resistance by transforming the dispersive soil to non-dispersive soil (Joga and Varaprasad 2020).

Erosion Function Apparatus (EFA) was used to observe soil erosion at different velocities of water flow (Kwon et al. 2019). Soil samples were loaded inside EFA and after the attainment of steady flow, the soil column was thrust above for it to project 1 mm into the water flow. P-waves were picked up by ultrasonic transducers and thus the eroded portion of soil could be determined. Soil treated by xanthan gum and guar gum exhibited higher resistance against erosion against the control specimens. A levee model scaled to 50 cm in length was kept on a steel plate before placing into a hydraulic flume testing channel of 60 × 90 × 900 cm dimensions (Lee et al. 2020). Provision to control flow rate was achieved by water gate, and an optical device was located to monitor the erosion of soil. When water overflowed the downscaled levee, detachment of surface in the upstream side was observed leading to rapid widening of hydraulic flow resulting in crumbling of the model. Xanthan gum-treated soil did not endure detachment of surface in the upstream portion. Even though toe erosion on the downstream portion was observed for the treated soil, the structure did not dissociate even after 4 h.

12.6 Field Implementation

Larson et al. (2016) used *Rhizobium tropici* biopolymer to retain the height of a protection berm, which resulted in enhanced slope stability, reduction in soil erosion and improvement in vegetation. Depending on the inclination of the slope, the area was categorized into different zones. Steeper slopes were provided with the deeper application of the *Rhizobium tropici* biopolymer whereas gentle slopes were stabilized by surface treatment of the biopolymer. A hydroseeder was employed for the implementation of biopolymer application with grass seeds. No maintenance activity was carried out for 3 years, and the treated slopes exhibited a 223% increase in plant vegetation and 11–28% reduction of surface roughness. The performance of deeper biopolymer application was better than the surficial application and the treated berm recorded a 33% reduction in cost compared to the traditional berm construction. Before field installation, long-term durability performance, suitable irrigation system, thermal resistance, UV resistance, etc. are to be considered and analyzed. One of the major factors in the selection of a biopolymer for practical purposes is the molecular weight of the biopolymer (Chang et al. 2015b). Appropriate methods in mixing, handling, installing the biopolymer mixture in various field conditions are to be ascertained.

The implementation could be done with aviation devices in remote, desert areas, spraying, hydroseeding, etc. (Chang et al. 2015b). The greater the molecular weight indicates a stronger the adsorption process. The functional group and length of the biopolymer chain influence the bonding of the biopolymer with the soil (Theng 1982). Scale effect plays a major role in the assessment of properties of a biopolymer with respect to counteracting shear forces. Laboratory mini-furrow tests yielded satisfactory results in the laboratory and below-par performance in field implementation. Formation of a stable soil—biopolymer network corresponding to field conditions may require enormous molecular size and striking a balance between binding and flocculation to reach the desired length of the furrow in field conditions (Orts et al. 2001). From the literature study, 0.5–2% biopolymer concentration is found to be optimum for effective stabilization. Polysaccharides such as xanthan gum, guar gum, β-glucan not only improve the strength characteristics but also increase water retention, vegetative growth, erosion resistance against various types of soils, problematic soils, etc. Suitability of appropriate erosion test is important for the evaluation of biopolymers' performance against erosion.

12.7 Summary

Biopolymers have the ability to change the soil morphology, structure and shear strength properties (i.e., cohesion and angle of internal friction) properties. These are the significant properties to enhance the erosional resistance to the problematic soils. The addition of biopolymers increases the particle size through particle aggregation and bonding. Also, the mass of the soil grains tends to increase. This change in size will improve the resistance of the soil against wind erosion particularly in case of loose soil. Researches on biopolymer enhanced soil also show that biopolymer addition enhances the soil aggregation property and its strength. The microstructural studies show the formation of new cementitious products that plug the soil particles together. The spraying of biopolymers to the target soil area at their optimum biopolymer content helps to improve the erosional resistance against the water. The hydrophilic nature of the biopolymers tends to improve the water retention behavior of the soil and stimulates vegetative growth. This, in turn, increases the resistance against the transportation of soil due to rainwater. A number of laboratory experiments were performed to show the efficacy of biopolymers against wind and water erosion. As biopolymers are organic materials, several chemical compatibility tests are to be done, to ensure their stability against various chemicals. Further, the long-term behavior of the biopolymer amended soil is still under discussion and investigation. Hence, the amendment of biopolymers will help in the resistance to soil erosion, and several field application studies are recommended to investigate the efficacy of the process under real-time construction projects.

References

Aguilar R, Nakamatsu J, Ramírez E, Elgegren M, Ayarza J, Kim S, Pando MA, Ortega-San-Martin L (2016) The potential use of chitosan as a biopolymer additive for enhanced mechanical properties and water resistance of earthen construction. Constr Build Mater 114:625–637. https://doi.org/10.1016/j.conbuildmat.2016.03.218

Almajed A, Lemboye K, Arab MG, Alnuaim A (2020) Mitigating wind erosion of sand using biopolymer-assisted EICP technique. Soils Found. https://doi.org/10.1016/j.sandf.2020.02.011

Anandha Kumar S, Sujatha ER (2021) An appraisal of the hydro-mechanical behaviour of polysaccharides, xanthan gum, guar gum and β-glucan amended soil. Carbohydr Polym 265:118083. https://doi.org/10.1016/j.carbpol.2021.118083

Arab MG, Mousa RA, Gabr AR, Azam AM, El-Badawy SM, Hassan AF (2019) Resilient behavior of sodium alginate-treated cohesive soils for pavement applications. J Mater Civ Eng 31:4018361. https://doi.org/10.1061/(asce)mt.1943-5533.0002565

Ayeldeen M, Negm A, El Sawwaf M, Gädda T (2018) Laboratory study of using biopolymer to reduce wind erosion. Int J Geotech Eng 12:228–240. https://doi.org/10.1080/19386362.2016.1264692

Ayeldeen MK, Negm AM, El Sawwaf MA (2016) Evaluating the physical characteristics of biopolymer/soil mixtures. Arab J Geosci 9:371. https://doi.org/10.1007/s12517-016-2366-1

Barani OR, Barfar P (2021) Effect of xanthan gum biopolymer on fracture properties of clay. J Mater Civ Eng 33:4020426. https://doi.org/10.1108/WJE-05-2020-0152

Bitar L (2020) Optimum mixing design of xanthan and gellan treated soils for slope stabilization for weathered shales and glacial tills in Nebraska. University of Nebraska

Bouazza A, Gates WP, Ranjith P (2009) Hydraulic conductivity of biopolymer-treated silty sand. Géotechnique 59:71–72. https://doi.org/10.1680/geot.2007.00137

Cabalar AF, Wiszniewski M, Skutnik Z (2017) Effects of Xanthan gum biopolymer on the permeability, odometer, unconfined compressive and triaxial shear behavior of a sand. Soil Mech Found Eng 54:356–361. https://doi.org/10.1007/s11204-017-9481-1

Caballero S, Acharya R, Banerjee A, Bheemasetti TV, Puppala A, Patil U (2016) Sustainable slope stabilization using biopolymer-reinforced soil. In: Geo-Chicago 2016 GSP 269, pp 458–466

Chang I, Cho G (2018) Shear strength behavior and parameters of microbial gellan gum-treated soils: from sand to clay. Acta Geotech 14:361–375. https://doi.org/10.1007/s11440-018-0641-x

Chang I, Cho G (2014) Geotechnical behavior of a beta-1,3/1,6-glucan biopolymer-treated residual soil. Geomech Eng 7:633–647. https://doi.org/10.12989/gae.2014.7.6.633

Chang I, Cho GC (2012) Strengthening of Korean residual soil with β-1,3/1,6-glucan biopolymer. Constr Build Mater 30:30–35. https://doi.org/10.1016/j.conbuildmat.2011.11.030

Chang I, Im J, Cho G (2018) Soil consistency and inter-particle characteristics of xanthan gum biopolymer containing soils with pore-fluid variation. Can Geotech J 56:1206–1213. https://doi.org/10.1139/cgj-2018-0254

Chang I, Im J, Cho G-C (2016) Geotechnical engineering behaviors of gellan gum biopolymer treated sand. Can Geotech J 53:1658–1670. https://doi.org/10.1139/cgj-2015-0475

Chang I, Im J, Cho GC (2016) Introduction of microbial biopolymers in soil treatment for future environmentally-friendly and sustainable geotechnical engineering. Sustainability 8:251. https://doi.org/10.3390/su8030251

Chang I, Im J, Prasidhi AK, Cho G (2015) Effects of Xanthan gum biopolymer on soil strengthening. Constr Build Mater 74:65–72. https://doi.org/10.1016/j.conbuildmat.2014.10.026

Chang I, Prasidhi AK, Im J, Shin HD, Cho GC (2015) Soil treatment using microbial biopolymers for anti-desertification purposes. Geoderma 253–254:39–47. https://doi.org/10.1016/j.geoderma.2015.04.006

Chen R, Ramey D, Weiland E, Lee I, Zhang L (2016) Experimental investigation on biopolymer strengthening of mine tailings. J Geotech Geoenviron Eng 142:6016017. https://doi.org/10.1061/(asce)gt.1943-5606.0001568

Choi SG, Chang I, Lee M, Lee JH, Han JT, Kwon TH (2020) Review on geotechnical engineering properties of sands treated by microbially induced calcium carbonate precipitation (MICP) and biopolymers. Constr Build Mater 246:118415. https://doi.org/10.1016/j.conbuildmat.2020.118415

Dehghan H, Tabarsa A, Latifi N, Bagheri Y (2018) Use of xanthan and guar gums in soil strengthening. Clean Technol Environ Policy 21:155–165. https://doi.org/10.1007/s10098-018-1625-0

Dejong JT, Soga K, Kavazanjian E et al (2013) Biogeochemical processes and geotechnical applications: Progress, opportunities and challenges. In: Bio- and Chemo- Mechanical Processes in Geotechnical Engineering-Geotechnical Sympoisum. Print 2013, pp 143–157. https://doi.org/10.1680/bcmpge.60531.014

Fatehi H, Bahmani M, Noorzad A (2019) Strengthening of dune sand with sodium alginate biopolymer. In: Geo-congress 2019: soil improvement, pp 157–166. https://doi.org/10.1061/9780784482117.015

Fatehi H, Mahdi S, Hashemolhosseini H, Mahdi S (2018) A novel study on using protein based biopolymers in soil strengthening. Constr Build Mater 167:813–821. https://doi.org/10.1016/j.conbuildmat.2018.02.028

Ghasemzadeh H, Modiri F (2020) Application of novel Persian gum hydrocolloid in soil stabilization. Carbohydr Polym 246:116639. https://doi.org/10.1016/j.carbpol.2020.116639

Gopika AS, Mohandas TV (2019) Soil Strengthening using caseinate: a Protein based biopolymer. Int J Res Eng Sci Manag 2:538–540

Ham SM, Chang I, Noh DH, Kwon TH, Muhunthan B (2018) Improvement of surface erosion resistance of sand by microbial biopolymer formation. J Geotech Geoenviron Eng 144:6018004. https://doi.org/10.1061/(ASCE)GT.1943-5606.0001900

Hataf N, Ghadir P, Ranjbar N (2018) Investigation of soil stabilization using chitosan biopolymer. J Clean Prod 170:1493–1500. https://doi.org/10.1016/j.jclepro.2017.09.256

Hataf N, Jamali R (2018) Effect of fine-grain percent on soil strength properties improved by biological method. Geomicrobiol J 35:695–703. https://doi.org/10.1080/01490451.2018.1454554

Horpibulsuk S, Rachan R, Chinkulkijniwat A, Raksachon Y, Suddeepong A (2010) Analysis of strength development in cement-stabilized silty clay from microstructural considerations. Constr Build Mater 24:2011–2021. https://doi.org/10.1016/j.conbuildmat.2010.03.011

Joga JR, Varaprasad BJS (2020) Effect of xanthan gum biopolymer on dispersive properties of soils. World J Eng 17:563–571. https://doi.org/10.1108/WJE-05-2020-0152

Kavazanjian E, Iglesias E, Karatas I (2009) Biopolymer soil stabilization for wind erosion control. In: Hamza M, Shahien M, El-Mossallamy Y (eds) Proceedings of the 17th international conference on soil mechanics and geotechnical engineering. IOS Press, Alexandria, Egypt, pp 881–884. https://doi.org/10.3233/978-1-60750-031-5-881

Khachatoorian R, Petrisor IG, Kwan CC, Yen TF (2003) Biopolymer plugging effect: laboratory-pressurized pumping flow studies. J Pet Sci Eng 38:13–21. https://doi.org/10.1016/S0920-4105(03)00019-6

Khan A, Adil M, Ahmad A, Hussain R, Zaman H (2018) Stabilization of soil using cement and bale straw. Int J Adv Eng Res Dev 5:44–49. https://doi.org/10.21090/IJAERD.39611

Khatami HR, Kelly BCO (2013) Improving mechanical properties of sand using biopolymers. J Geotech Geoenviron Eng 139:1402–1406. https://doi.org/10.1061/(ASCE)GT.1943-5606.0000861

Kumar SA, Sujatha ER (2020) Performance evaluation of β-glucan treated lean clay and efficacy of its choice as a sustainable alternative for ground improvement. Geomech Eng 21:413–422. https://doi.org/10.12989/gae.2020.21.5.413

Kumar SA, Sujatha ER, Pugazhendi A, Jamal MT (2021) Guar gum - stabilized soil : a clean , sustainable and economic alternative liner material for landfills. Clean Technol Environ Policy. https://doi.org/10.1007/s10098-021-02032-z

Kwon YM, Ham SM, Kwon TH, Cho GC, Chang I (2019) Surface-erosion behaviour of biopolymer-treated soils assessed by EFA. Geotech Lett 10:1–7. https://doi.org/10.1680/jgele.19.00106

Larson S, Newman JK, O Connor G, Griggs C, Martin A, Nijak G, Lord E, Duggar R, Leeson A (2013) Biopolymer as an alternative to petroleum-based polymers to control soil erosion. Iowa Army ammunition plant. Environmental security technology certification program office (dod) ARLINGTON VA

Larson S, Nijak Jr, G, Corcoran M, Lord E, Nestler C (2016) Evaluation of rhizobium tropici-derived biopolymer for erosion control of protective berms. field study. Iowa Army Ammunition Plant. US Army Engineer Research and Development Center Vicksburg United States

Latifi N, Horpibulsuk S, Meehan CL, Abd Majid MZ, Tahir MM, Mohamad ET (2016) Improvement of problematic soils with biopolymer—an environmentally friendly soil stabilizer. J Mater Civ Eng 29:1–11. https://doi.org/10.1061/(ASCE)MT.1943-5533.0001706

Latifi N, Horpibulsuk S, Meehan CL, Majid MZA, Rashid ASA (2016) Xanthan gum biopolymer: an eco-friendly additive for stabilization of tropical organic peat. Environ Earth Sci 75:825. https://doi.org/10.1007/s12665-016-5643-0

Lee S, Chang I, Chung MK, Kim Y, Kee J (2017) Geotechnical shear behavior of xanthan gum biopolymer treated sand from direct shear testing. Geomech Eng 12:831–847. https://doi.org/10.12989/gae.2017.12.5.831

Lee S, Chung M, Park HM, Song K, Chang I (2019) Xanthan gum biopolymer as soil-stabilization binder for road construction using local soil in Sri Lanka. J Mater Civ Eng 31:6019012. https://doi.org/10.1061/(ASCE)MT.1943-5533.0002909

Lee S, Kwon YM, Cho GC, Chang I (2020) Investigation of biopolymer treatment feasibility to mitigate surface erosion using a hydraulic flume apparatus. In: Geo-congress 2020: biogenetics. american society of civil engineers. Reston, VA, pp 46–52. https://doi.org/10.1061/9780784482834.006

Mishra AK, Ravindra V (2015) On the utilization of fly ash and cement mixtures as a landfill liner material. Int J Geosynth Gr Eng 1:1–7. https://doi.org/10.1007/s40891-015-0019-1

Muguda S, Lucas G, Hughes PN, Augarde CE, Perlot C, Bruno AW, Gallipoli D (2020) Durability and hygroscopic behaviour of biopolymer stabilised earthen construction materials. Constr Build Mater 259:119725. https://doi.org/10.1016/j.conbuildmat.2020.119725

Orts WJ, Sojka RE, Glenn GM, Gross RA (2001) Biopolymer additives for the reduction of soil erosion losses during irrigation. In: Biopolymers from polysaccharides and agroproteins, pp 102–116

Reddy NG, Nongmaithem RS, Basu D, Rao BH (2020) Application of biopolymers for improving the strength characteristics of red mud waste. Environ Geotech 1–20. https://doi.org/10.1680/jenge.19.00018

Saygili A, Dayan M (2019) Freeze-thaw behavior of lime stabilized clay reinforced with silica fume and synthetic fibers. Cold Reg Sci Technol 161:107–114. https://doi.org/10.1016/j.coldregions.2019.03.010

Smitha S, Rangaswamy K (2019) The potential use of biopolymers as a sustainable alternative for liquefaction mitigation-a review. In: Singh RM, Sudheer KP (ed) Advances in civil engineering, Lecture notes in civil engineering. Springer, Singapore, pp 25–34. https://doi.org/10.1007/978-981-15-5644-9_3

Smitha S, Rangaswamy K, Keerthi DS (2019) Triaxial test behaviour of silty sands treated with agar biopolymer. Int J Geotech Eng 1–12. https://doi.org/10.1080/19386362.2019.1679441

Smitha S, Sachan A (2016) Use of agar biopolymer to improve the shear strength behavior of sabarmati sand. Int J Geotech Eng 10:387–400. https://doi.org/10.1080/19386362.2016.1152674

Soldo A, Miletic M, Auad ML (2020) Biopolymers as a sustainable solution for the enhancement of soil mechanical properties. Sci Rep 10:1–13. https://doi.org/10.1038/s41598-019-57135-x

Sujatha ER, Sivaraman S, Subramani AK (2020) Impact of hydration and gelling properties of guar gum on the mechanism of soil modification. Arab J Geosci 13. https://doi.org/10.1007/s12517-020-06258-x

Swain K, Mahamaya M, Alam S, Das SK (2018) Stabilization of Dispersive Soil Using Biopolymer. In: Contemporary issues in geoenvironmental engineering, sustainable civil infrastructures, pp 132–147. https://doi.org/10.1007/978-3-319-61612-4_11

Theng, (1982) Clay-polymer interactions: summary and perspectives. Clay Miner 30:1–10. https://doi.org/10.1346/CCMN.1982.0300101

Tran ATP, Chang I, Cho GC (2019) Soil water retention and vegetation survivability improvement using microbial biopolymers in drylands. Geomech Eng 17:475–483. https://doi.org/10.12989/gae.2019.17.5.475

Tran TPA, Im J, Cho G-CIC (2017) Soil—water characteristics of xanthan gum biopolymer containing soils. In: Proceedings of the 19th international conference on soil mechanics and geotechnical engineering, pp 1091–1094

Viswanath SM (2019) Biopolymer stabilised earthen construction materials. Durham University, United Kingdom

Wade E, Zowada R, Foudazi R (2021) Alginate and guar gum spray application for improving soil aggregation and soil crust integrity. Carbohydr Polym Technol Appl 100114. https://doi.org/10.1016/j.carpta.2021.100114

Zhao W, Wei H, Jia L, Daryanto S, Zhang X, Liu Y (2018) Soil erodibility and its influencing factors on the loess plateau of China: a case study in the Ansai watershed. Solid Earth 9:1507–1516. https://doi.org/10.5194/se-9-1507-2018

Zhao Y, Zhuang J, Wang Y, Jia Y, Niu P, Jia K (2020) Improvement of loess characteristics using sodium alginate. Bull Eng Geol Environ 79:1879–1891. https://doi.org/10.1007/s10064-019-01675-z

Chapter 13
A Parameter to Assess the Strength of Fly Ash and GGBS-Based Geopolymer Concrete

Sumanth Kumar Bandaru and D. Rama Seshu

13.1 Introduction

When developing new-age building materials, "sustainability" and "energy efficiency" are key factors. The long-term view is to reduce exposure through the use of unwanted industrial by-products, which reduces the consumption of natural materials. In this direction, geopolymers have emerged as environmentally friendly substitutes for Portland cement, which, in many areas, not only reduce greenhouse gas emissions but also consume large volumes of industrial waste such as fly ash, mine tailings and metallurgical slag.

Geopolymer concrete (GPC) has recently been developed as an alternative to conventional Portland cement (OPC). GPC is produced by combining raw materials such as fly ash (FA), ground granular blast furnace slag (GGBS), which are rich in silicon (Si) and aluminum (Al), using highly alkaline liquids such as NaOH and/or sodium silicate solution (Na_2SiO_3). These alkaline liquids act as an activator and produce the binder needed to make concrete without the use of cement. In the recent past, several studies Davidovits (1999), Palomo et al. (1999), Lloyd and Van Deventer (2005) and Rangan (2008) have reported various parameters affecting the strength of GPC.

These parameters include the amount of source material, the ratio of the activator to the binder, the molarity of the activator solution. Published literature indicates that several variables influenced the strength of the GPC. The combined effect of various parameters on the strength of geopolymer concrete based on GGBS and fly ash was represented by the proposal of a parameter called "Binder Index (Bi)" Rama Seshu

S. K. Bandaru (✉)
Kakatiya Institute of Technology and Science, Warangal, India

D. R. Seshu
National Institute of Technology, Warangal, India
e-mail: drseshu@nitw.ac.in

K. R. Reddy et al. (eds.), *Advances in Sustainable Materials and Resilient Infrastructure*,
Springer Transactions in Civil and Environmental Engineering,
https://doi.org/10.1007/978-981-16-9744-9_13

et al. (2017). However, the proposed binder index takes into account the variation of few variations like molarity and ratio of source materials. Therefore, in this article, the previously proposed binder indicator is modified in such a way that it takes into account the main parameters that affect the strength of geopolymer concrete.

13.2 Factors Affecting Strength of Ggbs and Fly Ash-Based Geopolymer Concrete

In the recent past investigations, several researchers reported various parameters affecting strength are curing regime like temperature and duration of curing, sodium hydroxide solution concentration, ratio of sodium silicate to sodium hydroxide, alkaline activator solution to binder (FA + GGBS) ratio along with chemical composition of binder material.

It is concluded that curing regime and temperature influences the polymerization process in GPC, i.e. hardening process of GPC Hardjito and Rangan (2005), Lloyd and Van Deventer (2005), Rangan (2008, 2009). Various researchers have concluded that sodium silicate to sodium hydroxide ratio of 2.5 gives maximum compression strength for a constant binder content Lazarescu et al. (2017). Molarity of alkali (NaOH solution) is a vital role in strength of GPC, i.e. higher the concentration of NaOH solution, a higher compressive strength can be achieved.

The total water content (including in alkaline activators and extra water added if any) to total binding material also varies the strength of GPC. Higher fly ash content with higher alkaline activator content gives high compressive strength and vice versa, i.e. with a ratio of 1.4–2.3 gives healthy compressive strength Lazarescu et al. (2017). Alkaline to binder ratio impacts the strength of GPC, it reduces strength for a constant binder content. The strength of GPC was observed to increase with an increase in GGBS to FA ratio for a particular molarity of activator used, also noted that the increase in GGBS quantity setting time decreases along with an increase in strength Mallikarjuna Rao and Rao (2015), Jawahar and Mounika (2016). A comprehensive data of mixed proportions of constituents and strength properties have been extracted from Bhikshma and Kumar (2016), Annapurna and Kishore (2017), Ganapati Naidu et al. (2012), Rama Seshu et al. (2017, 2019), Gopalakrishnan (2019), Zende and Mamatha (2015), Rajini and Rao (2014), Krishnaraja et al. (2014), Rafeel et al. (2017), Prasanna et al. (2016), Ramamohana et al. (2019), Mallikarjuna Rao and Rao (2015), Mallikarjuna Rao et al. (2016), Mallikarjuna Rao and Gunneswara Rao (2018)—for geopolymer mortar. Each of the test data pointed plotted in various graphs corresponds to the strength of geopolymer concrete.

Previous studies and as per Fig. 13.1 showed that an increase in molarity/concentration of NaOH observed an increase in compressive strength. It is also observed that using an alkaline activator composed of sodium hydroxide solution and sodium silicate solution leads to better strengths. The compressive strength of

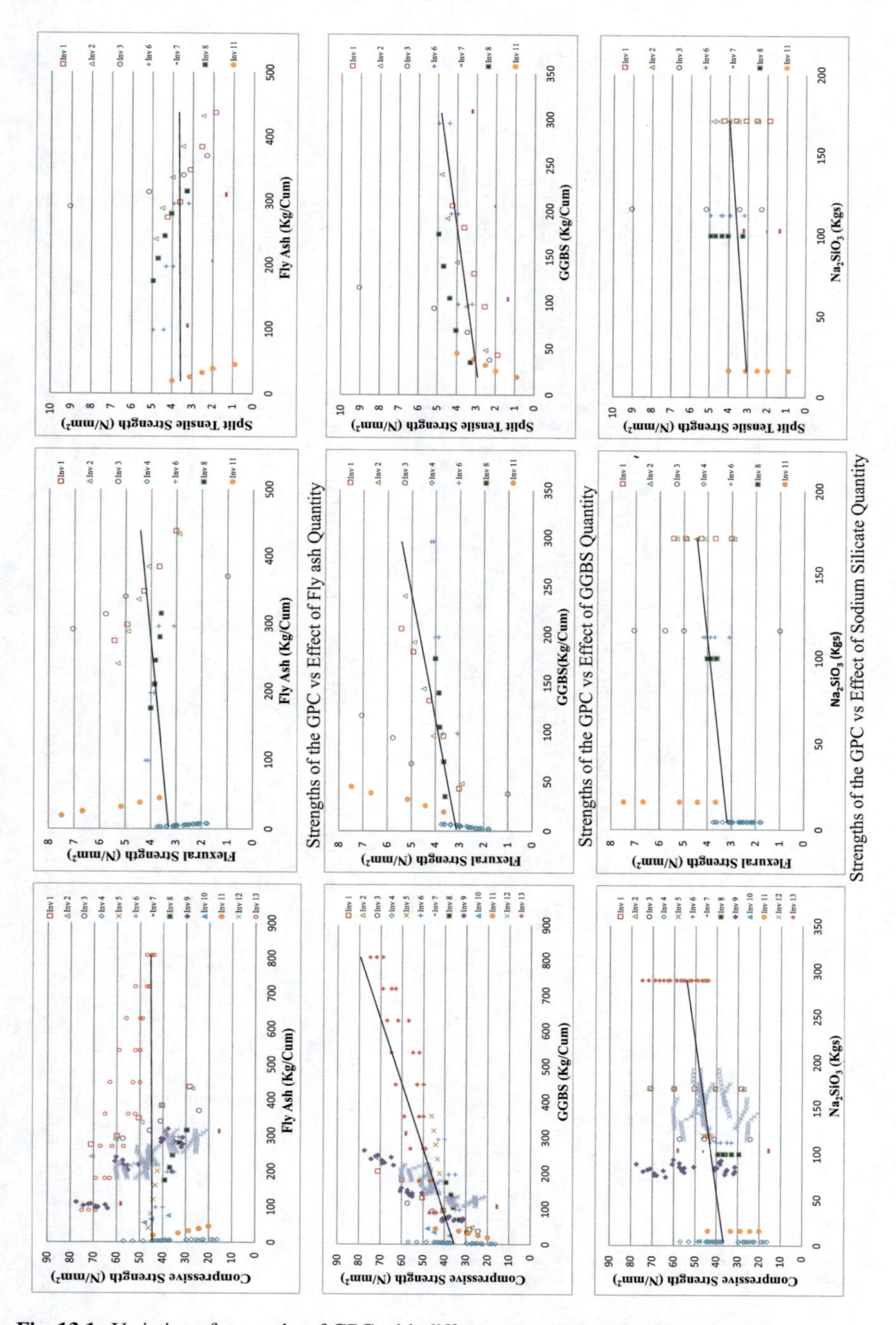

Fig. 13.1 Variation of strengths of GPC with different parameters of published works

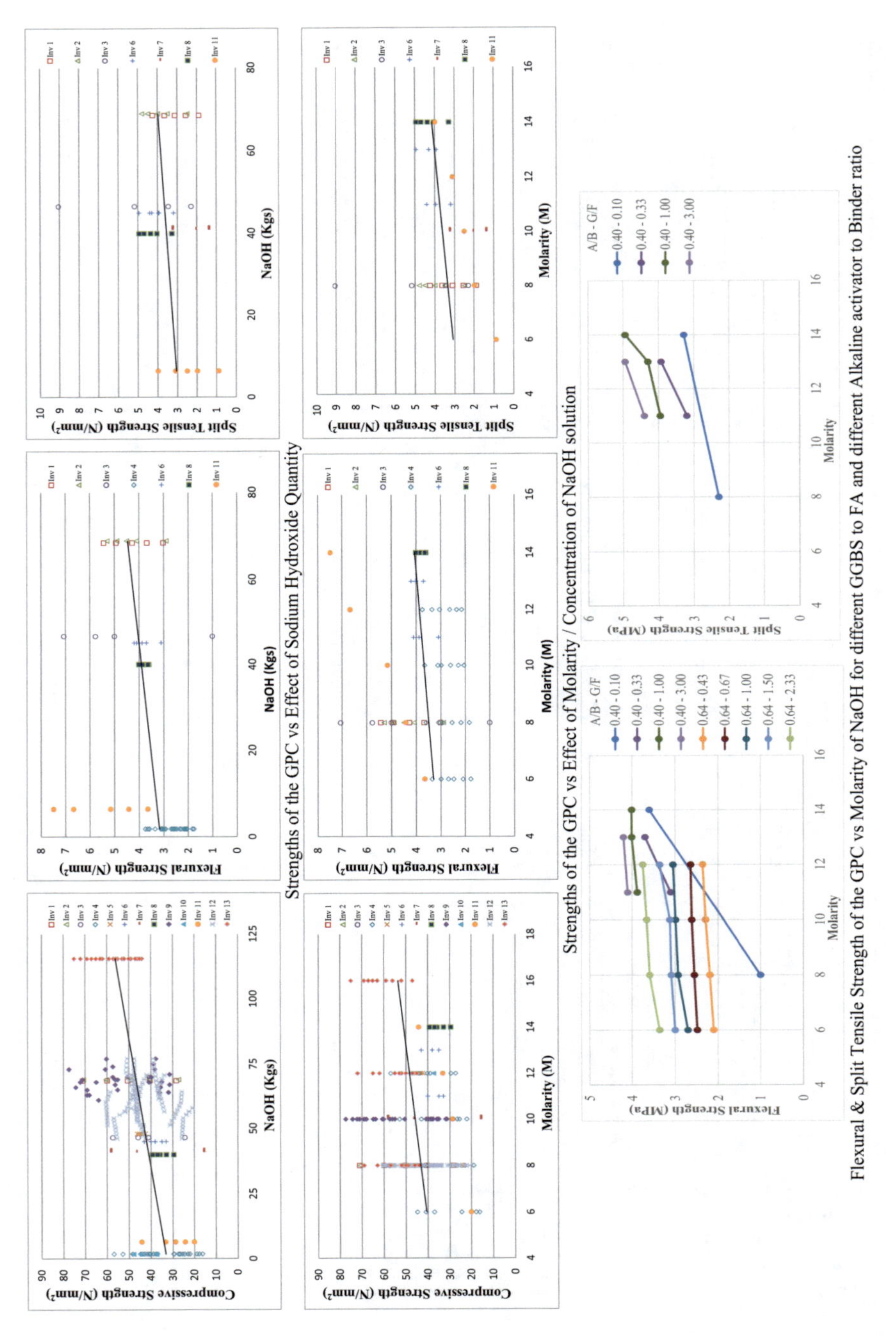

Fig. 13.1 (continued)

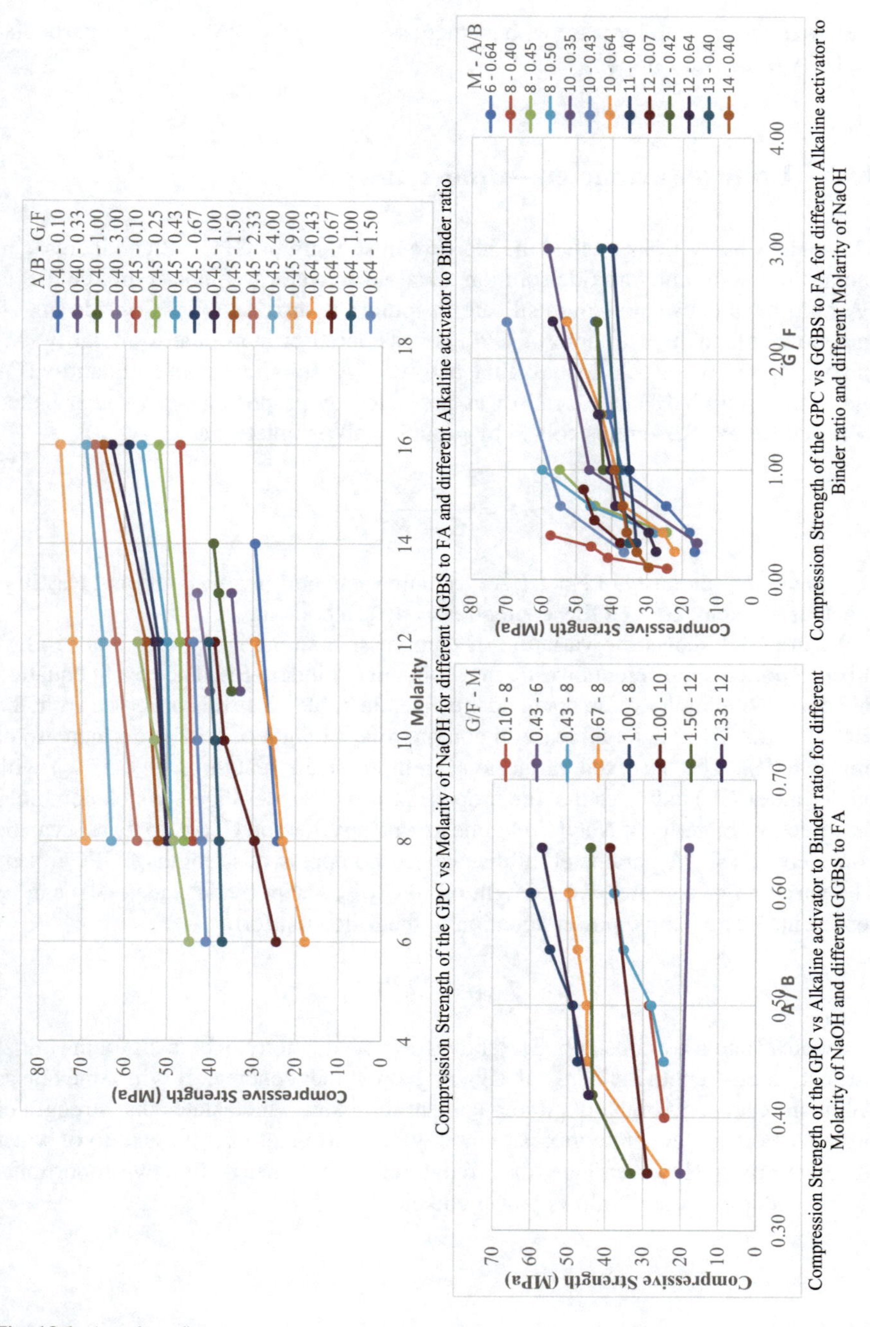

Fig. 13.1 (continued)

GPC was observed to increase with an increase in GGBS to FA ratio for a particular molarity of activator used.

13.3 Unified Parameter—Binder Index

The major observations include the increase in strength of GPC with an increase in molarity of sodium hydroxide solution, alkaline activator to binder ratio, GGBS to FA ratio for a constant sodium silicate to sodium hydroxide ratio. Considering all the above effects, the strength of GPC is considered proportional to molarity (M) of sodium hydroxide solution, alkaline activator (A) solution to binder quantity (FA + GGBS) and GGBS to FA ratio was presented by proposing a parameter called "**Binder Index (Bi)**" and is coined by grouping all parameters as

$$\boldsymbol{Bi} = \frac{\boldsymbol{MA}}{\boldsymbol{G}+\boldsymbol{F}}\left[\frac{\boldsymbol{G}}{\boldsymbol{F}}\right]$$

where, M = molarity of NaOH, A = alkaline activator (Both NaOH and Na_2SiO_3 together) content, G = GGBS content, F = fly ash content.

Figure 13.2 shows the variation of compressive strength (f_{gpc}) results of GPC mixes reported by different investigators with binder index (Bi). The best fit equation and corresponding R^2 value obtained are given in Table 13.1. It is observed from the Fig. 13.2 that there is an increase in compressive strength of GPC with increase of binder index. The observed variation of compressive strength of GPC (f_{gpc}) with binder index (Bi) indicates that the proposed form of binder index, which combines the effects of molarity of NaOH, alkaline to binder ratios and GGBS to fly ash, can be considered as single parameter influencing the compressive strength of GPC mixes. The variation of compressive strength of GPC (f_{gpc}) with binder index (Bi) can be represented by a simple power equation of the following form

$$\boldsymbol{f}_{\mathbf{gpc}} = \boldsymbol{N}[\boldsymbol{Bi}]^{\mathrm{L}}$$

where N and L are constants. The above form of equation can be basis for the initial estimation of strength in the mix design of geopolymer concrete. It is in same line as Abraham's law or Abraham's water–cement ratio law, which states that strength of ordinary Portland concrete mix is indirectly proportional to the mass ratio of water to cement ratio. However, here, the strength of geopolymer is directly proportional to the binder index and follows power equation.

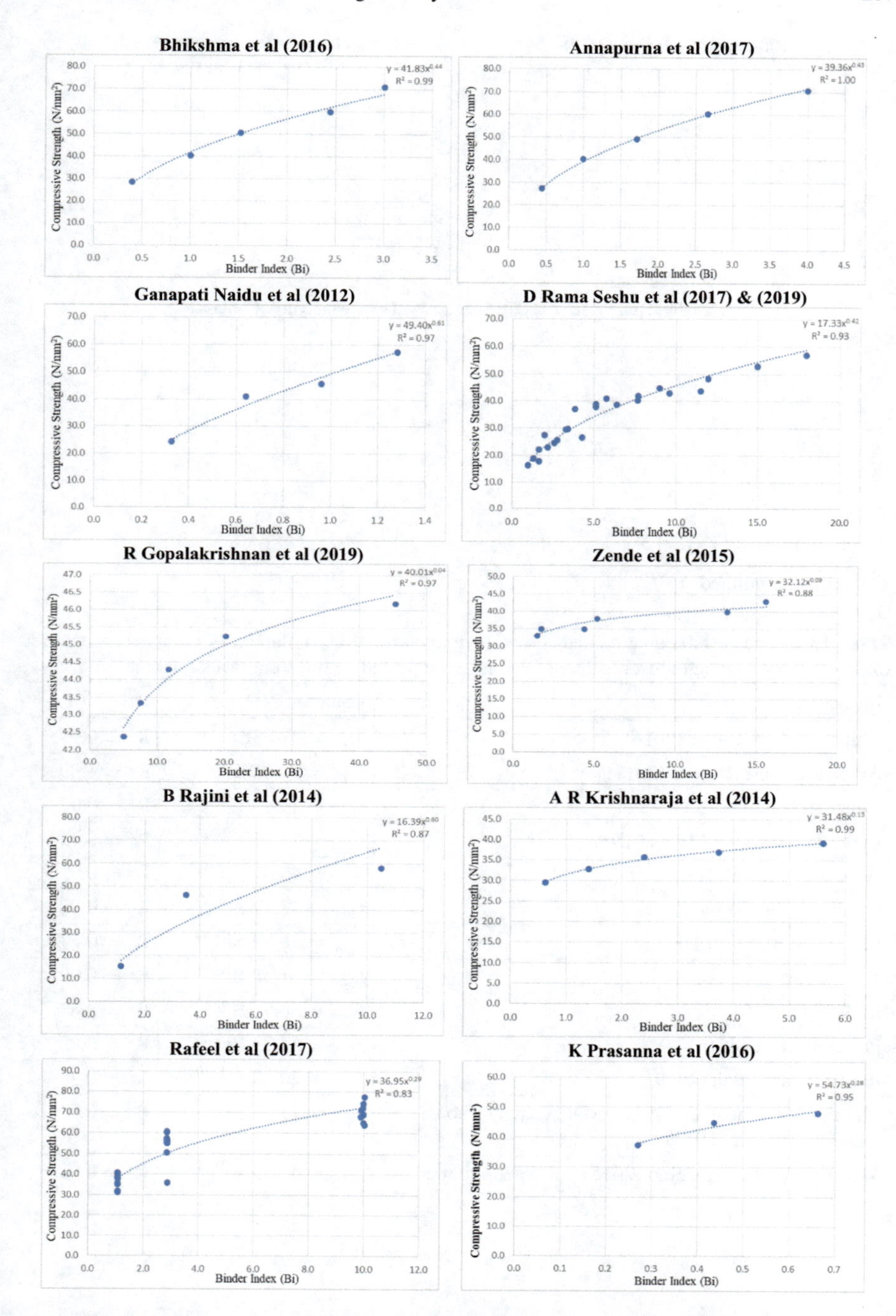

Fig. 13.2 Variation of compressive strength of GPC (f_{gpc}) with the proposed Binder index (Bi) for the experimental results of reported literature

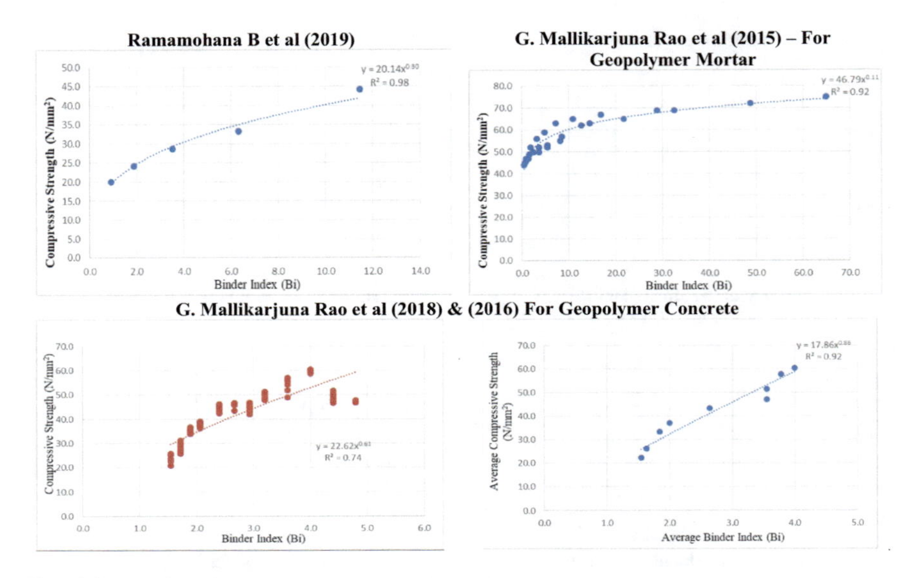

Fig. 13.2 (continued)

Table 13.1 The best fit equation and corresponding correlation coefficient (R^2) value obtained for the compressive strength test results of GPC mixes reported by different investigators

Investigator (s)	Equation	R^2
Bhikshma and Kumar (2016)	$f_{gpc} = 41.83Bi^{0.44}$	$R^2 = 0.99$
Annapurna and Kishore (2017)	$f_{gpc} = 39.36\ Bi^{0.43}$	$R^2 = 1.00$
Ganapati Naidu et al. (2012)	$f_{gpc} = 49.40Bi^{0.61}$	$R^2 = 0.97$
Rama Seshu et al. (2017, 2019)	$f_{gpc} = 17.33Bi^{0.42}$	$R^2 = 0.93$
Gopalakrishnan (2019)	$f_{gpc} = 40.01\ Bi^{0.04}$	$R^2 = 0.97$
Zende and Mamatha (2015)	$f_{gpc} = 32.12\ Bi^{0.09}$	$R^2 = 0.88$
Rajini and Rao (2014)	$f_{gpc} = 16.39\ Bi^{0.60}$	$R^2 = 0.87$
Krishnaraja et al. (2014)	$f_{gpc} = 31.48\ Bi^{0.13}$	$R^2 = 0.99$
Rafeel et al. (2017)	$f_{gpc} = 36.95\ Bi^{0.29}$	$R^2 = 0.83$
Prasanna et al. (2016)	$f_{gpc} = 54.73\ Bi^{0.28}$	$R^2 = 0.95$
Ramamohana et al. (2019)	$f_{gpc} = 20.14\ Bi^{0.30}$	$R^2 = 0.98$
Mallikarjuna Rao et al. (2016), Mallikarjuna Rao and Gunneswara Rao (2018)	$f_{gpc} = 17.86\ Bi^{0.86}$	$R^2 = 0.92$
Mallikarjuna Rao and Rao (2015)—For geopolymer mortar	$f_{gpc} = 46.79\ Bi^{0.11}$	$R^2 = 0.92$

13.4 Conclusions

The following are conclusions derived after the study of different variables affecting the compressive strength of geopolymer concrete mixes.

1. The compression strength of the GPC increased with an increase in the GGBS to FA ratio.
2. As the molarity of NaOH solution in alkaline activator increased, the compressive strength, split tensile strength and modulus of rupture of the GPC also increased. However, the increase in strength is not in proportion to the increase in molarity.
3. The compression strength of the GPC increased with an increase in alkaline / binder ratio for constant molarity of NaOH solution in alkaline activator.
4. The new proposed parameter called "Binder Index (Bi)", which combines the effects of alkaline to binder content ratio, GGPS to FA ratio and molarity of sodium hydroxide can be considered as single unique parameter to control the compressive strength of GPC.

$$\boldsymbol{Bi} = \frac{\boldsymbol{MA}}{\boldsymbol{G} + \boldsymbol{F}}\left[\frac{\boldsymbol{G}}{\boldsymbol{F}}\right]$$

where, M = molarity of NaOH, A = alkaline activator (both NaOH and Na_2SiO_3 together) content, G = GGBS content, F=fly ash content.

5. It is witnessed that increase in strength of GPC is with an increase of binder index.
6. The relation between compressive strength (f_{gpc}) and binder index (Bi) of GPC is non-linear and can be signified by a power equation.

$$f_{\text{gpc}} = \text{N}[\boldsymbol{Bi}]^{\text{L}}$$

7. The compressive strength of the GPC increased with an increase in the binder index. However, the increase in strength is not in proportion to the increase in the binder index.
8. A non-linear variation exists between the binder index and the compressive strength of GPC.

References

Annapurna D, Kishore R (2017) Evaluation of mechanical properties of fly ash and GGBS based geopolymer concrete. Int J Emerg Technol Innov Res 4(12):1028–1033. ISSN: 2349-5162
Bhikshma V, Kumar TN (2016) Mechanical properties of fly ash based geopolymer concrete with addition of GGBS. Indian Concrete J 64–68
Davidovits J (1999) Chemistry of geopolymeric systems, terminology. Geopolymer 99(292):9–39

Ganapati Naidu P, Adiseshu S, Satayanarayana PVV (2012) A study on strength properties of geopolymer concrete with addition of GGBS. Int J Eng Res Dev 19–28
Hardjito D, Rangan BV (2005) Development and properties of low-calcium fly ash-based geopolymer concrete. Curtin University of Technology
Jawahar JG, Mounika G (2016) Strength properties of fly ash and GGBS based geopolymer concrete. Asian J Civil Eng (BHRC) 17(1):127–135
Krishnaraja AR, Sathishkumar NP, Kumar TS, Kumar PD (2014) Mechanical behaviour of geopolymer concrete under ambient curing. Int J Sci Eng Technol 3(2):130–132
Lazarescu AV, Szilagyi H, Baera C, Ioani A (2017) The effect of alkaline activator ratio on the compressive strength of fly ash-based geopolymer paste. IOP Conf Ser: Matter Sci Eng 209012064
Lloyd RR, Van Deventer JSJ (2005) The microstructure of geopolymers synthesized from industrial waste. In: 1st International conference on engineering for waste treatment, Albi, France
Mallikarjuna Rao G, Rao TG (2015) Final setting time and compressive strength of fly ash and GGBS-based geopolymer paste and mortar. Arab J Sci Eng 40(11):3067–3074
Mallikarjuna Rao G, Rao TD, Seshu RD, Venkatesh A (2016) Mix proportioning of geopolymer concrete. Cement Wapno Beton 21(4):274–285
Mallikarjuna Rao G, Gunneswara Rao TD (2018) A quantitative method of approach in designing the mix proportions of fly ash and GGBS-based geopolymer concrete. Aust J Civ Eng 16(1):53–63
Palomo A, Grutzeck MW, Blanco MT (1999) Alkali-activated fly ashes: a cement for the future. Cem Concr Res 29(8):1323–1329
Prasanna K, Lakshminarayan B, Arun Kumar M, Dinesh Kumaran JR (2016) Fly ash based geopolymer concrete with GGBS. In: International conference on current research in engineering science and technology, India, pp 12–18. . E-ISSN: 2348-8352
Rafeel A, Vinai R, Soutsos M, Sha W (2017) Guidelines for mix proportioning of fly ash/GGBS based alkali activated concretes. Constr Build Mater 147:130–142
Gopalakrishnan R (2019) Durability of alumina silicate concrete based on slag/fly ash blends against corrosion. Eng Constr Archit Manag 26(8):1641–1651
Rajini B, Rao AN (2014) Mechanical properties of geopolymer concrete with fly ash and GGBS as source materials. Int J Innov Res Sci Eng Technol 3(9):15944–15953
Rama Seshu D, Shankaraiah R, Srinivas BS (2017) A study on the effect of binder index on compressive strength of geopolymer concrete. Cement Wapno Beton 84:211–215
Rama Seshu D, Shankaraiah R, Sesha SB (2019) The binder index—A parameter that influences the strength of geopolymer concrete. Slovak J Civil Eng 27(1):32–38
Ramamohana B, Gopinathan P, Chandrasekhar I (2019) Engineering properties of GGBS & fly ash synthesized geopolymer concrete at different environmental conditions by comparing with conventional concrete. Int J Recent Technol Eng (IJRTE) 7(5S4):399–407. ISSN: 2277-3878
Rangan BV (2008) Fly ash-based geopolymer concrete. Curtin University of Technology, Dept of Engineering, Tech. report No.GC4
Rangan BV (2009) Engineering properties of geopolymer concrete. In: Geopolymers. Woodhead Publishing, pp 211–226
Zende R, Mamatha A (2015) Study on fly ash and GGBS based geopolymer concrete under ambient curing. J Emerg Technol Innov Res 2(7):3082–3087

Chapter 14
Influence of Soft Drink Bottle Caps as Steel Fibre on Mechanical Properties of Concrete

P. Teja Abhilash, K. Tharani, and P. V. V. Satyanarayana

14.1 Introduction

Concrete plays an extensive role in the construction industry and infrastructure development. Its great strength, durability and veracity are the properties that are utilized in construction of Roads, Bridges, Airports, Railways, Tunnels, Port, Harbours and many other infrastructural projects. Concrete being weak in tension is reinforced with steel to obtain ultimate strength corresponding to the design (Kishore and Gupta 2019). There are various types of reinforcement out of which steel fibre is the one that can be obtained at a lower cost.

Research and design of steel fibre reinforced concrete (SFRC) began to increase about 40 years ago. The idea of adding the fibres to brittle materials to increase ductility goes back to old times (Felekoğlu et al. 2007). They differ in size, shape and surface structure. Fibre reinforcements are generally used for special concrete applications wherein short discrete fibres are used as additional reinforcing material. The fibres are added in low volume (less than $2\% \ V_f$) to that of the total volume of concrete in order to improve the hardened concrete properties (Anandan and Alsubih 2021). These fibres have different mechanical properties such as tensile strength, grade of mechanical anchorage and capability of stress distribution and absorption (Wang 2006). In ancient Egypt, straw was added to mud brick. Nowadays, various types of fibre are used; for example, in concrete structures the most popular is steel fibre. Many researches have been conducted on steel fibre reinforced concrete (SFRC) (Ranjbaran et al. 2018).

P. T. Abhilash (✉)
Kakatiya Institute of Technology and Science, Warangal, Telangana 506015, India

K. Tharani
National Institute of Technology, Warangal, Telangana 506004, India

P. V. V. Satyanarayana
Andhra University, Visakhapatnam, Andhra Pradesh 530003, India

K. R. Reddy et al. (eds.), *Advances in Sustainable Materials and Resilient Infrastructure*, Springer Transactions in Civil and Environmental Engineering,
https://doi.org/10.1007/978-981-16-9744-9_14

The SFRC is a compound material made of cement, fine aggregate, coarse aggregate and discrete steel fibres. The SFRC fails in tension only after steel fibres pull out of concrete. The mechanical strength properties of SFRC are closely related to the fibre parameters, matrix strength and their interaction (Thomas and Ramaswamy 2007). One of the most important properties of SFRC is its ability to transfer stresses across a cracked section which increases toughness of concrete in hardened state (Dahake and Charkha 2016).

The addition of steel fibres has shown a maximum increase in compressive strength up to 15% (Usman et al. 2020); however, an appreciable increase in bond strength and toughness was noted with the addition of steel fibres (Richardson and Landless 2009). The maximum volume fraction of fibres at the crack plane plays an important role in stable crack growth formation. Many investigations have proven the beneficial addition of high modulus fibre on the overall improvement in strength, ductility, toughness and durability (Susetyo et al. 2011). The basic fibre mechanism is to prevent the crack origination and propagation. The use of high strength fibres at a large volume fraction does possess relative advantages on the flexural and tensile strength (Chan et al. 2009; Kazmi et al. 2019).

Reliability of fibre reinforcement

Concrete being a brittle material which has low tensile strength and low strain capacity, as a result, the mechanical behaviour of concrete is critically influenced by crack propagation. Concrete in service may exhibit failure through cracks which are developed due to brittleness (Darshan et al. 2014). To improve properties of concrete like low tensile and low strain capacity fibre reinforced concrete (FRC) has been developed which is defined as concrete containing dispersed randomly oriented fibres. Use of Fibres in concrete experimented in 1910 and research on steel fibre addition in concrete started in early 1960s. By the 1960s, steel, glass (GFRC) and synthetic fibres such as polypropylene fibres were used in concrete.

1. ***Glass Fibre Reinforced Concrete (GFRC)***: Glass fibre reinforced concrete is found to be one of the best replacements for steel in the construction industry. GFRC is positioned properly either with direct anchors or slip anchors (Behera and Behera 2015).
2. ***Steel Fibre Reinforced Concrete (SFRC)***: Steel fibre reinforced concrete is made of hydraulic cement with random placement of steel fibre (Alsaif et al. 2018; Anandan and Alsubih 2021). Reinforcement of concrete by steel fibres is isotropic in nature that improves the resistance to fracture, disintegration and fatigue (Fang et al. 2020; Behera and Behera 2015).
3. ***Natural fibre reinforced concrete (NFRC)***: The cellulose fibres from pine trees are used to manufacture natural fibre reinforced concrete. Also, the waste carpet fibres prove to give good results (Behera and Behera 2015).
4. ***Polypropylene Fibre Reinforced (PFR) concrete***: Polypropylene has the capacity to resist because of its chemical reactions with the binding materials. Hence it is used abundantly (Syamsir et al. 2020; Behera and Behera 2015).

5. ***Asbestos Fibres***: The asbestos fibres are good at resisting temperatures, moderate at resisting strains but weak at handling the impact conditions. They are economic in nature and used only when cost is the prominent constraint (Behera and Behera 2015).
6. ***Carbon Fibres***: **Carbon fibres provide** high modulus of elasticity and flexural strength to concrete (Marcos et al. 2020). The strength and stiffness are optimum with the use of carbon fibres but are more susceptible to damage (Behera and Behera 2015).

Inclusion of above-mentioned fibres provide better tensile strength and crack arresting capacity to concrete. Higher quantities of steel waste fibres are generated from industries related to lathes, empty beverage metal cans and soft drink bottle caps. This is an environmental issue as steel waste fibres are non-biodegradable and involves processes either to recycle or reuse. The phenomenon of reuse from 3R's (Reduce, Reuse and Recycle) necessitates the sustainable usage of steel fibres. Preservation of environment and conservation of rapidly diminishing natural resources should be the essence of sustainable development. Bottle caps are the substitute for fibres and added to enhance the mechanical properties of concrete. The resulting compressive strength, split tensile strength and flexural strength of the mixture depends on the type of cement, size and type of aggregate, period and type of curing adopted. To investigate the increase in mechanical strength of normal concrete by addition of different percentage of bottle cap fibres, an experiment was set up. Plain Concrete with target mean strength of 33.75 N/mm^2 was designed and casted. The main variables in this experimental investigation are Bottle cap fibres with 0.5, 1.0, 1.5 and 2.0% (total wt. of concrete ingredients) which were added to above plain concrete for enhancement of mechanical properties.

The objective of current study is.

- To study the properties of steel fibre obtained from soft drink bottle caps
- To study the effect of bottle caps as fibre on mechanical properties of the concrete.

14.2 Materials

The materials used for preparation of M25 grade concrete are discussed in the following section. Ordinary Portland cement of 53 grade BIRLA brand confirming to B.I.S standards is used in this present investigation. The cement is tested for its various properties as per IS: 4031-1988 and found to be confirming to the requirements as per IS: 12,269-1987. Soft drink bottle caps are used as steel fibres. The caps are pressed by hammer to strips as shown in Fig. 14.1. The properties of steel fibre are given in Table 14.1. The sand obtained from Handri River near Kurnool is used as a fine aggregate in this project investigation. The sand is free from the clayey matter, silt and organic impurities, etc. The sand is tested for specific gravity, in accordance with IS: 383-1970 and it is 2.62, whereas its fineness modulus is 3.70. The sand confirms to zone-II. Crushed 20 mm maximum size was used. The specific

Fig. 14.1 Steel caps converted to fibre

Table 14.1 Properties of steel fibre

Property	Value
Length	20 mm
Diameter	0.75 mm
Aspect ratio	26.67
Tensile strength	

gravity was found to be 2.67. Fineness modulus is also determined and is found to be 2.5. The locally available potable water, which is free from concentration of acid and organic substance, is used for mixing the concrete.

14.3 Methodology

Initially, the physical properties of materials used for preparation of concrete are evaluated according to Indian standard codes. A mix design is calculated for the concrete grade of M25 and the quantities are mentioned in Table 14.1. The concrete is prepared by addition of steel fibres from 0 to 2.0% at the rate of 0.5%. Thus, five mixes are prepared and designated as R0, R0.5, R1.0, R1.5 and R2.0. The steel caps are pressed into sheets by means of hammer and cut into pieces. The trimmed pieces are added to concrete as reinforcement. Cubes, beams and cylinders are casted for each designation. Finally, the mechanical properties are evaluated after a curing period of 7 and 28 days. The results are analysed with the help of relative values of mechanical properties for varying percentages of fibre (Figs. 14.3, 14.4, 14.5).

Table 14.2 Mix proportions of M25 grade concrete

Cement: fine aggregate: coarse aggregate	1: 1.47: 2.88
Water	0.45

Table 14.3 Observed properties of materials used in concrete mix

Material	Property	Value
Cement (OPC 53)	Fineness	2.5%
	Standard consistency	29%
	Soundness	1 mm
	Specific gravity	3.12
	Initial and final setting time	43 and 172 min
	Compressive strength	34.17 N/mm^2
Fine aggregate	Fineness modulus	2.55
	Specific gravity	2.62
	Water absorption	1.23%
Coarse aggregate	Fineness modulus	3.70
	Specific gravity	2.67
	Water absorption	1.125%

14.4 Observations

The following (Table 14.3) shows the observations that were made on cement and aggregates. All the properties are evaluated according to IS code.

14.5 Tests on Concrete

The moulds are casted and cured for 7 and 28 days. The moulds are removed and air dried before testing the mechanical properties of concrete. The compressive strength of concrete is tested according to IS 516:1959. The split tensile strength of concrete is tested according to IS 5816:1999. The flexural strength of concrete is tested according to IS 516:1959. The test specimens before and after testing are shown from Fig. 14.2a to f.

14.6 Results and Discussions

The relative mechanical property is defined as the ratio of mechanical property at different percentage of fibre to mechanical property obtained without steel fibre. The graphical representation of relative mechanical property is shown from Fig. 14.3, 14.4 to 14.5. As the percentage of fibre increases, the relative compressive strength at 7-day curing increases which indicates that the early strength is not affected by addition of fibres (Dahake and Charkha 2016; Anandan and Alsubih 2021). Although, the compressive strength at 28-day curing is found to increase by 7 at 2% addition

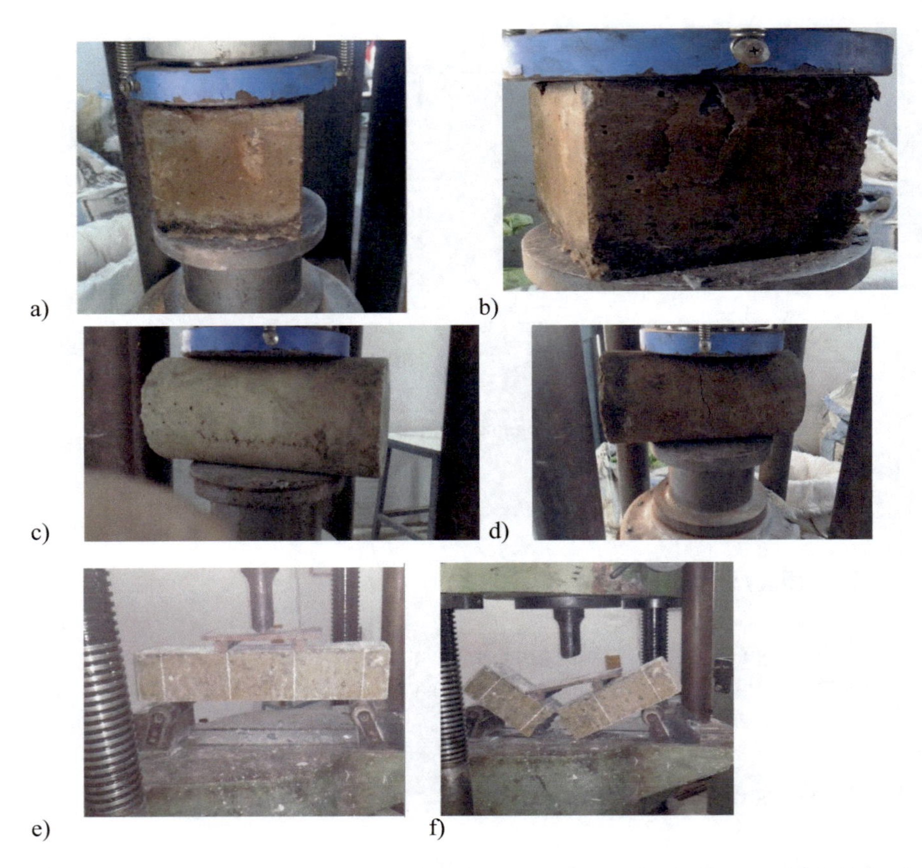

Fig. 14.2 Specimens of concrete for testing compressive strength, split tensile strength and flexural strength. **a**, **c**, **e** denotes before testing and **b**, **d**, **f** denotes after testing

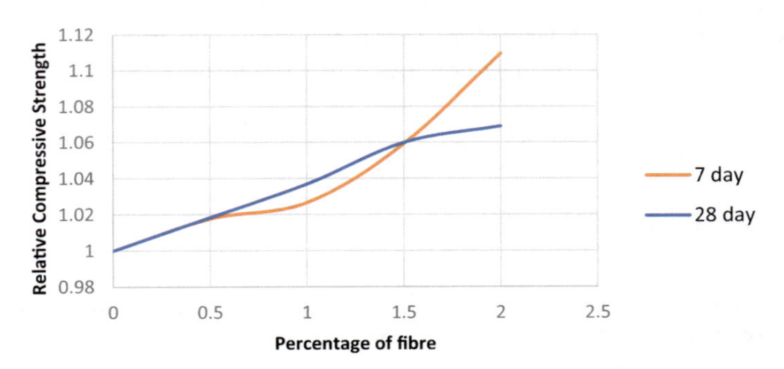

Fig. 14.3 Graphical representation of relative compressive strength with percentage of fibre

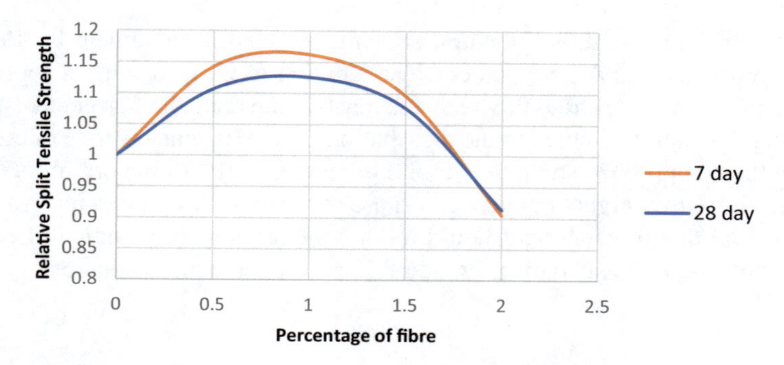

Fig. 14.4 Graphical representation of relative split tensile strength with percentage of fibre

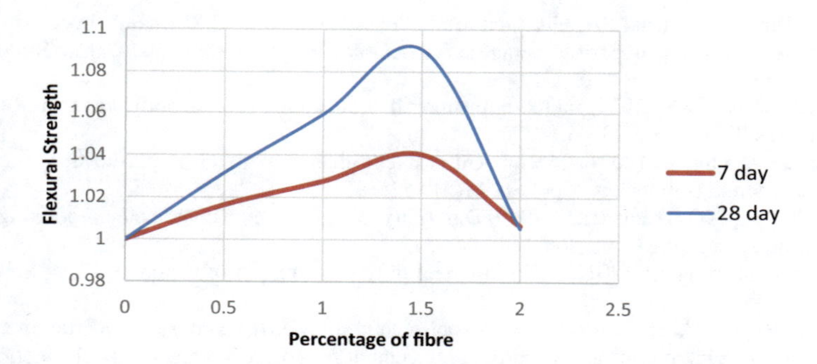

Fig. 14.5 Graphical representation of relative flexural strength with percentage of fibre

of fibre, the relative compressive strength is found to be less when compared to relative strength at 7-day curing. The relative split tensile strength is found to attain a maximum at 1% addition of fibre for both periods of curing. With respect to split tensile strength at 0% fibre, a maximum increase of 19% is found to be obtained at 1.5% fibre. The variation in relative and normal value indicates that the rate of increase in split tensile strength increases till 1.5% fibre content and then decreases. The flexural strength has attained a maximum at 1.5% fibre content which is also seen in relative strength values.

14.7 Conclusions

The addition of steel fibre to concrete has found to be advantageous in terms of improvising the mechanical properties of concrete. The mechanical properties are found to increase till 1.5% of fibre and then decline for further addition of fibre content. This increase is attributed to the effective load sharing capacity of fibres. The compressive strength is found to attain an increase of 7 at 2% of fibre content.

The split tensile strength and flexural strength is found to increment by 19% and 9% respectively at 1.5% fibre content. The addition of fibres results in appreciable increase in split tensile strength which ensures that the concrete has gained strength in tension. It can be attributed to the distribution of fibre in tension zone of concrete. Though the compressive strength is found to increase till 2%, the optimum content of fibre is 1.5% by weight of concrete because of the other two governing properties. It is essential that the concrete should resist compression and bending effectively. The prerequisite is satisfied by 1.5% addition of fibre content to concrete.

References

Alsaif A, Koutas L, Bernal SA, Guadagnini M, Pilakoutas K (2018) Mechanical performance of steel fibre reinforced rubberised concrete for flexible concrete pavements. Constr Build Mater 172:533–543

Behera GC, Behera RK (2015) Increase in strength of concrete by using bottle caps. Int Res J Eng Technol (IRJET) 2(03):1937–1942

Anandan S, Alsubih M (2021) Mechanical strength characterization of plastic fibre reinforced cement concrete composites. Appl Sci 11(2):852

Chan SYN, Feng NQ, Tsang MKC (2009) Durability of high-strength concrete incorporating carrier fluidifying agent. Mag Concr Res J 52:235–242

Dahake AG, Charkha KS (2016) Effect of steel fibres on strength of concrete. J Eng Sci Manag Educ 9(I):45–51

Darshan N, Rajani VA, Sharath BP (2014) Experimental study on the hardened properties of concrete by using soft drink bottle caps as partial replacement for coarse aggregates. Int J Emerg Trends Sci Technol 1(08):1335–1341

Fang C, Ali M, Xie T, Visintin P, Sheikh AH (2020) The influence of steel fibre properties on the shrinkage of ultra-high performance fibre reinforced concrete. Constr Build Mater 242:117993

Felekoğlu B, Türkel S, Altuntaş Y (2007) Effects of steel fibre reinforcement on surface wear resistance of self-compacting repair mortars. Cement Concr Compos 29(5):391–396

IS: 10262-2009. Recommended guidelines for concrete mix design, Bureau of Indian Standards, New Delhi, India

IS 12269-1987: specifications for 53 grade ordinary portland cement. New Delhi: IS

IS 5816: 1999 Spliting Tensile Strength of Concrete Method of Test. Bureau of Indian Standard, New Delhi

IS: 383-1970. Specifications for coarse and fine aggregates from natural sources for concrete, Bureau of Indian Standards, New Delhi, India

IS: 456-2000 Plain and reinforced concrete–code of practice. New Delhi, Bureau of Indian Standards

Kazmi SM, Munir MJ, Wu YF, Patnaikuni I, Zhou Y, Xing F (2019) Axial stress-strain behavior of macro-synthetic fibre reinforced recycled aggregate concrete. Cem Concr Compos J 97:341–356

Kishore K, Gupta N (2019) Experimental analysis on comparison of compressive strength prepared with steel tin cans and steel fibre. Int J Res Appl Sci Eng Technol 7(Iv):169–172

Marcos-Meson V, Fischer G, Solgaard A, Edvardsen C, Michel A (2020) Mechanical performance and corrosion damage of steel fibre reinforced concrete—A multiscale modelling approach. Constr Build Mater 234:117847

Ranjbaran F, Rezayfar O, Mirzababai R (2018) Experimental investigation of steel fibre-reinforced concrete beams under cyclic loading. Int J Adv Struct Eng 10(1):49–60

Richardson AE, Landless S (2009) Synthetic fibres and steel fibres in concrete with regard to bond strength and toughness. Built Environ Res 2:128–140

Susetyo J, Gauvreau P, Vecchio FJ (2011) Cracking behavior of steel fibre-reinforced concrete members containing conventional reinforcement. ACI Struct J 108:488–496

Syamsir A, Adnan W, Anggraini V, Zahari NM, Norhisham S, Beddu S, Itam Z (2020) Combine effect of soft drink tins and polypropylene fibres on mechanical properties of concrete. In: AIP conference proceedings, vol 2291, no 1. AIP Publishing LLC, , p 020101

Thomas J, Ramaswamy A (2007) Mechanical properties of steel fibre-reinforced concrete. J Mater Civ Eng 19(5):385–392

Usman M, Farooq SH, Umair M, Hanif A (2020) Axial compressive behavior of confined steel fibre reinforced high strength concrete. Constr Build Mater J 230:117043

Wang C (2006) Experimental investigation on behavior of steel fibre reinforced concrete MSc thesis. University of Canterbury, New Zealand

Wijatmiko I, Wibowo A, Nainggolan CR (2019) Strength characteristics of wasted soft drinks can as fibre reinforcement in lightweight concrete. Int J Geomate 17(60):31–36

Yang Y, Ren QW (2006) Experimental study on mechanical performance of steel fibre reinforced concrete. J Hohai Univ (Nat Sci) 1

Chapter 15
Effect of Recycled Asphalt Pavement (RAP) Aggregates on Strength of Fly Ash-GGBS-Based Alkali-Activated Concrete (AAC)

Hima Kiran Sepuri, Nabil Hossiney, Sarath Chandra, Yu Chen, Patrick Amoah Bekoe, and Vishnu Sai Nagavelly

15.1 Introduction

In India substantial growth in infrastructure projects has been observed due to recent urbanization, leading to the generation of waste in large quantities. Further, due to rising emission concerns, it has become imperative to adapt to environment-friendly measures to recycle the waste. Also, recycling waste in construction industry helps to mitigate various problems faced in construction industry (Chandra et al. 2021; Das et al. 2018; Hossiney et al. 2018; Pyngrope et al. 2021; Thejas and Hossiney 2020). There are different types of construction waste generated such as brick waste, concrete waste, stone waste, glass waste and recycled asphalt pavement (RAP), and this study primarily focuses on the recycling of RAP aggregates. Removal of existing deteriorated bituminous roads leads to RAP generation. Past studies show significant use of RAP in construction industry, and though there are successful findings (Avirneni et al. 2016; Mogawer et al. 2012; Zaumanis et al. 2014), there is still a need for novel ways of its utilization. Particularly, in developing countries where it is

H. K. Sepuri · N. Hossiney (✉) · S. Chandra
Department of Civil Engineering, CHRIST (Deemed To Be University), Bangalore 560074, India
e-mail: nabil.jalall@christuniversity.in

Y. Chen
School of Highway, Chang'an University, Middle Section of NanErhuan Road, Xi'an 710064, China

P. A. Bekoe
Senior Adjunct Lecturer, Department of Civil Engineering, Kwame Nkrumah University of Science and Technology, Kumasi, Ghana

V. S. Nagavelly
Department of Civil Engineering, School of Engineering and Technology, Central Queensland University, Melbourne Campus, Australia

K. R. Reddy et al. (eds.), *Advances in Sustainable Materials and Resilient Infrastructure*, Springer Transactions in Civil and Environmental Engineering,
https://doi.org/10.1007/978-981-16-9744-9_15

still downgraded in landfills, incurring loss to the construction industry. Further, due to practical challenges, government agencies find it difficult to completely recycle RAP in asphalt pavements. Therefore, innovative ways of managing RAP are often welcome. Recently, the use of RAP aggregates in cement concrete has become an attractive proposition (Shi et al. 2018; Singh et al. 2018; Su et al. 2013a, b, 2014; Tia et al. 2012). Mukhopadhyay and Shi (2019) studied the potential of RAP aggregates in PCC pavement and concluded that RAP aggregates which possess more intermediate size particles, improves aggregate gradation and thus PCC-mixes with dense aggregate gradation can be obtained. Further, RAP replaced for natural aggregates provides adequate concrete properties in terms of strength and durability and incurs various other benefits pertaining to social, economy and environment. Singh and Ransinchung (2020) studied the feasibility of RAP aggregates in concrete pavements. Laboratory and field studies were performed to ascertain the same. It was found that RAP aggregates reduce compressive strength of concrete, and lower reduction in flexural strength was found. This was further confirmed by structural evaluation with falling weight deflectometer. An increase in porosity of slab was observed with RAP incorporation and caused a lowering in temperature differential. Finally, concrete pavements with RAP aggregates showed comparable or slightly lower stresses than that of control slab without RAP aggregates.

Most of the studies from past with respect to RAP in concrete mixtures use Portland cement (Hossiney et al. 2008, 2010), and environment-friendly approach needs more attention. Since Portland cement contributes towards CO_2 emissions, its reduced usage will benefit the environment (Davidovits 1994). Geopolymers and alkali-activated binders have been studied extensively (Diaz et al. 2010; Fernandez-Jiménez and Palomo 2003; Shi et al. 2006) and have shown promise to achieve industrial ecology and reduce CO_2 emissions when compared to the usage of Portland cement (McLellan et al. 2011; Turner and Collins 2013). Therefore, this study is conducted to understand the feasibility of RAP aggregates in AAC, which will have multiple benefits such as; decrease burden on natural aggregates, increase the reusability of RAP aggregates in construction industry and reduce the usage of Portland cement.

15.2 Materials and Methods

Materials used for the preparation of AAC were fly ash (FLA), ground granulated blast furnace slag (GGBS), sodium hydroxide (SH) pellets of 98% purity, sodium silicate (SS) solution having $Na_2O = 15\%$, $SiO_2 = 29\%$ and $H_2O = 56\%$. Natural aggregates were acquired from local quarry, while RAP was obtained from the nearby state highway. RAP aggregates were reclaimed using milling process and fractionated in coarse and fine fraction using 4.75 mm sieve. Specific gravity of raw materials is seen in Fig 15.1a. RAP aggregates coarse and fine (C-RAP and F-RAP) showed lower values when compared to natural aggregates coarse and fine (CNA and FNA). One of the major reasons for such reduction has been due to asphalt in RAP aggregates.

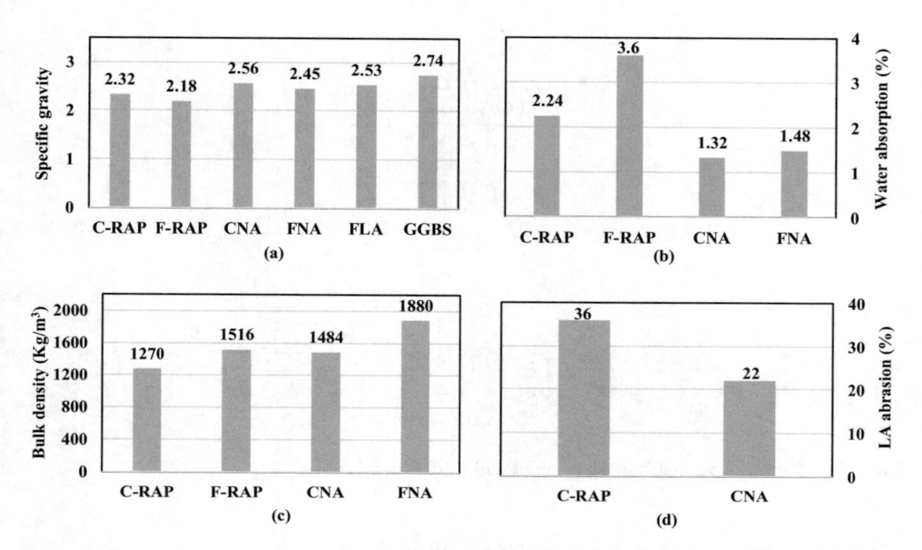

Fig. 15.1 Properties of aggregates, FLA, GGBS, **a** specific gravity, **b** water absorption, **c** bulk density, **d** abrasion resistance

Specific gravity of FLA ranges between 2.1 and 3.0 (Chesner et al. 1998), and similar results were obtained in this study. While, for GGBS in comparison to FLA, slightly higher specific gravity was observed. Figure 15.1b shows water absorption of aggregates. Higher water absorption for RAP aggregates was observed when compared to natural aggregates. This trend is similar to the past published findings (Singh et al. 2018). Figure 15.1c shows the bulk density of the aggregates. Crushed aggregate density depends on parent rock type and its porosity. While, in case of RAP aggregates, the presence of aged asphalt reduces the overall specific gravity causing a reduction in bulk density when compared to natural aggregates. Figure 15.1d shows LA abrasion value for coarse aggregates. A lower value is necessary to ensure that aggregates possess good resistance to frictional forces and wearing action. Higher value for C-RAP was observed and can be attributed to the presence of aged asphalt, which gets dislodged easily with fine particles, as exposed to abrasive charge. Particle size distribution of aggregates, FLA and GGBS are seen in Fig 15.2. FLA satisfied the fineness criteria of ASTM C 618 (2019), with more than 70% passing 45 μm sieve. Similarly, GGBS satisfied the fineness criteria of ASTM C989 (2018), with more than 80% passing 45 μm sieve. Particle size of F-RAP was coarser than FNA due to the presence of bitumen, which causes fine particles to agglomerate, while C-RAP was finer than CNA, and can be attributed to the milling operation, which breakdown bigger size aggregates into finer particles. Raw material chemical composition is presented in Table 15.1. Both RAP and natural aggregates have similar compositions. However, the silica content of natural and RAP aggregates is inert, in contrast to reactive silica in FLA and GGBS, which will contribute towards reaction with alkaline activators. Further, higher potassium concentrations were observed in natural and

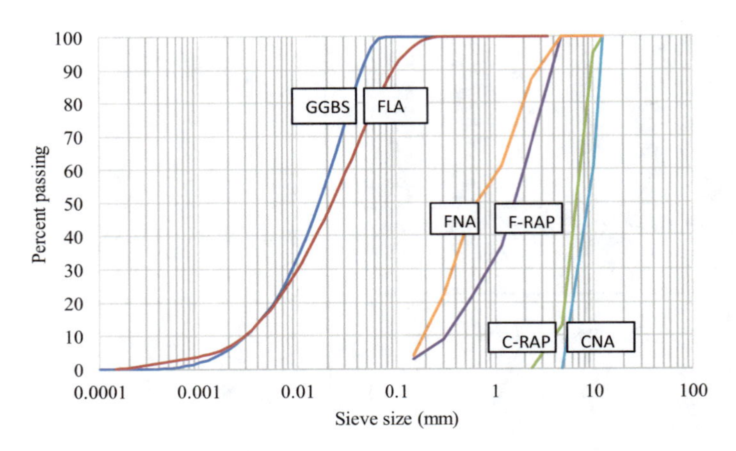

Fig. 15.2 Particle size distribution of aggregates, FLA, GGBS

Table 15.1 Chemical composition of aggregates, FLA, GGBS

Material	By weight, %							
	SiO_2	Al_2O_3	Fe_2O_3	CaO	MgO	K_2O	SO_3	LOI
Natural aggregates	52.9	11.1	7.5	8.3	–	14.1	–	–
RAP aggregates	53.6	7.1	7.8	7.9	–	16.5	–	–
FLA	46.5	19.5	13.5	7.6	1.9	–	–	2.4
GGBS	40.2	17.5	–	30	7.5	–	1.9	1.8

RAP aggregates due to local geological formations (Geological Survey of India 2006) FLA classified as class F as per ASTM C 618 (2019), since combined silicon dioxide, aluminum oxide and iron oxide was more than 70%, and GGBS satisfies the criteria's of BS 146 (2002) and ASTM C989 (2018), since the ratio of $(CaO + MgO)/(SiO_2)$ by mass exceeded 1.0, and sulfur content was less than 2.5%, respectively.

Details of evaluated mix proportions are presented in Table 15.2. Most constituents remain constant except for RAP varying from 0 to 75%, in increments of 25% by

Table 15.2 Mix proportions of AAC with RAP aggregates

Mix type	C-RAP	F-RAP	CNA	FNA	FLA	GGBS	SH sol	SS sol	SP	Water
	(Kg/m^3)									
AAC-RAP0	–	–	1201	647	286	122	41	103	8.1	60
AAC-RAP25	300	162	901	485	286	122	41	103	8.1	60
AAC-RAP50	601	324	600	323	286	122	41	103	8.1	60
AAC-RAP75	901	485	300	162	286	122	41	103	8.1	60

Note C-RAP and CNA were saturated surface dry and addition of sulphonated naphthalene formaldehyde as SP and extra water was to improve mix workability

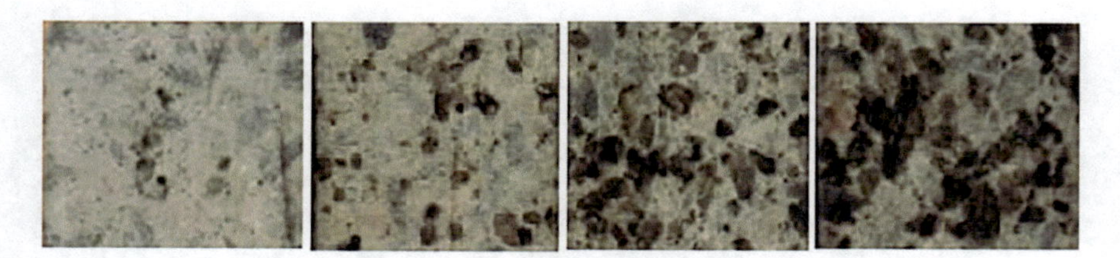

Fig. 15.3 Cross-section of hardened AAC. **a** AAC-RAP0, **b** AAC-RAP25, **c** AAC-RAP50, **d** AAC-RAP75

weight of the natural aggregates (coarse and fine). For instance, AAC-RAP0 indicates mix with no-RAP aggregates, while AAC-RAP25 indicates 25% replacement by weight of CNA, and 25% replacement by weight of FNA, by C-RAP and F-RAP aggregates, and so on for AAC-RAP50 and AAC-RAP75, respectively. Mixture constituents and proportions as well as mixing procedures were based on previous studies (Hardjito and Rangan 2005; Nath and Sarker 2014). Important mix design parameters used in this study were 8 molar SH solution, ratio of alkaline liquid to binder content is 0.35, ratio of SS solution to SH solution is 2.5, and FLA to GGBS proportion in total binder content is 70–30.

Replica of three cube specimens (150 mm × 150 mm × 150 mm) were casted for compressive strength, while cylinder specimens (150 mm diameter × 300 mm long) were casted for split tensile strength. Fresh AAC slump was determined in accordance with IS: 1199-1959 (1959). The compressive and splitting tensile strength were performed in accordance with IS: 516-1959 (1959) and IS: 5816-1999 (1999). This is a preliminary study, and only the strength properties of AAC were evaluated. Pan mixer was used for mixing, and the specimens prepared were air-cured at room temperature for a period of 7 and 28 days. Figure 15.3 shows the cross-section image of the developed AAC with RAP aggregates.

15.3 Results and Discussion

Fresh properties of AAC-RAP are shown in Fig 15.4. The unit weight and slump of AAC reduce with the addition of RAP aggregates, and this reduction was 2.7%, 5.3% and 8.4%, for AAC-RAP25, AAC-RAP50 and AAC-RAP75, when compared to AAC-RAP0, respectively. Behavior for such reduction can be attributed to lower specific gravity of RAP aggregates, when compared to natural aggregates. The slump is observed to be reduced by 12%, 21% and 32%, for AAC-RAP25, AAC-RAP50 and AAC-RAP75, when compared to AAC-RAP0, respectively. Reduction in slump is due to the presence of adhered dust around aged asphalt, and fine particles in RAP increase the demand for alkaline solution and water, thus influencing workability negatively. In another study (Hossiney et al. 2019, 2020), wherein RAP aggregates were used in alkali-activated paver blocks, the reduction trend in fresh properties

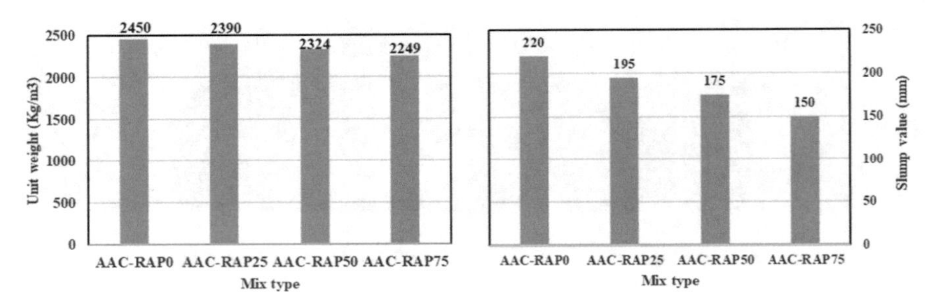

Fig. 15.4 Fresh properties of AAC-RAP mixes

was observed to be almost similar to the one reported here, indicating that RAP negatively influences the workability of AAC mixtures and proper use of admixtures is required to restore its workability.

Compressive and split tensile strength is considered as important property of hardened cement concrete to be known for most of the civil engineering applications. It will influence the performance of structural members and has bearing on its durability. The compressive and split tensile strength of hardened AAC mixes is shown in Fig 15.5. The addition of RAP aggregates has influenced the strength of hardened AAC negatively. The percentage reduction in 28 days' compressive strength was 14.3%, 22.9% and 37.1% for AAC-RAP25, AAC-RAP50 and AAC-RAP75, when compared to AAC-RAP0, respectively. The percentage reduction in 28 days split tensile strength was 10%, 16.7% and 22.2% for AAC-RAP25, AAC-RAP50 and AAC-RAP75, when compared to AAC-RAP0, respectively. The reduction rate in compressive strength is more sensitive to the RAP addition compared to that of split tensile strength as noticed in Fig 15.6. Such reductions in strength can be attributed to inability of RAP particles to adhere to the binder phase. Also, the presence of bitumen deteriorates the adhesion property and influences the strength properties of AAC-RAP mixes (Costa et al. 2021). Further, past literatures show similar trends in strength reduction for PCC-RAP mixes (Shi et al. 2017, 2019). Some factors contributing towards a reduction in strength of Portland cement concrete (PCC) with RAP aggregates have been due to higher porosity in the interfacial transition zone

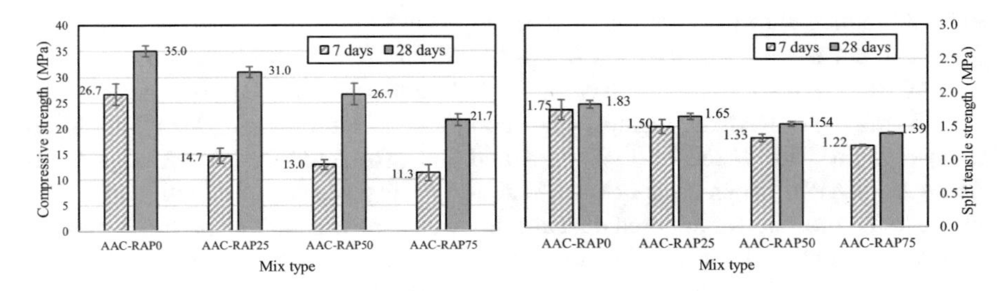

Fig. 15.5 Compressive and split tensile strength of AAC-RAP mixes

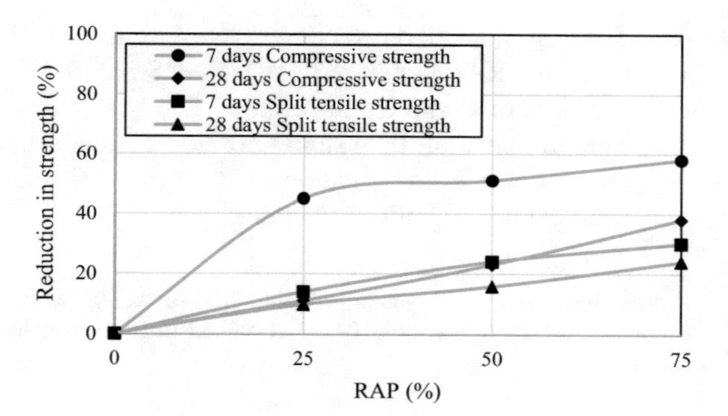

Fig. 15.6 Reduction in strength of AAC-with-RAP compared to AAC-without-RAP

(ITZ), due to lower amounts of calcium silicate hydrate at interface, as well lowering bulk modulus and leading to easier crack initiation (Brand and Roesler 2017a). Further, cohesive failure in asphalt (Brand and Roesler 2017b), and poor adhesion between RAP cementitious matrix (Said et al. 2017), reduces strength and bonding at the aggregate paste interface. Generally, the addition of FLA/GGBS improves the microstructure of cementitious matrices making it more compact and reducing voids. Also, in case of AAC, the improved bonding at aggregate paste interface is due to the presence of comparatively smaller particles that attach tightly, stronger ITZ, and absence of calcium hydroxide $Ca(OH)_2$, resulting in improved properties for AAC mixtures over PCC mixtures (de Toledo Pereira et al. 2018). This will further benefit AAC-RAP mixes in terms of durability, when compared to PCC-RAP mixes.

15.4 Conclusions

A comprehensive literature review indicates that the utilization of RAP aggregates in concrete is primarily based on Portland cement, and few efforts have been made to understand its effect in alkali-activated concrete (AAC). This preliminary investigation presents the effect of RAP on strength properties of fly ash/GGBS alkali-activated concrete. Based on laboratory experiments, the following conclusions are enumerated.

1. Inclusion of RAP aggregates inversely affects the workability of AAC-RAP concrete and was found to be highest for AAC-RAP75. The use of chemical admixtures is required to restore workability of AAC-RAP mixes.
2. Replacing natural aggregates by RAP aggregates in an AAC mix has reduced the compressive and splitting tensile strength and was found to be highest for AAC-RAP75.
3. The percentage reduction of compressive strength for AAC-RAP mixtures containing varying replacements of RAP aggregates is observed to be higher than its splitting tensile strength.

4. The minimum 28 days' compressive strength of 22 MPa for AAC-RAP75 indicates that RAP inclusive fly ash/GGBS-based AAC mixtures have potential in non-structural applications.
5. The findings of this study provide future scope to perform the detailed microstructure investigation of RAP in AAC with fly ash/GGBS and as well evaluate its durability properties.

Acknowledgements The authors would like to thank the lab staff of Civil and Mechanical Engineering Department at CHRIST (Deemed to be University) for providing support for this study.

References

ASTM C989/C989M-18a (2018) Standard specification for slag cement for use in concrete and mortars. ASTM International, West Conshohocken, PA. www.astm.org. https://doi.org/10.1520/C0989_C0989M-18A

ASTM C618-19 (2019) Standard specification for coal fly ash and raw or calcined natural Pozzolan for use in concrete. ASTM International, West Conshohocken, PA. www.astm.org. https://doi.org/10.1520/C0618-19

Avirneni D, Peddinti PRT, Saride S (2016) Durability and long term performance of geopolymer stabilized reclaimed asphalt pavement base courses. Constr Build Mater 121:198–209. Article first published online 3 June 2016. https://doi.org/10.1016/j.conbuildmat.2016.05.162

Brand AS, Roesler JR (2017a) Bonding in cementitious materials with asphalt-coated particles: part I-the interfacial transition zone. Constr Build Mater 130:171–181. Article first published online 12 Oct 2016. https://doi.org/10.1016/j.conbuildmat.2016.10.019

Brand AS, Roesler JR (2017b) Bonding in cementitious materials with asphalt-coated particles: part II-cement-asphalt chemical interactions. Constr Build Mater 130:182–192. Article first published online 12 Oct 2016. https://doi.org/10.1016/j.conbuildmat.2016.10.013

BS 146 (2002) Specification for Portland blast furnace cements. British Standard, London, UK

Chandra SK, Krishnaiah S, Reddy NG, Hossiney N, Peng L (2021) Strength development of geopolymer composites made from red Mud-Fly Ash as a subgrade material in road construction. J Hazard Toxic Radioact Waste 25:04020068, Article first published online 11 Nov 2020. https://doi.org/10.1061/(ASCE)HZ.2153-5515.0000575

Chesner WH, Collins RJ, MacKay MH (1998) Users guidelines for waste and by-product materials in pavement construction. Federal Highway Administration, McLean, VA, United States. https://rosap.ntl.bts.gov/view/dot/38365

Costa JO, Borges PHR, dos Santos FA, Bezerra ACS, Blom J, Van den bergh W (2021) The effect of reclaimed asphalt pavement (RAP) aggregates on the reaction, mechanical properties and microstructure of alkali-activated slag, CivilEng 2(3):794–810. Article first published online 4 Sept 2021. https://doi.org/10.3390/civileng2030043

Das P, Matcha B, Hossiney N, Mohan MK, Roy A, Kumar A (2018) Utilization of iron ore mines waste as civil construction material through geopolymer reactions. In: Geopolymers and other geosynthetics, Mazen Alshaaer and Han-Yong Jeon, Intech Open. https://doi.org/10.5772/intechopen.81709

Davidovits J (1994) Global warming impact on the cement and aggregate industries. World Resour Rev 6(2):263–278 de Toledo Pereira DS, da Silva FJ, Porto ABR, Candido VS, da Silva ACR, Filho FDCG, Monteiro SN (2018) Comparative analysis between properties and microstructures of geopolymeric concrete and portland concrete. J Mater Res Technol 7(4):606–611. Article first published online 22 Oct 2018. https://doi.org/10.1016/j.jmrt.2018.08.008

Diaz EI, Allouche EN, Eklund S (2010) Factors affecting the suitability of fly ash as source material for geopolymers. Fuel 89(5):992–996. Article first published online 30 Sept 2009. https://doi.org/10.1016/j.fuel.2009.09.012

Fernandez-Jiménez A, Palomo A (2003) Characterisation of fly ashes. Potential reactivity as alkaline cements. Fuel 82(18):2259–2265. Article first published online 24 Jun 2003. https://doi.org/10.1016/S0016-2361(03)00194-7

Geological survey of India (2006) Geology and mineral resources of the states of India, Part VII—Karnataka & Goa, Operation: Karnataka & Goa, Bangalore, India

Hardjito D, Rangan BV (2005) Development and properties of low calcium fly ash based geopolymer concrete. Curtin University of Technology, Perth, Australia. http://hdl.handle.net/20.500.11937/5594

Hossiney N, Wang G, Tia M, Bergin MJ (2008) Evaluation of concrete containing RAP for use in concrete pavement. In: Proceedings of the transportation research board annual meeting (Cdrom), transportation research board, Washington, DC. https://trid.trb.org/view/848720

Hossiney N, Tia M, Bergin MJ (2010) Concrete containing RAP for use in concrete pavement. Int J Pavement Res Technol 3(5):251–258

Hossiney N, Das P, Mohan MK, George J (2018) In-plant production of bricks containing waste foundry sand—A study with Belgaum foundry industry. Case Stud Constr Mater 9:e00170. Article first published online 15 May 2018. https://doi.org/10.1016/j.cscm.2018.e00170

Hossiney N, Sepuri HK, Mohan MK, HR A, Govindaraju S, Chyne J (2019) Alkali-activated concrete paver blocks made with recycled asphalt pavement (RAP) aggregates. Case Stud Constr Mater 12:e00322. Article first published online 7 Dec 2019. https://doi.org/10.1016/j.cscm

Hossiney N, Sepuri HK, Mohan MK, Chandra SK, Kumar SL, Thejas HK (2020) Geopolymer concrete paving blocks made with recycled asphalt pavement (RAP) aggregates towards sustainable urban mobility development. Cogent Eng 7(1):1824572. Article first published online 5 Oct 2020. https://doi.org/10.1080/23311916.2020.1824572

IS: 1199 (1959) Methods of sampling and analysis of concrete, Bureau of Indian Standards, New Delhi, India

IS: 516 (1959) Indian standard methods of tests for strength of concrete, Bureau of Indian Standards, New Delhi, India

IS: 5816 (1999) Indian standard for splitting tensile strength of concrete—Method of Test, Bureau of Indian Standards, New Delhi, India

McLellan BC, Williams RP, Lay J, Riessen AV, Corder GD (2011) Costs and carbon emissions for geopolymer pastes in comparison to ordinary portland cement. J Clean Prod 19(9):1080–1090. Article first published online 22 Feb 2011. https://doi.org/10.1016/j.jclepro.2011.02.010

Mogawer W, Bennert T, Daniel JS, Bonaquist R, Austerman A, Booshehrian A (2012) Performance characteristics of plant-produced high RAP mixtures, Road Mater. Pavement Des 13:183–208. Article first published online 18 April 2012. https://doi.org/10.1080/14680629.2012.657070

Mukhopadhyay A, Shi X (2019) Utilization of reclaimed asphalt pavement aggregates in Portland cement concrete for concrete pavement. ACI special publication, vol 334, pp 13–32, Article first published online 30 Sept 2019. https://www.concrete.org/publications/internationalconcreteabstractsportal.aspx?m=details&id=51720251

Nath P, Sarker PK (2014) Effect of GGBFS on setting, workability and early strength properties of fly ash geopolymer concrete cured in ambient condition. Constr Build Mater 66:163–171. Article first published online 12 June 2014. https://doi.org/10.1016/j.conbuildmat.2014.05.080

Pyngrope M, Hossiney N, Chen Y, Thejas HK, Chandra SK, Alex J, Kumar SL (2021) Properties of alkali-activated concrete (AAC) incorporating demolished building waste (DBW) as aggregates. Cogent Eng 8:1870791. Article first published online 11 Jan 2021. https://doi.org/10.1080/23311916.2020.1870791

Said SEEB, Khay SEE, Achour T, Loulizi A (2017) Modelling of the adhesion between reclaimed asphalt pavement aggregates and hydrated cement paste. Constr Build Mater 152:839–846. Article first published online 12 July 2017. https://doi.org/10.1016/j.conbuildmat.2017.07.078

Shi C, Krivenko PV, Roy DM (2006) Alkali-activated cements and concretes. Taylor and Francis, Abingdon, UK

Shi X, Mukhopadhya A, Liu KW (2017) Mix design formulation and evaluation of portland cement concrete paving mixtures containing reclaimed asphalt pavement. Constr Build Mater 152:756–768. Article first published online 12 July 2017. https://doi.org/10.1016/j.conbuildmat.2017.06.174

Shi X, Muhkhopadhyay A, Zollinger D (2018) Sustainability assessment for Portland cement concrete pavement containing reclaimed asphalt pavement aggregates. J Clean Prod 192:569–581. Article first published online 5 May 2018. https://doi.org/10.1016/j.jclepro.2018.05.004

Shi X, Mirsayar MM, Mukhopadhyay A, Zollinger D (2019) Characterization of two-parameter fracture properties of portland cement concrete containing reclaimed asphalt pavement aggregates by semicircular bending specimens. Cem Concr Compos 95:56–69. Article first published online 19 Oct 2018. https://doi.org/10.1016/j.cemconcomp.2018.10.013

Singh S, Ransinchung GD, RN, Debbarma S, Kumar P (2018) Utilization of reclaimed asphalt pavement aggregates containing waste from sugarcane mill for production of concrete mixes. J Clean Prod 174:42–52. Article first published online 19 October 2017. https://doi.org/10.1016/j.jclepro.2017.10.179

Singh S, Ransinchung RN, GD (2020) Laboratory and field evaluation of RAP for cement concrete pavements. J Transp Eng B: Pavements 146(2):04020011. Article first published online Feb 27 2020. https://doi.org/10.1061/JPEODX.0000162

Su YM, Hossiney N, Tia M (2013a) The analysis of air voids in concrete specimen using X-ray computed tomography. In: Proceedings of the SPIE, nondestructive characterization for composite materials, aerospace engineering, civil infrastructure, and homeland security, vol 8694. Article first published online 16 April 2013. https://doi.org/10.1117/12.2012267

Su YM, Hossiney N, Tia M (2013b) Indirect tensile strength of concrete containing reclaimed asphalt pavement using the superpave indirect tensile test. Adv Mat Res 723:368–375. Article first published online Aug 2013. https://doi.org/10.4028/www.scientific.net/AMR.723.368

Su YM, Hossiney N, Tia M, Bergin M (2014) Mechanical properties assessment of concrete containing reclaimed asphalt pavement using the superpave indirect tensile strength test. J Test Eval 42(4):912–920. Article first published online 1 July 2014. https://doi.org/10.1520/JTE20130093

Thejas HK, Hossiney N (2020) Use of waste foundry sand in precast concrete paver blocks—A study with belgaum foundry industry. In: Kanwar V, Shukla S (eds) Sustainable civil engineering practices, lecture notes in civil engineering, vol 72. Article first published online 30 April 2020. https://doi.org/10.1007/978-981-15-3677-9_1

Tia M, Hossiney N, Su YM, Chen Y, Do TA (2012) Use of reclaimed asphalt pavement in concrete pavement slabs. Florida Department of Transportation, Tallahassee, Florida, USA

Turner LK, Collins FG (2013) Carbon dioxide equivalent (CO_2-e) emissions: a comparison between geopolymer and OPC cement concrete. Constr Build Mater 43:125–130. Article first published online 14 Mar 2013. https://doi.org/10.1016/j.conbuildmat.2013.01.023

Zaumanis M, Mallick RB, Frank R (2014) 100% recycled hot mix asphalt: a review and analysis. Resour Conserv Recycl 92:230–245. Article first published online 1 Aug 2014. https://doi.org/10.1016/j.resconrec.2014.07.007

Chapter 16
An Index for Assessment of Onsite Waste Management Performance in Indian Construction Sites

Swarna Swetha Kolaventi, Tezeswi Tadepalli, and M. V. N. Siva Kumar

16.1 Introduction

The construction industries are considered to be the major greenhouse gas emitters (Semaan et al. 2017). The increase in urbanization leads to the corresponding increase in construction and demolition (C&D) waste generation in the last few years globally (Jain et al. 2020a). C&D wastes are adverse in comparison with domestic waste as they possess diverse characteristics and can ultimately lead to polluted environment (Bolyard et al. 2019; Baxter et al. 2019; Rao et al. 2020).

Inappropriate C&D waste management leads to tremendous urban menace by creating serious (i) air pollution—i.e. huge traces of particulate matter and can create severe dust nuisance, which can ultimately pose a severe threat to human health. (ii) Noise pollution—noise produced by the construction activities (demolition, usage of pneumatic hammers, concrete mixers, machinery operations, etc.) creates severe disturbances to day to day activities. (iii) Water pollution illegal disposal of C&D waste debris into water bodies creates serious water-logging scenarios. The utilization of non-renewable natural resources poses harmful threats to the environment globally. Therefore, there is an enormous requirement in monitoring the onsite C&D waste (Gupta 2018).

The global statistics of C&D waste generated are: USA-569 Million tons (MT), UK-136.2 MT, Australia-67 MT, China 1130 MT, Japan77 MT, Hong Kong-58 MT (Aliferova and Ncube 2017; Kabirifar et al. 2020; Kolaventi et al. 2021). The majority

S. S. Kolaventi (✉) · T. Tadepalli · M. V. N. Siva Kumar
NIT Warangal, Warangal, India
e-mail: swarnaswetha@student.nitw.ac.in

T. Tadepalli
e-mail: tezeswi@nitw.ac.in

M. V. N. Siva Kumar
e-mail: mvns@nitw.ac.in

K. R. Reddy et al. (eds.), *Advances in Sustainable Materials and Resilient Infrastructure*,
Springer Transactions in Civil and Environmental Engineering,
https://doi.org/10.1007/978-981-16-9744-9_16

of the waste material is usually disposed onto landfills or onto sides of the roads. The decrease in landfill sites and unavailability of raw materials like aggregates create a need for efficient management of C&D waste (Kolaventi et al. 2019). Construction waste management (CWM) system can recover the valuable material, which can conserve the existing natural resources. The first step for the recovery of valuable materials needs to start onsite (Kolaventi et al. 2021). However, the lack of quantified data makes onsite WM challenging.

In India, the amount of solid waste generated is 960 MT, out of which the construction waste is 100–400 MT, and this value may increase as it does not include non-documented waste (Kolaventi et al. 2019). Management of the wastes generated during construction is challenging in most of the Indian municipalities (Bolyard and Reinhart, 2016; Ghanimeh et al. 2016). Segregation, reuse and disposal of CW are given minimum importance. Regardless of possessing potential waste management infrastructure and policies, onsite CWM is non-existent in most of the construction sites (Kolaventi et al. 2020).

The current practice typically adopted by Indian construction companies to manage construction waste is illegal disposal of C&D waste onto low-lying areas, sides of the roads and incineration. Such procedures may tend to provide impermanent solution (Kolaventi et al. 2017). Therefore, there is an emerging need in developing alternative solutions for managing construction waste. The primary reason for non-implementation of CWM by Indian construction industries is due to the lack of standard documentation procedures at site (Koutamanis et al. 2018; Zhang et al. 2019; Islam et al. 2019). To mitigate this barrier, the current article attempts to develop an onsite tool to assess onsite CWM performance. Therefore, the objective of the current article is defined as—(i) rank the variables using relative importance index and evaluate the category weightages correspondingly and (ii) develop an index system to assess onsite WM performance of construction projects.

16.2 Literature Review

16.2.1 Definition of Construction Waste

CW is defined in previous literatures as the difference between the material utilized and that delivered at construction site (Cochran and Townsend 2007). Furthermore, it is defined as material loss that does not aid financial benefit to the project (Koshy and Apte 2012). It is a combination of construction materials such as rubble, steel, timber and tiles (Dania et al. 2007). Alternatively, it is also defined as an comprehensive and rational method to maintain environmental sustainability (Gilpin 1996; Townsend et al. 2004). Waste minimization is defined in existing literature as an action which produce minimal waste at source and which can be treated (Hopkinson et al. 2019; Charlson and Hons 2019). The most common constitutes of CW are poorly segregated construction materials (aggregate, concrete, timber, steel, bricks, earth, wires, etc.) and at times mixed with solid and hazardous waste.

16.2.2 Existing Studies on C&D Waste Management.

The typical studies available in CWM include—(i) influence factor identification (Owolabi et al. 2016; Wang et al. 2010; Wang and Zhengdao 2014; Weisheng and Yuan 2010), (ii) waste assessment (Havard Bergsdal et al. 2016; Shen et al. 2005; Solís-Guzmán et al. 2009), (iii) waste forecast (Banias et al. 2011; Blackwell et al. 2011; Cheng and Ma 2013), (iv) identification of barriers, benefits and measures studies on CWM (Yuan et al. 2011; Yuan 2017; Kolaventi et al. 2021), (v) attitude and behavior studies on CWM (Jain et al. 2020a; Yang et al. 2020; Friedrich 2021; Zhang et al. 2021) and (vi) C&D waste recycling (Jin et al. 2017; Ram et al. 2020; Zhang et al. 2020).

However, there are nominal studies available on onsite WM estimation using index-based system (Campbell 2019).

16.2.3 C&D WM in India

The Bureau of Indian Standards (BIS) is the National standard body of India established on 23 December 1986 for the harmonious development of standardization, quality certification of goods and the concerned affairs. Central pollution control board (CPCB 2016) of India is a statutory organization under the Ministry of Environment, Forest and Climate Change (Mo.E.F.C.C) (Ministry of Housing and Urban Affairs 2018). It co-ordinates the activities of SPCB by providing technical assistance with disputes resolving (CPCB 2017). Swachh Bharat Mission recognized the need for C&D waste management and doubled the points to 100 despite; there is improper management of C&D waste in India. It is estimated by Centre for Science and Environment (CSE), India recycles just 1% of its C&D waste that is being generated (Swachh Bharat Mission 2017). It is estimated that nearly 53 major cities were planning to set up recycling plants by 2020, of which 13 are established. The C&D waste recycling plants in India are running at loss, and it is reported that the material reaching the recycling plants is poorly segregated, the improper segregation of the C&D waste is due to lack of adoption of adequate onsite procedures. Therefore, in order to provide a framework for monitoring the CWMP, an index system is devised to measure onsite waste management performance of the construction sites, in the current article. A summary of C&D waste management guidelines in India is mentioned in appendix Table 16.6.

16.2.4 Software Developments in C&D Waste Management

Technological innovations are needed for the successful management of C&D waste (Oliveira et al. 2019). Earlier studies reported that the increase in societal development accelerated the usage of software technologies such as BIM (Building information and modeling), GIS (geographic information system), BDA (Big data analytics), RFID (Radio frequency identification tags), which have proven successful in solving C &D WM problems (Huang et al. 2018; Lu 2019). Similar studies are summarized in appendix (Table 16.7). Software advancements tend to solve C&D WM problems. However, the software extensions at site, which can quantify the onsite waste management performance of construction projects, are limited. The assessment of waste management performance onsite can aid in enhanced benefits like (i) identification of weak zones in the corresponding projects based on which measures can be proposed at management level, (ii) quantification of the total amount of waste that is being generated at their projects, (iii) recovery of valuable recyclable material, (iv) identification of the effective segregation system, (v) estimation of the alternative methods for material disposal. Therefore, the current study focused on developing an index system to assess onsite waste management performance of the individual projects.

16.3 Methodology

The article is divided into two principal sections (Fig. 16.1)—(i) ranking the variables using relative importance index and evaluate the category weightages correspondingly and (ii) developing an index system to assess onsite waste management performance of construction sites. Questionnaire and case studies are the survey tools used from the tested and validated scales of the concepts.

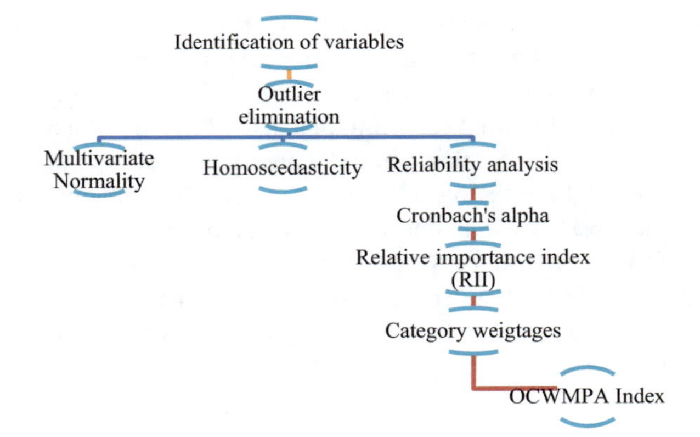

Fig. 16.1 Methodology

16.3.1 Identification and Ranking of OCWMPA Variables

The variables that can improve performance assessment onsite are identified based on previous literatures as well as interviews with the expert panel. At first a list of top-ranked variables (which are used to assess the CWM performance) are selected from previous literatures. The variables are further scrutinized by expertise professional (academicians, engineers with experience greater than 20 years) ahead of questionnaire drafting. Thus a questionnaire is drafted with 32 variables (Table 16.1). Seven-point Likert scale is used in the survey, which ranges from 1 to 7 (strongly disagree to strongly agree). The questionnaire is divided into two sections. The first half gathers the background data of the respondents, and the subsequent section focuses on the respondent's opinion on the level of significance of OCWMPA variables.

Google Forms are used for data collection from respondents, which include engineers, contractors, academicians and construction managers. The survey is conducted both virtually as well as offline. The respondents are selected based on three factors—(i) Academic background (B-tech), (ii) Knowledge-basic knowledge on CWM, and (iii) experience (minimum of 2 years of field experience), both government and private organizations are involved in the survey. Detailed demographics are shown below. The completed questionnaires are then exported to MS Excel and later to IBM SPSS 20.0 for further analysis. A total of 177 responses received of which 154 are online and 23 offline. The detailed demographics of the respondents are male respondents 87% and female respondents 13%, age below 30 years—40% and above 30 years—60%. Engineer—40%, manager—30%, contractor—20% and academicians—40%. Experience 1–5 years—35%, 6–10 years—55%, 11–15 years—5% and above 15 years—5%. To rank the variables based on the level of importance, relative importance index (RII) is used. The index ranges from $0 < \text{RII} < 1$, and the variables are ranked correspondingly.

$$RII = \frac{\sum W}{A \times N} \tag{16.1}$$

where

W weight given by the respondents to each variable (1–7)
A highest weight (7)
N the total no of people involved in survey.

16.3.2 Developing Onsite Construction Waste Management Performance Assessment (OCWMPA) Index

The level of CWM performance of the individual organizations can be assessed by utilizing the OCWMP variables. To devalue into implementation, top 25 ranked OCWMPA variables (based on RII values) are selected for the study to assess project

Table 16.1 OCWMPA variables

Factor category	Factor label	Factor name
Human resources	H1	Contractor involvement in construction waste management
	H2	Client involvement in construction waste management
	H3	Education of staff working on the construction site
	H4	Training programs at the construction site
	H5	Appointment of workers especially for segregation of waste
	H6	Supplier's involvement in construction waste collection
	H7	The management team for managing construction waste
Construction methods planning	C1	Supervision and control of the amount of construction waste generated
	C2	The practice of segregation, i.e. maintenance of separate bins for construction waste segregation
	C3	Cleaning up the site on a daily basis
	C4	Quantification of the amount of construction waste generated
	C5	RRR (reduce, reuse, recycle) strategy
	C6	Installation of information boards for segregation of construction waste
	C7	Allocation of separate space for material sorting at initial stages of construction
	C8	Informing methods to deal with remaining CW after recycling
	C9	Disposal of construction waste periodically by open dumping, incineration, etc
Materials and equipment	M1	Installation of recycling equipment at construction sites
	M2	Installation of equipment for waste sorting
	M3	Installation of mobile recycling plant at the construction site
	M4	Usage of recycled material at the construction site
	M5	Material transportation system for construction waste
Design and documentation	D1	Separate documentation (records) on recycling waste

(continued)

Table 16.1 (continued)

Factor category	Factor label	Factor name
	D2	Checklist on the execution of waste management plan
	D3	Database management system or any software technology for construction waste (BIM, etc.)
	D4	Maintenance of record on training programs, i.e. past, present and future schedules at construction sites
	D5	Changing of the design
Industry policy	I1	Awareness of government policies on construction waste generated
	I2	Following government norms on dealing with construction waste
	I3	Incentive for a contractor having a plan on CWM
	I4	Establish criteria for the quality and safety of recycled materials
	I5	Documentation of payment of taxes and penalties if the waste exceed permitted limits according to government policies
	I6	Practice of making money out of waste, i.e. selling, etc

performance. To analyze acceptance on the selected factors, they are modified to question answer format and the user opts for the most suitable answer. For instance, the question "we have controlled supervision and quantification of the construction waste generated", there are five choices available based on agreement, i.e. strongly disagree—1 to strongly agree—5. The scores are awarded with the help of expert panel members for 1–0, 2–0.25, 3–0.50, 4–0.75 and 5–1.00. This process is vital for variables, which are difficult to estimate in project management. Likewise, top 25 questions are converted to a question response format and distributed online.

Five construction organizations consisting of five respondents in each organization are used in the present study. The organizations are selected on the basis of the following parameters (i) knowledge on CWM, (ii) implementation of CWM, (iii) respondents experience greater than 5 years and (iv) with companies greater than 10 years of service life. Both government and corporate organizations are involved in the present study. A sample question response format for five questions is presented in Table 16.2.

The OCWMPA index indicates project performance towards CWM calculated using Eq. 16.2. The index ranges from 0 to 1000 and is further classified as OCWMPA index ranging from 0 to 250 as poor, 251 to 500 as fair, 501 to 750 as good and 751 to 1000 as excellent (Cha et al. 2009).

Table 16.2 Question–response format

Variable	Question	Response option	Score
C5	We adopt RRR strategy at construction sites	A,B,C,D,E	0,0.25,0.50,0.75,1
C1	We have controlled supervision and quantification of the construction waste generated		
I6	We have a practice of making money out of waste, i.e. selling, etc		
I5	We maintain documentation regarding Construction waste management (on payment of taxes, penalties) if the waste exceeds permitted limits based on government norms		
C2	We practice waste segregation at the construction site		

A—strongly disagree, B—somewhat disagree, C—neutral, D—somewhat agree, E—strongly agree

$$OCWMPA\,index = \sum_{i=4}^{4}\left(\sum_{j=1}^{1}\left(\sum_{k=1}^{m}(RS_{ijk} \times RW_{ijk}\right) \times FW_{ij} \times CW_i\right) \quad (16.2)$$

where

RS_{ijk} score awarded by kth response for jth factor in ith category;
RW_{ijk} weightage of kth response for jth factor in ith category $0 < RW_{ijk} \leq 1$;
CW_i weight of ith category $0 < CW_i \leq 7$;
FW_{ij} weight of jth factor in ith category $0 < FW_{ij} \leq 1$;
l number of factors in ith category; and
m number of responses for jth factor in ith category.

16.4 Analysis Results and Discussion

16.4.1 Outlier Elimination

The outliers and internal consistency of the data are assessed using various statistical tests.

16.4.1.1 Multivariate Normality

It is usually calculated using Mahalanobis distance. It is calculated by using IBM SPSS statistics. Mahalanobis distance for the total 32 variables is 106.001, which indicates a high distance. To reduce the distance, probability values are created and

Table 16.3 Mahlanobis distance before after outlier's elimination

Residuals statistics

	Minimum	Maximum	Mean	Std. deviation	N
Mahal. Distance	2.473	106.001	31.819	15.852	177
Mahal. Distance	2.708	71.083	31.808	13.387	167

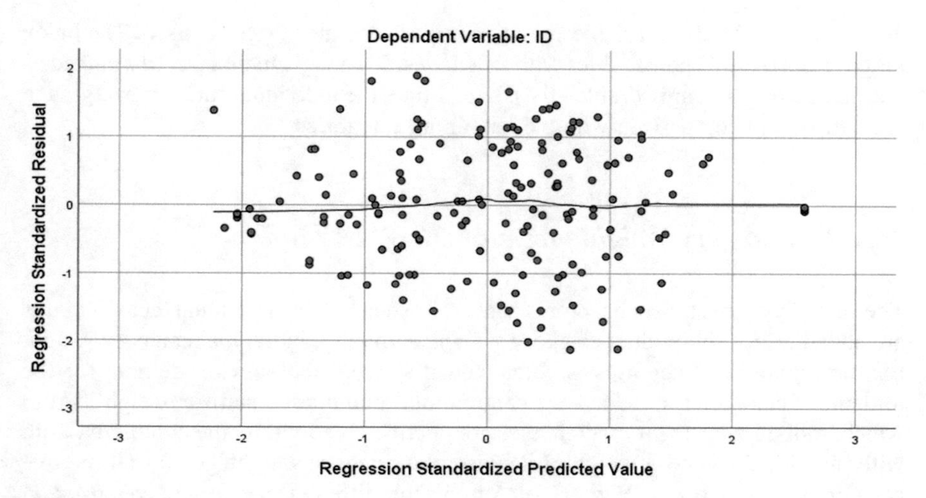

Fig. 16.2 Homoscedasticity

checked. The responses with probability values less than 0.01 are deleted. A total of 10 such responses were deleted and are eliminated from further analysis. The Mahalanobis value after deletion is 71.083 for a sample size of 167 respondents (Table 16.3).

16.4.1.2 Homoscedasticity

Homoscedasticity is calculated using IBM SPSS, the results are shown below. Loess line has been added to determine homoscedasticity for the data. From the scatter plot, it can be seen that the loess line is free from sharp curves; hence the data did not violate the assumption of homoscedasticity (Fig. 16.2).

16.4.1.3 Reliability Analysis

The internal consistency of the data is measured by using Cronbach's alpha. The value ranges from 0 (lower consistency) to 1 (higher consistency) (Olaniyi 2019).

The Cronbach's alpha for the entire variables is 0.929 and is acceptable with excellent internal consistency (Hair et al. 2010).

16.4.2 Relative Importance Index (RII)

To rank the OCWMPA variables, relative importance index (RII) is used. The index ranges from 0 to 1. The variables with the highest RII are given first priority ranked as one and correspondingly (Table 16.4). In addition, the individual category weightage is calculated in comparison with the remaining categories.

16.4.3 Category Weightages and OCWMPA Index

The sum of weights for each factor is analyzed. Then, all the individual factor weights are added to give individual category weight. Similarly, individual category weight for the remaining categories is calculated. Later on, the sum of weights for the remaining categories is added up to give total cumulative sum or weight. From which, individual category weightages are obtained by dividing the cumulative sum with individual category weight. Likewise, the category weightages for all the five categories are analyzed (Fig. 16.3). Among the different categories, *construction method and planning* occupies the highest weightage of 0. 286. In addition, the category weightage of *Human resources* is 0.216, *materials & equipment*—0.154, design and documentation—0.151 and industry policy—0.193. By using these category weightages, OCWMPA index is thus evaluated using Eq. 16.2.

It is therefore evident that one out of five case studies has shown excellent performance towards construction waste management respectively (Table 16.5).

16.4.4 Content Validity

Furthermore, the content validity of the variables used in the study is validated. For this, the subject matter experts (SME) panels are used to assess the content validity. The SME are questioned regarding the variables used and their relevance in the study. The organizations are selected on the basis of the following parameters (i) knowledge on CWM, (ii) implementation of CWM, (iii) respondents experience greater than 5 years and (iv) companies greater than 10 years of service life. A total of 15 SME are thus selected. To evaluate the SME responses, Lawshe's content validity test is used.

$$\mathrm{CVR} = [(\mathrm{ne} - \mathrm{N}) - \mathrm{N}/2]/2 \tag{16.3}$$

Table 16.4 Ranking of OCWMPA variables

OCWMPA	Coding	RII	Rank
RRR (reduce, reuse, recycle) strategy	C5	0.786	1
Supervision and control of the amount of construction waste generated	C1	0.781	2
Practice of making money out of waste, i.e. selling, etc	I6	0.772	3
Documentation of payment of taxes and penalties if the waste exceed permitted limits according to government policies	I5	0.769	4
The practice of segregation, i.e. maintenance of separate bins for construction waste segregation	C2	0.766	5
Usage of recycled material at the construction site	M4	0.763	6
Training programs at the construction site	H4	0.763	7
Contractor involvement in CW management	H1	0.756	8
Cleaning up the site on a daily basis	C3	0.754	9
Allocation of separate space for material sorting at initial stages of construction	C7	0.754	10
Incentive for a contractor having a plan on CWM	I3	0.754	11
Education of staff working on the construction site	H3	0.751	12
Following government norms on dealing with construction waste	I2	0.750	13
Awareness of government policies on construction waste generated	I1	0.747	14
A management team for managing CW	H7	0.744	15
Quantification of the amount of CW generated	C4	0.744	16
Material transportation system for construction waste	M5	0.740	17
Appointment of workers especially for segregation of waste	H5	0.735	18
Establish criteria for the quality and safety of recycled materials	I4	0.735	19
Checklist on the execution of waste management plan	D2	0.729	20
Installation of information boards for segregation of construction waste	C6	0.727	21
Informing methods to deal with remaining CW after recycling	C8	0.718	22
Maintenance of record on training programs, i.e. past, present and future schedules at construction sites	D4	0.711	23
Database management system or any software technology for construction waste (BIM, etc.)	D3	0.708	24
Installation of recycling equipment at construction sites	M1	0.704	25
Installation of mobile recycling plant at the construction site	M3	0.704	26
Client involvement in construction waste management	H2	0.701	27
Installation of equipment for waste sorting	M2	0.699	28
Disposal of construction waste periodically by open dumping, incineration, etc	C9	0.698	29
Changing of the design	D5	0.697	30
Separate documentation (records) on recycling waste	D1	0.692	31
Supplier's involvement in construction waste collection	H6	0.631	32

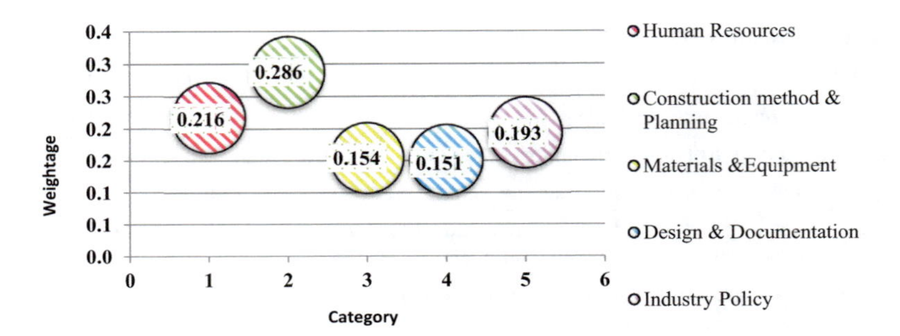

Fig. 16.3 Factor weightage

Table 16.5 OCWMPA index for the case studies selected

Case study	OCWMPA index	Performance assessment
1	495.8876	Fair (400–600)
2	847.342	Excellent (>800)
3	512.7306	Fair (400–600)
4	496.1329	Fair (400–600)
5	520.3007	Fair (400–600)

where CVR = content validity ratio, ne = number of experts in the panel given answer "yes, relevant"; and N = total number of experts in the panel.

The CVR for all the variables is calculated from which critical CVR (mean value of CVR) is calculated. The results indicate the critical CVR value of 0.63. Based on Lawshe's CVR critical table, for the panel size of 15 members, the acceptable limit is 0.49. Hence, the content validity of the items is verified and is acceptable.

16.5 Conclusion

16.5.1 Conclusions of the Analysis Results

The current article assesses the onsite waste management performance of the construction projects. For this, an index system is developed to assess CWM performance of projects. The RII of the variables is calculated to rank the corresponding variables. Among them *RRR (reduce, reuse, recycle* strategy*)* **(C5)** with RII of 0.786 is ranked as the topmost factor. *Supervision and control of the amount of construction waste generated* ***(C1)*** with RII 0.781 ranked second. *Practice of making money out of waste, i.e. selling* ***(I6)*** with RII-0.772 ranked third. *Documentation of payment of taxes and penalties if the waste exceed permitted limits according to government policies* ***(I5)*** with RII 0.769 ranked fourth and *the practice of segregation, i.e. maintenance of separate bins for construction waste* ***(C2)*** with RII-0.766 ranked fifth. In addition, the category weightage of all the five categories is evaluated. Among them, *Human resources* shared the highest category weightage of "0.216".

The OCWMPA index is used to assess onsite CWM of projects. The index ranges from 0 to 1000 and is further classified as 0–250 as poor, 2510–500 as fair, 501–750 as good and 751–1000 as excellent. The results of five construction projects conclude one out of five projects (Project 4) has excellent performance, i.e. OCWMPA index of 847. The remaining projects perform weak towards CWM, i.e. with index value between 400 and 600. The OCWMPA index helps the project manager or engineers to infer CWM performance of their respective projects along with the identification of weak areas which need improvement.

16.5.2 Suggestions Based on Categorization:

Human resources: Adequate training to the workforce serves as an appropriate solution for effective onsite C&D waste management. In addition,

- Training manuals on effective C&D waste minimization and management.
- Checklists for C&D WM activities.
- Appointment of separate technical team for devising effective solutions to deal with C&D waste.
- Inclusion of C&D waste management in tender documents.
- Building plan approvals with mandated C&D waste management.

Construction methods and planning: Usage of advanced software technologies like GIS, BIM for effective onsite monitoring.

- Onsite mobile recycling plants.
- Eco points for temporary storage of materials.
- Planning for selective building deconstruction.
- Usage of advanced software applications for waste recovery, recycling and circularity
- Providing friendly competitive programs among workforce regarding waste management.

Materials and Equipment:

- Mandatory recycled material usage in government construction projects.
- Tax levies on recycled products.
- Establishment of confidence in recycled products usage.
- Developing quality standards for recycled materials.

Design and Documentation:

- Establishing database management system.
- Minimizing design and documentation errors.

- Quantifying the waste that is being generated onsite through the entire life cycle of construction project.
- Maintaining documentation records on training programs, i.e. past, present and future schedules at construction sites.

Industry policy:

- Incentive schemes and promotions based on effective management of C&D waste.
- Developing strategies for marketing C&D products.
- Developing revenue generation-based solutions.
- Appointment of waste inspectors are few solutions suggested in Indian context.

16.5.3 Implications of the Research on Academia, Industry and Policy Developers

The current article theoretically analyzes the construction waste management performance onsite using an OCWMA index. The study identifies various parameters that are required to assess onsite waste management performance of the construction projects using the relative importance index. The construction industry should look for sustainable solutions to conserve existing natural resources. The construction organizations should promote knowledge about CWM and create awareness of the detrimental effects of C&D when left unmanaged. The study has implications for Indian Construction Industry and aims to usher in an era of onsite CWM.

The government/non-government organizations should host several campaigns to increase awareness on CWM. Therefore, it is evident that conducting seminars, workshops on CWM onsite and effective solutions such as including wall of fame and shame boards onsite, assessment of WM performance of construction sites can reduce illegal disposal of CW. The findings from the study provide a basis for the government as well as regulators to establish enhanced strategies towards sustainable C&D waste management in India, which includes:

- Establishment of contractual clauses that are requisites for implementing CWM.
- Framing of legislations, which includes deconstruction plan at planning phase.
- Recruiting and training workforce.
- Setting up of recycling target for every project.
- Creation of societal awareness.
- Improving employee motivation for implementing CWM.

In expanding this line of research to other CW-oriented guidelines, researchers and practitioners may design more effective and human-oriented CWM programs. Specifically, the proposed extensive framework aids in improved insights of individual, corporate and regulatory factors, which governs individual behavior towards material efficiency and sustainability.

Appendix

Table 16.6 C&D waste regulations in India (CPCB 2017; Ministry of housing and urban affairs 2017; BMTPC 2018)

Year	Guidelines	Description	Issuing authority/Government of India
2000	Schedule-3 of MSWM handling rules	Representation on C&DW	Ministry of Environment and forests
2000	Manual on MSWM	Preliminary guidelines on C&DW handling	Ministry of Urban Development (MoUD),
2006	C &D WM and Disposal rules	C&D WM	Municipal Corporation of Greater Mumbai
2010	Report of the management committee on MSWM	Recommendation of special focus on C&DW	Working Committee on MSWM
2012	Circular to establish C&DW recycling facilities under Swachh Bharat Mission (SBM)	Establishment of C&DW recycling plants in cities with residents greater than one million	MoUD
2014	Guidelines regarding sustainable Habitat	Guidelines and precautions on reuse, recovery of materials during deconstruction	Central Public Works Department (CPWD) & Ministry of Housing & Urban Affairs (MoHUA)
2015	Guidelines in using recycled C&DW materials	Addition of 2%–10% of recycled C&D material in buildings and roads projects	Public Works Department, Govt. of NCT of Delhi
2015	Draft SWM rules	Inclusion of C&D WM chapter	Ministry of Environment, Forest and Climate Change (MoEFCC),
2016	C&D W	Establishment of rules managing waste originated from C&D restoration of any structure	MoEFCC
2016	Guidelines in usage of C&DW in construction projects of Government of India	Addressing the shortage of construction materials based on increased demand	Building Materials and Technology Promotion Council (BMTPC), MoHUA

(continued)

Table 16.6 (continued)

Year	Guidelines	Description	Issuing authority/Government of India
2016	Notification on usage of recycled C&DW	Mandatory usage of recycled C&DW in construction projects if it is available within 100 km radius	CPWD, MoHUA
2016	Indian Standard (IS) 383: 2016-Coarse & Fine Aggregate for Concrete-Specification' (III-revision)	Addition of fine and coarse aggregate produced by recycling of C&D waste in IS 383:2016	Bureau of Indian Standards (BIS), IS 383:2016, Ministry of Consumer affairs, Food and Public Distribution
2016	National Building Code of India (Volume-II)	Recommendation on replacement of coarse aggregate up to 30%, in fresh concrete. The percentage can be increased up to 50% for pavements subjected to pure compression	(BIS), Ministry of Consumer Affairs, Food and Public Distribution
2017	Guidelines on Environmental Management of C &DW	Environmental management of C&DW in compliance with C&D WM Rules, 2016 (CPCB 2017)	Central Pollution Control Board (CPCB), MoEFCC
2017	Indian Roads Congress (IRC) 121: 2017	Specifications for usage of C&D W in roads sector	IRC

Table 16.7 Software applications in construction waste management

Author	Description	Software technology
Wang et al. (2018)	Life cycle assessment of carbon emissions for a building	Building information and modeling (BIM)
Cheng and Ma (2013)	Automated estimation and planning of demolition and renovation waste	BIM
Seror and Portnov (2018)	Identification of illegal C&D waste dumping yards	Geographic information system (GIS)
Wu et al. (2016)	Quantification of the demolition waste from origin to final disposal	GIS
Chen and Lu (2017)	Identification of key factors affecting demolition waste generation	Big data analytics
Lu et al. (2016)	Comparison of CWM performance of several public and private projects	Big data analytics
Zainun et al. (2015)	Quantification and mapping of illegal dumping of construction waste	GPS and GIS
Yu et al. (2019)	Prediction of demolition waste generation	Image recognition technology
Di et al. (2016)	Assessment of distribution size of recycling C&D aggregate	Image analysis
Li et al. (2005)	Construction waste reduction	GPS and GIS
Li et al. (2003)	Tracking of real-time data regarding used and unused materials onsite and waste debris for incentive reward program	Barcode technology
Wang et al. (2019)	Identification of nails and screws onsite for recycling	Computer vision technology

References

Aliferova TE, Ncube A (2017) Construction waste generation in Malaysia construction industry: illegal dumping activities. Mater Sci Eng 271(012040):1–8, Article First Published Online:2017, https://doi.org/10.1088/1757-899X/271/1/012040

Banias G, Achillas C, Vlachokostas C, et al (2011) A web-based decision support system for the optimal management of construction and demolition waste. Waste Manage 31(12):2497–2502, Article First Published Online:2011, https://doi.org/10.1016/j.wasman.2011.07.018

Baxter LK, Dionisio K, Pradeep P, et al (2019) Human exposure factors as potential determinants of the heterogeneity in city-speci fi c associations between PM 2.5 and mortality. J Expo Sci Environ Epidemiol 29(4):1–11, Article First Published Online:2019, https://doi.org/10.1038/s41370-018-0080-7

Bergsdal H, Bohne RA, HB (2016) Projection of construction. J Ind Ecol 11(3):27–39, Article First Published Online:2016

Blackwell M, Hobbs G, Adams K (2011) Understanding and predicting construction waste. Proc ICE - Waste Resour Manage 164(4):239–245, Article First Published Online:2011, https://doi.org/10.1680/warm.2011.164.4.239

BMTPC (2018) Utilisation of recycled produce of construction & demolition waste: a ready reckoner

Bolyard SC, Reinhart DR (2016) Application of landfill treatment approaches for stabilization of municipal solid waste. Waste Manage, https://doi.org/10.1016/j.wasman.2016.01.024

Bolyard SC, Reinhart DR, Richardson D (2019) Conventional and fourier transform infrared characterization of waste and leachate during municipal solid waste stabilization. Chemosphere 227:34–42, Article First Published Online:2019, https://doi.org/10.1016/j.chemosphere.2019.04.035

Campbell A (2019) Mass timber in the circular economy: paradigm in practice? Proc Inst Civ Eng Eng Sustain 172(3):141–152, Article First Published Online:2019, 03090569610105762

Cha HS, Kim J, Han J-Y (2009) Identifying and assessing influence factors on improving waste management performance for building construction projects. J Constr Eng Manage 135(7):647–656, Article First Published Online:2009, https://doi.org/10.1061/(ASCE)0733-9364(2009)135:7(647)

Charlson A, Hons M (2019) Briefi ng: embedding circular thinking in a major UK infrastructure project. Proc Inst Civil Eng Eng Sustain 172(3):115–118, Article First Published Online:2019

Chen X, Lu W (2017) Identifying factors in fl uencing demolition waste generation in Hong Kong. J Clean Prod 141:799–811, Article First Published Online:2017, https://doi.org/10.1016/j.jclepro.2016.09.164

Cheng JCP, Ma LYH (2013) A BIM-based system for demolition and renovation waste estimation and planning. Waste Manage 33(6):1539–1551, Article First Published Online:2013, https://doi.org/10.1016/j.wasman.2013.01.001

Cochran K, Townsend T (2007) Estimation of regional building-related C & D debris generation and composition: case study for Florida, US. Waste Manage 27:921–931, Article First Published Online:2007, 10.1016/j.wasman.2006.03.023

CPCB (Central Pollution Control Board) (2017) Guidelines on environmental management of construction & demolition waste. Delhi, India

Dania AA, Kehinde JO, Bala K (2007) A study of construction material waste management practices by construction firms in Nigeria. In: Proceedings of the 3rd Scottish conference for postgraduate researchers of the built and natural environment, pp 121–129

Di F, Bianconi F, Micale C, et al (2016) Quality assessment for recycling aggregates from construction and demolition waste: an image-based approach for particle size estimation. Waste Manage 48:344–352, Article First Published Online:2016, https://doi.org/10.1016/j.wasman.2015.12.005

Friedrich D (2021) Consumer and expert behaviour towards biobased wood-polymer building products: a comparative multi-factorial study according to theory of planned behaviour. Archit Eng Des Manage 18(1), Article First Published Online:2021

Ghanimeh S, Jawad D, Semaan P (2016) Quantification of construction and demolition waste: a measure toward effective modeling. In: Third international conference on advances in computational tools for engineering applications (ACTEA) quantification, pp 83–86

Gilpin A (1996) Dictionary of environment and sustainable development. John Wiley & Sons, Chichester and New York, Newyork, Unites States

Gupta S (2018) The impact of C & D waste on Indian Environment: a critical review. Civil Eng Res J 5(2):57–63, Article First Published Online:2018, https://doi.org/10.19080/cerj.2018.05.555658

Hair JF, Black WC, Babin BJ, Anderson RE (2010) Mulitivariate data analysis: a global perspective. Pearson Pranctise, Cham, https://doi.org/10.1016/j.ijpharm.2011.02.019

Hopkinson P, Chen HM, Zhou K, et al (2019) Recovery and reuse of structural products from end-of-life buildings. Proc Inst Civil Eng Eng Sustain 172(3):119–128, Article First Published Online:2019, https://doi.org/10.1680/jensu.18.00007

Huang B, Wang X, Kua H, et al (2018) Construction and demolition waste management in China through the 3R principle. Resour Conserv Recycl 129(4):36–44, Article First Published Online:2018, https://doi.org/10.1016/j.resconrec.2017.09.029

Islam R, Nazifa TH, Yuniarto A, et al (2019) An empirical study of construction and demolition waste generation and implication of recycling. Waste Manage 95(11):10–21, Article First Published Online:2019, https://doi.org/10.1016/j.wasman.2019.05.049

Jain S, Singhal S, Jain NK, Bhaskar K (2020a), Construction and demolition waste recycling: Investigating the role of theory of planned behavior, institutional pressures and environmental consciousness. J Clean Prod 263:121405, Article First Published Online:2020, https://doi.org/10.1016/j.jclepro.2020.121405

Jin R, Li B, Zhou T, et al (2017) An empirical study of perceptions towards construction and demolition waste recycling and reuse in China. Resour Conserv Recycl, 2017 126(6):86–98, Article First Published Online:2017, https://doi.org/10.1016/j.resconrec.2017.07.034

Kabirifar K, Mojtahedi M, Wang C, Tam VWY (2020) Construction and demolition waste management contributing factors coupled with reduce , reuse , and recycle strategies for effective waste management: a review. J Clean Prod 263:121265, Article First Published Online:2020, https://doi.org/10.1016/j.jclepro.2020.121265

Kolaventi SS, Tezeswi TP, Kumar MVNS, Momand H (2021) Implementing site waste-management plans, recycling in India: barriers, benefits, measures. ICE-Engineering Sustain, https://doi.org/10.1680/jensu.21.00032|10.1680/jensu.21.00032

Kolaventi SS, Tezeswi TP, Siva Kumar MVN (2019) An assessment of construction waste management in India: a statistical approach. Waste Manage Res 1–16, Article First Published Online:2019, https://doi.org/10.1177/0734242X19867754

Kolaventi SS Hikmatulla Momand Tezeswi TP Sivakumar MVN (2020) Construction waste in India: a structural equation model for identification of causes. Proc Inst Civil Eng – Eng Sustain 1–10, Article First Published Online:2020, https://doi.org/10.1680/jensu.19.00047

Kolaventi SS, Tezeswi TP, Sivakumar MVN (2017) A modeling approach to construction waste management. In: Proceedings of urbanization challenges in emerging economies. ASCE, Cham, pp 11–20

Koshy R, Apte EMR (2012) Waste minimization of construction materials on a bridge site (Cement and Reinforcement Steel) - a regression and correlation analysis. Int J Eng Innov Technol 2(1):6–14, Article First Published Online:2012

Koutamanis A, Reijn B Van, Bueren E Van (2018) Urban mining and buildings: a review of possibilities and limitations. Resour Conserv Recycl 138(June):32–39, Article First Published Online:2018, https://doi.org/10.1016/j.resconrec.2018.06.024

Li H, Chen Z, Wong CTC (2003) Barcode technology for an incentive reward program to reduce construction wastes. Comput Civ Infrastruct Eng 18(4):313–324, Article First Published Online:2003, https://doi.org/10.1111/1467-8667.00320

Li H, Chen Z, Yong L, Kong SCW (2005) Application of integrated GPS and GIS technology for reducing construction waste and improving construction efficiency. Autom Constr 14:323–331, Article First Published Online:2005, https://doi.org/10.1016/j.autcon.2004.08.007

Lu W (2019), Big data analytics to identify illegal construction waste dumping: a Hong Kong study. Resour Conserv Recycl 141(August 2018):264–272, Article First Published Online:2019, https://doi.org/10.1016/j.resconrec.2018.10.039

Lu W, Chen X, Ho DCW, Wang H (2016) Analysis of the construction waste management performance in Hong Kong: the public and private sectors compared using big data. J Clean Prod 112:521–531, Article First Published Online:2016, https://doi.org/10.1016/j.jclepro.2015.06.106

Ministry of housing and urban affairs (2017) Guidelines for Swachh Bharat Mission

Ministry of Housing and Urban Affairs (2018) Strategy for Promoting Processing of Construction and Demolition (C&D) Waste and Utilisation of Recycled Products

Olaniyi AA (2019) Scholar Journal of Applied Sciences

Oliveira MS, Oliveira E, Wanderley A, Campos A, Fonseca A (2019) Smart management of waste from construction sites: mobile application technology in the city manaus, amazonas, Brazil. In: *XIII CTV 2019 Proc XIII Int Conf Virtual City Territ "Challenges Paradig Contemp city"*, 2019, 8426–8440, Article First Published Online:2019, https://doi.org/10.5821/ctv.8426

Owolabi HA, Oyedele LO, Ajayi SO, et al (2016) Evaluation criteria for construction waste management tools: towards a holistic BIM framework. Int J Sustain Build Technol Urban Dev 7(1):3–21, Article First Published Online:2016, https://doi.org/10.1080/2093761x.2016.1152203

Ram VG, Kishore KC, Kalidindi SN (2020) Environmental benefits of construction and demolition debris recycling: evidence from an Indian case study using life cycle assessment. J Clean Prod 255:20258, Article First Published Online:2020, https://doi.org/10.1016/j.jclepro.2020.120258

Rao ST, Luo H, Astitha M, et al (2020) On the limit to the accuracy of regional-scale air quality models. Atmos Chem Phys 20(3):1627–1639, Article First Published Online:2020

Semaan P, Mirella Abdel M, Mario C, et al (2017) Life cycle assessment of concrete industry in developing nations. In: International conference on sustainable infrastructure, pp 197–206

Seror N, Portnov BA (2018), Identifying areas under potential risk of illegal construction and demolition waste dumping using GIS tools. Waste Manage 75:22–29, Article First Published Online:2018, https://doi.org/10.1016/j.wasman.2018.01.027

Shen LY, Lu WS, Yao H, Wu DH (2005) A computer-based scoring method for measuring the environmental performance of construction activities. Autom Constr 14(3):297–309, Article First Published Online:2005, https://doi.org/10.1016/j.autcon.2004.08.017

Solís-Guzmán J, Marrero M, Montes-Delgado MV, Ramírez-de-Arellano A (2009) A Spanish model for quantification and management of construction waste. Waste Manage 29(9):2542–2548, Article First Published Online:2009, https://doi.org/10.1016/j.wasman.2009.05.009

Townsend T, Tolaymat T, Leo K, Jambeck J (2004) Heavy metals in recovered fines from construction and demolition debris recycling facilities in Florida. Sci ofthe Total Environ 332:1–11, Article First Published Online:2004, https://doi.org/10.1016/j.scitotenv.2004.03.011

Wang J and Zhengdao L (2014) Critical factors in effective construction waste minimization at the design stage: a Shenzhen case study, China. Resour Conserv Recycl 82(1):1–7, Article First Published Online:2014

Wang J, Wu H, Duan H, et al (2018) Combining life cycle assessment and building information modelling to account for carbon emission of building demolition waste: a case study. J Clean Prod 172:3154–3166, Article First Published Online:2018, https://doi.org/10.1016/j.jclepro.2017.11.087

Wang J, Yuan H, Kang X, Lu W (2010) Critical success factors for on-site sorting of construction waste: a china study. Resour Conserv Recycl 54(11):931–936, Article First Published Online:2010, https://doi.org/10.1016/j.resconrec.2010.01.012

Wang Z, Li H, Zhang X (2019) Construction waste recycling robot for nails and screws: computer vision technology and neural network approach. Autom Constr 97(November 2018):220–228, Article First Published Online:2019, https://doi.org/10.1016/j.autcon.2018.11.009

Weisheng L, Yuan H (2010) Exploring critical success factors for waste management in construction projects of China. Resour Conserv Recycl 55(2):201–208, Article First Published Online:2010, https://doi.org/10.1016/j.resconrec.2010.09.010

Wu H, Wang J, Duan H, et al (2016) An innovative approach to managing demolition waste via GIS (geographic information system): a case study in Shenzhen city, China. J Clean Prod 112:494–503, Article First Published Online:2016, https://doi.org/10.1016/j.jclepro.2015.08.096

Yang B, Song X, Yuan H, Zuo J (2020) A model for investigating construction workers' waste reduction behaviors. J Clean Prod 265:121841, Article First Published Online:2020, https://doi.org/10.1016/j.jclepro.2020.121841

Yu B, Wang J, Li J, et al (2019) Prediction of large-scale demolition waste generation during urban renewal: a hybrid trilogy method. Waste Manage J 89:1–9, Article First Published Online:2019, https://doi.org/10.1016/j.wasman.2019.03.063

Yuan H (2017) Barriers and countermeasures for managing construction and demolition waste: a case of Shenzhen in China. J Clean Prod 157(4):84–93, Article First Published Online:2017, https://doi.org/10.1016/j.jclepro.2017.04.137

Yuan H, Shen L, Wang J (2011) Major obstacles to improving the performance of waste management in China's construction industry. Facilities 29(5):224–242, Article First Published Online:2011, https://doi.org/10.1108/02632771111120538

Zainun NY, Othman W, Zainun NY, Othman W (2015) Quantification and mapping of construction waste generation in Parit Raja quantification and mapping of construction waste generation in Parit Raja. https://doi.org/10.4028/www.scientific.net/AMM.773-774.1032|10.4028/www.scientific.net/AMM.773-774.1032

Zhang A, Venkatesh VG, Liu Y, et al (2019) Barriers to smart waste management for a circular economy in China. J Clean Prod 240:118198, Article First Published Online:2019, https://doi.org/10.1016/j.jclepro.2019.118198

Zhang C, Hu M, Yang X, et al (2020) Upgrading construction and demolition waste management from downcycling to recycling in the Netherlands. J Clean Prod 3:121718, Article First Published Online:2020, https://doi.org/10.1016/j.jclepro.2020.121718

Zhang G, Zhang Y, Tian W, et al (2021) Bridging the intention–behavior gap: effect of altruistic motives on developers' action towards green redevelopment of industrial brownfields. Sustain 13(2):1–16, Article First Published Online:2021, https://doi.org/10.3390/su13020977

Chapter 17
Production of Lightweight Aggregates for Construction Industry from Industrial Byproducts: A Review

Manu S. Nadesan and Abin Joy

17.1 Introduction

Industrialization is significant for the economic growth and development of a society, and can also be detrimental to the environment. The increment in the generation of industrial wastes and depletion of natural resources are considered as the present concerns of society. Most of the industrial byproducts should be disposed of properly, otherwise, they may cause contamination to the environment. Proper disposal of these industrial wastes is highly energy-intensive and has financial requirements. The production of lightweight aggregate (LWA) from industrial byproducts is one of the methods for safe and massive disposal. The production and use of lightweight aggregate (LWA) in the construction industry is not a new concept. The evidence regarding the utilization of lightweight aggregate can be obtained from ancient civilizations. Ancient time volcanic originated natural rocks such as pumice, scoria, etc. were used as LWA for various applications. Some of these ancient structures like Hagia Sofia and Pantheon still exist without any serious damages. Apart from this, it was noticed that the structural and insulation efficiency of the LWA concretes are better than that of normal aggregate concrete (Nadesan and Dinakar 2017).

Concrete is considered the second most consumed manufactured material on planet earth. Each concrete mix generally contains 60–75% of aggregates by its total volume. So, it is clearly understood that the fresh and hardened properties of the developed concretes will be governed by the properties of the aggregates that have been used. Natural aggregates are derived from naturally occurring geological sources and have been processed such as crushing, washing, and sizing. And the use of natural

M. S. Nadesan (✉) · A. Joy
ASIET, Kalady, Kerala 683574, India
e-mail: msn10@iitbbs.ac.in

A. Joy
e-mail: abin.ce@adishankara.ac.in

K. R. Reddy et al. (eds.), *Advances in Sustainable Materials and Resilient Infrastructure*,
Springer Transactions in Civil and Environmental Engineering,
https://doi.org/10.1007/978-981-16-9744-9_17

aggregates will cause depletion of the resources with time. Minimizing the aggregate mining from natural resources is highly necessary for sustainable development. The effective utilization of various industrial wastes such as fly ash, municipal solid waste, palm oil shell, and water treatment sludge requires a proper understanding of the behavioural characteristics of these materials. The present study emphasises various possibilities of resources, production techniques, and the properties of the produced aggregates. Presently, the most common LWAs that have been widely used are expanded clay, shale, and slate. All of these aggregates were produced using naturally occurring resources. This cannot be considered as a perfect example of sustainable development. To provide a solution to this, the present paper made an attempt to review various existing techniques and the properties of the lightweight aggregates that are produced from various industrial wastes.

17.2 Lightweight Concrete

Lightweight concrete is widely used for various structural as well as insulation applications. It provides a significant reduction in dead load which is considered a major contributor to the total load of the structure. It is possible to use lightweight concrete having a density less than 400 kg/m^3 for insulation purposes, whereas structural grade concretes possess a density starting from 1600 kg/m^3. Lightweight concrete (LWC) is a broad term that mainly consists of aerated concrete, no-fines concrete and lightweight aggregate concrete (LWAC). LWC can be produced by inducing aeration or omitting fine content or by using lightweight aggregates. Lightweight aggregates are mainly obtained in two ways. One is a naturally occurring state, most of them will be volcanic rocks; pumice, perlites, and scoria are examples of this. And the other one is artificial aggregates; can be produced either from natural resources or from industrial byproducts. Lightweight aggregates for concrete manufactured from natural resources like clay, shale, and slate have been produced over many decades. Recently, many researches have been carried out in the direction of production of lightweight aggregate from industrial byproducts.

Most of these aggregates are light in weight because of the pore structure developed within them during the production stage. Aggregate having a particle density less than 2000 kg/m^3 or dry loose bulk density of less than 1200 kg/m^3 is considered as lightweight aggregates (BSI 92/17688). Lightweight aggregate concrete has been produced by replacing the normal aggregate with lightweight aggregates. Often the coarse aggregates are replaced with LWA and the fine aggregates will be the same normal density aggregates. This lightweight concrete is not only reduced weight but have better insulation properties against heat and sound. However, the strength of the lightweight aggregates is less compared to the normal aggregates. The most common uses of LWC are floor, roof, bridge deck, and others that include pavements. The American Concrete Institute defines structural lightweight concrete as having a characteristic compressive strength of more than 17.2 N/mm^2 and air dry density less than 1840 kg/m^3. However, the British Code of practice BS8110 recommends maximum density up to 2000 kg/m^3 and a minimum characteristic compressive strength of 15 N/mm^2 after 28 days of curing.

17.3 Methodology

The present study aims to evaluate the aggregates that have been produced from various industrial wastes. Also, check the properties of the aggregates and concretes with the standard norms as mentioned in the previous section. This study will also provide some insights about various aggregate production techniques which may provide innovative solutions to the disposal of various other wastes or byproducts that are not incorporated in the present study.

17.4 Lightweight Aggregate Production and Properties

Artificially produced aggregates are generally spherical in nature and the size fraction will be uniform. Also the porous nature of the aggregate cause certain difficulties in the mix proportioning stage. These can be overcome either by saturating these aggregates before undergoing mixing or compensating for the absorbed water by adding extra water along with the mixing water. Studies show that the percolation of the paste into the aggregates will be beneficial to the concrete in terms of its hardened properties (Zhang and GjØrv 1990). Also, these porous aggregates will exhibit internal curing, which is favourable in many ways as far as the quality of the concrete is considered. Generally, curing is only possible externally for normal aggregates concretes. But in the case of lightweight aggregate, it is possible internally also. The water absorbed during the mixing time will facilitate curing during the hardening of the surrounding paste. This will refine the ITZ around the aggregates and rearrange the tortuosity of the concrete. This means the paste matrix of the lightweight aggregate concrete could be more uniform compared to normal aggregate concrete. Also, it was suggested that these porous aggregates may act as a buffer zone for the harmful chemical that ingress into the concrete.

The production of aggregates from various industrial byproducts mainly undergoes the basic steps such as mixing of raw materials, agglomeration and binding or hardening of the fly as particles (Bijen 1986). Agglomeration means fine particles are joined together and form a colloidal system, which can be simply broken by mechanical forces. The common techniques adopted for agglomeration are agitation granulation and compaction. Agitation granulation can be performed with the help of a disc, pan, drum, cone, or mixer. Similarly, compaction is carried out using processes such as unidirectional piston type, roll pressing, extrusion, or pellet mills.

In the granulation stage, a proper binding agent is also added along with the powder materials. The commonly used binding agents are water, starch solution, emulsions of mineral oil or wastewater of certain industries, etc. With the increase in the amount of liquid content the cohesion between the powders also increases. To improve the strength of the green pellets, agents such as bentonite, lime, Portland cement, or clay can be used. For the hardening of green pellets, treatments such as cold bonding, autoclaving, and sintering have been used. Among these, the sintering process facilitates the maximum amount of waste disposal in a short duration through the production of aggregates.

The pelletization efficiency and the strength of artificial aggregates can be increased by the use of higher fine fractions during the mixing stage. Among the different hardening methods, sintering requires the high content of SiO_2 and Al_2O_3 in raw materials, while the cold-bonded requires cementitious raw materials and a long curing time (Ren et al. 2021). The sintering temperature is more than 1100 °C, so it is considered as one of the energy-intensive processes, but it is a rapid hardening technique. The accelerated carbonation method was also considered as a sustainable hardening method for artificial aggregates. A study shows that while using basic oxygen furnace slag as the raw material of artificial aggregates, the application of carbonation during the curing after the granulation process caused the property enhancement of aggregates (Jiang and Ling 2020).

17.5 Potential Industrial Byproducts

17.5.1 Fly Ash

Fly ash is the byproduct that generates from the coal-based thermal power plants after the combustion of pulverized fuel ash (PFA). It was found that lightweight aggregates can be produced from fly ash using various techniques. Various methods associated with the production of these aggregates were mentioned in various literature works (Bijen 1986). It was found that the optimum dosage of water content for the fly ash aggregate production is around 20–25%. Presently most of the researches are focused on the aggregates that have been produced from fly ash through cold bonding and sintering. From earlier studies, it is noticed that the physical properties of the produced aggregates are influenced by the fineness of the fly ash. Various mixing ratios and the aggregate production parameters have already been mentioned in detail elsewhere (Nadesan and Dinakar 2017; Tajra et al. 2019). Some of the properties of the sintered aggregates were tabulated in Table 17.1 for a better understanding of the aggregate properties

LBD: Loose bulk density, RBD: Roded bulk density

The aggregates produced using the sintering technique appear spherical in shape and have a specific gravity between 1.33 and 2.35 (Nadesan and Dinakar 2017). The loose bulk density and water absorption varied between 765 and 936 kg/m^3 and 1–34%, respectively. The density of the concrete produced using sintered fly ash aggregates varied from 1650 to 2010 kg/m^3. Also from the studies, it is found that the compressive strength is between 24 and 74 MPa. These concretes exhibited

Table 17.1 Physical and mechanical properties of sintered fly ash aggregates

Specific gravity	LBD (kg/m^3)	RBD (kg/m^3)	Water absorption (%)	Crushing value N	Crushing strength (MPa)	10% fine (tonnes)	References
1.7–2.35	–	–	16–22	–	–	1.75–4.25	Harikrishnan and Ramamurthy (2006)
1.51–1.93	–	–	0.7–18.4	–	5.1–19.3	–	Kockal and Ozturan (2011)
1.8–1.92	–	–	19–20	–	–	2.9–4.2	Geetha and. Ramamurthy (2011)
1.33	–	–	2.7	–	6	–	Wu et al. (2013)
1.46	840	–	15	–	8.7	–	Wegen and Bijen. (1985)
1.57–1.60	933–936	993–999	0.8–19.3		5.1–23.1		Kockal and Ozturan (2010)
1.57	900	1000	1.75		18.34		Gomathi and. Sivakumar (2015)

mechanical and durability properties sufficient enough to consider them as structural grade concrete. The durability properties of these concretes were comparable with the normal aggregate concrete having the same strength. Due to lower density these concretes exhibit higher structural efficiency than the normal aggregate concrete.

The hardened properties of these concretes are associated with the behaviour of its interfacial transition zone (ITZ). Most of these aggregates have uniformly distributed pores. Studies show that the thickness and hardness of the ITZ around the sintered fly ash aggregate are superior to a normal aggregate matrix. In some cases the hardness of the ITZ was observed better than that of the cement paste; this phenomenon is not observed in the case of normal aggregate concrete. Also if the sintering temperature increases the pozzolanic reactivity of the aggregate also found to increase. Sometimes the paste matrix also intruded into the porous structure of the LWA and seals some of the pores (Fig. 17.1). This will again lead to the pore refinement and successively reduce permeability.

Fly ash possesses sufficient pozzolanic reactivity, and this will be utilized to form a water-resistant bonding material in the presence of calcium hydroxides. Normally, lime or Portland cement are the sources of calcium hydroxide for the pozzolanic reaction. Polymerization is another technique that can be used in the case of fly ash. It is also evident that calcium sulphate, calcium sulphite, and alkaline substances may have a favourable impact on the mechanical properties of the concrete. The

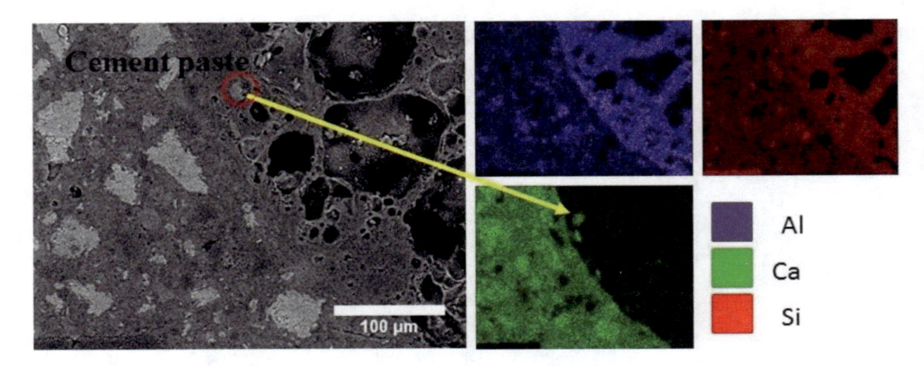

Fig. 17.1 Intrusion of cement paste into the aggregate pore (Nadesan and Dinakar 2018)

specific gravity and loose bulk densities of cold-bonded aggregate reported in various literature works are in the range of 0.88–2.3 and 510–1460 kg/m^3, respectively. Like sintered aggregate cold-bonded aggregates also have a spherical shape, and this will be beneficial in terms of workability of the concrete. The literature works show that the cold-bonded aggregates are able to produce concrete with a compressive strength varied from 7 to 70 MPa, Young's modulus from 6 to 62 GPa, and a density varying between 1120 and 2350 kg/m^3. The mechanical properties of the concrete produced using cold-bonded aggregates clearly show that this aggregate can be used for the development of structural grade concrete also. It is also possible to produce concrete having a thermal conductivity less than 0.75 W/m.K using cold-bonded aggregate. Mechanical and chemical interlocking between cold-bonded aggregate and cement matrix may lead to the formation of a superior ITZ around the aggregates. It is also possible to produce concrete having water penetration depth less than 30 mm and water absorption less than 10% using cold-bonded fly ash aggregates.

Another method to produce fly ash aggregate is through the crushing technique. Initially, geopolymer cubes or slabs will be cast, and then crush them into angular aggregates. In this process fly ash, granulated blast furnace slag (GGBS), and anhydrous sodium meta-silicate were used as the ingredients. Results showed that the produced GLAs with different mixes were all lightweight with a loose bulk density below 800 kg/m^3 and an apparent density of 1450–1750 kg/m^3. The produced GLAs were further used as coarse aggregates to produce geopolymer aggregate concrete (GAC) (Xu et al. 2021).

Apart from these commonly used aggregate production techniques, Flashag is another patent aggregate manufactured from fly ash. These aggregates are not pelletized but sintered, and the process is considered simple and inexpensive compared to the pelletized sintered method. The concrete produced using Flashag aggregates is found to be 21% lighter than the granite aggregate concrete and 15% stronger than the high strength concrete produced using normal aggregate. It is also noticed that the concrete produced using these aggregates possesses 30% less drying

Fig. 17.2 Flashag (on the left), granite (in the middle), sintered fly ash aggregate (on the right) (Kayali 2005)

shrinkage compared to normal aggregate concrete (Kayali 2005). Fig. 17.2 exhibited the aggregates that are produced from fly ash using various techniques.

17.5.2 Municipal Solid Waste

Rapid urbanization and enhancement in living standard is the reason behind the sudden growth in the generation of municipal solid wastes (MSW) all over the world. Incineration is one of the widely used procedures to dispose of the generated municipal solid waste. Municipal solid waste incineration ash (MSWIA) are continually generated in large amounts as a result of this incineration. MSWIA may contain a large amount of toxic or hazardous elements. In recent years, stabilization/solidification treatment of MSWIA is considered as one of the best practices to treat hazardous material in an efficient and convenient way (Chen et al. 2020). It is possible to produce lightweight aggregate from MSWIA by pelletizing the ash and then high-temperature sintering. The flowchart of the production process has been shown in Fig. 17.3.

Sun et al. (2021) observed that it is possible to produce aggregate having crushing strength varied from 8 to 22 MPa using MSWIA. In that study red mud is also used as one of the ingredients. And the results indicate that as the percentage replacement of red mud increases the strength decreases. The bulk density and the water absorption of these aggregates varied from 1209 to 1241 kg/m^3 and 0.36–17%, respectively. The alkalinity of the produced aggregate ranges from 9.4 to 9.9. But the individual alkalinity of both the ingredients is less than that of this value. This could be due to

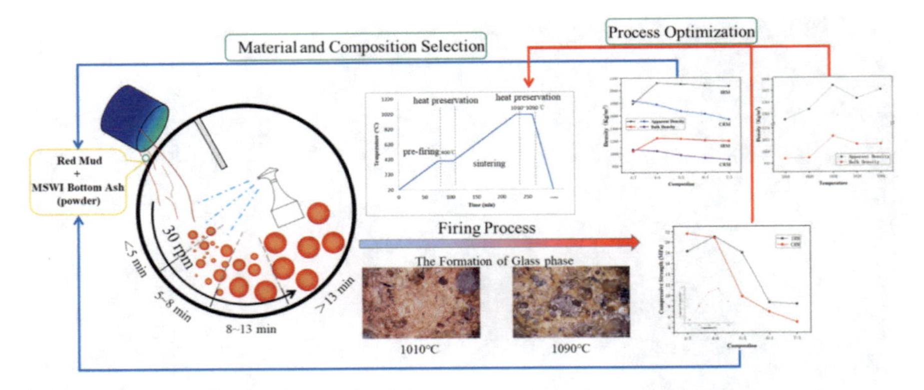

Fig. 17.3 Aggregate produced from municipal solid waste (Sun et al. 2021)

the existence of stable alkaline substances in the red mud after mineralization and precipitation (Whittington et al. 1998).

17.5.3 Palm Oil Shell

The production of palm oil is increasing due to the demand of the consumer from many parts of the world. Malaysia, one of the largest producers of palm oil, generates approximately a total of 61.1 MT of palm oil solid wastes such as empty fruit branches, fibres, and kernels generated per year (Mohammad et al. 2016). Oil palm shell (OPS) is one of the byproducts from palm oil mills, which is obtained after the extraction of the oil from the fruit. Fig. 17.4 provides the basic information about the shape and size of the oil palm shell. Previous studies (Mo et al. 2015; Basri et al. 1999) already accepted that OPS is a potential material that can be incorporated in concrete as an aggregate.

It has been reported that the addition of supplementary cementitious material like fly ash has adverse effects on the porosity and compressive strength of the concrete (Muthusamy et al. 2021). Since 1985, OPS has been used as a lightweight aggregate in research studies to manufacture lightweight aggregate concrete (Salam and Abdullah 1985). In many researches the compressive strength of the OPS LWAC has been reported between 13 and 22 MPa. Also with the inclusion of mineral admixtures 28-day compressive strength up to 48 MPa with a dry density of about 1870 kg/m^3 has been reported for crushed OPS lightweight aggregate concrete (Alengaram et al. 2013; Shafigh et al. 2014). It is also possible to produce concrete containing a certain percentage of OPS shell as coarse aggregate with normal aggregate and these concretes will fall under the category of semi lightweight aggregate. If the curing condition of the concrete is improper, then concrete containing OPS aggregates experiences a higher loss in compressive strength compared to normal aggregate concrete. It is also recommended that for concrete having less water absorption and minimum

Fig. 17.4 Oil palm shell (Muthusamy et al. 2021)

drying shrinkage the optimum amount of coarse aggregate must be limited to 50% of the total volume of the coarse aggregate (Maghfouri et al. 2018).

17.5.4 Water Treatment Sludge

Water is considered as most consumed material on planet earth. Water treatment is an unavoidable stage in water supply system all over the world. This water treatment process will generate organic waste matter known as sludge. Domestic households produce an average of 200–300 L of wastewater per person every day. Disposal of this generated wastewater safely is a necessity of the community. As a result of wastewater treatment, sewage sludge is an organic waste produced from the water treatment plants. It is expected that 13 million tonnes of dry matter has been produced in 2020 and the amount increases with time (Durdevic et al. 2019). Many studies have shown that this sewage sludge can be used for the production of lightweight aggregate (Abdul et al. 2004, Montero et al. 2009).

Normally, sewage sludge is used as a partial replacement at a low substitution (<15%) level for the production of expanded clay aggregates. A flowchart of the manufacturing process of lightweight aggregates from MSW has been exhibited in Fig. 17.5. The inclusion of the sewage sludge having less expanding properties minimizes the expandability and the porosity of the aggregates (Tuan et al. 2013). This will lead to an increase in the density and reduce the water absorption of the aggregates. Results indicate that concrete produced with aggregate containing 7% of

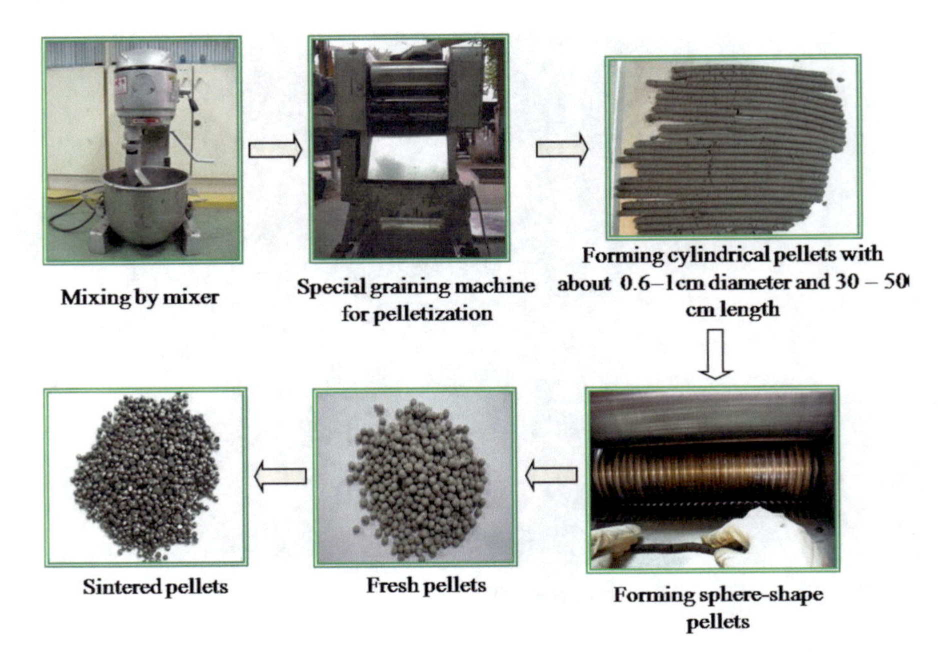

Fig. 17.5 Flowchart of the manufacturing process of lightweight aggregates (Li et al. 2020)

the sludge and expanded clay aggregates exhibited similar results (Tuan et al. 2013). X-ray fluorescence spectrometry studies show that the sewage sludge is primarily composed of SiO_2 followed by Al_2O_3 and Fe_2O_3 (Li et al. 2020). Most of the water treatment sludge have high Al_2O_3 due to the use of flocculants. Due to the presence of Al_2O_3, it is possible to produce ultra-lightweight aggregate having a density of less than 0.8 g/m^3.

17.6 Conclusions

From the review of various literature works, it was noticed that all the industrial wastes considered in the present study have the potential to produce the LWA.

Among the reviewed industrial wastes, OPS does not require much treatment and energy-intensive process. Whereas the production of aggregates from fly ash facilitates large disposal of the generated fly ash. It was also found that for the quick production of aggregate from fly ash, MSW, and water treatment sludge, the sintering technique is used.

And the literature works showed that the density of the concretes produced from these aggregates was less than 2000 kg/m^3. It indicates that it is possible to produce LWAs from these industrial wastes.

The compressive strength results indicated that most of the concretes exhibited strength above 20 MPa. So it is possible to produce structural grade concretes by the inclusion of these aggregates. Proper durability assessment has to be made before using them in the reinforced cement concrete.

References

Abdul GL, Samad AA, Wong CHK, Jaafar MS, Baki AM (2004) Reusability of sewage sludge in clay bricks‖, J Mater Cycles Waste Manage 6:41–47

Alengaram UJ, Muhit BAA, Jumaat MZ (2013) Utilization of oil palm kernel shell as lightweight aggregate in concrete—a review. Constr Build Mater 38:161–172

Basri HB, Mannan MA, Zain MFM (1999) Concrete using waste oil palm shells as aggregate. Cem Concr Res 29:619–622

Bijen JMJM (1986) Manufacturing processes of artificial lightweight aggregates from fly ash. Int J Cem Compos Lightweight Concr 8(3):191–199

BSI Document 92/17688: European Draft Standard Specification for Lightweight Aggregates. CEN/TC154/SC5, Sub-Committee Lightweight Aggregates, October 1992

Chen L, Wang YS, Wang L, Zhang Y, Li J, Tong L, Hu Q, Dai JG, Tsang DCW (2020) Stabilization/solidification of municipal solid waste incineration fly ash by phosphate-enhanced calcium aluminate cement. J Hazard Mater 124404

Durdevic D, Blecich P, Juric Z (2019) Energy recovery from sewage sludge: the case study of Croatia. Energies 12:1927

Geetha S, Ramamurthy K (2011) Properties of sintered low calcium bottom ash aggregate with clay binders. Constr Build Mater 25:2002–2013

Gomathi P, Sivakumar A (2015) Accelerated curing effects on the mechanical performance of cold bonded and sintered fly ash aggregate concrete. Constr Build Mater 77:276–287

Harikrishnan KI, Ramamurthy K (2006) Influence of pelletization process on the properties of fly ash aggregates. Waste Manage 26:846–852

Jiang Y, Ling TC (2020) Production of artificial aggregates from steel-making slag: influences of accelerated carbonation during granulation and/or post-curing. J CO2 Util l36:135–144

Kayali O, Flashag-New lightweight aggregate for high strength and durable concrete. In: 2005 World coal Ash (WOCA), April 11–15 2005, Lxington, Kentucky, USA

Kockal NU, Ozturan T (2011) Optimization of properties of fly ash aggregates for high-strength lightweight concrete production. Mater Des 32:3586–3593

Kockal NU, Ozturan T (2010) Effects of lightweight fly ash aggregate properties on the behavior of lightweight concretes. J Hazard Mater 179:954–965

Li X, He C, Lv Y, Jian S, Liu G, Jiang W, Jiang D (2020) Utilization of municipal sewage sludge and waste glass powder in production of lightweight aggregates. Constr Build Mater 256:119413

Maghfouri M, Shafigh P, Aslam M (2018) Optimum oil palm shell content as coarse aggregate in concrete based on mechanical and durability properties. Adva Mater Sci Eng 2018:14, Article ID 4271497

Mo KH, Alengaram UJ, Jumaat MZ, Yap SP (2015) Feasibility study of high volume slag as cement replacement for sustainable structural lightweight oil palm shell concrete. J Clean Prod 91:297–304

Mohammad MUI, Mo KH, Alengaram UJ, Jumaat MZ (2016) Mechanical and fresh properties of sustainable oil palm shell lightweight concrete incorporating palm oil fuel ash. J Clean Prod 115:307–314

Montero MA, Jordán MM, Hernández-Crespo MS, Sanfeliu T (2009) The use of sewage sludge and marble residues in the manufacture of ceramic tile bodies. Appl Clay Sci 46:404–408

Muthusamy K, Budiea AMA, Azhar NW, Jaafar MS, Mohsin SMS, Arifin NF, Yahaya FM (2021) Durability properties of oil palm shell lightweight aggregate concrete containing fly ash as partial cement replacement. Mater Today: Proc l41:56–60

Nadesan MS, Diankar P (2018) Micro-structural behaviour of interfacial transition zone of the porous sintered fly ash aggregate. J Buil Eng 16:31–38

Nadesan MS, Dinakar P (2017) Structural concrete using sintered fly ash lightweight aggregate: a review. Constr Build Mater l154:928–944

Ren P, Ling TC, Mo KH (2021) Recent advances in artificial aggregate production. J Clean Prod 291:125215

Salam SKAAA, Abdullah A (1985) Lightweight concrete using oil palm shells as aggregates. In: Proceedings of national symposium on oil palm by-products for agro-based industries, Kuala Lumpur, Malaysia

Shafigh P, Mahmud HB, Jumaat MZB, Ahmmad R, Bahri S (2014) Structural lightweight aggregate concrete using two types of waste from the palm oil industry as aggregate. J Clean Prod 80:187–196

Sun Y, Li JS, Chen Z, Xue Q, Sun Q, Zhou Y, Liu XCL, Poon CS (2021) Production of lightweight aggregate ceramsite from red mud and municipal solid waste incineration bottom ash: mechanism and optimization. Constr Build Mater 287:122993

Tajra F, Abd Elrahman M, Stephan D (2019) The production and properties of cold-bonded aggregate and its applications in concrete: a review. Constr Build Mater 225:29–43

Tuan BLA, Hwang CL, Lin KL, Chen YY, Young MP (2013) Development of lightweight aggregate from sewage sludge and waste glass powder for concrete. Constr Build Mater 47:334–339. https://doi.org/10.1016/j.conbuildmat.2013.05.039

Wegen GJL, Bijen JMJM (1985) Properties of concrete made with three types of artificial PFA coarse aggregates. Int J Cem Compos Lightweight Concr 7(3):159–167

Whittington BI, Fletcher BL, Talbot C (1998) The effect of reaction conditions on the composition of desilication product (DSP) formed under simulated Bayer conditions. Hydrometallurgy 49(1–2):1–22

Wu X, Wu Z, Zheng J, Zhang X (2013) Bond behaviour of deformed bars in Self compacting Lightweight concrete subjected to lateral pressure. Mag Concr Res 65(23):1396–1410

Xu LY, Qian LP, Huang BT, Dai JG (2021) Development of artificial one-part geopolymer lightweight aggregates by crushing technique. J Clean Prod 315:128200

Zhang MH, GjØrv E (1990) Microstructure of the interfacial zone between lightweight aggregate and cement paste. Cem Concr Res 20:610–618

Chapter 18
Sulfate Resistant Mortar Using Coarse Fraction of Red Mud as Fine Aggregate

Anshumali Mishra, Bajaya K. Das, Shamshad Alam, and Sarat Kumar Das

18.1 Introduction

Red mud (RM) is one of the major industrial wastes generated from the aluminium industry with a generation rate of about 120 million tons per annum worldwide (Nath et al. 2015). The present scenario of red mud management is disposal into the sea, storage in dry form, and the storage in slurry form in the pond (Agrawal et al. 2004). Disposal into the sea endangers the aquatic life whereas the storage on the ground takes vast tracts of usable land and create problems like soil contamination, ground and surface water pollution due to its alkaline nature (pH > 11). Occasional failure of dike results in spreading the red mud slurry over large vicinity, which causes causalities, polluting the surface water (Mayes et al. 2011). As the sand mining is becoming prohibitive and for the sustainable development of the industries, it is important to utilize the industrial waste as a value-added material. Several studies have been made on industrial waste like fly ash and slag but the study on the red mud is limited. Gordon et al. (1996) studied the compressive strength of composites developed using Jamaican red mud, hydrated lime, condensed silica fume, and limestone without using Portland cement as binder and found a compressive strength of 15–18 MPa. Singh et al. (1996) prepared three different types of

A. Mishra · S. K. Das (✉)
Indian Institute of Technology (ISM), Dhanbad, India
e-mail: saratdas@iitism.ac.in

A. Mishra
e-mail: anshumali.18dr0036@cve.iitism.ac.in

B. K. Das
KIIT University, Bhubaneswar, India

S. Alam
Chandigarh University, Mohali, India

K. R. Reddy et al. (eds.), *Advances in Sustainable Materials and Resilient Infrastructure*, Springer Transactions in Civil and Environmental Engineering,
https://doi.org/10.1007/978-981-16-9744-9_18

cement (aluminoferrite-belite, aluminoferrite-ferrite-aluminates and sulfoaluminate-aluminoferrite-ferrite) using red mud collected from HINDALCO, whose strength was found very comparable to ordinary Portland cement (OPC). Zhihua et al. (2003) studied the properties and microstructural behaviour of the alkali-activated slag-red mud cementitious material (ASRC) by conducting strength test, microstructural studies (MIP & SEM) and special properties like carbonation resistance, simulated sea water, diluted acid, sulfate solution and freeze thaw cycles and found a very compacted and integrated structure, less coarse crystallized structure, high early and ultimate strength and considerable resistance to chemical attack. Tsakiridis et al. (2004) investigated the feasibility of red mud utilization in the production of cement. Zhang et al. (2009) prepared the red mud based cementitious material using RM-coal gangue mix, blast furnace slag, clinker and gypsum and presented the result in term of chemical, physical and mechanical properties. Najar et al. (2014) studied the physical and chemical parameters of RM based lightweight-foamed bricks by keeping in mind the potential utilization of RM as a building and construction material. Molineux et al. (2016) in their study, focused on the potential use of different percentage of red mud in the manufacturing of lightweight aggregate as a replacement of pulverized fuel ash (PFA). Nikbin et al. (2018) studied the effect of replacement of cement by 25% red mud in lightweight concrete on mechanical properties. However, in all the cases the red mud has been utilized to replace cement but the study on the coarse fraction (>0.075mm) of red mud (Red Sand (RS)) to replace sand is limited to Alam et al. (2017), which discussed the possible use of RS in mortar based on the particle shape analysis. Another aspect is the durability of mortar/concrete under the sulfate environment. Few studies have been performed on the effect of sulfate environment on the mortar/concrete prepared using natural sand. Ouyang et al. (2014) studied the effect of water to cement ratio and solution concentration on the surface hardness of concrete exposed to sulfate attack. The mortar/concrete partially submerged in the sulfate solution undergoes dual sulfate attack (chemical and physical) resulting in the development of microcracks in the submerged portion (Nehdi et al. 2014). However, the physical sulfate attack on mortar/concrete can be controlled by applying the epoxy and silane at the surface (Suleiman et al. 2014). However, this type of study was not found for the mortar prepared from RS. So, in the present study, the work by Alam et al. (2017) has been extended by performing laboratory investigations on the mortar prepared using the RS and the results are presented in term of compressive strength, sulfate resistance and corrosion resistance which is further correlated with the microstructural changes.

18.2 Material and Methodology

The red mud was collected from HINDALCO, Muri, Jharkhand, India. The red mud sample was washed on 0.075 mm sieve to separate the coarse fraction (>0.075 mm) (RS) from red mud. To compare the results, the river sand was used with particle size varying from 1 mm–0.500 mm and the ordinary Portland cement (OPC) of 53 grade was used to prepare the mortar.

The cubical specimens of 50mm were prepared as per ASTM C109/C109M (2021) using the river sand as well as RS in 1:3 and 1:3.5 instead of 1:2.75 binder to the aggregate ratio and 0.6 water-cement ratio as obtained based on trial and previous work by the authors. Three sets each containing three specimens was prepared to study the compressive strength. The compressive strength of one set was studied after 28 days water curing. The other two sets were exposed to sulfate environment for 90 days, in which one set was exposed to sulfate environment after 28 days water curing and other without water curing. The sulfate environment was provided by submerging the specimen into the mixture of $MgSO_4$ (2.5%) and Na_2SO_4 (2.5%) and only $MgSO_4$ (5%) solution. The solution was changed with the fresh solution to maintain the environment aggression at interval of 30 days. The morphology and surface chemistry of the sand samples are studied using SEM and EDX analysis of crushed mortar matrix obtained after the compressive strength test. The mineralogical composition of RS and RS mortar matrix was studied using X-ray diffraction (XRD) test performed using Rigaku Japan/Ultima-IV model with C_uK_α radiation.

18.3 Results and Discussion

18.3.1 Physical Characteristics

RS is separated from red mud by washing it on 0.075 mm sieve. The grain size distribution curves of RS along with the river sand are shown in Fig. 18.1. It is observed that RS is finer than that of river sand. The RS is found to have 82.5% fine sand (0.425 mm–0.075 mm) and 17.5% medium sand (2 mm–0.425 mm) (ASTM D2487-11). The coefficient of uniformity (C_u) and coefficient of curvature (C_c) (Table 18.1) shows that both the sand (RS and river sand) used in the present study are poorly graded sand (SP) as per Unified soil classification system (USCS) (ASTM D2487 2011). Although the RS used in present study is poorly graded, the RS from another source may show different grading as it depends on the method and degree of grinding of bauxite. The specific gravity (G_s) of RS is found as 3.21 (Table 18.1) which is much higher than that of the specific gravity of river sand (2.49) due to the presence of iron oxide in RS. The maximum and minimum void ratios (e_{max} and e_{min}) of RS is found as 1.46 and 1.05 respectively, which is higher than the maximum (0.96) and minimum (0.66) void ratios of river sand. However, the maximum variation in the void ratio of RS (0.41) is higher than that of river sand (0.27) which is the controlling factor for compactness of the material, (Alam et al. 2017).

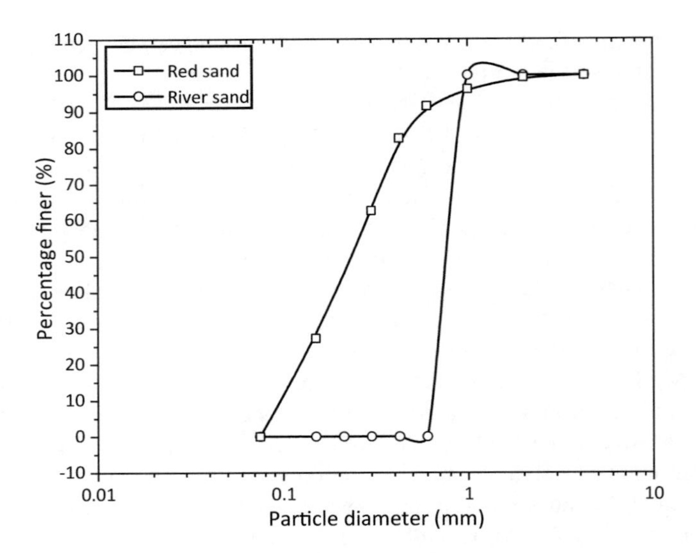

Fig. 18.1 Grain size distribution curve (Modified from Alam et al. 2017)

Table 18.1 Physical characteristics of RS and river sand (*Source* Alam et al. 2017)

Parameter	RS	River sand
G_s	3.21	2.49
e_{min}	1.05	0.66
e_{mzax}	1.46	0.96
γ_{dmax} (kN/m^3)	15.30	14.72
γ_{dmin} (kN/m^3)	12.85	12.46
C_u	2.89	1.46
C_c	0.98	0.94

18.3.2 Compressive Strength

The test samples were prepared as discussed in previous section (material and methodology). Figure 18.2 is showing the typical photo of the test specimens (Fig. 18.2a), specimens under sulfate solution (Fig. 18.2b) and the dry specimen after sulfate attack (Fig. 18.2c). The white layer at the surface after sulfate attack (Fig. 18.2c) may be due to the salt deposit and formation of glauberite ($Na_2Ca(SO_4)_2$). Figure 18.3a and b show the brittle failure of the river sand-based mortar and ductile failure of RS-based specimen under the load, which was loaded with a strain rate of 1.25 mm/min. After the compressive strength test, the strain-stress curve for all the specimens is plotted by considering the average value of three replicates.

Figures 18.4 and 18.5 shows the strain-stress curve of the cube prepared using the river sand in the binder to aggregate ratio of 1:3 and 1:3.5. Figures 18.4 and

Fig. 18.2 Typical photo of **a** test specimen **b** specimen in sulfate environment **c** specimen exposure of 90 days sulfate environment

Fig. 18.3 Typical photo showing the failure pattern of **a** river sand and **b** RS-based mortar

18.5 presents a comparison between the compressive strength of specimen cured in water for 28 days and the specimen subjected to 90 days sulfate environment before and after water curing. The 28 days compressive strength of the river sand mortar with a binder to the aggregate ratio of 1:3 and 1:3.5 are found as 33 MPa and 27 MPa for two different exposure condition, respectively (Figs. 18.4 and 18.5). The compressive strength of the specimens that were water cured for 28 days and exposed to the different sulfate environments was found to decrease by 34% and 46% in mixture of $MgSO_4$ & Na_2SO_4 and by 59% and 67% in $MgSO_4$ for the binder to the aggregate ratio of 1:3 and 1:3.5, respectively.

The decrease in strength after sulfate attack shows the negative impact of sulfate environment on the mortar prepared using the river sand which can be attributed to formation of ettringite (Ouyang et al. 2014). The same Figs. 18.4 and 18.5 are

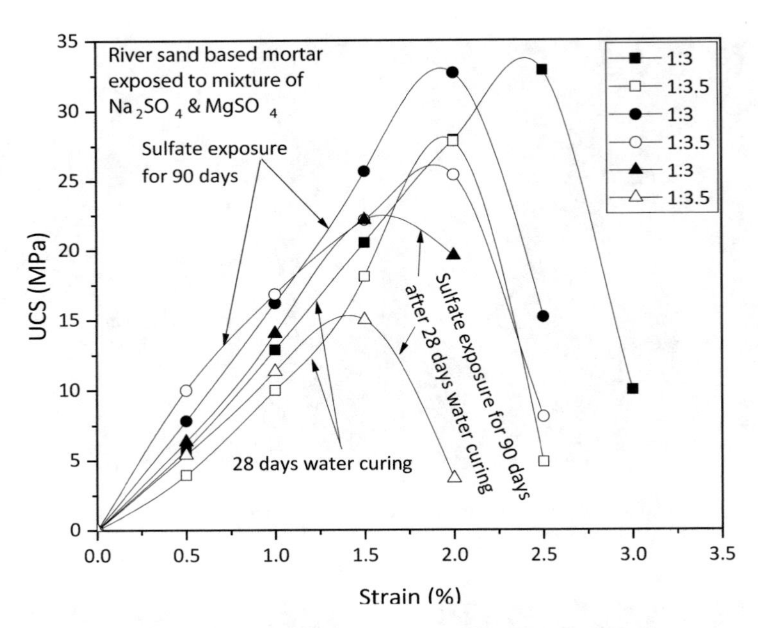

Fig. 18.4 Strain–stress curve of river sand-based mortar exposed to mixture of Na_2SO_4 & $MgSO_4$

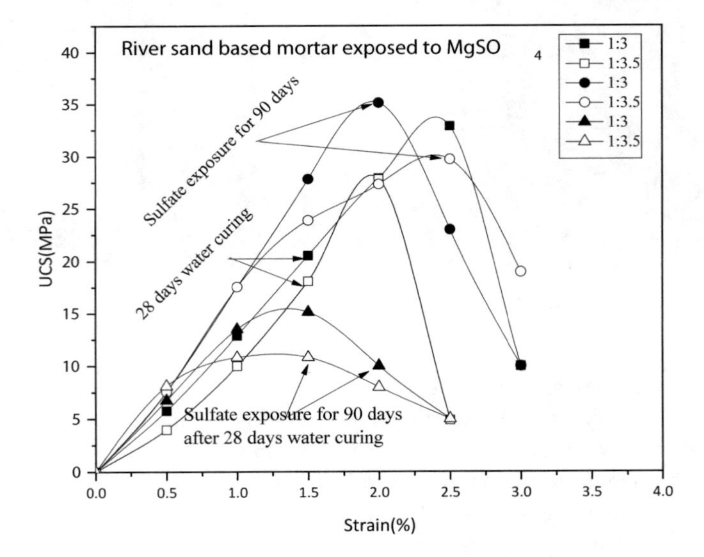

Fig. 18.5 Strain–stress curve of river sand-based mortar exposed to $MgSO_4$

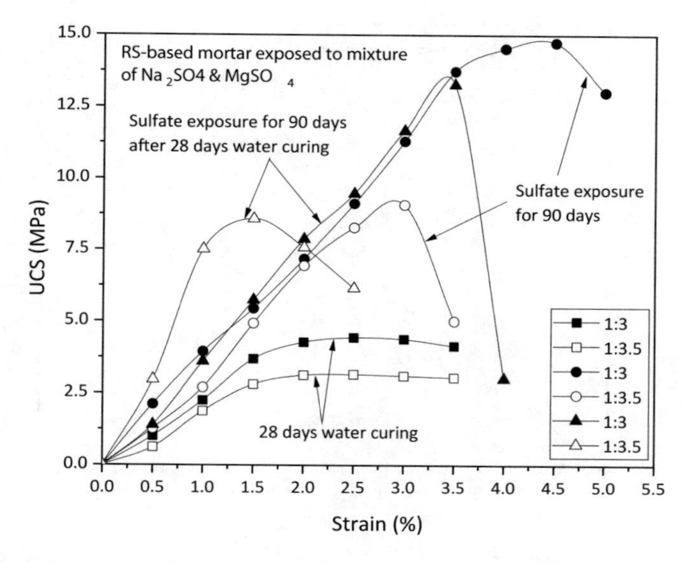

Fig. 18.6 Strain–stress curve of RS-based mortar exposed to mixture of Na_2SO_4 & $MgSO_4$

showing the effect of water curing on the sulfate attack. It can be observed that the specimen went through 90 days sulfate exposure without water curing shows higher strength as compared to the specimen went through sulfate environment after 28 days water curing irrespective of the binder to aggregate ratio and sulfate solution. A negligible difference between the compressive strength of 28 days cured specimen and specimen went through sulfate environment without water curing is observed (Figs. 18.4 and 18.5). The negligible difference in strength may be due to the early ettringite formation, whereas the delay ettringite formation causes microcracks due to early surface hardening and decrease the compressive strength (Ouyang et al. 2014).

Figures 18.6 and 18.7 presents the strain–stress curve of the specimen prepared using the RS in the binder to the aggregate ratio of 1:3 and 1:3.5. The compressive strength of mortar prepared using RS in the binder to aggregate ratio of 1:3 is found as 4.37 MPa whereas for binder to aggregate ratio of 1:3.5, it is found as 2.63 for 28 days water cured specimens (Figs. 18.6 and 18.7) which is much lower than that of river sand-based mortar for the similar binder to aggregate ratio (Figs. 18.4 and 18.5). It is found that the sulfate attack on the red mud based mortar increases its compressive strength, which may be due to the formation of ettringite, which causes compatible expansion due to the ductile nature of the RS matrix (Ouyang et al. 2014). The compressive strength of mortar subjected to sulfate attack for 90 days after 28 days water curing are found as 13.75 MPa for mixture of $MgSO_4$ & Na_2SO_4. Whereas the compressive strength of 10.67 MPa and 5.8 MPa is observed in $MgSO_4$ for the binder to the aggregate ratio of 1:3 and 1:3.5, respectively as also shown in Table 18.2. However, for the same condition, the river sand-based mortar shows

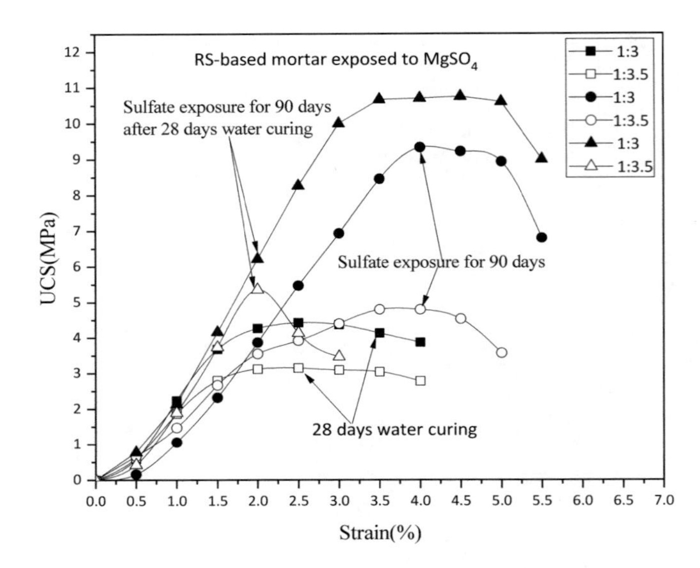

Fig. 18.7 Strain–stress curve of RS-based mortar exposed to $MgSO_4$

Table 18.2 Compressive strength comparison of RS-based mortar

	Sample description		28 days water curing	90 days sulfate curing	90 days sulfate curing + 28 days water curing
Red Sand based mortar	$Na_2SO_4 + MgSO_4$	1:3	4.37	15	13.75
		1:3.5	2.63	8.75	8
	$MgSO_4$	1:3	4	9.5	10.67
		1:3.5	3	4.8	5.8
River Sand based mortar	$Na_2SO_4 + MgSO_4$	1:3	32	31	22
		1:3.5	27	25	15
	$MgSO_4$	1:3	32	33	15
		1:3.5	27	30	10

higher strength as compared to the RS-based mortar. Figures 18.6 and 18.7 presents a comparison between the compressive strength of specimen went through sulfate environment before and after water curing. It is observed that the sulfate attack on the uncured specimen increases its compressive strength. The compressive strength of uncured specimen of RS-based mortar after sulfate attack is found as 15 MPa (1:3) and 8.75 MPa (1:3.5) when exposed to mixture of $MgSO_4$ & Na_2SO_4. Whereas the compressive strength of 9.5 MPa (1:3) and 4.8 MPa (1:3.5) was observed for $MgSO_4$ (Table 18.2). In both the cases, the compressive strength observed is very similar to

Table 18.3 Surface chemistry of RS-based mortar after sulfate attack. (Na_2SO_4 & $MgSO_4$)

Element	O	Mg	Al	Si	S	Cl	K	Ca
Percentage by weight	60.16	0.05	4.13	0.17	8.69	0.01	0.18	26.61

the RS-based mortar specimen that went through sulfate environment after 28 days of water curing (Figs. 18.4 and 18.5).

The gain in strength in the RS-based mortar after sulfate attack is due to the formation of ettringite ($Ca_6Al_2(SO_4)_3(OH)_{12} \cdot 26H_2O$), magnesite ($MgCO_3$) and plaster of paris ($2CaSO_4 \cdot H_2O$) as also identified in XRD analysis (Figs. 18.8 and 18.9). The gain in strength may also be due to the formation of dense matrix as observed in SEM image, but further detail study is required in this regard. Along with the ettringite and plaster of paris, sulfate attack also causes the formation of sodium-calcium sulfate (glauberite) at the surface of the specimen as white patches as shown in Fig. 18.2c. The formation of ettringite is also identified in the SEM image (Fig. 18.10) of RS-based mortar after sulfate attack. However, the presence of ettringite was not significant enough in case of $MgSO_4$ (Fig. 18.11).

The surface chemistry of the mortar prepared using RS is studied using EDX (Figs. 18.12 and 18.13) and is presented in Tables 18.3 and 18.4 for mixture of $MgSO_4$ and Na_2SO_4 and only $MgSO_4$ solutions, respectively. The surface was found to be rich in Ca (26.61%) followed by 8.69% sulphur along with 4.13% Al. The majority of Ca, S and Al may be due to the formation of ettringite and plaster of paris.

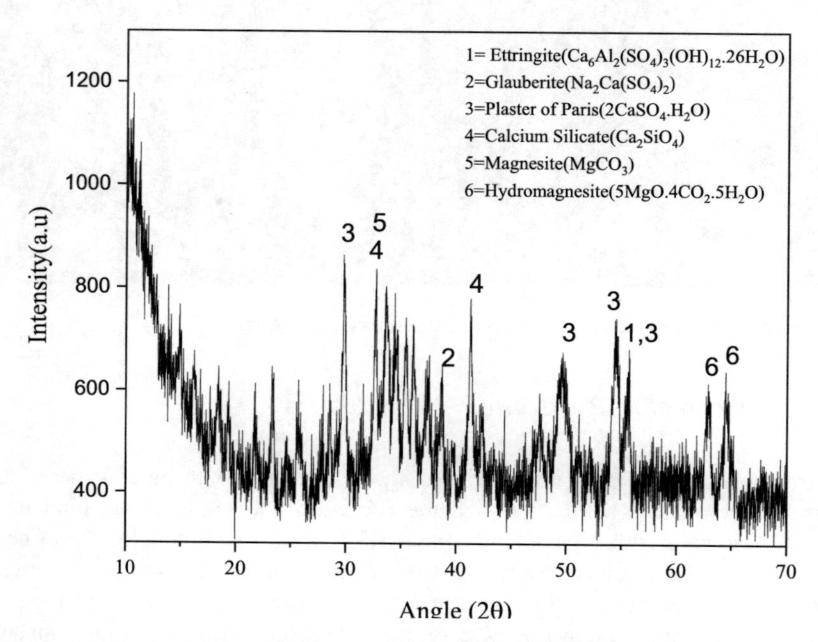

Fig. 18.8 XRD plot of RS-based mortar after sulfate attack (Na_2SO_4 & $MgSO_4$)

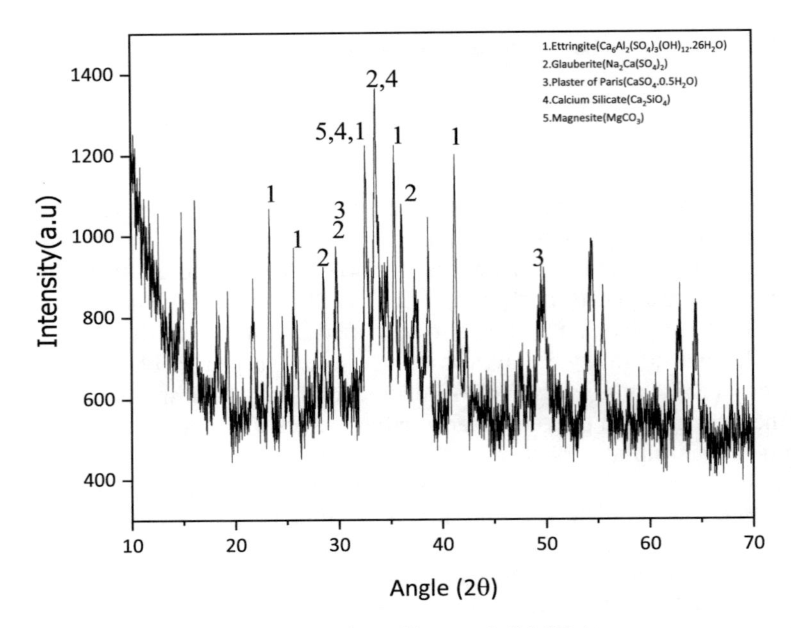

Fig. 18.9 XRD plot of RS-based mortar after sulfate attack ($MgSO_4$)

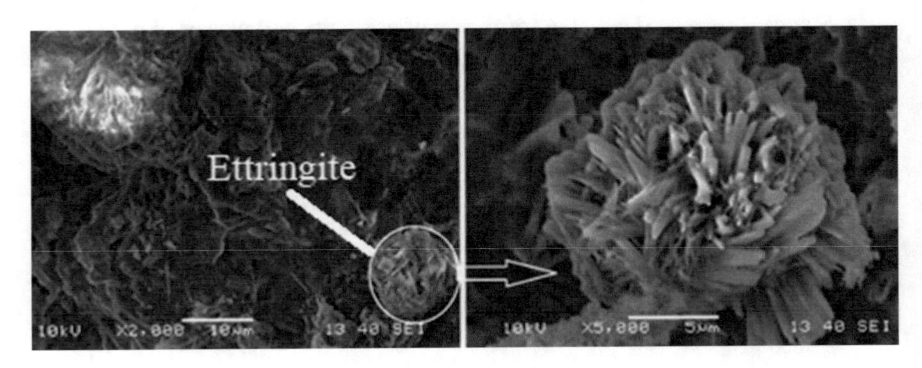

Fig. 18.10 SEM image of RS-based mortar after sulfate attack (Na_2SO_4 & $MgSO_4$)

18.3.3 Corrosion Resistance

Corrosion resistance is an important parameter of a matrix to be considered as a mortar. Ribeiro et al. (2012) studied the corrosion behaviour of red mud based concrete. However, this type of study on the RS has not been found. So, the effect of RS on the corrosion resistance has been studied on specimen with binder to aggregate ratio of 1:2 at w/c of 0.4 using the polarization resistance technique developed by Miranda et al. (2005). Figure 18.14 is showing the setup for the corrosion resistance test. Figure 18.14a is showing the series connection between AC to DC converter,

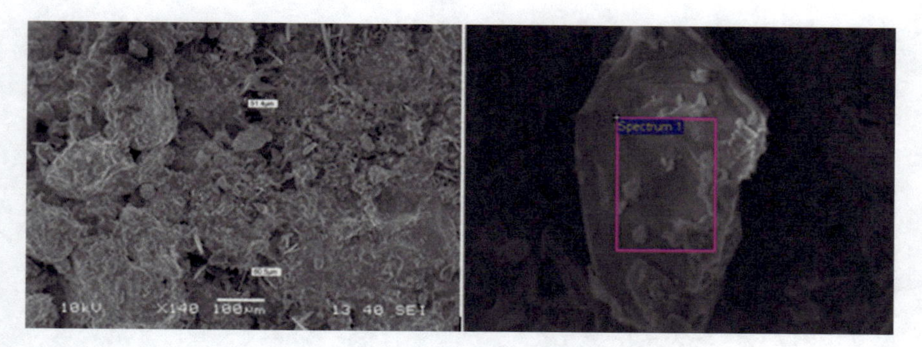

Fig. 18.11 SEM image of RS-based mortar after sulfate attack ($MgSO_4$)

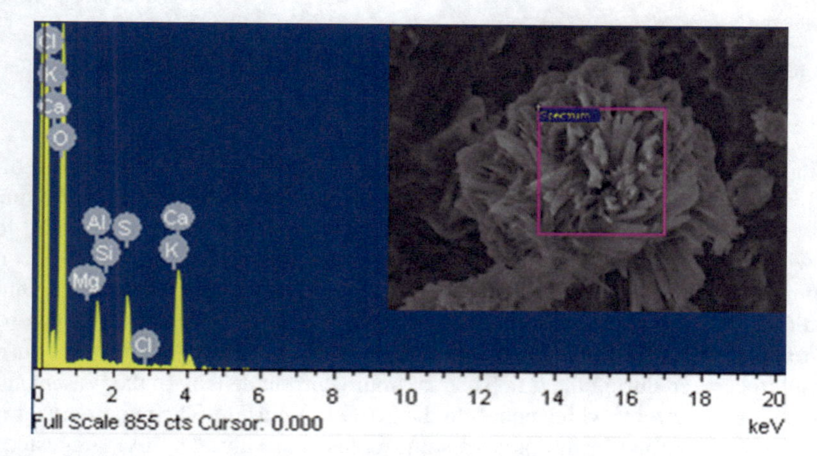

Fig. 18.12 EDX of RS-based mortar after sulfate attack (Na_2SO_4 & $MgSO_4$)

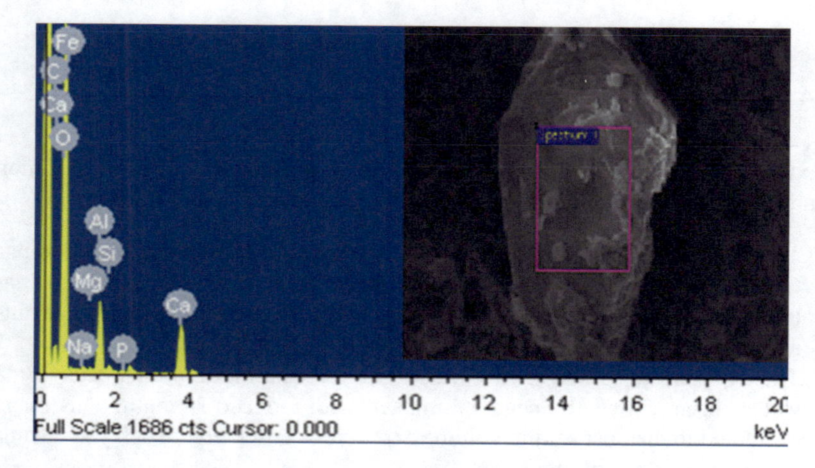

Fig. 18.13 EDX of RS-based mortar after sulfate attack ($MgSO_4$)

Table 18.4 Surface chemistry of RS-based mortar after sulfate attack ($MgSO_4$)

Element	O	Mg	Al	Si	C	Na	Fe	Ca
Percentage by weight	64.32	0.37	6.89	0.69	5.54	0.26	4.66	17.27

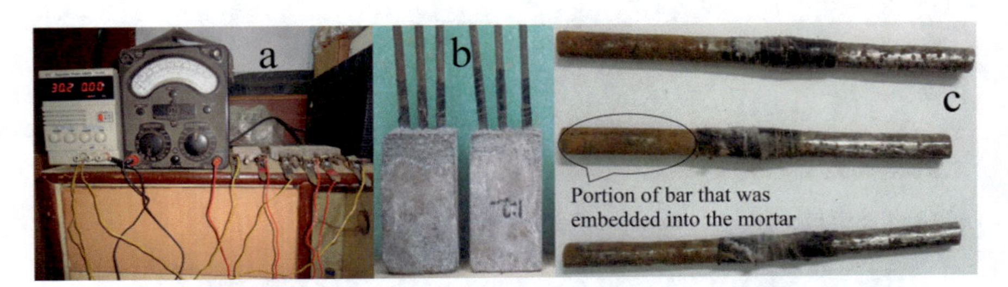

Fig. 18.14 Typical photo showing the **a** corrosion resistance setup, **b** specimen, and **c** bar after the test

multi-metre and specimens to measure the current in the circuit. A DC voltage of 30 Volt was applied through the circuit and the current was measured using the multi-metre. Figure 18.14b is showing the samples after the test whereas Fig. 18.14c is showing the reinforcing bar after 6 months of the test period. No trace of corrosion on the bar surface is found but imprint due to red colour of red sand can be seen (Fig. 18.14c) after 6 months, which confirm that the RS matrix can passivate the reinforcement from the corrosion. Figure 18.15 is showing the time vs current density measured during the test. The maximum current density in the present study is found as 18 $\mu A/cm^2$ which equals to the 0.169 mA current. The maximum current (0.169 mA) recorded in the present study is less the value (1 mA) suggested by Miranda et al. (2005) for passivating the reinforcement from corrosion.

18.4 Conclusions

Based on the laboratory tests performed to study the resistance of RS-based mortar towards the sulfate attack and corrosion, following conclusions can be drawn:

1. The mortar prepared using RS shows lower compressive strength as compared to that with river sand. However, the exposure to sulfate environment increases the compressive strength of RS-based mortar due to the formation of ettringite and plaster of Paris.
2. The exposure to sulfate environment for river sand-based mortar after 28 days water curing shows the negative impact, with reduced strength. This may be attributed to the post setting volume expansion caused due to delayed ettringite formation during sulfate curing.

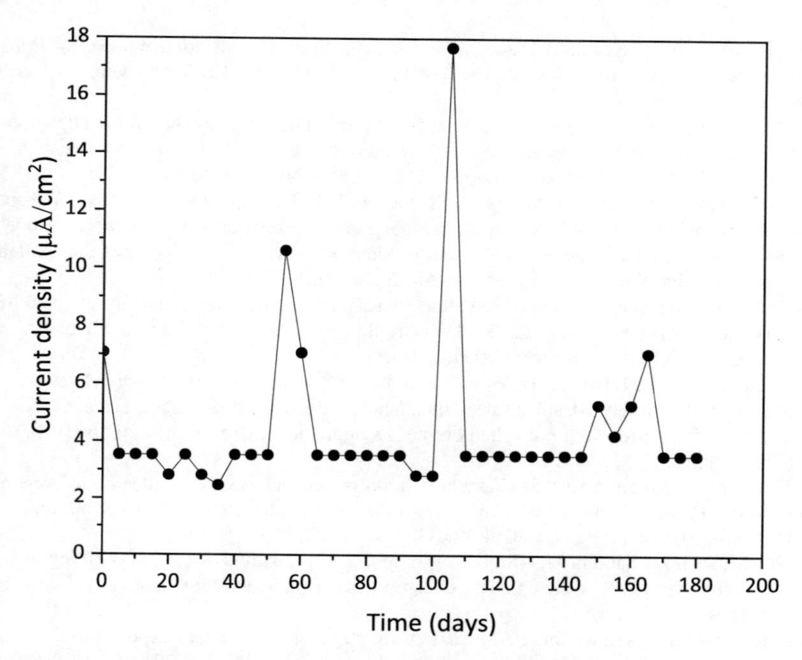

Fig. 18.15 Time versus current density plot for RS-based mortar

3. The current resistance test shows that the mortar prepared using the RS is effective in controlling the corrosion of reinforcing bar, which may attributed to the alkaline condition caused due to high pH of RS.

Acknowledgements The authors would like to acknowledge the partial financial support for present work from Department of Science and Technology (DST), India under the SERB programme SB/S3/CEE/036/2013 and CRG/2020/006383.

References

Agrawal A, Sahu KK, Pandey BD (2004) Solid waste management in non-ferrous industries in India. Resour Conserv Recycl 42(2):99–120. Article first published online: 2 June 2004. https://doi.org/10.1016/j.resconrec.2003.10.004

Alam s, Das SK, Rao H (2017) Characterization of coarse fraction of red mud as a civil engineering construction material. J Clean Prod 168:679–691. Article first published online: 30 August 2017. https://doi.org/10.1016/j.jclepro.2017.08.210

ASTM C109/C109M (2021), Standard Test Method for Compressive Strength of Hydraulic Cement Mortars (Using 2-in. or [50 mm] Cube Specimens), ASTM International, West Conshohocken, PA, www.astm.org

ASTM D2487 (2017), Standard Practice for Classification of Soils for Engineering Purposes (Unified Soil Classification System), ASTM International, West Conshohocken, PA, www.astm.org
Gordon JN, Pinnock WR, Moore MM (1996) A preliminary investigation of strength development in Jamaican red mud composites. Cem Concr Compos 18(6):371–379. Article first published online: 12 February 1999. https://doi.org/10.1016/S0958-9465(96)00027-3
Mayes WM, Jarvis AP, Burke IT, Walton M, Gruiz K (2011) Trace and rare earth element dispersal downstream of the Ajka red mud spill, Hungary, In: Wolkersdorfer C, Freund A, Rude TR, (eds) 11th Inetrnational mine water association congress-mine water – Managing the challenges. International Mine Water Association, 29–34. ISBN 9781897009475
Miranda JM, Jimenez AF, Gonzailez JA, Palomo A (2005) Corrosion resistance in activated fly ash mortars. Cem Concr Res 35(6):1210–1217. Article first published online: 10 November 2004. https://doi.org/10.1016/j.cemconres.2004.07.030
Molineux CJ, Newport DJ, Ayati B, Wang C, Connop SP, Green JE (2016) Bauxite residue (Red mud) as a pulverised fuel ash substitute in the manufacture of lightweight aggregate. J Clean Prod 112(1):401–408. Article first published online: 18 September 2015. https://doi.org/10.1016/j.jclepro.2015.09.024
Nath H, Sahoo P, Sahoo A (2015) Characterization of red mud treated under high temperature fluidization. Powder Technol 269:233–239. Article first published online: 16 September 2014. https://doi.org/10.1016/j.powtec.2014.09.011
Najar PA, Mukesh J, Chaddha M, Nimje T, Agnihotri AA, Satpathy BK (2014) Building materials from aluminium industry wastes. In: National seminar on green building materials & construction technologies, New Delhi
Nehdi ML, Suleiman AR, Soliman AM (2014) Investigation of concrete exposed to dual sulfate attack. Cem Concr Res 64:42–53. Article first published online: 10 July 2014. https://doi.org/10.1016/j.cemconres.2014.06.002
Nikbin IM, Alianghazadeh M, Charkhtab SH, Fathollahpour A (2018) Environmental impacts and mechanical properties of lightweight concrete containing bauxite residue (red mud). J Clean Prod 172:2683–2694. Article first published online: 21 November 2017. https://doi.org/10.1016/j.jclepro.2017.11.143
Ouyang W, Chen J, Jiang M (2014) Evolution of surface hardness of concrete under sulfate attack. Constr Build Mater 53:419–424. Article first published online: 29 December 2013. https://doi.org/10.1016/j.conbuildmat.2013.11.107
Ribeiro DV, Labrincha JA, Morelli MR (2012) Effect of red mud addition on the corrosion parameters of reinforced concrete evaluated by electrochemical methods. Ibracon Struct Mater J 5(4):451–467. Article first published online: 6 October 2011. https://doi.org/10.1016/j.cemconres.2011.09.002
Singh M, Upadhayay SN, Prasad PM (1996) Preparation of special cement from Red Mud. Waste Manage 16(8):665–670. Article first published online: 26 February 1999. https://doi.org/10.1016/S0956-053X(97)00004-4
Suleiman AK, Soliman AM, Nehdi ML (2014) Effect of surface treatment on durability of concrete exposed to physical sulfate attack. Constr Build Mater 73:674–681. Article first published online: 27 October 2014. https://doi.org/10.1016/j.conbuildmat.2014.10.006
Tsakiridis PE, Leonardou SA, Oustadakis P (2004) Red mud addition in the raw meal for the production of Portland cement clinker. J Hazard Mater 116(1–2):103–110. Article first published online: 22 October 2004. https://doi.org/10.1016/j.jhazmat.2004.08.002
Zhang N, Sun H, Liu X, Zhang J (2009) Early-age characteristics of red mud–coal gangue cementitious material. J Hazard Mater 167(1–3):927–932. Article first published online: 30 January 2009. https://doi.org/10.1016/j.jhazmat.2009.01.086
Zhihua P, Dongxu L, Jian Y, Nanru Y (2003) Properties and microstructure of the hardened alkali-activated red mud–slag cementitious material. Cem Concr Res 33(9):1437–1441. Article first published online: 12 April 2003. https://doi.org/10.1016/S0008-8846(03)00093-0

Chapter 19
Condition Assessment and Repair Strategy for RCC Chimney of Thermal Power Station Located in Semi-Arid Region in India

T. V. G. Reddy, P. N. Ojha, Brijesh Singh, Rizwan Anwar, and Vikas Patel

19.1 Introduction

The first step in concrete repair is to evaluate the current condition of the concrete structure. This evaluation may include a visual inspection of the structure, a review of available design and construction documents, a review of records of any previous repair work, review of maintenance records before any repair work is put in hand, and the cause of damage must be identified as clearly as possible. Because many deficiencies are caused by more than one mechanism, a basic understanding of the causes of concrete deterioration is essential to determine what has actually happened to a particular concrete structure and why. It is always useful to ask questions of as many as possible of the people who were concerned with the design or construction. The next step must be to consider the objective of the repair, which will generally be to restore or enhance one or more of the following:

- Durability: Repair is carried out on structure that has appeared to be less durable.
- Structural strength: Increase of structural strength to make structure serviceable.
- Function: Repair is considered for the restoration of fitness for use.
- Appearance: Improvement in appearance is often required to enhance the overall aesthetic of the structure.

Of these four requirements, restoration of durability is by far the most common in repair work. One must also consider whether the repair is to be structural or cosmetic. Only after deciding on the most likely cause of damage and the purpose of the work, the method of repair should be chosen. The third step that must be considered during repair is Quality assurance/Quality control in order to evaluate the performance of repair materials to achieve a satisfactory repair, with adequate strength and durability. Many practitioners have a lot of confusion regarding the

T. V. G. Reddy · P. N. Ojha · B. Singh (✉) · R. Anwar · V. Patel
National Council for Cement & Building Materials, Ballabgarh, Haryana 121004, India

K. R. Reddy et al. (eds.), *Advances in Sustainable Materials and Resilient Infrastructure*, Springer Transactions in Civil and Environmental Engineering,
https://doi.org/10.1007/978-981-16-9744-9_19

property that needs to be evaluated and the minimum acceptable values to achieve an effective repair for different situations. Rebound hammer test, pull-out and pull-off tests, Ultrasonic pulse velocity (UPV) test, Core sampling and testing, cover survey and carbonation test are mostly used for the assessment of existing concrete structures (Bungey 1989; Malhotra and Carino 2004; Bhaskar and Rajeev 1999). It is important to note that almost all the NDT methods indirectly estimate the concrete strength and strength obtained by these methods, in most of the cases, is comparable (4). Even then, no single method can be said to be fully reliable and therefore, more than one method should be performed and results should be correlated (Goyal and Mahesh 1997; Kenneth et al. 2012; Ramachandran 1996). Tarek et al. (2003) carried out the study to determine the corrosion of steel in RC structures when exposed to a marine environment by evaluating physical and chemical properties of corrosion, presence of chloride ion and permeability properties of concrete. Study indicated that the w/c ratio has great influence on the magnitudes of corrosion. As the narrow cracks heals considerably fast in the marine environment, chances of reduction in corrosion rate can be seen. Sharma et al. (2015) reported the distress assessment, repair and strengthening of RCC members of turbo generator foundation of Anpara Thermal Power Station at Uttar Pradesh (India) indicating need of detailed investigation for understanding cause of distress and selection of appropriate repair methodology for effective repair system.

Presently, there are very few standard code available specifically for repair materials because of diverse nature of the materials and varying requirements. However, professional societies or institutes like American Concrete Institute (ACI) provide some guidelines, which are quite useful. The European Standard EN 1504 entitled: Products and systems for the repair and protection of concrete structures, provides an improved framework for achieving successful and durable repair systems. In India, only CPWD handbook on repair and rehabilitation of RCC buildings provides specifications based on performance properties of repair materials and also suggests the repair methodologies. Hence it is necessary to have a field quality plan giving details of the testing and frequency in order to execute the work in proper manner.

19.2 Research Significance

RCC structures constructed during early 70s and late 80s of the last century in India are found to be under distressed conditions due to inadequate specifications and poor construction practices. The main purpose of this research work is to firstly briefly describe how to carry out the condition assessment of RCC structures before taking up repair work and secondly it also describes evaluation of repair materials including Quality Assurance (QA) & Quality Control (QC) measures undertaken during execution of repair to evaluate the performance of repair material and repair work as whole for obtaining the desired service life.

19.3 Case Study on Condition Assessment of RCC Chimney

The investigated structure is 40 years old 150 m high RCC Chimney in Thermal Power Station located in semi-arid region. The condition assessment study was carried out due to distress condition of the structure with ageing. During the visual inspection, rust strains were observed at many locations, which gave an indication of corrosion on rebars (Figs. 19.1 and 19.2). The concrete cover to the reinforcement was delaminated, cracked and spalled. Also the concrete cover was found to be inadequate at many locations. Inspection of the concrete showed honeycombing at few locations that may be due to inadequate compaction during construction. The damage to exposed surface of RCC Chimney was mostly at the higher elevations i.e., above 65 m. No apparent visual distress was observed at lower elevations of chimney.

Fig. 19.1 Showing distress in terms of spalling of concrete, corrosion of rebars in RCC Chimney

Fig. 19.2 Showing distress in terms of spalling of concrete, corrosion and reduction in diameter of rebars in RCC Chimney

19.3.1 Findings on Cause of Damage Using Non Destructive Evaluation Techniques

Non-destructive testing was carried out in RCC Chimney and the test results were analyzed to work out the cause of damage and find out the suitable repair methodology. Based on ultrasonic pulse velocity and rebound hammer testing done by random sampling technique, overall quality of concrete was graded as 'Good'. The test results of concrete cores were found to meet the specified characteristic compressive strength of M25 grade concrete. Based on the half-cell potential measurement, corrosion was found to be 'uncertain'. Carbonation was found to vary from 5 to 15 mm at lower elevations and 30 mm at higher elevations. The concrete cover was found to be inadequate at many locations with values varying from 20 to 60 mm at lower elevations and 10 to 40 mm at higher elevations. The corrosion was observed at those locations where the concrete cover was less. Chloride and sulphate content was found to be within the specified limit. Hence the cause of damage for distress in structure is found to be carbonation, leading to corrosion and inadequate concrete cover.

19.3.2 Repair and Strengthening Measures

1. Removal of all soft and loose concrete from the visible distressed concrete
2. Cleaning of rust from corroded reinforcing steel and providing anticorrosive treatment.
3. Applying Concrete Penetrating Corrosion Inhibitor (CPCI) over the concrete substrate
4. Applying two component epoxy bond coat conforming to specifications of ASTM C882 type II to ensure the effective bond old substrate obtained after chisel cutting of cover concrete and new mortar.
5. Building up the profile using Polymer Modified Mortar (PMM) in 15–20 mm thick layers with application of bond coat between each layer.
6. Strengthening the concrete around cracked portions by wrapping chimney stack (0.5 m wide) using single layer non-metallic composite glass fiber polymer wrapping system
7. Applying protective coating on surface of RCC Chimney to protect the concrete from further deterioration. Chemical and UV resistant polyurethane coating for the top 50 m and acrylic anti-carbonation coating with silane siloxane primer at the lower elevations were recommended.

19.4 Quality Assurance/Quality Control

19.4.1 Material Testing

Testing for the performance of repair materials play an important role to achieve a satisfactory repair, with adequate strength and durability. Material testing including solid content, bond strength using slant shear test, compressive strength, density, specific gravity, pH value should be carried out prior to commencement of work. Initially Quality Assurance Plan (QAP) was prepared for material testing frequency and type of test to be carried out. Accordingly, repair material samples were collected and tested in laboratory. Also the Manufacturer's Test Certificate (MTC's) of all materials were also reviewed. The materials testing and the acceptance criteria was carried out based on the available relevant codes and also by manufacturer's test certificate, wherever required. The test results obtained for the repair materials are as follows:

19.4.1.1 Rust Remover

The rust remover was tested for specific gravity and pH value. The specific gravity test was carried out as per IS 9162 was found to be 1.17 which was meeting the requirement of manufacturer's test certificate. The rust remover was found to be slightly acidic in nature.

19.4.1.2 Anti-corrosive Coating

The properties of the coating that were evaluated at author's laboratory were tack free time and density as per ASTM C679 and IS 9162, respectively. The results obtained for tack free time was 26 min and for density at 30 °C was 2.05 g/cm^3. Also the aspect of the coating was greyish liquid. All results obtained were found to be meeting the requirements of MTCs.

19.4.1.3 Corrosion Inhibitor

The properties of corrosion inhibitor that were evaluated at author's laboratory were pH value and specific gravity as per IS 9103 and IS 9162, respectively. The results obtained for pH value was 10.59 and for specific gravity at 30 °C was 1.14. Also aspect of the inhibitor was colorless. All results obtained were found to be meeting the requirements of MTCs.

19.4.1.4 Bonding Agent (To Bond Substrate to New Mortar)-Epoxy A&B

The testing of compressive strength as per IS 9162, bond strength as per ASTM C882 and density as per IS 9162 were done at author's laboratory. As per the specifications provided by NCB, the bonding agent should meet the specifications of ASTM C881 type II epoxy. The specimens were prepared for bond test as per ASTM C882 (Fig. 19.3). The epoxy A was not meeting the requirements of type II epoxy although the data sheet provided by the manufacturer says that the bonding agent conforms to type II as per ASTM C881. Hence the client was suggested not to use the epoxy A as a bonding agent, therefore, epoxy B was used that was found to meet the requirements of type II epoxy. The results are shown in Table 19.1.

Fig. 19.3 Testing of bond strength by slant shear test as per ASTM C882

Table 19.1 Test results of physical properties of Epoxy bonding agent

Property	Physical requirements of type II Bonding agent as per ASTM C881	Test results of epoxy A	Whether epoxy A meeting the requirement	Test results of epoxy B	Whether epoxy B meeting the requirements
Compressive strength @ 7 days (MPa)	Min. 35	37	Yes	82	Yes
Bond strength @ 14 days (MPa)	Min. 10	3.51	No	11.04	Yes
Density at 30 °C	–	1.55	Yes	1.90	Yes
Aspect	–	Concrete grey	Yes	Concrete grey	Yes

Table 19.2 Test results of physical properties of SBR latex

Property of SBR Latex	Test results of SBR latex	Physical requirements of type I latex as per ASTM C1059	Quality requirement as per CPWD handbook	Quality requirement as per MTC's	Whether meeting the requirements
Bond strength @ 14 days (Mpa)	4.73	Min. 2.8	–	–	Yes
Solid content (%)	40.30		Not less than 35	39 to 41	Yes
pH	7.10		–	7 to 9	Yes
Specific gravity	1.02		–	1.01 to 1.03	Yes
Aspect	Milky white		–	Milky white	Yes

19.4.1.5 SBR Latex

The testing of solid content as per IS 9103:1999, bond strength as per ASTM C1042 and specific gravity as per IS 9162:1979 and pH value as per IS 9103:1999 were done at author's Laboratory. As per the specifications provided by NCB, the SBR latex should meet the specifications of ASTM C1059 type I Latex. The bond strength was evaluated based on Table 19.1 of ASTM C1059. As per Table 5.5 of CPWD handbook, solid content in latex should not be less than 35% and within ±1% of the value marked by the manufacturer. The test results are shown in Table 19.2. Based on results, SBR latex was found to meet the quality requirements of CPWD handbook, MTC's and ASTM C1059.

19.4.1.6 Epoxy Grout

The properties of epoxy grout that were evaluated at author's laboratory were compressive strength as per IS 9162:1979, density as per IS 9162:1979 and pot life. The quality requirements of epoxy grout were verified based on the values given in Sl. No., 9.3.2.1,a,ii of CPWD handbook. The test results obtained are shown in Table 19.3. All the results obtained were found to be meeting the requirements of CPWD handbook and Manufacturer's Test Certificate.

19.4.1.7 Polymer Modified Mortar (PMM)

Ordinary Portland Cement (OPC) of 43 grade conforming to IS269, crushed sand conforming to Zone II of IS 383, mixing water conforming to IS 456, polymer conforming to type I latex of ASTM C1059 and polypropylene fibers were used

Table 19.3 Test results of physical properties of Epoxy grout material

Property	Test results	Quality requirement as per CPWD handbook	Whether meeting the requirements
Compressive strength @ 7 days (Mpa)	67.50	Min. 60	Yes
Bond strength @ 14 days (Mpa)	11	Min. 3.5	Yes
Pot life (minutes)	40	Min.30	Yes
Density at 30 °C	1.11	-	Yes
Aspect	Pale yellow	-	Yes

for casting of polymer modified mortar. Three samples each consisting of different ingredients were cast and tested for compressive strength at author's laboratory. Firstly, conventional mortar (designated as S1), secondly, polymer modified mortar consisting of latex only (designated as S2), thirdly, polymer modified mortar consisting of both latex and fibers (designated as S3). As per CPWD handbook, minimum compressive strength after 28 days should be 20 MPa. As per ASTM C1438, minimum compressive strength of polymer modified mortar shall be 70% of reference material (conventional mortar). Test results obtained for sample S1, S2 & S3 are shown in Table 19.4:

The curing under wet conditions, such as water immersion or moist curing, applicable to ordinary cement mortar or concrete, is detrimental to polymer modified mortar. Normally, PMM requires a different curing regime because of the incorporation of polymer latexes. Almost optimum properties of the modified systems are achieved by a combined moist and dry cure, i.e., moist cure for 1 to 3 days, followed by dry cure at ambient temperature[7]. Testing with three different curing regimes were carried out at author's laboratory and the test results obtained are shown in Table 19.5. The maximum strength is achieved when combined wet and dry curing is used. Hence the same was recommended at site.

Table 19.4 Test results of compressive strength of different mortars

Sample identification	Compressive strength @ 28 days (MPa)	Quality requirement as per CPWD handbook	Quality requirement as per ASTM C1438
S1 (Conventional mortar)	48	–	–
S2 (Polymer mortar with SBR latex)	44.25	Min. 20 MPa	Min.33.6 MPa
S3 (Polymer mortar with SBR latex and fibers)	42	Min. 20 MPa	Min.33.6 MPa

Table 19.5 Test results of compressive strength of polymer modified mortar with different curing regimes

Sample identification	Curing regime	Compressive strength @ 7 days (MPa)
S4	7 days air curing	28.5
S5	7 days moist curing	26.2
S6	3 days moist and 4 days air curing	30.5

Table 19.6 Test results of unidirectional woven glass fiber fabric

Description	Characteristics of glass fiber wrap as per data sheet	Test results	Whether meeting the requirements
Density of wrap	935 g/sqm ± 47 g/sqm	900 g/sqm	Yes
Fiber density	2.56 g/cc	2.58 g/cc	Yes
Effective fiber wrap thickness	0.363 mm	0.47 to 0.53 mm	Yes
Tensile modulus	76,000 N/mm^2	–	–
Tensile strength	3400 N/mm^2	–	–

19.4.1.8 Unidirectional Woven Glass Fiber Fabric

The areal weight, effective fiber wrap thickness and fiber density were measured at author's laboratory. The test results obtained are shown in Table 19.6.

19.4.1.9 Protective Coating to Concrete

The properties of the coating that were evaluated at author's laboratory were solid content as per IS 9103:1999 and density as per IS 9162:1979, respectively. The results obtained for solid content was 72.67% and for density at 30 °C was 1.24 g/cm^3. Also the aspect of the coating was colored free flow liquid. All the results obtained were found to be meeting the requirements of MTCs.

19.4.2 Execution of Work

During the execution of repair work, quality inspection was carried out by qualified/experienced engineers to follow the approved repair methodology and QAP, etc. Since some repair materials may be hazardous to the workers involved in their application, hence the safety measures were reviewed before the commencement of work.

19.4.2.1 Surface Preparation and Sand Blasting

All the apparently cracked/loose/honeycombed/unsound surfaces of concrete were chipped/chiseled off. Cleaning off existing loose concrete was done carefully by manual or low impact frequency hammer. The entire surface was checked by light hammering (using 2 lb/1 kg hammer) for detection of loose/unsound concrete. Edges of chipped patch will bear vertical/reverse cut with minimum depth at edges as 10 mm. Sand blasting was done to clean the prepared concrete surface of all loose, lightly sticking materials, including the foreign materials, loose concrete, aggregate, etc. so as to provide a good bond with the applied mortar. Coarse sand conforming to Zone II as per IS 383 was sprayed over the substrate using air compressor having pressure of 4–7 kg/cm^2.

19.4.2.2 Reinforcement Treatment

After carrying out the chipping, chemical rust remover was applied by brush over the corroded reinforcement surface thoroughly. After 24 h of application the surface was cleaned with wire brush and all the loose particles were removed. Since rust remover was slightly acidic in nature, the reinforcement was washed with water and was allowed to dry. Visual inspection was carried out to verify whether the rust was properly removed or not. Two coats of anti-corrosion were applied over the cleaned reinforcement by brush @ 5 to 6 sqm per liter. Also two coats of corrosion inhibitors were applied over the concrete @ 4 sqm per liter.

19.4.2.3 Epoxy Grouting

V-cuts of size 10 mm $\times$ 10 mm and fixing of Poly Vinyl Chloride (PVC) nipples after drilling holes @ 300 mm c/c was done along the cracks. Subsequently sealing/closing the v-cut using epoxy mortar was done one day prior to carrying out injection grouting (refer Fig. 19.4). Epoxy was then grouted into the cracks with pump at a pressure of 3–5 kg/cm^2(refer Fig. 19.5). In case of vertical cracks injection was started at the lowest nipple and continued until the injected grout begins to flow out at the next higher nipple. Whereas in case of horizontal locations, injection was started from one nipple and continued until the injected grout begins to flow out at the other nipple. After minimum of one day, when the system was cured, the nipples were cut.

After the completion of epoxy grouting, cores were extracted on cracked portion to physically verify the depth of ingress of grout material (refer Fig. 19.6). Wherever the epoxy grout was not found to be filled in the crack, the same was re-grouted. The UPV test using surface probing was also carried out over the grouted cracked portion to verify the ingress of grout material (refer Fig. 19.7). UPV values were found to vary from 3.70 km/sec to 4.76 km/sec with an average velocity of 4.34 km/sec that indicated Good quality as per IS 13311:1992 (Part I). The photograph showing the core extracted and UPV test over the crack is shown below.

Fig. 19.4 Application of epoxy mortar to seal the crack in RCC Chimney

Fig. 19.5 Epoxy grouting of cracks using pump in RCC Chimney

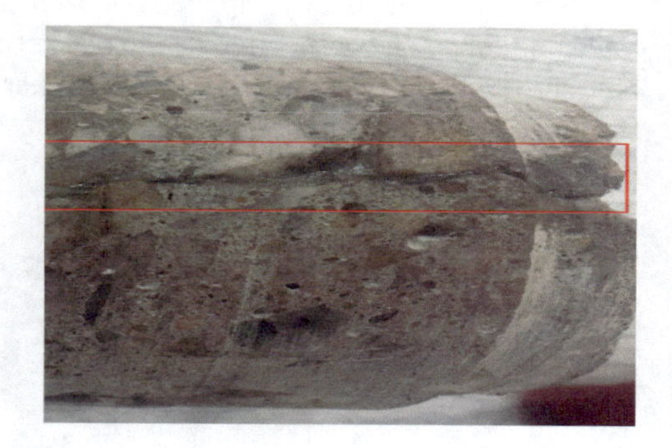

Fig. 19.6 Core extraction from grouted crack portion of RCC Chimney

19.4.2.4 Coat

Epoxy bond coat was applied by brush over the concrete substrate. It was ensured that the PMM was applied while the bond coat was in tacky condition, however if the epoxy bond coat loses its tacky condition, the same was removed or slightly abraded and the second coat of epoxy was applied before placing the PMM. It was ensured

Fig. 19.7 UPV testing around grouted crack by surface probing method

that during the application of bond coat the atmospheric temperature was below 40 °C. The material was consumed @ 0.5–0.6 kg/sqm.

19.4.2.5 Polymer Modified Mortar

Building up the profile of structural member up to required depth by using SBR latex conforming to ASTM C1059 type-I in damaged areas (1 cement-3 part graded cleaned river sand + 15% latex by weight of cement) with 0.35 w/c ratio, in 15–20 mm or 10–15 mm thick layers by applying bond coat between successive/each layers was done. Polypropylene fibers were also added to reduce shrinkage in PMM. Proper curing was done for the repair work. Gunny bags were used for effective curing if needed. 3 days of wet curing followed by air curing was done. A bonding coat of Styrene Butadiene Rubber (SBR) polymer (@10% of cement weight) modified cementitious bond coat was applied between each layer of polymer modified mortar. Such bonding coat was not allowed to dry before application of new layer of mortar. Non-destructive testing i.e., rebound hammer and UPV test was carried out using random sampling technique over the PMM (refer Figs. 19.8 and 19.9). The test results as obtained over the PMM are shown in Table 19.7.

Fig. 19.8 Rebound hammer testing over polymer modified mortar used in repair of RCC Chimney

Fig. 19.9 UPV testing over polymer modified mortar used in repair of RCC chimney by surface probing method

Table 19.7 Test results of rebound hammer and UPV testing on PMM

Location	Compressive strength based on rebound hammer test (N/mm^2)	UPV test results in km/s (surface probing)
L1 @ EL + 105 m	41.50 to 43.00	2.86 to 3.42
L2 @ EL + 105 m	38.00 to 41.00	2.68 to 3.01
L3 @ EL + 100 m	39.00	2.60 to 3.42

Note As per clause 5.4.1 of IS 13311:1992 (Part I), surface probing in general gives lower pulse velocity than in case of cross probing and depending on number of parameters, the difference could be of the order of about 1 km/sec

Since the UPV values did not give the clear idea about the uniformity in PMM, hence the 50 mm dia cores were extracted from PMM (refer Figs. 19.10 and 19.11). As seen in figure the quality of core extracted from one of the location of PMM was not good, hence the PMM was recommended to dismantle and recast once again wherever the quality was found poor.

Fig. 19.10 Core extraction over polymer modified mortar used in repair of RCC Chimney

Fig. 19.11 Polymer modified mortar core showing poor quality of repair

19.4.2.6 Pull Off Test

Pull off testing was done to check the bonding of new material with substrate. Locations for the pull off testing were randomly identified and 50 mm dia cores were drilled and metal disc was adhered at two different locations (refer Fig. 19.12). The results of Pull off test indicate that the bond strength was greater than the tensile strength of the overlay mortar. The test setup of the pull off test is shown in figure below. Also the core was extracted once again over the surface where the PMM was recast and the quality of the mortar was found to be satisfactory by this time (refer Fig. 19.13).

Fig. 19.12 Pull off testing over polymer modified mortar used in repair of RCC Chimney

Fig. 19.13 Polymer Modified Mortar core showing good quality of repair

Fig. 19.14 Finished surface after application of GFRP in periphery of RCC Chimney

19.4.2.7 Fibre Wrapping

Procedure for application of GFRP system included: (a) surface preparation of all affected area of RCC column using mechanical grinder to smoothen the surface was carried out. (b) application of primer over the prepared surface, followed by. (c) application of thixotropic putty to fill the holes and uneven surface (d) application of 900 gsm GFRP wrapping to strengthen the concrete around the cracked portions by wrapping chimney stack in desired orientation using tamping roller to avoid any air voids. (e) after application of GFRP Wrap, a coat of epoxy saturant was applied after a minimum of 12 h. (g) finishing the surface was done using 10 mm thick PMM (refer Fig. 19.14).

19.4.2.8 Protective Coating

Excessive moisture in concrete at the time of coating can destroy the performance of protective coating systems by causing blistering and detachment of the film. Hence it is necessary to measure moisture content in concrete prior to application of any coating. ASTM D4263-2005 describes a method for indicating moisture in concrete by the plastic sheet method. This is a nondestructive test that requires firmly taping the perimeter of a sheet of plastic to the concrete and allowing it to remain in place for a minimum of 16 h. At the end of the exposure, the underside of the sheet and surface of the concrete was visually examined for the presence of moisture. After the sheet was removed from the concrete surface, no apparent excessive moisture was noticed. Acceptance criteria are not explicitly stated in standard, but coatings are typically not applied if the test indicates that moisture is visibly present as per ASTM C-882. Before applying the protective coating, the moisture content was also measured using Humidity meter as shown in Fig. 19.15. The value of moisture content is obtained with the help of graph between specific resistance and moisture content as shown in Fig. 19.16. The moisture content obtained from the graph was 5.26%. As per Cement Concrete & Aggregates Australia (CCAA), the concrete is deemed dry enough when

Fig. 19.15 Humidity meter to measure moisture content on RCC Chimney

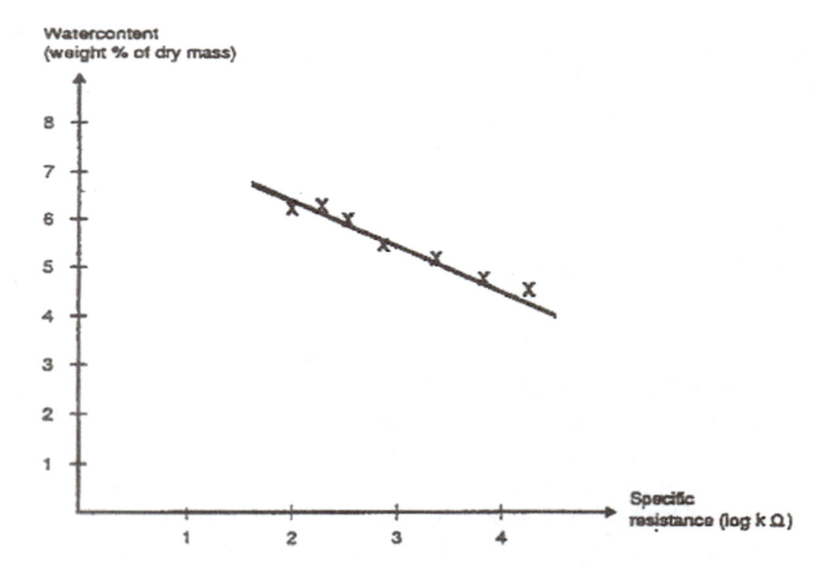

Fig. 19.16 Graph between specific resistance and moisture content as provided by the manufacturer

the moisture content is not more than 5.5%. Based on the test results, the concrete was deemed dry enough; hence there was no problem in applying the protective coating over concrete.

The protective coating work was not started yet. The chemical resistance coating has to be applied over the top 50 m height of chimney in three layers to achieve a total Dry Film Thickness (DFT) of 185–215 microns. At lower elevations, anti-carbonation coating has to be applied with one primer coat of silane siloxane and two anti-carbonation coats to achieve total DFT of 225–240 microns. The thicknesses of the coatings will be measured using wet film thickness gauge. Dry film thickness will then be calculated by multiplying solid content in coating to wet film thickness.

19.5 Conclusion

1. Testing for the performance of repair materials play an important role to achieve a satisfactory repair, with adequate strength and durability.
2. Various tests were carried out on different repair materials like rust remover, anticorrosive coating, corrosion inhibitors, bond coats, grout materials, Polymer Modified Mortar (PMM), Glass Fibre Reinforced Polymer (GFRP) wrap, protective coatings, etc. The test results of repair materials were compared with available codes, manuals & manufacturers test certificate.
3. During the execution of repair work, quality inspection was carried out by qualified/experienced engineers to follow the approved repair methodology and QAP, etc.
4. Non Destructive Evaluation Techniques like rebound hammer, ultrasonic pulse velocity, core extraction and pull off test, etc. was carried out on repaired portion to check the quality.
5. Formulation of accurate repair strategy through timely condition assessment and adoption of latest available techniques/materials for repair/rehabilitation of RCC structures is key in achieving desired service life for trouble free operations.

Acknowledgements This paper pertains to a R&D work carried out by Construction Development and Research Centre at National Council for Cement and Building Materials. The Authors acknowledge the contribution of engineers and lab assistants of Institute in carrying out the investigation work.

References

ACI Committee 546R-04, Concrete repair guide, American Concrete Institute

ACI Committee 548.1R-2008, Guide for the use of polymers in concrete, American Concrete Institute

ACI Committee 548.3R-2003, Polymer modified concrete, American Concrete Institute

ACI Committee 548.4R-93, Standard specification for latex modified concrete (LMC) overlays, American Concrete Institute

ASTM C1059/C1059M 2021, Standard specification for latex agents for bonding fresh to hardened concrete

ASTM C1438–2017, Standard specification for latex and powder polymer modifiers for use in hydraulic cement concrete and mortar

ASTM C881/C881M 2020a, Standard specifications for epoxy resin based bonding systems for concrete
ASTM C882/C882M–2020, Standard test method for bond strength of epoxy resin systems used with concrete by slant shear
ASTM D4263-2018, Standard test method for indicating moisture in concrete by the Plastic sheet method
Bungey JH (1989) Testing of concrete in structures. Surrey University Press, New York
Bhaskar S, Rajeev G (1999) Assessment and rehabilitation of concrete structures, In: Proceedings of international conference on structural engineering, Ghaziabad (U.P) India
Cement Concrete & Aggregates Australia (CCAA) (2007) Moisture in concrete and moisture sensitive finishes and coatings
EN 1504 (Part 1 to 10) 2017, Products and systems for the repair and protection of concrete structures Handbook on Repair and Rehabilitation of RCC Buildings, published by CPWD, Govt. of India
Goyal BK, Chandra M (1997) Damage assessment of concrete structures by NDT techniques. In: International conference on maintenance & durability of concrete structures, Hyderabad, India
IS 269:2015 (Reaffirmed: 2020) Ordinary Portland cement—specifications
IS 383:2016, Coarse and fine aggregate for concrete—specification
IS 456-2000 (Reaffirmed: 2021) Plain and reinforced concrete—code of practice
IS 9103:1999 (Reaffirmed: 2018) Specification for concrete admixtures
IS 9162:1979 (Reaffirmed: 2016) Methods of tests for epoxy resins, hardeners and epoxy resin compositions for floor topping
IS: 13311 – 1992 (Part I) (Reaffirmed: 2018) Non destructive testing of concrete—methods of test, part—I ultrasonic pulse velocity
IS: 13311 – 1992 (Part II) (Reaffirmed: 2018) Non destructive testing of oncrete—methods of test, part—II rebound hammer
Trimbler KA, Brown KJ (2012) Measuring moisture in walls, pp 29–36. RCI-Inc.
Malhotra VM,Carino NJ (2004) Handbook on nondestructive testing of concrete. CRS Press, Washington DC
Sharma S, Arora VV (2015) Distress assessment, repair and strengthening of RCC members of turbo generator foundation of anpara thermal power station at Uttar Pradesh (India). In: International conference on the regeneration and conservation of concrete structures, Nagasaki, Japan
Mohammed TU, Otsuki N, Hamada H (2003) Corrosion of steel bars in cracked concrete under marine environment. 15(5):460. 10.1061/(ASCE0899-1561)
Ramachandran VS (1996) Concrete admixtures handbook: properties, science and technology. 2nd edn

Chapter 20
An Experimental Study of Using Biopolymer for Liquefaction Mitigation of Silty Sand—A Sustainable Alternative

S. Smitha and K. Rangaswamy

20.1 Introduction

In the present day due to increase in urbanization, the construction activities has spurred and that has led to deterioration of the environment. According to UN habitat even though cities account for less than 2% of the earth's surface they produce 60% of world's greenhouse gas emission. Also, construction industry is held responsible for about 18% of the carbon emissions. Soil is an imperative construction material that takes up the load from the super structure through foundation. So it has to be adequate to take up any kind of loading that may act upon it including earthquake loading and dynamic loading. But certain types of soil like silty sand and fine sand are incapable of sustaining dynamic loading when they are in saturated condition due to a phenomenon called liquefaction. The soil loses its shear strength and start behaving like a liquid in the event of any cyclic loading. Thus it would have to be improved. The conventional methods of liquefaction mitigation mostly make use of unsustainable practices like cement stabilization that increases the carbon footprint immensely. About 0.95 ton of CO_2 is produced per ton of cement production (Larson 2011). Other methods like grouting using cement or chemicals also adversely affect the groundwater and the soil ecosystem (Kim et al. 2017; Benhelal et al. 2013). Another traditional method is the use of dynamic compaction that is acceptable from environmental point of view, but is not so economic when done in small scale and it cause disturbance to the nearby structures when done in built environment (Hayden 1994).

However from the past few years people have started becoming aware about the deteriorating environmental scenario and rigorous research is being done on usage of innovative sustainable materials even in the realm of ground improvement. For the purpose of liquefaction mitigation new methods like soil desaturation by

S. Smitha (✉) · K. Rangaswamy
National Institute of Technology Calicut, Calicut, India

K. R. Reddy et al. (eds.), *Advances in Sustainable Materials and Resilient Infrastructure*, Springer Transactions in Civil and Environmental Engineering,
https://doi.org/10.1007/978-981-16-9744-9_20

introduction of N_2 gas or air bubbles (O'Donnell et al. 2017; Kavazanjian et al. 2015; He et al. 2013), microbial-induced calcite precipitation (Xiao et al. 2018; Feng and Montoya 2017) and use of bio- and nano-materials are becoming popular nowadays (Thomas and Rangaswamy 2020; Smitha and Rangaswamy 2021; Smitha et al. 2021; Rangaswamy et al. 2021; Huang and Wang 2016). The studies concerning use of biopolymers for soil strengthening (Reddy et al. 2020; Smitha and Sachan 2016; Chang et al. 2015), erosion control (Kwon et al. 2020; Reddy et al. 2018; Ham et al. 2018), permeability reduction (Biju and Arnepalli 2020; Kumar and Sujatha 2021; Dora et al. 2021), etc. are widely carried out. In the current research the use of biopolymer as a stabilization material against mitigation of liquefaction have been explored by performing a series of consolidated undrained cyclic triaxial tests. The tests were performed at different strain amplitudes ranging from 0.3 to 1.2% to study the effect of level of cyclic loading on silty sand as well as biopolymer stabilized silty sand. To simulate earthquake condition by strain-controlled cyclic triaxial testing method, the range of axial strain amplitude adopted in previous studies like that of Pandya and Sachan (2019) and Vijayasri et al. (2016) and Thomas and Rangaswamy (2020) is similar to the strain range employed in the current study. Also, for stress controlled studies in saturated loose silty sands subjected to cyclic loading, the development of axial strain with pore pressure buildup is considerably slow up to about 0.5%. Further, the axial strain is significantly increased with developed pore pressures to reach the state of initial liquefaction at about 1 to 1.5% strain level. After triggering the liquefaction state, flow failure occurs with continuous deformations within a few loading cycles that could not be predicted (Boominathan et al. 2010). Hence this range of strain levels was adopted for the current study. The main objective of this paper is to study the effect of variation in strain levels on cyclic strength properties of biopolymer treated and untreated silty sand. The effect of stabilization on pore water pressure response and stress path behavior is also analyzed in detail.

20.2 Materials and Methods

The materials used for the study include silty sand collected from Kalpetta, Wayanad, Kerala and agar biopolymer that was purchased from the manufacturer Urban platter. The soil was found to have 72% sand, 22% silt and 3% clay having maximum and minimum density of 1.42 g/cc and 1.24 g/cc, respectively. It has a specific gravity of 2.65 major portion of the soil fell in the category of liquefiable soil range given by Tsuchida (1970) and the grain size distribution of the soil is given in Fig. 20.1. The agar biopolymer, that was procured from Urban Platter was of food grade quality and had 700 g/cm^2 of gel strength, a pH of 7.2 and has a molecular weight of 336.3. The value of gel strength and molecular weight were obtained from the manufacturer. It was creamy white in appearance and was available as a powder form. It forms a solution with water only if the temperature rises to above 85 °C.

Air pluviation method was demonstrated to produce reconstituted soil samples with least soil degradation (Cresswell et al. 1999). As reported by Raghunandan

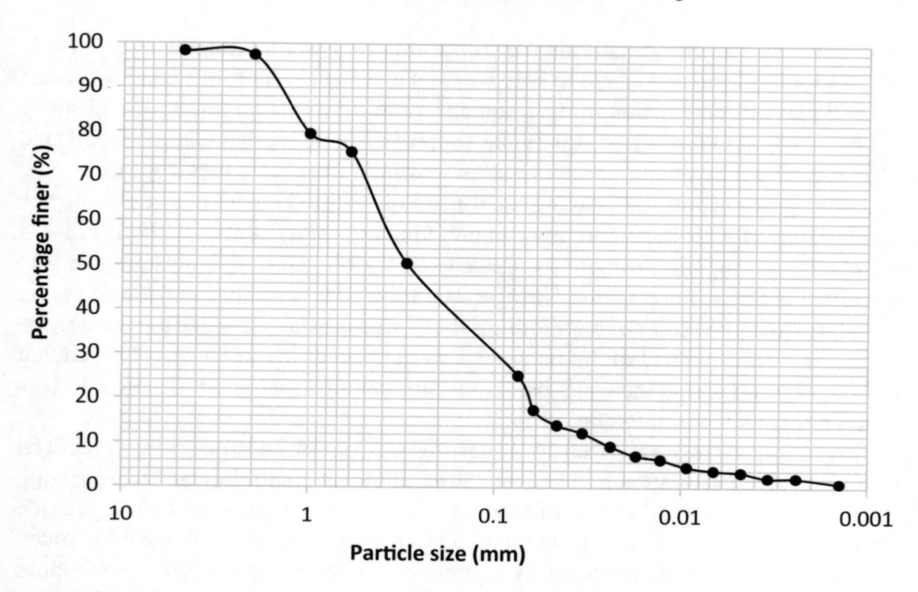

Fig. 20.1 Particle size distribution curve of the silty sand from Kalpetta

et al. (2012) samples prepared by moist tamping always showed softening behavior, whereas samples prepared by pluviation could soften or harden depending on its relative density. Also, air pluviation method mimicked natural depositional processes more closely, to create a grain structure similar to that of naturally deposited river sands than other methods like moist tamping or water pluviation. Therefore, untreated silty sand specimens were prepared by using air pluviation. The soil samples were made at a relative density of 30% to depict a loose condition of soil. The height for air pluviation was fixed at 5cm after some trial and error. The agar was added to soil in a solution form by wet mixing method. It was available in a powder from the manufacturer. In this study 2% agar biopolymer by weight of dry soil was used i.e., soil-biopolymer ratio adopted is 100:2. (For 100 g of soil 2 g of biopolymer is used). Beyond 2% dosage there were problems in mixing of biopolymer solution with the soil since the viscosity of biopolymer solution increases with increase in concentration. For preparing the agar solution it was added to water and the mixture was heated to about 90°C so that agar forms a homogenous solution in water. The quantity of water required was calculated by finding the saturation water content. The hot solution was then poured to the oven dry soil and mixed well and cast in cylindrical split mould that had an internal diameter of 5 cm and a height of 10 cm. For this particular biopolymer agar, since it has a property of gelation hysteresis wet mixing method is more suitable. There is significant difference between its melting and gelling temperature. Agar forms solution in water only at around 90°C and as soon as the temperature falls down the viscosity of the agar solution increases and it sets to form gel when the temperature drops to below 40°C. If dry mixing was done the agar biopolymer would not be able to form gel in between the soil particles. It would just act as filler densifying the soil. But during wet mixing the agar is made into a solution form and added to soil, where it sets to gel that forms coating around the soil particle, connection bridges and also pore-filling within the soil. The sticky nature of the gel holds the soil particles together. The specimen could be extracted from the mould by about 20 minutes from casting. They were then kept for curing in a temperature controlled environment till 7 days at a temperature of 30 °C. The photographs of sample preparation can be found in Smitha et al. (2019). There was weight loss during the curing of the treated soil samples since the moisture loss was permitted in the curing process. The agar was allowed to dehydrate and thereby form crosslinks with the soil structure during the curing period. In order to monitor the curing, the water content of the specimens were noted down and it was checked that at 7 days curing period the water content for the samples were within $2.5 \pm 1\%$. This ensured that the curing was uniform.

Strain-controlled method of cyclic testing was performed for all the tests. The testing involved saturation and consolidation of the sample followed by subjecting it to repeated sinusoidal cycles of loading. The saturation involved carbon dioxide circulation, water circulation from top to bottom which was then followed by forced pressure saturation accompanied by subsequent checking of the Skempton's pore pressure parameter, B-value (B-value is the ratio of change in pore pressure to the change in confining stress). The specimens were assumed to be fully saturated when B- value greater than 0.95 was obtained. It was followed by consolidation of the sample in an isotropic condition and then subjecting it to cyclic loading.

20.3 Results and Discussions

20.3.1 Pore Pressure Response

Pore pressure was noted with the help of a pore pressure transducer connected to the bottom of the soil sample. Since the samples were saturated before cyclic loading stage, the pore water took up the load coming on to the sample and the pore water pressure (PWP) also varied cyclically corresponding to the loading given. PWP obtained was represented as the excess pore water pressure (EPWP) ratio by dividing the PWP by the effective confining pressure. The soil is said to liquefy when PWP becomes equal to the confining pressure. It was seen that in untreated soil the PWP reached a value equal to the applied confining pressure within few cycles as it got liquefied. In treated soil also it increased with increase in number of cycles, that was rapid in the initial few cycles. But as the cycles progressed, the increase in EPWP was slower and it was almost constant by about 40 cycles for all the tests. In the figures the EPWP ratio upto 50 cycles of loading is plotted. The EPWP ratio of untreated soil is plotted with respect to the cyclic axial strain in Fig. 20.2. It can be seen that the EPWP rises and subsequently reaches a value of 1 at all the amplitudes signifying that it liquefied. The number of cycles to liquefaction was 30, 23, 19, 14 & 9 for

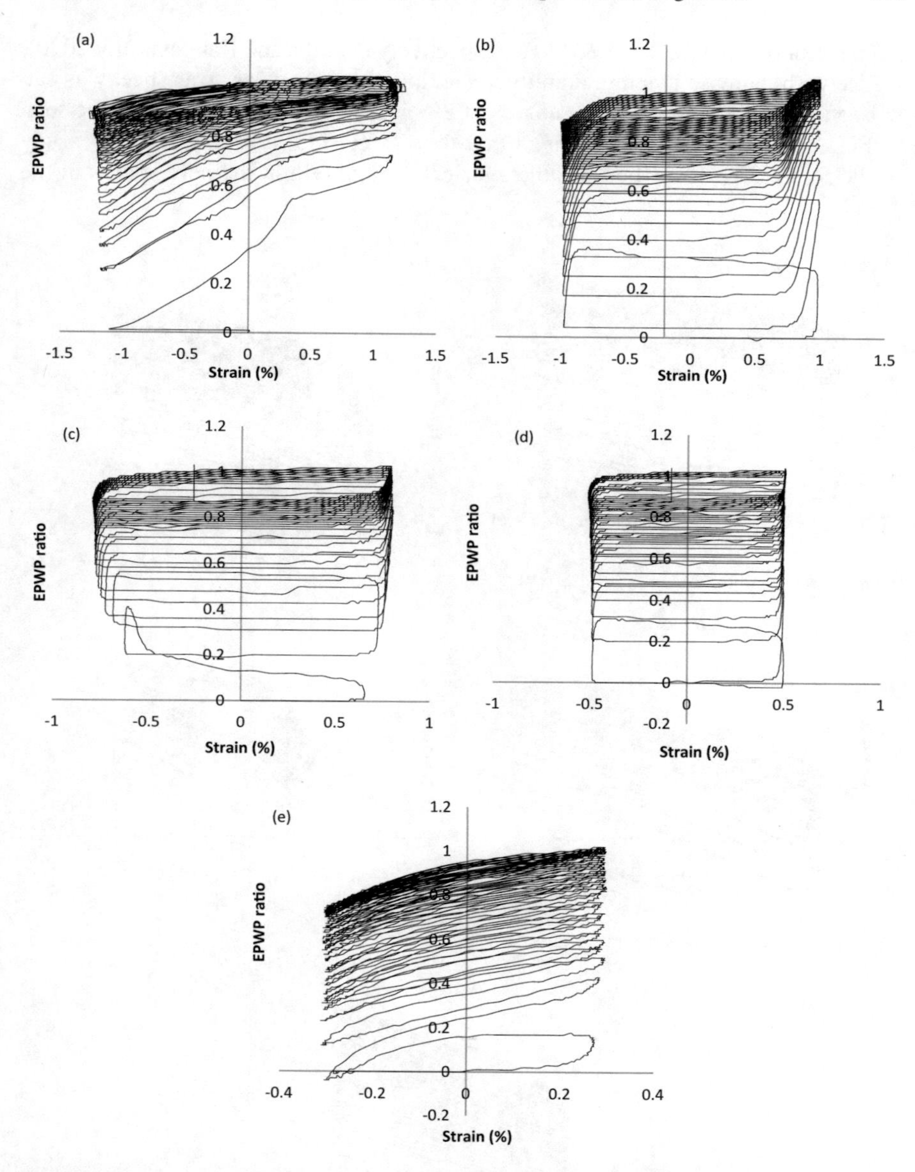

Fig. 20.2 Pore pressure response of untreated silty sand at strain amplitude of **a** 1.2%, **b** 1%, **c** 0.8%, **d** 0.5% and **e** 0.3%

axial strain of 0.3, 0.5, 0.8, 1 & 1.2%, respectively (Smitha and Rangaswamy 2020). Hence with increase in strain amplitude the liquefaction process was speedy as can be seen from the graph. Similar curves of 2% agar treated soil samples can be seen in Fig. 20.3a–e. It can be perceived that the EPWP ratio does not reach 1 in any of the figures. This was because the agar gel present within the pore spaces in the

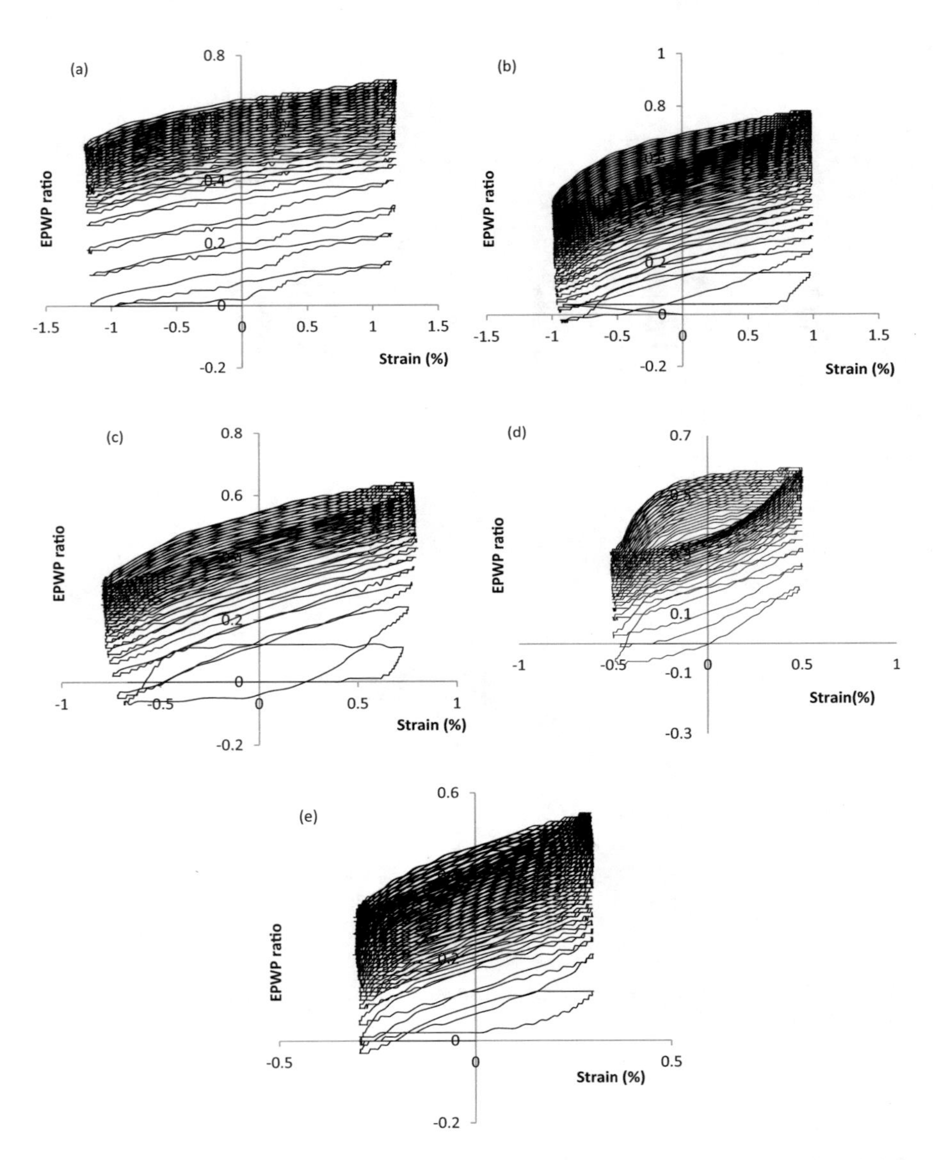

Fig. 20.3 Pore pressure response of 2% agar stabilized silty sand at strain amplitude of **a** 1.2%, **b** 1%, **c** 0.8%, **d** 0.5% and **e** 0.3%

soil sample take up the pressure due to cyclic loading instead of the pore water. This reduced the PWP buildup significantly. Furthermore, the maximum EPWP buildup at the end of 50 cycles was compared for studying the variation caused due to different strain amplitudes and it was seen to be increasing with increase in the amplitude. The maximum EPWP ratio in 50th cycle was 0.47, 0.59, 0.64, 0.67 & 0.71 for 0.3, 0.5, 0.8, 1 & 1.2% amplitudes. Thus biopolymer treatment could serve as a sustainable liquefaction mitigation alternative since it reduced the PWP buildup.

20.3.2 Stress Path Response

Stress path represents the states of stresses a material undergo during its loading stage. Each point in the stress path represents the stress states during cyclic loading. Stress path is obtained by finding the effective mean stress (p') and effective peak shear stress (*q* or *q*') and plotting *q* versus *p*'. Here *p*' and *q* was found from the equations based on MIT definition which is given by:-

$$p' = \frac{(\sigma_1' + \sigma_3')}{2} \tag{20.1a}$$

$$q' = \frac{(\sigma_1' - \sigma_3')}{2} \tag{20.1b}$$

where, σ_1' and σ_3' are effective major and minor principle stresses. The stress paths are indicative of the stress degradation with number of cycles. The stress paths of untreated soil at different strain amplitude have been plotted in Fig. 20.4a–e. From the start of the test, the range of q can be found to decrease for all the curves. The decrease is least for lower strain amplitude (0.3%) and highest for 1.2% strain amplitude. *p*' decreased steadily from 100 kPa at the start of the test to 0 kPa by the end. In similar stress path plots for 2% agar treated soil (Fig. 20.5a–e) the decrease in *p* and *q*' values from the start of the test is very minimal as compared to that in untreated silty sand. The stress path loops can be seen to overlap with each other and its propagation towards the origin is constrained. This attribute implies that the pore pressure generation is restricted due to agar treatment. The decrease in *p*'-value from the start of the test is least for 0.3% amplitude and maximum for 1.2% strain. This corresponds to the result obtained in EPWP plots. The width of the q indicated the resistance to stress taken up by the specimens. The range of q-value slightly decreased for higher amplitude from the start of the test whereas it remained same till the end of the test for lower amplitudes of testing. This might be because at higher amplitudes the connection between the soil and gel might be weakened to some extent that caused the resistance to take up cyclic loads to decrease.

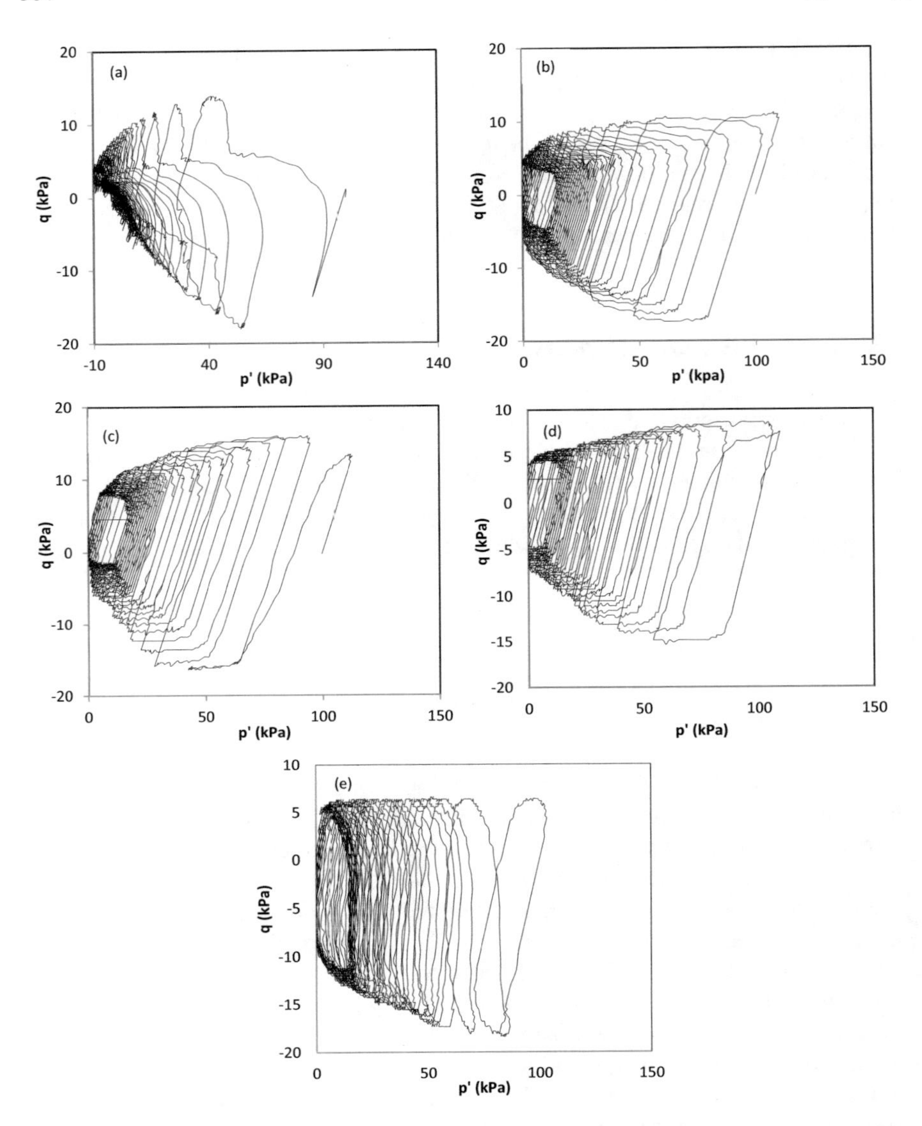

Fig. 20.4 Stress path response of untreated silty sand at strain amplitudes of **a** 1.2%, **b** 1%, **c** 0.8%, **d** 0.5% and **e** 0.3%

20.4 Summary

The chapter deals with testing the efficiency of agar biopolymer treatment on silty soil against liquefaction at different magnitudes of cyclic strain amplitude. The experimental study consisting of a series of strain-controlled consolidated undrained cyclic triaxial test at different amplitudes of strain yielded the following conclusions:-

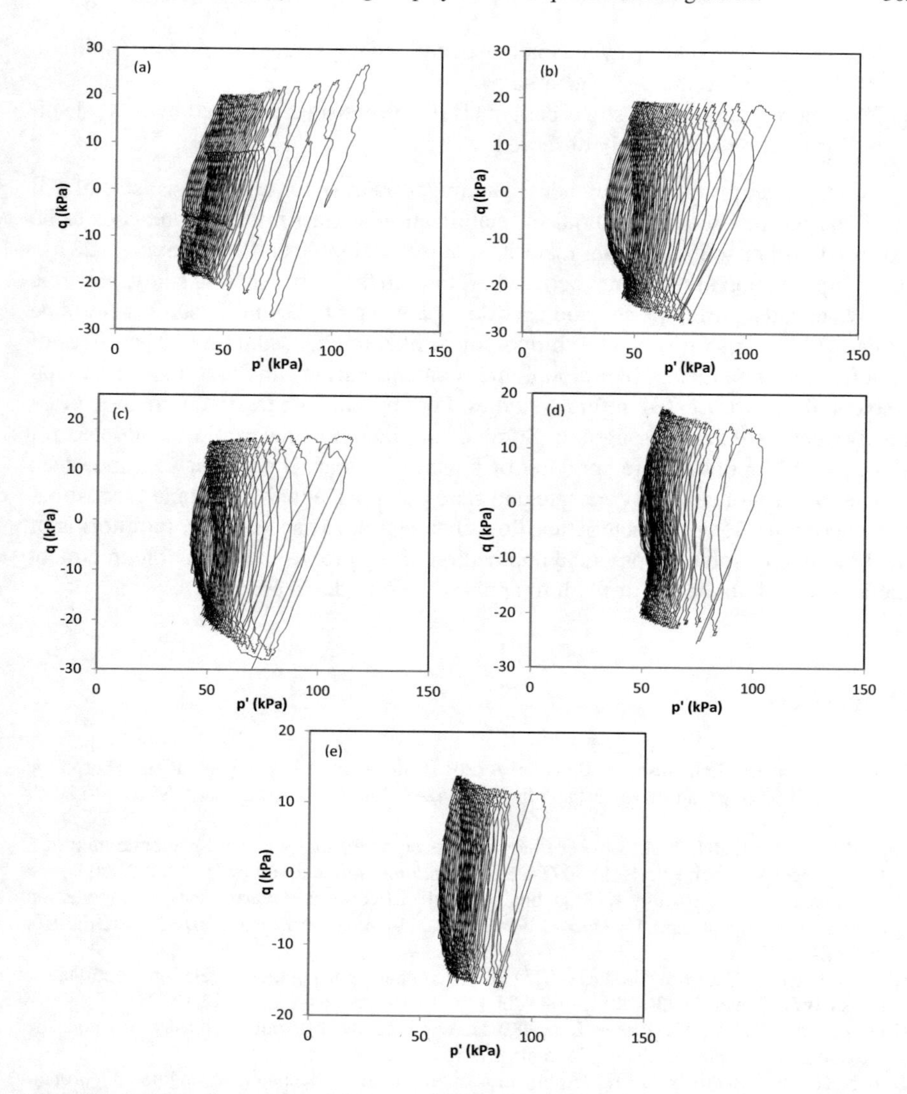

Fig. 20.5 Stress path response of 2% agar stabilized silty sand at strain amplitudes of **a** 1.2%, **b** 1%, **c** 0.8%, **d** 0.5% and **e** 0.3%

- Pore pressure buildup reduced significantly upon treatment with agar biopolymer and the EPWP decrease was higher for lower strain amplitude of loading. But even at 1.2% axial strain the maximum EPWP ratio decreased by 29%.
- The EPWP ratio for untreated soil rose to 1 within few loading cycles itself whereas for treated soil it rose in the initial cycles and became steady after some time.

- The resistance to take up cyclic loading decreased with increasing strain amplitude as observed from the peak shear stress.
- From pore pressure and stress path analysis the resistance offered by treated soil to liquefaction is explicitly observed.

Soil treatment using biopolymer is gaining popularity for different aspects of soil stabilization. Its use as a material for stabilization against liquefaction needs to be explored further so that a better material that is sustainable and efficient is made use of for liquefaction mitigation practices in future. In this experimental study, only one relative density, curing period and agar dosage is explored. Similar studies could be extended to using various other biopolymers like xanthan, guar, etc. The effect of various other parameters like biopolymer content, curing time and mixing method also could be investigated in future studies. For applying the treatment method practically, many studies are ongoing. Some of the methods that could be adopted for wet mix of biopolymer are spraying or injection of the biopolymer solution, deep mixing by grouting, biopolymer treated sand columns, etc. With detailed investigations about rheology characteristics, flow characteristics and viscosity monitoring of the biopolymer solution, practical application of the proposed sustainable treatment method would not be too difficult to apply in field in the near future.

References

Benhelal E, Zahedi G, Shamsaei E, Bahadori A (2013) Global strategies and potentials to curb CO_2 emissions in cement industry. J Cleaner Prod 51:142–161. https://doi.org/10.1016/j.jclepro.2012.10.049

Biju MS, Arnepalli DN (2020) Effect of biopolymers on permeability of sand-bentonite mixtures. J Rock Mech Geotech Eng 12(5):1093–1102. https://doi.org/10.1016/j.jrmge.2020.02.004

Boominathan A, Rangaswamy K, Rajagopal K (2010) Effect of non-plastic fines on liquefaction resistance of Gujarat sand. Int J Geotech Eng 4(2):241–253. https://doi.org/10.3328/IJGE.2010.04.02.241-253

Chang I, Prasidhi AK, Im J, Cho GC (2015) Soil strengthening using thermogelation biopolymers. Constr Build Mater 77:430–438. https://doi.org/10.1016/j.conbuildmat.2014.12.116

Cresswell A, Barton ME, Brown R (1999) Determining the Maximum Density of Sands by Pluviation. Geotechnical Testing Journal, ASTM 22(4):324–328

Dora N, Nanda P, Reddy NG (2021) Application of biopolymers for enhancing engineering properties of problematic soils and industrial wastes: a review. In: Biswas S, Metya S, Kumar S, Samui P (eds) Advances in sustainable construction materials. Lecture notes in civil engineering, vol 124, Springer, Singapore. https://doi.org/10.1007/978-981-33-4590-4_20

Feng K, Montoya BM (2017) Quantifying level of microbial-induced cementation for cyclically loaded sand. J. Geotech. Geoenviron. Eng. 143(6):06017005. https://doi.org/10.1061/(ASCE)GT.1943-5606.0001682

Ham S-M, Chang I, Noh D-H, Kwon T-H, Muhunthan B (2018) Improvement of surface erosion resistance of sand by microbial biopolymer formation. Journal of Geotechnical and Geoenvironmental Engineering 144:06018004. https://doi.org/10.1061/(ASCE)GT.1943-5606.0001900

Hayden PF, Baez, JI (1994) State of practice of liquefaction mitigation in North America. In: Proceedings of international workshop on remedial treatment of liquefiable soils. Tsukuba Science City, Japan

He J, Chu J, Ivanov V (2013) Mitigation of liquefaction of saturated sand using biogas. Géotechnique 63(4):267–275. https://doi.org/10.1680/geot.SIP13.P.004

Huang Y, Wang L (2016). Laboratory investigation of liquefaction mitigation in silty sand using nanoparticles. Eng Geol 204: 23–32. https://doi.org/10.1016/j.enggeo.2016.01.015.

Kavazanjian E, Donnell STO, Hamdan N (2015) Biogeotechnical mitigation of earthquake-induced soil liquefaction by denitrification: a two-stage process. In: Proceedings of 6th international conference on earthquake geotechnical engineering. New Zealand, ISSMGE and Christchurch

Kim AR, Chang I, Cho GC, Shim SH (2017) Strength and dynamic properties of cement-mixed Korean marine clays. KSCE J Civil Eng 1–12. https://doi.org/10.1007/s12205-017-1686-3

Kumar SA, Sujatha ER (2021) An appraisal of the hydro-mechanical behaviour of polysaccharides, xanthan gum, guar gum and β-glucan amended soil. Carbohydr Polym 265. https://doi.org/10.1016/j.carbpol.2021.118083

Kwon Y-M, Ham S-M, Kwon T-H, Cho G-C, Chang I (2020) Surface-erosion behaviour of biopolymer treated soils assessed by EFA. Géotechnique Letters 10:1–7. https://doi.org/10.1680/jgele.19.00106

Larson A (2011) 'Sustainability, Innovation, and Entrepreneurship'; University of Virginia: Charlottesville. VA, USA

O'Donnell ST, Kavazanjian E, Rittmann BE (2017) MIDP: liquefaction mitigation via microbial denitrification as a two-stage process. II: MICP. J Geotech Geoenviron Eng 143(12):04017095.https://doi.org/10.1061/(ASCE)GT.1943-5606.0001806

Pandya S, Sachan A (2019) Effect of frequency and amplitude on dynamic behaviour, stiffness degradation and energy dissipation of saturated cohesive soil. Geomech Geoeng 6025. https://doi.org/10.1080/17486025.2019.1680885.

Raghunandan MA, Juneja A, Benson Hsiung BC (2012) Preparation of reconstituted sand samples in the laboratory. Int J Geotech Eng 6(1):125–131. https://doi.org/10.3328/IJGE.2012.06.01.125-131

Rangaswamy K, Thomas G, Smitha S (2021) Influence of bio and nano materials on dynamic characterisation of soils. In: Sitharam T, Jakka R, Kolathayar S (eds) Latest developments in geotechnical earthquake engineering and soil dynamics. Springer transactions in civil and environmental engineering, Springer, Singapore. https://doi.org/10.1007/978-981-16-1468-2_25

Reddy NG, Rao BH, Reddy KR (2018) Biopolymer amendment for mitigating dispersive characteristics of red mud waste. Géotechnique Letters. 8(3):201–207. https://doi.org/10.1680/jgele.18.00033

Reddy NG, Nongmaithem RS, Basu D, Rao BH (2020) Application of biopolymers for improving the strength characteristics of red mud waste. Environmental Geotechnics. https://doi.org/10.1680/jenge.19.00018

Smitha S, Rangaswamy K (2021) Experimental Study on Unconfined Compressive and Cyclic Triaxial Test Behavior of Agar Biopolymer Treated Silty Sand. Arabian Journal of Geoscience 14:590. https://doi.org/10.1007/s12517-021-06955-1

Smitha S, Rangaswamy K (2020) Effect of biopolymer treatment on pore pressure response and dynamic properties of silty sand. J Mater Civ Eng 32(8):04020217. https://doi.org/10.1061/(ASCE)MT.1943-5533.0003285

Smitha S, Rangaswamy K, Balaswamy Naik P (2021). Liquefaction mitigation of silty sands using xanthan biopolymer. In: Sitharam TG, Parthasarathy CR, Kolathayar S (eds) Ground Improvement Techniques. Lecture Notes in Civil Engineering, vol 118. Springer, Singapore. https://doi.org/10.1007/978-981-15-9988-0_23 (Scopus indexed, Best paper award in 7th ICRAGEE)

Smitha S, Rangaswamy K, Keerthi DS (2019) Triaxial test behavior of silty sands treated with agar biopolymer. Int J Geotech Eng 1–12. https://doi.org/10.1080/19386362.2019.167944

Smitha S, Sachan A (2016) Use of agar biopolymer to improve the shear strength behavior of Sabarmati sand. International Journal of Geotechnical Engineering, Taylor & Francis 10(4):387–400. https://doi.org/10.1080/19386362.2016.1152674

Thomas G, Rangaswamy K (2020) Dynamic soil properties of nanoparticles and bioenzyme treated soft clay. Soil dynamics and earthquake engineering. Elsevier Ltd, 137(October 2019), pp 106324. https://doi.org/10.1016/j.soildyn.2020.106324.

Tsuchida H (1970) Prediction and remedial measures against liquefaction of sandy soil. [In Japanese.] In: Vol. 3 of Annual Seminar of Port and Harbor Research Institute, 1–33. Tokyo, International Association of Ports and Harbors

Vijayasri T, Patra NR, Raychowdhury P (2016) Cyclic behavior and liquefaction potential of Renusagar pond ash reinforced with geotextiles. J Mater Civ Eng 28(11):04016125. https://doi.org/10.1061/(ASCE)MT.1943-5533.0001633

Xiao P, Liu H, Xiao Y, Stuedlein AW, Evans TM (2018) Liquefaction resistance of bio-cemented calcareous sand. Soil Dyn. Earthquake Eng. 107:9–19. https://doi.org/10.1016/j.soildyn.2018.01.008

Chapter 21
Durability Based Service Life Estimation of RC Structural Components

Bhaskar Sangoju

21.1 Introduction

Concrete is one of the most commonly used construction materials due to low cost, versatility and adaptability. Extensive damages due to rebar corrosion are commonly noticed for reinforced concrete (RC) structures. ACI (American Concrete Institute) 201.2R (2016) defines durability of concrete as the resistance to chemical attack, weathering action and other degradation processes, etc. BS (British Standard) 8110 (1997) says that a durable concrete is the one, that protects rebar from corrosion and will perform satisfactorily for its design life. It can be said that rebar corrosion in concrete is mainly due to the ingress of aggressive agents like chlorides, carbon dioxide (CO_2), etc. through the cover concrete. Hence, the durability of RC structure can be related to the transport properties and the chemical composition of concrete being used. The transport properties of concrete, viz., permeability (air/fluid), sorptivity, diffusivity, migration, etc., are a function of concrete penetrability and is affected by parameters such as the materials used for construction, water-to-cement (w/c) ratio, degree of hydration, etc. In general, it is assumed that concrete durability can be expected to achieve through strength, however, there is no simple relationship between strength and the durability (Swamy 2008). Neville (1987) suggests the possible reasons for deterioration of concrete structures are: (i) lack of understanding of deterioration mechanisms, (ii) inadequate acceptance criteria of site concrete, (iii) changes in cement materials and (iv) changes in construction practices.

It can be said that the current construction practices is to follow 'prescriptive specifications', where limiting values are provided for mix design parameters such as w/c ratio, cement quantity and the construction procedures that are to be followed. The major deficiency in following 'prescriptive specifications' is that there is no

B. Sangoju (✉)
CSIR-Structural Engineering Research Centre, Chennai 600 113, India
e-mail: bhaskar@serc.res.in

K. R. Reddy et al. (eds.), *Advances in Sustainable Materials and Resilient Infrastructure*, Springer Transactions in Civil and Environmental Engineering,
https://doi.org/10.1007/978-981-16-9744-9_21

flexibility to take into account the effect of new materials that improve the properties of concrete. Sometimes, the improved properties can be achieved by using locally available materials at cheaper prices. In recent times, the trend is changing towards 'performance based specifications', where in the as-built concrete performance can be evaluated. This will ensure adequate quality and durability of the as-built finished structure (Bhaskar et al. 2020). In 'performance based specifications/criteria', performance of finished product only will be specified. Therefore, there will be a flexibility in the selection of materials, which enables the constructor to take advantage of the low-cost and locally available materials.

As it is already mentioned that the deterioration due to corrosion is one of the growing problems all over the world. Reliable service life prediction is necessary for ensuring long term safety and also for planning inspection/maintenance activities. Generally, service life refers to the period/time during which a structure meets or exceeds the limits/requirements set for it. These limits/requirements can be technical, functional or economical (ACI 365.1R 2017). Technical service life is the period in service until a defined unacceptable state is reached, for example, limiting the loss of rebar cross-section, spalling of concrete, crack width, etc. As shown in Fig. 21.1, the deterioration due to rebar corrosion is considered in two phases: (i) initiation phase and (ii) propagation phase (Tuutti 1982). In the initiation phase, the time of ingress of deteriorating agents (chloride ion/CO_2) to the rebar is the main concern. It can be seen from Fig. 21.1 that propagation phase can be further divided into stable propagation and unstable propagation periods. The corrosion process at later stages of propagation period could differ significantly from that of during initial stages (may be before the cover concrete cracking) (Austroads 2000). Reliable estimation of initiation and propagation periods is the basis for service life prediction. On safer side, in the propagation phase, the time duration can be considered up to the end

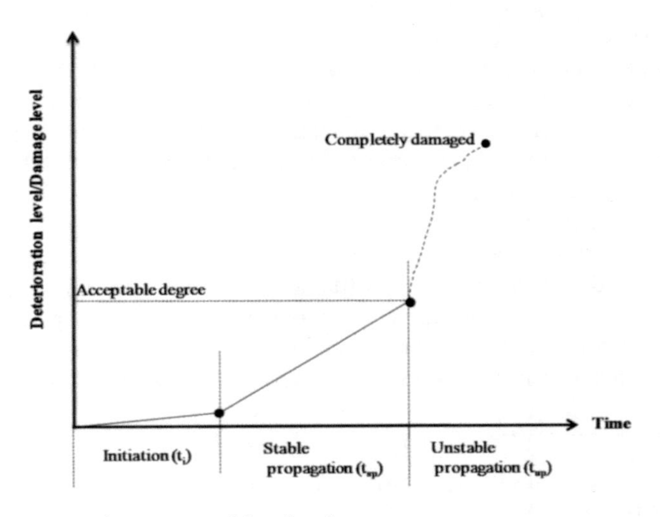

Fig. 21.1 Rebar corrosion—stages of deterioration

of stable propagation period, i.e., the time from corrosion initiation to the expected damage level (i.e., reduction in rebar cross-section, cracking, spalling of concrete, etc.).

It can be said that the service life predictions are always complicated and difficult due to the fact that it is influenced by many factors that are mostly probabilistic or uncertain in nature (viz., the quality of cover concrete, thickness of cover concrete, mechanical loading conditions, exposure conditions, etc.) (Austroads 2000; Anoop et al. 2002; Life-365 2013; Bhaskar et al. 2015; ACI 365.1R 2017). Some researchers, on the conservative side, approximated the service life as the length of the corrosion initiation phase, while others suggest the service life as the sum of corrosion initiation and stable propagation periods. Nevertheless, the stable propagation period depends on the limiting value of the damage level (viz., rebar section loss or surface crack width/spalling, etc.). Hence, the main objectives of this chapter are to familiarize the reader on the concepts of:

- rebar corrosion in RC structures and dominant mechanisms responsible for rebar corrosion.
- performance based specifications for long term durability of RC structures, and
- service life estimation of RC structures/components.

21.2 Rebar Corrosion in RC Structures

Rebars are generally protected from corrosion due to the high alkalinity of surrounding concrete. However, with the ingress of deteriorating agents such as chlorides and carbon dioxide (CO_2) through the cover concrete, the passivity is lost, resulting in corrosion of rebar. A brief discussion on (i) chloride induced corrosion and (ii) carbonation induced corrosion are given below.

21.2.1 Chloride Induced Corrosion

The rebar corrosion due to chloride ions is the most common forms of deterioration with huge costs and effects. However, the actual mechanism of chloride induced corrosion in concrete is not yet completely known (ACI 222R 2019). The possible sources of chloride ions (Cl^-) in concrete are: surrounding environment (marine), aggregates, mixing water, admixtures, salts used for de-icing, etc. It is reported that chloride ions react with the ferrous/ferric oxide (Fe^{2+}/Fe^{3+}, a thin passive layer) to form a soluble complex that dissolves in the surrounding solution causing local failure of the passive layer. The passivation breakdown occurs when the chlorides at the rebar surface exceed a certain amount, called threshold level, typically 0.4–1% by mass of binder (Montemor et al. 2003; Mackechnie et al. 2004). The possible reactions that occur in chloride induced corrosion are presented in Table 21.1 (Bentur et al. 1998; Callister 2000).

Table 21.1 Possible reactions that occur in chloride induced corrosion

Description	Possible reaction
Depassivation of protective layer	$Fe^{2+} + 2C1^{-} \rightarrow FeC1_2$(soluble in pore solution)
Corrosion of steel	$Fe \rightarrow Fe^{2+} + 2^{e-}$(oxidation) $\frac{1}{2}O_2 + H_2O + 2e^{-} \rightarrow 2(OH)^{-}$(reduction)
Formation of different rust products	$Fe^{2+} + 2(OH)^{-} \rightarrow Fe(OH)_2$ $2Fe(OH)_2 + \frac{1}{2}O_2 + H_2O \rightarrow 2Fe(OH)_3$ $\rightarrow Fe_2O_3.3H_2O$ (Rust)

Factors influencing the rate of chloride induced corrosion are:

- aggressive environment, concentration of chloride ions
- availability of oxygen and moisture
- type of steel rebar
- concrete material constituents

21.2.2 Carbonation Induced Corrosion

In carbonation induced corrosion, CO_2 in the atmosphere diffuses through cover concrete to reach the rebar and reduce the pH of concrete to 8 or 9, at which the passivating/oxide film over the rebar is no longer stable (Broomfield 2006). With adequate supply of oxygen and moisture, corrosion can initiate. The concrete carbonation is a slow process, the rate of carbonation is determined by the rate of penetration of CO_2 into the concrete that primarily depends on the porosity and permeability of the concrete. The carbonation process involves the following:

Initially, the atmospheric CO_2 reacts with water available in concrete pores to form carbonic acid (H_2CO_3).

$$H_2O + CO_2 \rightarrow H_2CO_3 \tag{21.1}$$

This is followed by reaction of the carbonic acid with calcium hydroxide ($Ca(OH)_2$) to form calcium carbonate ($CaCO_3$).

$$H_2CO_3 = Ca(OH)_2 \rightarrow CaCO_3 + 2H_2O \tag{21.2}$$

The above process leads to the depassivation of the rebar in contact with the carbonated zones. The water released during the chemical reactions can sustain both the formation of carbonic acid and the carbonation process. The carbonates formed are bigger molecules than the hydroxides, thereby increasing the density of the cement paste and also the strength, locally (Neville 1996). However, the carbonic acid formation lowers the pH of the concrete pore solution (8 to 9). As the carbonation front reaches the rebar, the corrosion process begins with the availability of moisture/water and oxygen. Carbonation induced corrosion occurs rapidly when there is insufficient concrete cover over rebar and also in highly permeable concretes. Factors influencing the rate of carbonation of concrete (Parrott 1990) are:

- amount of CO_2 in air
- relative humidity of concrete
- amount of precipitation of $CaCO_3$
- carbonation resistance of concrete (concrete permeability, amount of $Ca(OH)_2$ in concrete)

21.3 Durability Criteria—Performance Based Specifications

It is known that durability of a structure is the ability of the structure to withstand the design service conditions without significant deterioration. Recent experiences indicate that RC structures do not provide adequate resistance to corrosion under aggressive environments (Grantham 2011; Bhaskar et al. 2020). As discussed earlier, the current construction practice assumes that the measured compressive strength on standard cube samples will ensure compliance with the design requirements/specifications. However, it is well known that this assumption does not consider the variability of site concrete due to the concreting practices such as placing, consolidation, finishing, curing, etc. (Bhaskar et al. 2021). Therefore, emphasis shall be given to the concrete durability besides the strength.

Many a times, concrete satisfy the strength criteria, however, there can be a deficiency in the surface region, in terms of honeycombing, voids, inadequate cover thickness, etc. causing easy penetration of deteriorating agents. To address this, there is a need to measure the surface quality of concrete, in terms of easily measurable durability performance indicators (Hooton et al. 2005). It is already discussed that for corrosion to occur chloride ions/CO_2, moisture and O_2 must be available in a sufficient quantity at rebar level. That means, the corrosion resistance of RC structure/structural member depends mainly on the near surface quality of concrete and adequate rebar cover thickness. So, it is always better to specify the durability performance criteria of as-built structure based on the deterioration process, besides the commonly used strength criteria (RILEM TC 230 PSC 2016; Bhaskar et al. 2020, 2021) This criterion shall be established with due care based on the exposure conditions and variability of test methods and importance/requirements of the structure/structural members. If the evaluated durability parameters are not complying with the acceptance criteria, corrective measures can be taken for the remaining construction and repair measures can be taken up for the completed works.

Table 21.2 Durability performance criteria—test parameter specification

RCPT (Coulombs)	RCMT ($\times 10^{-12}$ m^2/s)	OPI	Cover thickness; variation of w.r.t prescribed (%)	Durability performance
100 to 1000	2 to 8	>10.0	$\pm$ 10	Excellent
1000 to 2000	2 to 8	9.5 to 10.0	$\pm$ 20	Very good
2000 to 4000	8 to 16	9.0 to 9.5	$\pm$ 25	Good
> 4000	>16	< 9.0	> $\pm$ 25	Poor

The durability performance criteria for RC structures/structural members, the relevant test parameter and its specification, when exposed to chloride or carbonation induced environments is very important. The durability parameters evaluation based on tests such as rapid chloride penetration (RCP); rapid chloride migration (RCM) and oxygen permeability index (OPI) tests can ensure good quality of cover concrete (ASTM C 1202, 2016; NT Build 492, 1999; SANS 3001, 2015). Required cover thickness to the rebar can be ensured by means of cover metre survey (Bungey et al. 2006). Table 21.2 presents the details of test parameters and suggested durability performance criteria (Bhaskar et al. 2021). It can be said that for long term durability of RC structural components, the minimum durability performance can be at least 'very good'. If the durability performance falls below the specified, necessary protective/repair measures need to be carried out.

21.4 Service Life Estimation

Chloride induced corrosion

As discussed earlier, the total service life (t_{total}) can be the sum of (i) corrosion initiation period (t_i) and (ii) stable propagation period (t_{sp}) (Tuutti 1982; Austroads 2000; Bhaskar et al. 2015):

$$t_{total} = t_i + t_{sp} \tag{21.3}$$

For the estimation of t_i, Fick's second law of diffusion is being used after applying specific boundary conditions, and can be expressed as (Life-365 2013; ACI 365.1R 2017; Andrade 1993; Gjorv 1994):

$$C(x, t) = C_s\left[1 - erf\left(\frac{x}{2\sqrt{(D_c t_i)}}\right)\right] \tag{21.4}$$

where, $C(x,t)$—amount of chlorides at depth x (or Cl_{th}) after time t_i; C_s—surface chloride concentration; *erf*—Gaussian error function; x—clear cover thickness; and D_c—diffusion coefficient.

The t_i is the time required to reach threshold chloride concentration, Cl_{th} at the rebar surface. The parameters Cl_{th}, C_s and D_c are dynamic and can vary considerably. Appropriate values can be assumed based on the exposure conditions, materials and mix proportions used. However, it is to be noted that Cl_{th} depends on number of factors which include the rebar properties, surrounding concrete properties, etc. (Alonso et al. 2000; Pillai and Trejo 2005; Angst et al. 2009). Cl_{th} can be taken as 0.4% by mass of cement (i.e., 0.05% by mass of concrete), which is being used commonly (Austroads 2000; Life-365 2013; ACI 365.1R 2017). In general, C_s increases with the exposure time and tends to reach a maximum value of the range 5–6% by mass of cement (1–2% by mass of concrete). Based on Eq. (21.4), the t_i can be estimated as:

$$t_i = \frac{x^2}{4D_c\left[erf^{-1}\left(1 - \frac{Cl_{th}}{C_s}\right)\right]^2} \quad (21.5)$$

The propagation period, t_{sp} can be estimated based on the Rodriguez et al. (1996) model:

$$\theta(t) = \theta(0) - p(t) \quad (21.6)$$

$$p(t) = 0.0116\,\alpha\, i_{corr} t_{sp} \quad (21.7)$$

where $\theta(t)$- reduction in rebar diameter (mm) with time t (years); $\theta(0)$—original rebar diameter (mm); $p(t)$—corrosion loss (mm); i_{corr}—corrosion current density ($\mu A/cm^2$); and 0.0116—conversion factor which converts $\mu A/cm^2$ to mm/year; α—accounts for pitting effects ($\alpha = 5$ to 10) (González et al. 1995).

Carbonation induced corrosion

It is reported in literature that the carbonation depth roughly follows parabolic relationship and is expressed as (Soutsos 2010; Neves et al. 2018; Ramesh and Bhaskar 2019):

$$x_c = kt^n \quad (21.8)$$

where, x_c = carbonation depth, mm.

k = carbonation rate/coefficient, mm/year^{n} and.

t = exposure time/duration, years and n is an exponent.

The exponent n can vary in the range from 0.4 to 0.6, depending upon the type of concrete and exposure conditions. Usually n is taken as 0.5 and hence, Eq. (21.8) can be written as

$$x_c = kt^{0.5} \tag{21.9}$$

In Eq. (21.9), the carbonation coefficient, k depends on number of factors such as the penetrability, CO_2concentration of in the atmosphere, amount of carbonatable material, the aggressiveness of environment viz., temperature, RH, rain, solar exposure, etc. Equation (21.9) can also be used to estimate the service life (corrosion initiation period) of RC structures exposed to carbonation induced environment. The carbonation depth (x_c) obtained through drilled cores/fresh pieces taken out from the structure, at time t_0, allow to know k and therefore, to estimate the corrosion initiation time t_i at which the carbonation front reach the rebar and depassivate it (i.e., x_c equals the clear cover depth, x). x_c is typically measured by spraying phenolphthalein solution on the freshly broken concrete surface.

21.5 Service Life Estimation—A Parametric Study

In this section, the results of a parametric study carried out under accelerated chloride induced environment is presented (Bhaskar et al. 2015). Two types of concretes, ordinary portland cement (OPC) and portland pozzolana cement (PPC) concretes with three w/c ratios with different rebar cover thicknesses (x) were considered. In the study, diffusion coefficient (D_c), a durability parameter, which reflects the resistance to chloride ion penetration of concrete was evaluated for OPC and PPC concretes (NT Build 355 1997). Appropriate values for Cl_{th} (0.05% by weight of concrete) and C_s (1% by mass of concrete) were assumed (Life-365 2013). By using Eq. (21.5), t_i has been estimated for concretes with three *w/c*s (0.57, 0.47, and 0.37) and four cover thicknesses (x = 20, 30, 40 and 60 mm). The estimated t_i values for different concretes were presented in Table 21.3. It can be seen that for a cover thickness of 20 mm, the t_i of OPC-0.57 concrete is 2.1 years, whereas for the corresponding PPC-0.57 concrete, it is 4.2, which is double. Similarly, for a cover thickness of 40 mm, the t_i of PPC-0.57 concrete is nearly double to that of the corresponding OPC-0.57 concrete. It can also be seen from Table 21.3 that as w/c decreases, the

Table 21.3 Predicted t_i values for OPC and PPC concretes with different D_c and x

Concrete	Diffusion coefficient, D_c(m^2/sec)	t_i, in years, for different cover thicknesses (x)			
		20 (mm)	30 (mm)	40 (mm)	60 (mm)
OPC-0.57	7.67×10^{-13}	2.1	4.8	8.6	19.4
PPC-0.57	3.92×10^{-13}	4.2	9.5	16.8	37.9
OPC-0.47	4.40×10^{-13}	3.8	8.4	15.0	33.8
PPC-0.47	1.25×10^{-13}	13.2	29.7	52.8	118.9
OPC-0.37	2.40×10^{-13}	6.9	15.5	27.5	61.9
PPC-0.37	5.20×10^{-14}	31.7	71.4	127.0	285.7

t_i increases, as expected. Nevertheless, the increase in $t_i$is significant in PPC-0.47 and PPC-0.37 concretes when compared to that of the corresponding OPC-47 and OPC-0.37 concretes.

For the estimation of stable propagation period, t_{sp}, 10% loss in cross-sectional area of the steel rebar was considered as the damage limit (CEB 1983; Andrade et al. 1990). The corrosion current density (i_{corr}) values of selected concretes were estimated through accelerated corrosion tests. The i_{corr}-values were presented in Table 21.4. It can be seen that i_{corr} for a rebar in the OPC concrete is higher than that of the corresponding i_{corr} for PPC concrete. This could be due to the lower chloride ion permeability and increased resistivity of PPC concrete (Scott and Alexander 2007; Bhaskar et al. 2011). This shows the beneficial effect of fly ash in terms of lower i_{corr}-values for PPC concretes when compared to that of the corresponding OPC concretes.

It is to be noted that the i_{corr}—values obtained were under accelerated conditions, hence, the values are very high than those observed in the field (atmospheric conditions). Andrade et al. (1990) and González et al. (1995) reported i_{corr}—values of 1 to 3 $\mu A/cm^2$ in the case of active corrosion and values of 10 $\mu A/cm^2$ or higher were observed, rarely, in extreme conditions. Therefore, in the parametric study, the field i_{corr}—values of 1, 5 and 10 $\mu A/cm^2$ were considered to OPC-0.57 concrete. The corresponding field i_{corr}—values for other selected concretes were assumed to vary proportionally to the i_{corr}-values (see Table 21.4) evaluated in the accelerated corrosion tests (Bhaskar et al. 2015; Morinaga et al. 1994). Figure 21.2 shows the schematic plot of % loss in rebar cross-section with time for OPC-0.57 and PPC-0.57 concretes. It can be observed that: (i) longer corrosion initiation period (t_i) for PPC concrete when compared to that of OPC concrete, (ii) higher rate of corrosion with higher i_{corr}-values, and (iii) longer stable propagation periods (t_{sp}) for PPC concrete.

The estimated service lives (t_{total}) for OPC and PPC concretes are presented in Table 21.5. It can be seen that the service life of PPC-0.57 concrete is 2.6 times to that of the service life of OPC-0.57 concrete when the i_{corr} for the OPC-0.57 is 1 $\mu A/cm^2$. Also, as can be expected, the ratio between the t_{total}-estimates of the PPC and OPC concretes increases as the w/c decreases. The parametric study results reveal that PPC concretes perform better corrosion resistance than OPC concretes in chloride induced environments.

With reference to carbonation induced corrosion, Otieno et al. (2020) reported that the carbonation rates are relatively more in some blended cement concretes (such as OPC + Flyash; OPC + Blast furnace slag) than that of OPC concretes with the same water-to-binder ratio because of the depletion of portlandite due to pozzolanic

Table 21.4 i_{corr}—values for OPC and PPC concretes

Concrete	i_{corr} ($\mu A/cm^2$)	Concrete	i_{corr} ($\mu A/cm^2$)
OPC-0.57	768	PPC-0.57	289
OPC-0.47	581	PPC-0.47	212
OPC-0.37	500	PPC-0.37	191

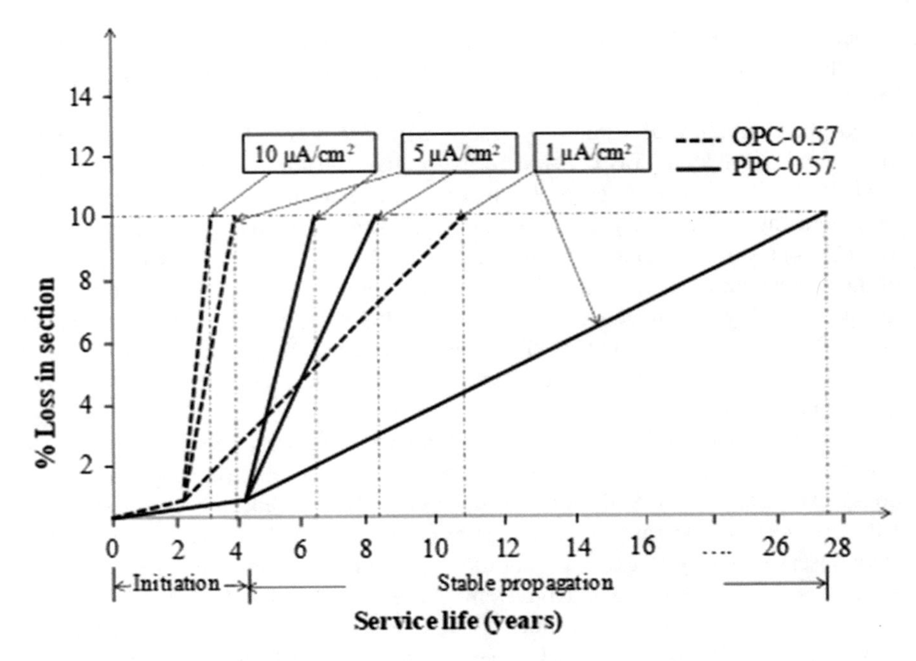

Fig. 21.2 Loss in rebar cross-sectional area with time for different i_{corr}—values

reaction. However, this may be to some extent offset by their low penetrability prior to carbonation.

21.6 Conclusions

Long term durability is one of the major concerns due to the early deterioration of RC structures caused by the rebar corrosion. In this chapter, basics of rebar corrosion and dominant corrosion mechanisms, performance based specifications and service life estimation of RC structures/components are discussed. Following are the major conclusions:

- The corrosion resistance of RC structure depends mostly on the cover concrete quality and its adequate thickness. Therefore, for long term durability of as-built RC structure, there is a need to verify the properties of cover concrete such as its penetrability to aggressive agents and specified cover thickness.
- The penetrability of cover concrete can be evaluated by test methods such as RCPT, RCMT, OPI, etc., depending on the exposure condition and specific durability requirements.

Table 21.5 Estimated service lives for OPC and PPC concretes

Type of concrete	(t_i), years	(t_{sp}), years	Service life (t_{total}), years
Field i_{corr} for OPC-0.57 concrete is 1 μA/cm^2			
OPC-0.57	2.1	8.6	10.7
PPC-0.57	4.2	23.3	27.5
OPC-0.47	3.8	12.3	16.1
PPC-0.47	13.2	28.7	41.9
OPC-0.37	6.9	14.4	21.3
PPC-0.37	31.7	43.1	74.8
Field i_{corr} for OPC-0.57 concrete is 5 μA/cm^2			
OPC-0.57	2.1	1.7	3.8
PPC-0.57	4.2	4.6	8.8
OPC-0.47	3.8	2.4	6.2
PPC-0.47	13.2	5.8	19.0
OPC-0.37	6.9	2.8	9.7
PPC-0.37	31.7	8.6	40.3
Field i_{corr} for OPC-0.57 concrete is 10 μA/cm^2			
OPC-0.57	2.1	0.9	3.0
PPC-0.57	4.2	2.3	6.5
OPC-0.47	3.8	1.2	5.0
PPC-0.47	13.2	2.9	16.1
OPC-0.37	6.9	1.4	8.3
PPC-0.37	31.7	4.3	36.0

- A durability performance criterion including test parameter specification is suggested for RC structures/components exposed to chloride or carbonation induced environments.
- The parametric study on durability based service life in chloride induced environments reveal that the PPC based concretes performed superior to that of the corresponding OPC based concretes.

References

ACI 201.2R-16 (2016) Guide to durable concrete.ACI Publication, USA

ACI 222R-19 (2019) Guide to protection of reinforcing steel in concrete against corrosion. ACI Publication, USA

ACI 365.1R-17 (2017) Report on service life prediction – State of the art report, ACI Publication, USA

Alonso C, Andrade C, Castellote M, Castro P (2000) Chloride Threshold Values to Depassivate Reinforcing Bars Embedded in a Standardized OPC Mortar. Cem Concr Res 30(7):1047–1055

Andrade C, Alonso MA, Gonzalez JA (1990) An initial effort to use the corrosion rate measurements for estimating rebar durability. In: Berke NS, Chaker V, Whiting D (eds) Corrosion rates of steel in concrete, ASTM STP 1065, Philadelphia, pp 29–37

Andrade C (1993) Calculation of chloride diffusion coefficients in concrete from ionic migration measurements. Cem Concr Res 23:724–742

Angst U, Elsener B, Larsen CK, Vennesland Ø (2009) Critical chloride content in reinforced concrete – A review. Cem Concr Res 39:1122–1138

Anoop MB, Balaji Rao K, Appa Rao TVSR (2002) Application of fuzzy sets for estimating service life of reinforced concrete structural members in corrosive environments. Engng. Structures 24(9):1229–1242

Austroads (2000) Service life prediction of reinforced concrete structures, Austroads Project No. N.T&E.9813, Austroads Publication No. AP-T07/00, Sydney, Australia

ASTM C1202 (2017) Standard test method for electrical indication of concrete ability to resist chloride ion penetration. American Society for Testing and Materials

Bentur A, Diamond S, Berke S (1998) Steel corrosion in concrete: fundamentals and civil engineering practice, E & FN Spon

Bhaskar S, Gettu R, Bharatkumar BH, Neelamegam M (2011) Chloride-induced corrosion of steel in cracked OPC and PPC concretes: Experimental study. J. Mater. Civil Engng. 23(7):1057–1066

Bhaskar S, Pillai RG, Gettu R, Bharatkumar, BH, Iyer NR (2015) Use of portland pozzolana cement to enhance the service life of reinforced concrete exposed to chloride attack. J Mater Civil Eng 27(11):04015031–1–8.

Bhaskar S, Ramesh G, Bharatkumar BH (2020) A review on performance based specifications towards concrete durability. Struct Concr https://doi.org/10.1002/suco.201900542

Bhaskar S, Kanchanadevi A, Sivasubramanian K, Ramanjaneyulu K (2021) Durability criteria for pre-cast RC box units and repair measures based on NDT&E. J Perform Constr Facil https://doi.org/10.1061/(ASCE)CF.1943-5509.0001646

Broomfield JP (2006) Corrosion of steel in concrete: understanding, investigation and repair, 2nd edn. Taylor and Francis Ltd, UK

BS 8110 (1997) Structural use of concrete—part 1: code of practice for design and construction. British Standards Institution, London

Bungey JH, Millard SG, Grantham MG (2006) Testing of concrete in structures, 4th edn. Taylor & Francis, NY, USA

Callister WD, Jr (2000) Materials science and engineering: an introduction. John Wiley & Sons, Inc, NY

CEB (1983) Assessment of Concrete structures and design procedures for upgrading. Bulletin No 162:87–90

Gjørv OE (1994) Important test methods for evaluation of reinforced concrete durability.. In: Mehta PK (ed) Concrete in the 21st century: past, Present and future. ACI SP 144, pp 545–574

González JA, Andrade C, Alonso C, Feliu S (1995) Comparison of rates of general corrosion and maximum pitting penetration on concrete embedded steel reinforcement. Cem Concr Res 25(2):257–264

Grantham MG (2011) Understanding defects, testing and inspection, Concrete repair: a practical guide. In: Grantham MG (ed), Spon Press, (Taylor & Francis), Milton Park, Abingdon, Oxon, UK, pp 1–55

Hooton RD, Hover K, Bickley JA (2005) Performance standards and specifications for concrete: Recent Canadian developments. The Ind. Con. J. 79(12):31–37

Life-365 Service life prediction model and computer program for predicting the service life and life-cycle cost of reinforced concrete exposed to chlorides, Version 2.2, Life-365 Consortium II, 2013

Mackechnie JR, Alexander MG, Heiyantuduwa R, Rylands T (2004) The effectiveness of organic corrosion inhibitors for reinforced concrete. Research monograph No. 7, Department of Civil Engineering, University of Cape Town, S.A

Montemor MF, Simoes AMP, Ferreira MGS (2003) Chloride-induced corrosion on reinforcing steel: from the fundamentals to the monitoring techniques. Cem. Con. Comp. 25:491–502
Morinaga S, Irino K, Ohta T, Arai H (1994) Life prediction of existing reinforced concrete structures determined by corrosion. In: Swamy RN (ed) Corrosion and Corrosion Protection of Steel in Concrete. Academic Press, Sheffield, UK, pp 603–618
Neves R, Torrent R, Imamoto K (2018) Residual service life of carbonated structures based on site non-destructive tests. Cem Concr Res 109:10–18
Neville AM (1987) Why we have concrete durability problems. ACI SP-100, Katherine and Bryant Mather International Conference on Concrete Durability, (American Concrete Institute, Detroit, pp. 21–48
Neville AM (1996) Properties of concrete, Addison Wesley Longman Ltd
NT Build 355 (1997) Concrete, mortar and cement based repair materials: chloride diffusion coefficient from migration cell experiments. Nordtest Method, Finland
NT Build 492 (1999) Chloride migration coefficient from non-steady-state migration experiments. Nordtest Method, Finland
Otieno M, Ikotun J, Ballam Y (2020) Experimental investigations on the effect of concrete quality, exposure conditions and duration of initial moist curing on carbonation rate in concretes exposed to urban, inland environment. Const Build Mat 246 118443
Parrott LJ (1990) Damage caused by carbonation of reinforced concrete. Mater Struct 23(3):230–234
Pillai RG, Trejo D (2005) Surface condition effects on critical chloride threshold of steel reinforcement. ACI Mater J 102(2):103–109
Ramesh G, Bhaskar S (2019) Carbonation-induced corrosion: a breif review on prediction models. J Inst Eng (India):series A https://doi.org/10.1007/s40030-020-00434-8.
Richardson M (2002) Fundamentals of durable reinforced concrete. Spon press, London, Londres e Nova Iorque
RILEM TC 230 PSC (2016) Performance based specifications and control of concrete durability. State-of-the-art report, Beushausen H, Fernandez LL (eds)
Rodriguez J, Ortega LM, Casal J, Diez JM (1996) Assessing structural conditions of concrete structures with corroded reinforcement. In: Dhir RK, Jones MR, (eds), Concrete repair, rehabilitation and protection, E & FN Spon, London, pp 65–78
SANS 3001-CO3–2 (2015) Concrete durability index testing–oxygen permeability test. South African National Standard, South Africa
Scott A, Alexander MG (2007) The influence of binder type, cracking and cover on corrosion rates of steel in chloride contaminated concrete. Mag Concr Res 59(7):495–505
Soutsos M (ed) (2010) Concrete durability: a practical guide to the design of durable concrete structures. Thomas Telford Ltd., London, UK
Swamy, R. N., (2008) Sustainable concrete for the 21 century concept of strength through durability. Japan Society of Civil Engineers Concrete Committee Newsletter, 13, http://www.jsce.or.jp/committee/concrete/e/newsletter/newsletter13/Paper1.pdf
Tuutti K (1982) Corrosion of steel in concrete, CBI, research report 4, Stockholm

Chapter 22
Composite Cement: A Sustainable Binding Material for Real Time Construction Practice in India

Chandra Sekhar Karadumpa and Rathish Kumar Pancharathi

22.1 Introduction

India is the second major producer of cement in the world after China (CMA 2018) and third largest emitter of carbon dioxide after USA and China (Mishra et al. 2015). The per capita cement consumption in India is about 190 kg against world's average consumption of 350 kg (IBEF 2016). The cement production in India in 2014 was 275 MT and is expected to grow at a much faster rate to reach 1000 MT by 2050 (Imbabi et al. 2012). There exists a huge demand for cement in the years to come. Cement manufacturing process is involved with high energy demand of 710 kcal/kg clinker for fuel and 78 kWh/tonne of cement for electrical energy (12th five year plan 2012–2017, GoI 2013) resulting in emission of carbon dioxide (0.82 tonne/tonne of cement) during calcination process where lime stone gets converted into calcium oxide with release of CO_2. The combustion of fuels for the calcination process also releases CO_2. Indian Bureau of Mines (IBM) reports that the availability of high grade lime stone is about 89,862 million tonnes and could last for 35 years (Indian minerals year book, 2014). Bureau of Energy Efficiency (BEE) reported that, in India, the energy consumed by the cement industry in 2007 was about 607 PJ (BEE, 2007) and the total energy (both electrical and thermal) required to meet the demand for cement production in the year 2030 at an increasing rate of 10% is projected to be 6190 PJ (Krishnan et al. 2012). International Energy Agency (IEA) reports that the consumption of energy in India with respect to cement industry is projected to be between 42 and 48 million tonnes of oil equivalent (MTOE) and carbon dioxide

C. S. Karadumpa (✉) · R. K. Pancharathi
Department of Civil Engineering, National Institute of Technology, Warangal, Telangana 506004, India
e-mail: chandukv19@student.nitw.ac.in

R. K. Pancharathi
e-mail: rateeshp@nitw.ac.in

K. R. Reddy et al. (eds.), *Advances in Sustainable Materials and Resilient Infrastructure*, Springer Transactions in Civil and Environmental Engineering,
https://doi.org/10.1007/978-981-16-9744-9_22

emission is between 422 and 483 million tonnes by the year 2050 (Trudeau et al. 2011). The annual GHG emission from the cement industry alone is about 129 MT in the year 2007 which is 6% of total GHG emission in the country (Midha et al. 2015). For developing countries like India, with rapid increase in population and demand for infrastructure, it is required to use low energy alternative mineral additives in cement to minimize the emission of GHG, to reduce energy consumption and also to restore the lime stone resources for many years to come. IEA reported that for energy production, India mainly depends on coal (33%) and oils (23%) as main source and the share of biomass and waste usage in energy production is 19% which is large compared to other countries (Trudeau et al. 2011). Confederation of Indian Industry (CII) reported that, about 1.6 million tonnes of alternative fuels are used in production of cement which is very high when compared to other countries in alternate fuel consumption as on year 2018 (CII 2014). India reserves 306.6 billion tonnes of coal and it occupies fifth place in the world in the availability of coal (BP Energy outlook 2017). At present, the power generation is mainly coal-based in India and it accounts for 55% of total power generation followed by, hydro (21%), renewable energy sources (11%), gas (10%) and nuclear (2%) (Planning Commission, GoI 2013). About 85% of coal burned power plants in India uses subcritical technology having efficiency rate of less than 35% (IEA 2015). If the technology is upgraded with Clean Coal Technology (CCT) and High Efficiency Low Emission (HELE) like supercritical and ultra-super critical technology and in burning of coal found to reduce emission of greenhouse gases between 20 and 25% and restoring of coal for future (Mishra et al. 2015). The carbon emission in India has reduced from 1.12 to 0.719 t CO_2/tonne of cement from the year 1996 to 2010 (Trudeau et al. 2011). This is due to increase in the production of blended cements with upgraded technology by reducing clinker content. By 2014, OPC, PPC and PSC contributed to 27%, 66% and 7% respectively of the total cement production (CII 2014). In view of energy requirement, carbon emission and moving towards sustainability, production of blended cements having partial replacements of mineral additives with OPC has increased in recent times. In 2015, Bureau of Indian Standard has released a code of practice IS as per 16,415:2015 (BIS, 2015a) on composite cement to recommend the combined usage of FA and GBFS as partial substitution to OPC. The optimum replacement levels of FA and GBFS are determined based on the requirement of strength and effective packing of the binder (Karadumpa and Pancharathi 2021a; Karadumpa and Pancharathi 2021b). Replacement of Supplementary Cementitious Materials (SCMs) in cement are found to have several benefits compared to OPC in terms of ecological benefits, performance and durability aspects of concrete such as reduction in heat of hydration (Cheah et al. 2019; Mehta and Monteiro 2014), emission of CO_2 (Hendriks et al. 1998; Worrell et al. 2001; Rehan and Nehdi 2005), conservation of raw materials, reduction in energy demand (García-Segura et al. 2014; Sobolev 2003), lowering the cost of production (Hendriks et al. 1998; Zhang et al. 2013), consumption of industrial wastes (Zhang et al. 2013; Bayraktar 2019), enhanced workability (Cheah et al. 2019; Mehta and Monteiro 2014; Gholampour and Ozbakkaloglu 2017) and long term strength (Ghrici et al. 2007; Bapat et al. 2001), resistance to corrosion (Guneyisi et al. 2007; Ogirigbo and Black 2019; Papadakis

2000), improved density and reduction in concrete permeability (Kayali and Ahmed 2013; Atis and Bilim 2007) resistance to alkali aggregate reaction (Angulo Ramirez et al. 2018) drying shrinkage (Hu et al. 2017; Bouzoubaa et al. 2001). In this paper, ternary blended composite cement, binary blended PPC and PSC cements and plain OPC are investigated from the points of view of energy consumption, carbon dioxide emission and cost involved in preparation of cement and thus concretes.

22.2 Materials and Methodology

22.2.1 Materials

22.2.1.1 Cement

OPC of 53 grade as per IS 269:2015 (BIS 2015b), PPC comprising 70% of OPC and 30% of fly ash (class F) confirming to IS 1489 (Part 1):1999 (BIS 1999a), PSC with OPC and GBFS contents 55% and 45% respectively as per IS 455:2015 (BIS 2015c), obtained from southern part of India, Composite Cement (CC) with 50% of OPC, 20% FA and 30% GBFS prepared by interblending process and confirming to IS 16415:2015 are used in the study. The permissible replacement levels of fly ash, GBFS and/or pozzolana in the preparation of composite cement prescribed by different international standards are shown in Table 22.1. The physical properties of cements are shown in Table 22.2.

Table 22.1 Comparison of standards of different countries on composite cement

Standard	Name	Clinker/OPC	GBFS	Pozzolana	Fly ash	Performance Additives
BIS	IS 16415:2015	35–65	20–50		15–35	As per required
EN 197–1	CEM V/A	40–64	18–30	18–30		0–5
	CEM V/B	20–38	31–50	31–50		0–5
ASTM C-595–14	Type IT (S20)P(10)	70	20	–	10	–
	Type IT (P25)P(10)	65	–	10	25	–
	Type IT (S20)P(10)	60	20	–	20	–
CSA	CSA A3000	40	60			–

Table 22.2 FA, GBFS, OPC, PPC, PSC and CC physical properties

Property	FA	GBFS	OPC	PPC	PSC	CC
Specific gravity	2.20	2.40	3.12	2.87	2.96	2.88
Standard consistency (%)	–	–	29	31	32	29.5
Surface area (fineness) (m^2/Kg)	325	350	300	300	285	330
IST (minutes)	–	–	90	100	95	110
FST (minutes)	–	–	250	285	305	280
Soundness by Le-chatelier method (mm)	–	1.6	1.0	1.2	0.8	1.0
Soundness by Autoclave method (%)	0.35	–	0.074	0.08	0.06	0.02
Drying shrinkage (%)	–	–	0.15	0.1	0.09	0.08
Loss on ignition (%)	0.8	1.1	1.42	3.24	1.75	1.42

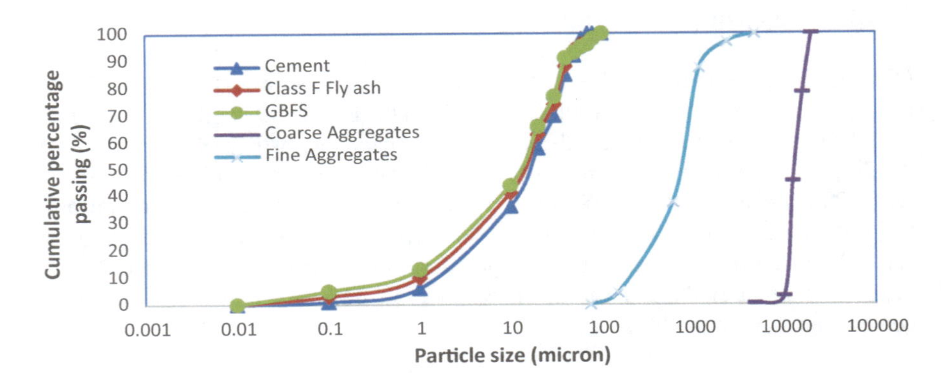

Fig. 22.1 Particle size distribution of cement, FA, GBFS and aggregates

22.2.1.2 Fine and Coarse Aggregates

River sand falling in Zone II as per IS 383:2016 is considered as fine aggregate in this study. The particle size distribution (PSD) of fine and coarse aggregates is shown in Fig. 22.1 and the physical properties are as indicated in Table 22.4.

22.2.1.3 Secondary Cementitious Materials (SCMs)

GBFS as per IS 12089:1987 (BIS 1987) and FA as per IS 3812:2013 (BIS 2013) obtained from industries located in southern part of India are used as mineral additives to OPC. The PSD of FA and GBFS is shown in Fig. 22.1. The physical properties of aggregates are determined and tabulated in Table 22.2. The oxide composition of OPC, FA, GBFS and CC is determined using X-ray fluorescence Spectroscopy (XRF) and is shown in Table 22.3.

Table 22.3 Chemical composition of FA, GBFS, OPC, PPC, PSC and CC (XRF analysis)

Composition	FA	GBFS	OPC	PPC	PSC	CC
Chemicals						
CaO	3.245	40.146	64.561	38.88	56.64	41.635
SiO_2	61.107	37.518	21.03	38.266	26.82	33.943
Al_2O_3	27.825	17.100	3.422	14.121	10.057	15.023
Fe_2O_3	3.045	0.541	4.332	4.031	2.438	2.662
SO_3	0.549	1.792	3.418	2.142	2.681	3.009
MgO	1.120	1.091	1.811	1.456	1.538	1.603
TiO_2	1.442	0.837	0.231	0.898	1.006	0.926
K_2O	1.433	0.380	0.752	1.403	0.547	0.034
MnO	0.041	0.496	0.087	0.072	0.551	0.105

Table 22.4 Aggregate properties

Properties	Coarse aggregates	Fine aggregates
Specific gravity (G)	2.72	2.65
Bulk density (kg/m^3)	1468	1495
Void content (%)	41.3	34.6
Water absorption (%)	0.71	1.36
Fineness modulus	6.96	2.36

22.2.1.4 Water

Potable water was used for all the mixtures.

22.2.1.5 Superplasticizer

A polycarboxylate-ether (PCE) based superplasticizer (PolyHEED-8950) conforming to IS 9103:1999 (BIS 1999b) is used in this study.

22.2.2 *Compressive Strength*

Concrete cube samples of size 150 mm side were tested for the compressive strength as per IS 516:2004 (BIS 2004).

22.2.3 Estimation of Embodied Energy and CO_2 Equivalent

The total embodied energy (E_T) and carbon footprint (C_T) of OPC, PPC, PSC and CC concretes of different grades is estimated using the Eqs. 22.1 and 22.2 below, respectively.

$$E_T = W_w E_w + W_c E_c + W_F E_F + W_G E_G + W_s E_s + W_{ca} E_{ca} + W_{sp} E_{sp} \quad (22.1)$$

$$C_T = W_w C_w + W_c C_c + W_F C_F + W_G C_G + W_s C_s + W_{ca} C_{ca} + W_{sp} C_{sp} \quad (22.2)$$

W_w, W_c, W_F, W_G, W_s, W_{ca} *and* W_{sp} are the weight fractions multiplied with corresponding energy and emission factors of water, cement, fly ash, GBFS, sand, coarse aggregates and super plasticizer respectively per cubic meter of concrete. The energy and emission factors considered in this study are as shown in Table 22.5.

22.3 Experimental Methodology

Three grades of concrete mixes M20, M30 and M40 are designed using four different types of cements (OPC, PPC, PSC and CC) by IS 10262:2019 (BIS 2019) guidelines as shown in Table 22.6 to study the fresh, hardened, energy and CO_{2e} emission properties of different concrete mixes. Figure 22.2 shows the methodology adopted for assessing the behaviour of different concrete mixes.

Table 22.5 Energy, emission and price factors of concrete making materials

Material	Energy (kJ/kg)	CO_{2e} (g/kg)	Price per kg (INR)	References
Cement (OPC)	5500	830	6.8	Prakasan et al. (2020), Reddy and Jagadish (2003), Collins (2010), Flower and Sanjayan (2007)
Fly ash	100	27	0.08	Aysha et al. (2014)
GBFS	1600	143	2.0	Aysha et al. (2014)
Coarse aggregates	83	45.9	1.0	Flower and Sanjayan (2007), Aysha et al. (2014)
Fine aggregates	80	13.9	1.5	Flower and Sanjayan (2007), Aysha et al. (2014)
Water	10	0.001	0.05	Aysha et al. (2014)
Superplasticizer	10,000	600	100	Aysha et al. (2014)

Table 22.6 Mix design of different types of concrete as per IS 10262:2019

Mix id	W/B	W (kg)	C (kg)	FA (kg)	GBFS (kg)	S (kg)	G	SP (kg)	Compressive strength (MPa)			WD (g/cc)	DD (g/cc)	VS (mm)
									7 days	28 days	90 days			
OPCM20	0.58	191.5	330.8	–	–	696.8	1228.4	–	21.56	30.45	31.63	2.41	2.40	95
PPCM20	0.53	191.5	278.3	119.3	–	773.3	998.9	–	17.54	26.82	29.24	2.37	2.35	100
PSCM20	0.53	191.5	198.8	–	162.6	781.8	1009.9	–	18.28	27.56	30.12	2.40	2.38	90
CCM20	0.50	191.5	198.8	79.1	119.5	639.9	1068.7	–	15.82	25.42	28.95	2.38	2.37	95
OPCM30	0.50	180.4	360.1	–	–	701.2	1212.2	–	26.56	38.46	40.06	2.41	2.40	75
PPCM30	0.42	147.5	270.4	115.9	–	683.4	1211.9	3.86	21.54	34.12	37.22	2.37	2.36	85
PSCM30	0.42	147.5	193.1	–	158.1	660.0	1170.4	3.51	23.23	35.67	38.18	2.40	2.39	80
CCM30	0.48	147.5	193.2	77.3	115.9	640.6	1212.6	3.86	20.24	32.48	36.44	2.37	2.36	85
OPCM40	0.41	147.5	359.8	–	–	651.1	1164.5	3.58	34.18	49.44	50.23	2.41	2.40	80
PPCM40	0.33	155.2	332.1	142.1	–	719.2	1069.4	4.74	27.52	43.88	47.24	2.38	2.37	85
PSCM40	0.36	158.6	265.2	–	172.2	770.8	1101.2	4.31	30.27	45.21	48.11	2.39	2.38	70
CCM40	0.35	147.4	236.5	94.8	142.2	628.3	1062.8	4.74	26.58	42.16	46.95	2.37	2.36	75

W = Water; GBFS = Granulated blast furnace slag; C = Cement; FA = Fly ash; S = Sand; G = Gravel; SP = Superplasticizer; WD = Wet density; DD = Dry density; VS = Vertical slump

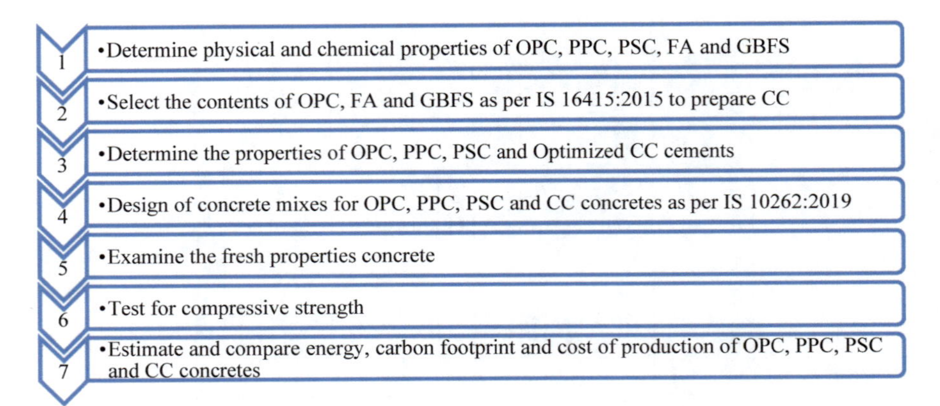

Fig. 22.2 Methodology opted for comparison of cements

22.4 Results and Discussions

22.4.1 Compressive Strength

Three grades of concretes are designed using OPC, PSC, PPC and CC cements as per IS 10262:2019. Table 22.6 shows the composition of concrete mixture, strength and fresh properties of different concretes prepared using these cements. Figure 22.3 shows the compressive strength of different grades of concretes at the curing ages of 7, 28 and 90 days. The concretes made of PPC cement on an average exhibited 19%, 11.48% and 6.86% lower compressive strength compared to OPC concrete at the end of 7, 28 and 90 days, respectively. The concretes designed using PSC cements depicted an average of 13.06%, 8.43% and 4.56% less compressive strength than the reference OPC concrete at the end of 7, 28 and 90 days, respectively. In the same way, the composite cement concretes have shown 24.22%, 15.6% and 8.01% lower compressive strength than reference OPC concrete at curing durations 7, 28 and 90 days, respectively. In general, the compressive strength achieved by different cement concretes follows the relation OPC > PSC > PPC > CC. This is attributed to the slower rate of pozzolanic action of fly ash and GBFS compared to the hydration of OPC. It is noticed that, concretes designed using PSC exhibited higher compressive strength than PPC and CC concretes at all ages of curing. This may be due to the hydraulic nature of GBFS which adds to the faster reaction along with OPC hydration process. It is obvious that, concretes designed using CC have shown lower compressive strength than all other cement concretes due to high (50%) replacement of OPC with 30% GBFS and 20% fly ash. Hence, in case of CC, the compressive strength is mainly governed by the pozzolanic activity of mineral additives. From Fig. 22.3, it can be noticed that, the compressive strength of PPC, PSC and CC concretes increased significantly beyond 28 days till 90 days compared to reference OPC concretes. This is due to maximum completion of hydration process at 28 days

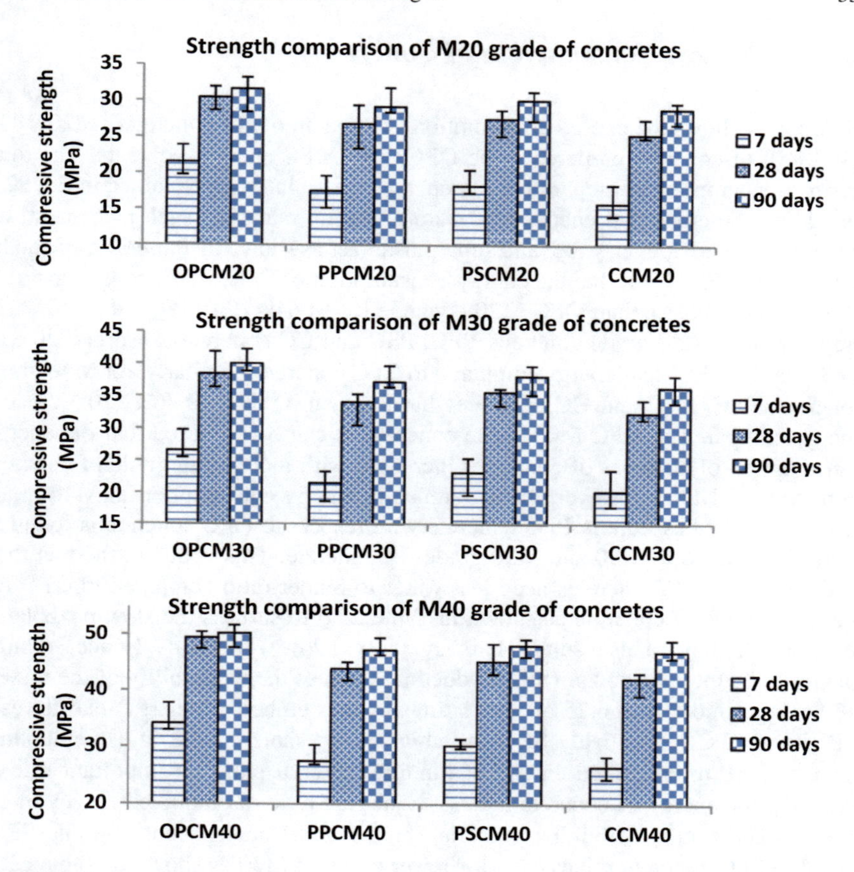

Fig. 22.3 Compressive strength of concretes prepared using OPC, PPC, PSC and CC as per IS 10262:2019

in OPC and incomplete pozzolanic reaction of FA and GBFS at 28 days in PPC, PSC and CC. It can also be noticed that CC concretes exhibited rise in compressive strength between 28 to 90 days compared to all other cement concretes. This is due to availability of more quantity of mineral additives in CC to participate in pozzolanic action compared to other cements. At 28 days, though concretes designed using CC resulted in less compressive strength than other cement concretes, it has achieved the characteristic compressive strength for all the grades of concrete. The systematic design by considering the precise material properties of CC concrete constituents would reduce the uncertainty or standard deviation of compressive strength results which makes safe, economic and sustainable concrete mix design.

22.4.2 Energy Consumption of Concretes

Figure 22.4 shows the energy consumption of three grades of concretes (M20, M30 and M40) of concrete prepared using OPC, PSC, PPC and CC cements. The total energy consumption in preparation of concrete is estimated as explained in Eq. 22.1 by adding the embodied energy of concrete making materials such as cement, fly ash, GBFS, sand, gravel, water and superplasticizer as shown in Table 22.5. It can be noticed from Fig. 22.4, that the energy consumption of PPC, PSC and CC concrete of M20 grade is less than OPC M20 concrete by 14.64%, 24.11% and 27.52%. In preparation of M30 grade concrete, PPC, PSC and CC concretes absorbed 20.8%, 29.8% and 32.32% less energy compared to OPC concrete. Similarly, for M40 grade concrete, PPC, PSC and CC concretes have shown 5.95%, 10.76%, 20.32% less energy compared to OPC M40 grade concrete. It can be observed that the energy consumption of all types of concretes increased with increase in grade of concrete from M20 to M40. However, the difference in energy consumption in M40 grade of PSC, CC and especially PPC concrete with respect to OPC concrete is found to reduce compared to M20 and M30 grades of concrete. This is due to the fact that, PPC, PSC and CC concretes need less water to binder ratio compared to OPC for the same grade of concrete as shown in Table 22.6 to surpass the slow pozzolanic reaction at 28 days. This results in harsh mixes with low workability. Hence, suitable quantity of super plasticizer (SP) is added to increase the workability of the mixes. SP having almost 100, 6.25 and 1.81 times higher embodied energy than fly ash, GBFS and OPC, respectively. SP being a high energy material, the ratio of embodied energy of SP to embodied energy of binding material plays an important role in assessing the energy of concrete mixes. Higher the ratio of embodied energy of SP to embodied energy of binding material, higher is the energy consumption. SP is added to PPC concrete mixes even for lower grade of M30 as shown in Table 22.6, to enhance the concrete workability. These results increase in energy consumption

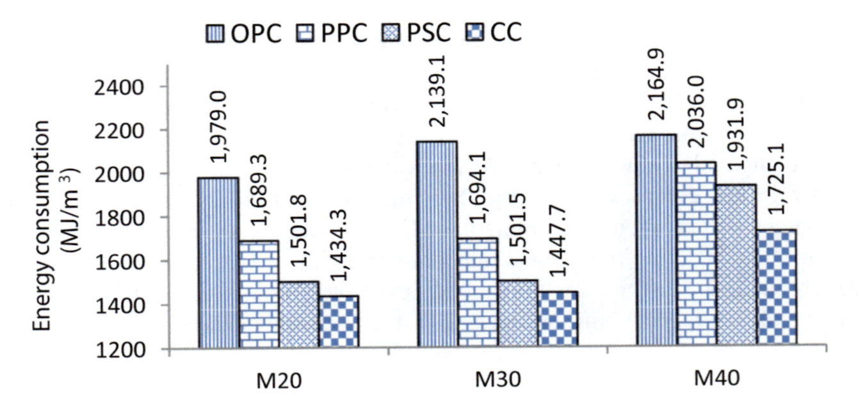

Fig. 22.4 Embodied energy of OPC, PPC, PSC and CC concretes

of PPC concrete mixes. Also, the percentage of fly ash (not to exceed 30%) in PPC is also a parameter that affects the energy consumption of PPC mixes. In PSC, the amount of GBFS is 45% with 55% OPC and with energy ratio of SP to GBFS 6.25, makes PSC a low energy binder compared to OPC and PPC. It is obvious that CC having 50% of OPC is replaced with fly ash and GBFS that makes it a low energy binder compared to OPC, PPC and PSC. It can also be noticed that, with increase in grade of concrete, the requirement of SP also increased in blended cement concretes compared to the reference OPC concrete.

22.4.3 Carbon Footprint Analysis of Concretes

Figure 22.5 shows the carbon dioxide equivalent to emission potential of three grades of concretes (M20, M30 and M40) prepared using OPC, PSC, PPC and CC cements. The total carbon emission is estimated as explained in Eq. 22.2 by adding the embodied carbon of concrete making materials such as cement, fly ash, GBFS, sand, gravel, water and superplasticizer as shown in Table 22.5. From Fig. 22.5, it can be noticed that the carbon emission increased with increase in grade of concrete for all type of concretes. The carbon emission of M20 grade of PPC, PSC and CC concretes is 14.63%, 27.93% and 28.9% less than carbon emission of reference OPC concrete. In preparation of M30 grade concrete, PPC, PSC and CC concretes showed 19.02%, 31.95% and 32.5% less carbon emission compared to OPC concrete. Similarly, for M40 grade concrete, PPC, PSC and CC concretes exhibited 6.02%, 15.06%, 23.05% less carbon emission compared to OPC M40 grade concrete. It can be noticed that the difference in carbon emission of PPC, PSC and CC concretes reduced from M30 to M40 grade with respect to reference OPC concrete. The M20 and M30 mixes of OPC are designed without the addition of SP. The carbon emission factor of cement (830 g/kg) is higher than emission factor of SP (600 g/kg). The

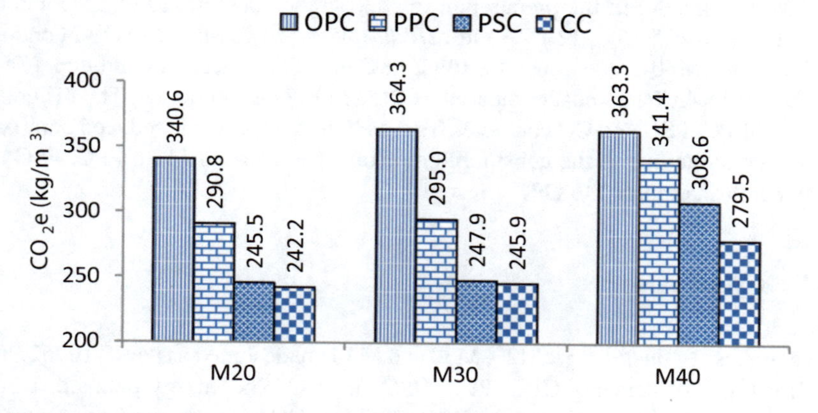

Fig. 22.5 Carbon emission of OPC, PPC, PSC and CC concretes

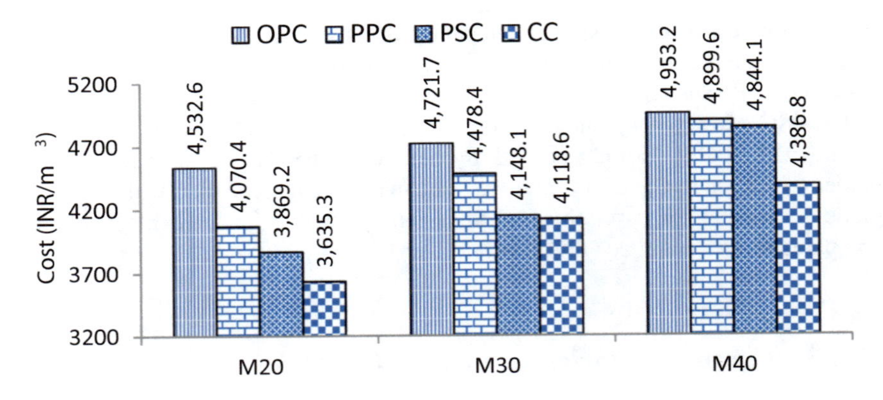

Fig. 22.6 Cost analysis of OPC, PPC, PSC and CC concretes

addition of SP enhances the workability and reduces the water demand by almost 23% thereby reducing cement content significantly to maintain the required water to binder ratio without compromising on strength. The low demand for cement is attributed to reduction in the rate of carbon emission.

22.4.4 Cost Comparison of Different Types of Concrete

Figure 22.6 shows the cost comparison of three grades of concrete prepared using four different cements. The total cost is evaluated by summing up the cost of concrete making materials such as cement, FA, GBFS, sand, gravel, water and superplasticizer as shown in Table 22.5. From Fig. 22.6, it can be observed that the cost increased with increase in grade of concrete for all type of concretes. The cost of M20 grade of PPC, PSC and CC concretes is 10.2%, 14.64% and 19.8% less than the cost of OPC M20 concrete. In the preparation of M30 grade concrete, PPC, PSC and CC concretes showed 5.15%, 12.15% and 12.77% lower cost compared to OPC concrete. Similarly, for M40 grade concrete, PPC, PSC and CC concretes exhibited 1.08%, 2.20% and 11.43% less cost compared to OPC M40 grade concrete. The difference in cost of PPC, PSC and CC concretes from M30 to M40 grade is reduced compared to OPC concrete due to the consumption of more quantity of SP in PPC, PSC and CC concretes compared to OPC concrete.

22.5 Conclusions

Three grades of concrete viz. M20, M30 and M40 are designed as per IS 10262:2019 for four types of cements OPC, PPC, PSC and CC. The energy demand, carbon footprint and cost are estimated for PPC, PSC and CC concretes and are compared with the reference OPC concrete. Some important conclusions drawn from the current study are listed below.

- The composite cement concretes exhibited 24.22%, 15.6% and 8.01% lower compressive strength than the reference OPC concrete at 7, 28 and 90 days, respectively. However, all grades of CC concretes have achieved the minimum required strength at the end of 28 days. Similar scenario was observed with PPC and PSC concretes also following the relation OPC > PSC > PPC > CC.
- The embodied energy of CC concrete is 27.52%, 32.32% and 20.32% lower than OPC concrete for M20, M30 and M40 grades, respectively. The energy consumption for all the grades of PPC and PSC concretes is less than OPC concrete but more than CC concrete. The ratio of embodied energy of SP to embodied energy of binding material plays a vital role in assessing the energy of concrete mixes.
- Carbon emission increased with increase in grade of concrete for all type of cements. The carbon emission of CC concretes is 28.9%, 32.5% and 23.05% less compared to OPC concrete for M20, M30 and M40 grades, respectively. Similar trend is noticed with other blended cements also with emission rate varying as OPC > PSC > PPC > CC.
- The cost of CC concretes is 19.8%, 12.77% and 11.43% lower than the cost of OPC concrete for M20, M30 and M40 grades, respectively.
- Overall, it can be concluded that the composite cement concretes consume less energy, emits low carbon and can be produced at less cost compared to OPC, PPC and PSC concretes without compromising on the required compressive strength. Hence, composite cement has clearly come out as an effective sustainable binding material for construction practice. The use of High Efficiency Low Energy (HELE) system in the manufacturing process of cement also adds to further reduction energy consumption and emission of GHG.

References

Angulo-Ramírez DE, de Gutiérrez RM, Medeiros M (2018) Alkali-activated Portland blast furnace slag cement mortars: performance to alkali-aggregate reaction. Constr Build Mater 179:49–56. https://doi.org/10.1016/j.conbuildmat.2018.05.183

Atiş CD, Bilim C (2007) Wet and dry cured compressive strength of concrete containing ground granulated blast-furnace slag. Build Environ 42(8):3060–3065

Aysha H, Hemalatha T, Arunachalam N, Ramachandra Murthy A, Nagesh RI (2014) Assessment of embodied energy in the production of ultra high performance concrete (UHPC). Int J Students Res Technol Manage 2(03):113–120

Bapat JD (2001) Performance of cement concrete with mineral admixtures. Adv Cem Res 13(4):139–155

Bayraktar OY (2019) The possibility of fly ash and blast furnace slag disposal by using these environmental wastes as substitutes in portland cement. Environ Monit Assess 191(9):560

Bouzoubaa N, Zhang MH, Malhotra VM (2001) Mechanical properties and durability of concrete made with high-volume fly ash blended cements using a coarse fly ash. Cem Concr Res 31(10):1393–1402

BP. Energy outlook energy (2017) BP stat Rev World Energy 2017

Cement Manufacturers' Association (CMA) (2018) Facts about cement industry. http://www.cmaindia.org/industry/facts-about-cement-industry.html

Cheah CB, Tiong LL, Ng EP, Oo CW (2019) The engineering performance of concrete containing high volume of ground granulated blast furnace slag and pulverized fly ash with polycarboxylate-based superplasticizer. Constr Build Mater 202:909–921

CO_GEN NGB (2007) Bureau of energy efficiency. Government of India, Ministry of power. https://beeindia.gov.in/

Collins F (2010) Inclusion of carbonation during the life cycle of built and recycled concrete: influence on their carbon footprint. Int J Life Cycle Assess 15(6):549–556

Confederation of Indian Industry (CII) (2014). Cement Vision 2025: Scaling new heights, India. https://www.cii.in/PublicationDetail.aspx?enc=WXkAtxdUuP1i9IA0aQeKZk1qYIamOjUgx00bWzKmWJ0

Flower DJ, Sanjayan JG (2007) Green house gas emissions due to concrete manufacture. Int J Life Cycle Assess 12(5):282–288

García-Segura T, Yepes V, Alcalá J (2014) Life cycle greenhouse gas emissions of blended cement concrete including carbonation and durability. Int J Life Cycle Assess 19(1):3–12

Gholampour A, Ozbakkaloglu T (2017) Performance of sustainable concretes containing very high volume Class-F fly ash and ground granulated blast furnace slag. J Clean Prod 162:1407–1417

Ghrici M, Kenai S, Said-Mansour M (2007) Mechanical properties and durability of mortar and concrete containing natural pozzolana and limestone blended cements. Cement Concr Compos 29(7):542–549

Güneyisi E, Özturan T, Gesoglu M (2007) Effect of initial curing on chloride ingress and corrosion resistance characteristics of concretes made with plain and blended cements. Build Environ 42(7):2676–2685

Hendriks CA, Worrell E, De Jager D, Blok K, Riemer P (1998) Emission reduction of greenhouse gases from the cement industry. In: Proceedings of the fourth international conference on greenhouse gas control technologies (pp 939–944). Interlaken, Austria, IEA GHG R&D Programme

Hu X, Shi Z, Shi C, Wu Z, Tong B, Ou Z, De Schutter G (2017) Drying shrinkage and cracking resistance of concrete made with ternary cementitious components. Constr Build Mater 149:406–415

IEA. India Energy outlook (2015) World energy outlook spec Rep. https://www.iea.org/publications/freepublications/publication/africa-energy-outlook.html

Imbabi MS, Carrigan C, McKenna S (2012) Trends and developments in green cement and concrete technology. Int J Sustain Built Environ 1(2):194–216

Indian Brand Equity Foundation (IBEF) (2016) Report on cement, India. https://www.ibef.org/download/Cement-January-2016.pdf

IS 12089:1987 Indian standard specification for granulated slag for the manufacture of portland slag cement, Bureau of Indian Standards, New Delhi

IS 1489 (part 1):1991 Portland Pozzolana Cement- Specification, Bureau of Indian Standards, New Delhi

IS 16415:2015 Indian standard specification for composite cement, Bureau of Indian Standards, New Delhi

IS 269:2015 Indian standard specification for ordinary portland cement, Bureau of Indian Standards, New Delhi.

IS 3812 (part 1):2013 Indian standard specification for pulverized fuel ash, Bureau of Indian Standards, New Delhi.

IS 383:2016 Indian standard specification for coarse and fine aggregates for concrete, Bureau of Indian Standards, New Delhi

IS 455:2015 Portland Slag Cement- Specification, Bureau of Indian Standards, New Delhi
IS 516:2004 Indian Standard methods of tests for strength of concrete, Bureau of Indian Standards, New Delhi
IS 9103:1999 Specification for concrete admixtures, Bureau of Indian Standards, New Delhi
IS 10262:2019 Concrete mix proportioning – Guidelines, Bureau of Indian Standards, New Delhi
Karadumpa CS, Pancharathi RK (2021) Influence of particle packing theories on strength and microstructure properties of composite cement-based Mortars. J Mater Civ Eng 33(10):04021267. https://doi.org/10.1061/(ASCE)MT.1943-5533.0003848
Karadumpa CS, Pancharathi RK (2021) Developing a novel mix design methodology for slow hardening composite cement concretes through packing density approach. Constr Build Mater 303:124391.https://doi.org/10.1016/j.conbuildmat.2021.124391
Kayali O, Ahmed MS (2013) Assessment of high volume replacement fly ash concrete–Concept of performance index. Constr Build Mater 39:71–76
Mishra MK, Khare N, Agarwal AB (2015) Scenario analysis of the CO_2 emission reduction potential through clean coal technology in India's power sector: 2014–2050. Elsevier, Energy strategy reviews
Mehta PK, Monteiro PJ (2014) Concrete: microstructure, properties, and materials. McGraw-Hill
Ministry of Mines (2014) Indian minerals year book India. https://ibm.gov.in/?c=pages&m=index&id=481
Ogirigbo OR, Black L (2019) The effect of slag composition and curing duration on the chloride ingress resistance of slag-blended cements. Adv Cem Res 31(5):243–250
Papadakis VG (2000) Effect of supplementary cementing materials on concrete resistance against carbonation and chloride ingress. Cem Concr Res 30(2):291–299
Planning commission, Government of India (2013) Constitution of task force on cement industry for the 12th five year plan
Prakasan S, Palaniappan, S, Gettu R (2020) Study of energy use and CO_2 Emissions in the Manufacturing of Clinker and Cement. J Inst Eng (India): Series A 101(1):221–232
Reddy BV, Jagadish KS (2003) Embodied energy of common and alternative building materials and technologies. Energy Build 35(2):129–137
Rehan R, Nehdi M (2005) Carbon dioxide emissions and climate change: policy implications for the cement industry. Environ Sci Policy 8(2):105–114
Krishnan SS, Murali Ramakrishnan A, Venkatesh V, Shyam Sunder P, Ramakrishna G (2012) Plant specific energy efficiency modeling and analysis of the Indian cement industry for robust policy implementation. Centre for study of science, technology and policy (CSTEP), Bangalore, India
Midha S, Kumar K, Mathew J, Maurya RK, Kaur A, Panwar RK, Sokhi K (2015) Statistics related to climate change-India. Central Statistics office, Ministry of statistics and programme implementation, Government of India
Sobolev K (2003) Sustainable development of the cement industry and blended cements to meet ecological challenges. TheScientificWorldJOURNAL 3
Trudeau N, Tam C, Graczyk D, Taylor P (2011) Energy transition for industry: India and the global context. Int Energy Agency (IEA). France, Paris
URL: https://niti.gov.in/planningcommission.gov.in/docs/aboutus/committee/index.php?about=11strindx.htm
Worrell E, Price L, Martin N, Hendriks C, Meida LO (2001) Carbon dioxide emissions from the global cement industry. Annu Rev Energy Env 26(1):303–329
Zhang T, Gao P, Gao P, Wei J, Yu Q (2013) Effectiveness of novel and traditional methods to incorporate industrial wastes in cementitious materials—an overview. Resour Conserv Recycl 74:134–143
Zhang T, Yu Q, Wei J, Li J, Zhang P (2011) Preparation of high performance blended cements and reclamation of iron concentrate from basic oxygen furnace steel slag. Resour Conserv Recycl 56(1):48–55

Chapter 23
Leaching Methods for the Environmental Assessment of Industrial Waste Before Its Use in Construction

Mercedes Regadío, Julia Rosales, Manuel Cabrera, Steven F. Thornton, and Francisco Agrela

23.1 Introduction

In recent decades the use of waste and by-products alone or substituting part of raw material in construction, energy production, agriculture, pollution control barriers, landfill covers, soil remediation, water treatment or landscape reconstructions has gained increased attention (Naik and Kraus 2003; Monte et al. 2009; Azad and Samarakoon 2021). The recovering of (non-hazardous) solid waste as alternative substitute materials for raw materials is often referred to as **beneficial use** or **valorization**. It is important to find safe possibilities for the valorization of waste and by-products because it would: (1) decrease the risk of environmental pollution from stockpiling of wastes, (2) avoid the use of specialized landfill sites, (3) provide a low cost readily available material and (4) conserve natural resources and raw materials, which would

Notes: (1) chemical elements are written in abbreviated form, (2) standard methods are cited in footnotes.

M. Regadío (✉) · S. F. Thornton
The University of Sheffield, Sheffield, UK
e-mail: mregadio@mregadio.com

S. F. Thornton
e-mail: s.f.thornton@sheffield.ac.uk

J. Rosales · M. Cabrera · F. Agrela
Universidad de Córdoba, Córdoba, Spain
e-mail: jrosales@uco.es

M. Cabrera
e-mail: manuel.cabrera@uco.es

F. Agrela
e-mail: ir1agsaf@uco.es

K. R. Reddy et al. (eds.), *Advances in Sustainable Materials and Resilient Infrastructure*, Springer Transactions in Civil and Environmental Engineering,
https://doi.org/10.1007/978-981-16-9744-9_23

otherwise be consumed. The most important types of waste and by-products (hereafter referred to as waste) currently available as promising alternatives for use in construction can be found in Table 23.1. In the last decades, their reuse and application are increasing over time, while their landfilling is decreasing.

The technical viability and feasibility of using different processed waste in new applications must be accompanied by acceptable low environmental impacts (Coelho and de Brito 2012; Jamshidi and White 2019). The environmental and public health repercussions of using such wastes are assessed with different standardized methodologies, mainly leaching tests (van der Sloot et al. 1991). After these tests, if the waste has been shown not to produce negative effects, they are classified as either inert or non-hazardous waste and can finally be used, for example, in construction. Otherwise, efforts could be made to reduce the pollutant load of the waste to levels suitable for the new application by converting it to inert and less hazardous waste (e.g., alkaline-activated geopolymers in Kiventerä et al. 2018a). Hazardous waste can be **stabilized** with cement or pozzolans, leading to its solidification within a cementitious matrix and converting the heavy metals into less soluble salts, which do not leach at appreciable rate (Chang et al. 1999; Cabrera et al. 2016). In some cases, if the waste is used in the production of cement, pollutants can be eliminated due to the high temperatures used in cement manufacture (Ferreira et al. 2003; Lederer et al. 2017). The incineration of sewage sludge also reduces pollutants (and their potential accumulation) through the disinfection and detoxification that occurs at those temperatures (Chakraborty et al. 2017).

Most studies have analyzed the use of waste in civil infrastructure and building works from the physical, chemical and durability standpoints (references within Table 23.1). However, the environmental impact caused by the use of this waste in the long-term, and assessed through leachate analysis, is neglected. Such analysis is, nevertheless, essential because heavy metals and other elements may be released into the soil by infiltrating rainwater at higher levels than permitted. This can result in potential contamination of soil and (ground)water resources and damage to ecosystems. In such cases, the waste is classified as hazardous waste and is not suitable for use, e.g., as isolated aggregate in civil engineering. Since the inception of environmental policies, the ecological impact generated by the use of waste is as or more important than finding possibilities for its valorization. The present chapter discusses the different leaching methods used for the environmental assessment of wastes in Europe and the United States.

23.2 Standard Leaching Tests for Environmental Impact Assessment

Historically different leaching tests have been used to evaluate the release of chemicals from waste materials and identify if they meet relevant safety requirements for disposal or, more recently, for a new application. However, there is little awareness

Table 23.1 Waste and by-products readily available in large quantities for potential use, especially as construction material

Waste / by-product generated	Potential use
Coal ash from combustion in thermal power plants*	In mortars directly (Wyrzykowski et al. 2016), in cementitious materials after a pre-grinding treatment (Carmona et al. 2010), in concrete and concrete blocks (Esquinas et al. 2018), in glass and ceramic products, in soil remediation and agriculture (Jayaranjan et al. 2014)
Biomass ash from combustion in thermal power plants*	In agriculture as fertilizer (Ondrasek et al. 2021), in road base or sub-base as both a granular material treated with cement and a stabilizing agent (Cabrera et al. 2016), in mortars as a cement substitute (Maschio et al. 2011; Modolo et al. 2013) and in lightweight concrete (Rosales et al. 2016)
Steel slag from combustion in blast furnaces: basic oxygen furnace (BOF) slag, electric arc furnace (EAF) slag and ladle furnace refining slag	In high-tracked road layers as aggregate, in hydraulic constructions and in agriculture as fertilizer (Gao et al. 2020; Liu et al. 2020). In asphaltic surface layers that resisted deformation, cutting and polishing for > 25 years (Motz and Geiseler 2001), in metal recovery (Shen et al. 2004)
Incineration ash from municipal solid waste (households and small industrial plants)*	Manufacture of cement (Tang et al. 2015; Lederer et al. 2017) with similar properties to Ordinary Portland Cement after substituting 40–60% of the clinker (Berg and Neal 1998; Guo et al. 2014). Manufacture of concrete, for all three: as a direct substitute for cement (Aubert et al. 2004; Jurič et al. 2006), as reinforcement fibres (Malagavelli and Patura 2011) and as a replacement for fine aggregate (Rashid and Frantz 1992; Chen et al. 2008). In highways or roads (Pandeline et al. 1997; Forteza et al. 2004; respectively), in backfill materials (Lin et al. 2012) and in soils as a stabilizer material with high shear strength and low compressibility replacing lime or cement (Ferreira et al. 2003)
Incineration ash from sewage sludge (sewage systems from households and small industrial)*	Manufacture of cement (Pan et al. 2003; Donatello and Cheeseman 2013), of concrete mixes (Tay 1987; Tay et al. 1991), of bricks and tiles (Alleman and Berman 1984; Anderson 2002), of asphalt paving mixes (Al Sayed et al. 1995) and of mortars and concrete in 15–30% cement substitutions (Monzó et al. 2004)

(continued)

Table 23.1 (continued)

Waste / by-product generated	Potential use
Construction and demolition waste (ceramic wastes, concrete waste, asphalt material wastes, …)	Although they may contain hazardous products such as Cr, Sb or sulphates, most recycled aggregates are suitable for use as construction material from a physico-chemical and environmental viewpoint if there is selective collection at source (Agrela et al. 2021). Manufacture of landfill barriers substituting 5–20% of the clay (Regadío et al. 2020)
Pulp/paper sludge and ash from pulp and paper production	In agriculture, combustion with energy recovery, and brick and ceramic construction (Camberato et al. 2006; Sutcu and Akkurt 2009; Martínez et al. 2012). Sludge was specifically applied in: cement and concrete as a direct substitute for fine aggregates (Srinivasan et al. 2010; Frías et al. 2015), ceramic bricks (Muñoz et al. 2016), clinker substituting the limestone (Castro et al. 2009) and fibres in fibre-cement and concrete pavements of asphalt concrete mixes (Modolo et al. 2007, 2011). Ash was specifically applied in agriculture as soil amendment and carbon sequestrator (Ram and Masto 2014), and in the restoration of acidic soils with heavy metals (Liu et al. 2017; Alvarenga et al. 2019)
Waste from mining and mineral processing (mine tailings, rocks, soils, oil sands or loose sediments from various waste minerals and natural rocks)	In construction industry (Almeida et al. 2020), as filler for embankments (Kuranchie et al. 2013), as aggregates in roads, dams, buildings, or embankments (Chesner et al. 2002), in ceramic materials (Lemeshev et al. 2004), in the manufacture of cement and concrete (Khudyakova et al. 2020) after treating sulphidic minerals to prevent sulphate attack from worsening the final mechanical properties or to prevent acidic contamination (Kiventerä et al. 2018a, 2018b; Capasso et al. 2019)

*Normally the fly ash is more reusable and less environmentally problematic in construction than the bottom ash, due to more suitable chemical properties and very low amount of organic matter

and understanding of the various leachability tests available, which is essential to make informed decisions on waste practices and pathways for the reuse of wastes.

Leaching tests were initially developed for waste classification and applied for beneficial use studies. The tests were devised to provide a **waste classification** needed to identify the appropriate waste management scenario or waste disposal environment. By studying the **toxicity** of an extract which is representative of the actual leachate produced from a waste in the field, the tests assigned a hazard level to the waste, to inform its final disposal option. Later, the tests were also employed to classify the toxicity level of the waste, but in this case using the classification to decide whether the waste is safe to use in a given application, rather than its disposal option. By analyzing the water composition in contact with the waste, the tests assign it a hazard level which, if low enough, enables the waste to be reused under safety and security standards. The standardized leaching methods of the US EPA and European Commission create a uniform framework on assessment methodology for waste characterization and compliance (footnotes 1–16). Authorities may apply to them minor modifications as they deem appropriate. Further guidance on available leaching tests can be found in BS EN ISO 18772:2014,[1] which also discusses the difference between analysis for characterization/toxicity purposes and quality control/compliance purposes. These leaching procedures may also be used to assess the impact of a contaminated site on groundwater, or the safety of the site for a new activity or use.

Leaching tests examine which chemicals, especially potential contaminants, are transferred from the solid waste to the aqueous phase or form soluble compounds when both phases interact with each other. At equilibrium the contaminants are partitioned between the waste and aqueous phase, the liquid being the mobile phase transporting the soluble pollutants. A higher partition coefficient for a chemical means a lower concentration in the liquid, and therefore higher sorption or retention of the chemical by the waste. The leaching processes are dominated by physical mechanisms, such as surface washing, and chemical mechanisms, such as diffusion and dissolution of minerals controlled by the mineral aqueous solubility and solution pH.

Three types of leaching tests are commonly used with and without pH variation, to determine if the waste complies with safety criteria for different potential applications or disposal methods: batch leaching tests, column percolation tests and monolithic tank tests. In addition, *field-scale leaching tests* can also be conducted by constructing and running waste cells that simulate waste leaching processes under real field conditions. Although these tests provide valuable information on the field-scale leaching behaviour of the waste, they are time-consuming, expensive and labour-intensive. Therefore, very little field leaching data on waste compliance and characterization is available. Hence, field studies were not included in the following review of different leaching tests. Upon completion of the leaching test, the analyses on the components released from the waste to the aqueous phase will inform the disposal and reuse of waste materials according to the environmental, health and safety requirements. This is undertaken by a risk assessor using the parameters provided by the tests (estimation of emission concentrations, exposure, release factors, etc.) to evaluate the risk potential from the exposure caused by a different use of the waste.

[1] BS EN ISO 18772:**2014**, Soil quality. Guidance on leaching procedures for subsequent chemical and ecotoxicological testing of soils and soil materials. pp. 1–40.

23.2.1 Batch Leaching Tests (with/without pH Test)

These tests are the preferred choice at laboratory-scale for regulatory assessment due to their simple implementation, good reproducibility, short time requirements (hours to days), wide availability and low cost. In addition, the results can be applied to different applications depending on the type of batch test, and the pH is easy to control and vary if desired. However, they are less representative because the procedure does not simulate actual site-specific leaching conditions, such as rain percolation. Also the materials are not tested under the potential engineering design or configuration and application. They simply estimate the maximum amount of contaminants that can be released, which may exceed the actual amount that would be released under field conditions.

Materials in a granular state, such as soil (land use change), waste (classification) or sand and aggregates (to be added to cement), are commonly characterized with these tests, but not cements or similar components. The material can be limited to a maximum particle size of 4 or 10 mm and is tumbled in a bottle containing water in a specific liquid-to-solid ratio depending on the standard (usually 2 or 10 L/kg). One (single step) or different liquid-to-solid (cumulative) steps can be carried out to determine the leaching potential of the materials. After mixing the samples with the extraction solution, the suspension is filtered and the chemistry is analysed. The release of chemicals is expressed as the mass of each one per unit (dry) mass of the material in mg/kg dried mass (Europe), or per unit volume of the leachate in mg/L (USA). Based on these concentrations, the material is classified for its disposal or potential use. In Europe, the corresponding limit values for inert and non-hazardous materials are shown in Table 23.2 (mg/kg). When these leaching levels are exceeded, the materials are classified as hazardous waste, i.e., they should be disposed in special facilities instead of in municipal landfills and are not allowed to be reused.

In the USA both the EPA SW-846 Test Method **1311**:1992 TCLP[2] (**landfill leachate application**) and the EPA SW-846 Test Method **1312**:1994 SPLP[3] (**beneficial use application**) are among the most used laboratory leaching tests. The first (known as Toxicity Characteristic Leaching Procedure or TCLP) was developed to simulate contaminant leaching resulting from waste in a municipal solid waste landfill environment. The extraction solution to be used is previously determined based on the alkalinity of the waste. The method determines if waste is characteristically hazardous (classified as "D" by the US EPA) and therefore poses unacceptable environmental hazards if improperly disposed. This occurs when the component concentration in the leachate is above the corresponding limit values (Table 23.3). The second (known as Synthetic Precipitation Leaching Procedure or SPLP) was developed to simulate leaching of rainfall through soil or materials from a contaminated site that is intended for beneficial use and that may affect groundwater or

[2] EPA SW-846 Test Method 1311:**1992**, Toxicity Characteristic Leaching Procedure (TCLP). pp. 1–18.

[3] EPA SW-846 Test Method 1312:**1994**, Synthetic Precipitation Leaching Procedure (SPLP). pp. 1–30.

Table 23.2 Maximum metal concentrations as leaching limit values (mg/kg dried mass) measured in the extraction solution after 24 h of batch leaching of a representative sample of waste at a liquid-to-solid ratio of 10 L/kg dry (Limits by: Council Decision 2003/33/EC,[4] Leaching tests by: CEN—EN **12,457**:2002 Part 4[5])

	Inert waste	Non-hazardous waste[a]
Cr	<0.5	<10
Ni	<0.4	<10
Cu	<2	<50
Zn	<4	<50
As	<0.5	<2
Se	<0.1	<0.5
Mo	<0.5	<10
Cd	<0.04	<1
Sb	<0.06	<0.7
Ba	<20	<100
Hg	<0.01	<0.2
Pb	<0.5	<10
SO_4^{2-}	<6,000	<20,000

[a]When the leaching limit values of non-hazardous waste are exceeded, the waste is classified as hazardous and is not accepted in municipal landfills

Table 23.3 Maximum metal concentrations as leaching limit values (mg/L) measured in the extraction solution after 16–18 h of batch leaching of a representative sample of waste at a liquid-to-solid ratio of 19–21 L/kg wet (Limits by: 40 CFR 261.24,[6] Leaching tests by: EPA SW-846 Test Method **1311**:1992[4])

Metals (classification of EPA hazardous waste)	Regulatory level of TCLP Max mg/L
Arsenic (D004)	<5
Barium (D005)	<100
Cadmium (D006)	<1
Chromium (D007)	<5
Lead (D008)	<5
Mercury (D009)	<0.2
Selenium (D010)	<1
Silver (D011)	<5

other receptors. The extraction solution is a diluted sulphuric / nitric acid mixture of pH 4.2 (60/40% weight). The method determines whether soil or material can be revalorized in new beneficial uses exposed to weathering, without endangering the

[4] Council Decision **2003**/33/EC of 19 December 2002 establishing criteria and procedures for the acceptance of waste at landfills pursuant to Article 16 of and Annex II to Directive 1999/31/EC (OJL 11, 16.1.2003, pp. 27–49) last update 22.07.2020.

[5] CEN—EN 12457:**2002** (Part 1 to 4), Characterisation of Waste—Leaching—Compliance Test for Leaching of Granular Waste Materials and Sludges with high solid content. pp. 1–28 (1), 1–29 (2), 1–34 (3), 1–30 (4).

[6] CFR (Code of Federal Regulations), Title 40, Volume 28, Chapter I, Subchapter I, Part 261, Subpart C, Sect. 261.24.

environment. Nevertheless, in the case of contamination by oily or similar petroleum waste, TCLP is the preferred method. To improve the limitations of both methods (Kosson et al. 2002; US EPA 2004 Appendix F), the US EPA recently developed the **Leaching Environmental Assessment Framework (LEAF, 2017)**. This is a new leaching evaluation system that describes four methods to assess the release of inorganic constituents of potential concern for a wide range of solid material (two batch tests, one column test and one monolithic test). The batch tests are parallel batch extractions rotated between 24 and 72 h (from smaller to larger particles size). The first (known as EPA SW-846 LEAF Method **1316**:2017[7]) is a screening test at five different liquid-to-solid ratios (0.5–10 L/kg dry, i.e., 20–400 g of dry material with 150–200 mL of water). The second (known as EPA SW-846 LEAF Method **1313**:2017[8]) is the same as the previous test but at a single liquid-to-solid ratio (10 L/kg dry), and with acid/base addition to achieve nine final pH values (from 2 to 10.5) in the extracts. This makes it a more time-consuming method. There are other standards[9,10,11] used for batch leaching tests in the USA and Europe.

23.2.2 Column Percolation Tests

These leaching tests are time-consuming (days to weeks), relatively expensive, involve a more sophisticated implementation (e.g., use of pump, tubing, monitoring, accumulated parameters, data logging, etc.) and can be problematic (e.g., channeling due to non-uniform packing or clogging of the packed column). They are not applicable to materials that biodegrade biologically or generate excessive gas, heat or solidification when reacting with the liquid, which would make the final hydraulic conductivity outside the specified range. The pH is difficult to control and it is dictated by the material itself. In addition, these tests may not be reproducible due to flow channeling, clogging and biological activity. However, they can **quantify contaminant retention in the matrix** and provide data at very low liquid-to-solid ratios that are close to field conditions (0.1–0.5 L/kg) with low dilution of chemical concentrations (thereby improving analytical sensitivity). The leaching behaviour data provided by these tests are more realistic and can be applied to more specific scenarios for longer time periods, because they simulate the **release progress** (and potential retardation) of a contaminant in the long-term (van der Sloot et al. 1996).

[7] EPA SW-846 LEAF Method 1316:**2017**, Liquid–Solid Partitioning as a Function of Liquid–Solid Ratio Using a Parallel Batch Extraction Procedure. pp. 1–20.

[8] EPA SW-846 LEAF Method 1313:**2017**, Liquid–Solid Partitioning as a Function of Extract pH Using a Parallel Batch Extraction Procedure. pp. 1–30.

[9] EPA SW-846 Test Method 1310B:**2004**, Extraction procedure (EP) toxicity test method and structural integrity test. pp. 1- 18.

[10] CEN—EN 14,429:**2015** (replaces de one of 2005), Characterization of waste—Leaching behaviour test—Influence of pH on leaching with initial acid/base addition. pp. 1–30.

[11] ASTM D3987—12(**2020**), Standard Practice for Shake Extraction of Solid Waste with Water, ASTM International, West Conshohocken, PA, 2020, www.astm.org, DOI: 10.1520/D3987-12R20.

Table 23.4 Maximum metal concentrations as leaching limit values (mg/L) measured in the first eluate of the extraction solution percolating through a representative sample of waste column at a liquid-to-solid ratio of 0.1 L/kg (Limits by: Council Decision 2003/33/EC[2], Leaching tests by: CEN–EN 14405:2017[12])

	Inert waste	Non-hazardous waste[a]
Cr	<0.1	<2.5
Ni	<0.12	<3
Cu	<0.6	<30
Zn	<1.2	<15
As	<0.06	<0.3
Se	<0.04	<0.2
Mo	<0.2	<3.5
Cd	<0.02	<0.3
Sb	<0.1	<0.15
Ba	<4	<20
Hg	<0.002	<0.03
Pb	<0.15	<3
SO_4^{2-}	<1,500	<8,500

[a]When the leaching limit values of non-hazardous waste are exceeded, the waste is classified as hazardous and is not accepted in municipal landfills

Granular materials which, once confined in a columnar vessel, allow liquid to percolate through them, can be characterized in column tests (e.g., waste or soils looser, made up of smaller particles like sand or pebbles). Both CEN–EN **14405**:2017[12] (replacing the one of 2004) and EPA SW-846 LEAF Method **1314**:2017[13] are the European and USA guidelines, respectively, that establish the specifications for percolation tests. For method consistency, the column radius is designed in proportion to the particle size of the material, approximately 5–20 times the nominal maximum particle diameter. An upward and continuous flow of water is slowly pumped through the column, at a rate of a few centimetres a day, thus simulating groundwater or rainwater gradually circulating through the material, as it would do, for example, through a road sub-base. Upward flow is used to minimize air entrainment and preferential fluid flow. Chemicals are released (or "extracted") from the material into the fluid as it moves through the column. This process can require 20–30 days before the leachate samples are analysed for contaminants as a function of time (i.e., at different **accumulated** liquid-to-solid ratios). According to the chemical concentrations, the material is classified for its disposal or potential use. In Europe, the corresponding limit values (mg/L) for inert and non-hazardous materials are shown in Table 23.4. When these leaching values are exceeded, the materials are classified as hazardous waste, i.e., they should be disposed in special

[12] CEN–EN 14405:**2017** (replaces the one of 2004), Characterization of Waste—Leaching behaviour tests—Up-flow percolation test (under specified conditions). pp. 1–46.

[13] EPA SW-846 LEAF Method 1314:**2017**, Liquid–Solid Partitioning as a Function of Liquid–Solid Ratio for Constituents in Solid Materials Using an Up-Flow Percolation Column Procedure. pp. 1–30.

facilities instead of in municipal landfills and are not allowed to be reused. There are standards for column percolation tests in Europe[12], the USA[13] and OECD.[14] In these tests, the release of chemicals is expressed in terms of mass per volume of the leachate (mg/L), both in Europe and USA.

23.2.3 Monolithic Tank Tests

These relatively new styles of tests are inexpensive, simply implemented and applicable for the assessment of materials under conditions which more closely resemble a field-scale context. They are the most important to apply for concrete and cement-stabilized wastes to determine the potential for contamination by leaching that could occur in a water medium, and the success of solidification in trapping contaminants within the waste over long timescales (Cabrera et al. 2019). The tank tests provide long-term assessment of the leaching of inorganic components by the enclosed water. The leaching occurs mainly through diffusion, by simulating in real situations the release under aerobic and defined conditions as a function of time. They are more realistic and sometimes provide lower leaching concentrations (higher data resolution) compared with batch and column tests. This helps to avoid the misinterpretation of false negatives. However, tank tests are very long tests (typically months) and the pH value at which the leaching occurs is determined by the material.

Tank tests are applied to **water-saturated** monolithic forms (e.g., cements, solidified wastes and concrete mortar) or compacted granular (<2 mm) materials with monolithic behaviour (e.g., soils, sediments and stacked granular wastes), rather than to unbound granular materials. The monoliths can be obtained by appropriate manufacturing, curing and drying. The compacted granular materials can be obtained by modified Proctor methods at optimum moisture content in cylindrical moulds. Both CEN–EN **15863**:2015[15] and EPA SW-846 LEAF Method **1315**:2017[16] are the European and USA guidelines, respectively, that establish the specifications for tank tests. The saturated hydraulic conductivity of the solid sample must be $\leq 10^{-8}$ m/s to avoid water circulating through the monolith rather than around it. The solid sample is fully submerged in a vessel containing water with an adjusted pH, where it is left without being agitated or tumbled. The predominant water flow is around the material and the release of constituents is controlled by diffusion to the boundary. The water is sampled, drained and replaced at pre-set times at increasing regular intervals ($\geq$9), with the entire test lasting between 4 and 77 days. The eluate pH, electrical conductivity and the mass of eluent absorbed into the solid matrix are measured for

---

[14] OECD TG 312:**2004**, Leaching in Soil Columns, OECD Guidelines for the Testing of Chemicals. pp. 1–15, DOI: 10.1787/9789264070561-en.

[15] CEN—EN 15863:**2015**, Characterization of waste—Leaching behaviour test for basic characterization—Dynamic monolithic leaching test with periodic leachant renewal, under fixed conditions. pp. 1–66.

[16] EPA SW-846 LEAF Method 1315:**2017**, Mass Transfer Rates of Constituents in Monolithic or Compacted Granular Materials Using a Semi-Dynamic Tank Leaching Procedure. pp. 1–37.

Table 23.5 Maximum concentrations as limits values (mg/L) for drinking water consumption established by the United States Environmental Protection Agency (US EPA), World Health Organization (WHO) and European Commission (EC)

	US EPA	WHO	EC
Zn	5.0	3.0	–
Cr	0.1	0.05	0.05
Pb	0.015	0.01	0.01
As	0.05	0.01	0.01
Cd	0.005	0.003	0.005
Ni	–	0.02	0.02
Ba	2.0	0.3	–
Sr	4	–	–

each leaching time-interval. The extraction of the components is by (one- to three-dimensional) mass transfer from a regular, geometric, external surface area of the monoliths (the sample size is $\geq$ 4–5 cm in all directions), to the eluent (the volume of the holding vessel is $\approx$ 2–5 times the volume of the sample). The aqueous extracts are analyzed to determine the concentrations of soluble constituents at different times (i.e., different liquid-to-solid ratios). Thus, the liquid-to-solid ratio is calculated in terms of the "volume of liquid" to the "exposed **surface area** of solid", e.g., 9–12 mL/cm^2 (the space between any exposed surface and the vessel wall is $\geq$ 2 cm). According to the chemical concentrations, the test quantifies whether the treatment (compaction, cementation, blending, …) is effective in decreasing the release of pollutants to desirable concentrations, at the corresponding spatial–temporal scales.

There are monolithic tank standards used in Europe[15] and the USA[16]. There are no clearly established limits for leaching from monolithic specimens in tank tests. Therefore, the limits established for drinking water consumption can be used to ensure the protection of public health (Table 23.5).

23.2.4 PH Leaching Tests

The pH can vary widely between samples (e.g., pH 4 for acid rain, pH 9–10 for aggregates, pH 12.5–14 for concrete matrices) and affects the behaviour of leachates. To study this, leaching tests can be performed using extraction aqueous solutions at different pH values (from pH 2 to 13) to attain specified endpoint pH values (from pH 2 to 10.5). These methods (known as *pH dependence tests*) determine the influence of pH in the extraction of dissolved components[7,9,10]. The shape of the **concentration-pH** curve indicates the speciation of the pollutant in the solid phase (Kosson et al. 2002). **Cationic species** (e.g., Cd) usually have a maximum concentration in the acidic pH range that decreases to lower values at alkaline pH, while **oxyanionic species** (e.g., $[AsO_4]^-$, $[MnO_4]^-$) usually have a maximum concentration in the neutral to slightly alkaline range. **Amphoteric species** (e.g., Pb, Cr(III) and Cu) often show a maximum concentration in the acidic pH greater than that of cationic

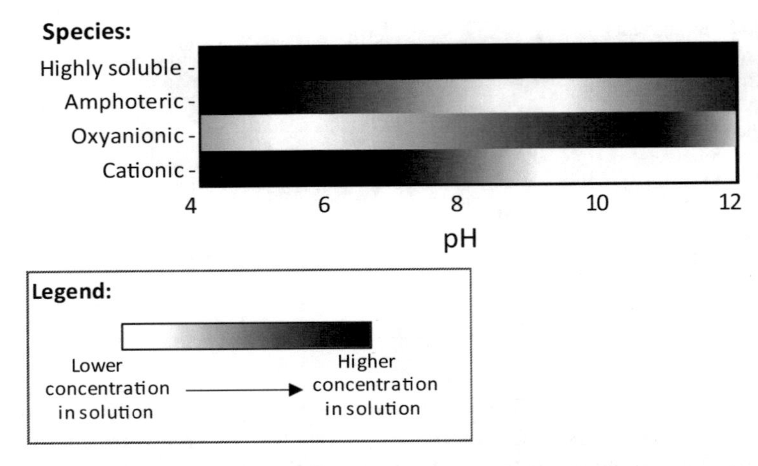

Fig. 23.1 Scheme of eluate concentration of cationic, oxyanionic, amphoteric and highly soluble species as a function of pH (liquid-to-solid ratio: 10 L/kg dry)

species, and a minimum concentration in the near-neutral to slightly acid pH range, increasing again for alkaline pH values typically due to the solubility of hydroxide complexes (e.g., $[Pb(OH_3)]^-$). In the case of **highly soluble species** (e.g., Na^+, K^+ and Cl^-), the solution pH has little effect on their aqueous concentration (Fig. 23.1).

These concentrations may be used as inputs in geochemical speciation models to infer the mineral phases, adsorption reactions and soluble complexes that control the release of contaminants (van der Sloot et al. 2008). This is important as the release of some contaminants can increase significantly with a decrease or increase in pH, which could ultimately lead to concerns regarding the dissolved loads of elements (e.g., the leaching of Pb under high pH values is enhanced and could reach hazardous levels, even in cases where this metal is present in the matrix in low amounts).

23.3 Forward Look

Testing techniques as well as environmental, health and safety considerations have constantly changed over the years and continue to develop, as does the associated legislation. While parameters for assessing pollution and thresholds have been applied across the board, the current trend is to consider and investigate each case individually. In the past, leaching tests were not designed to simulate real field conditions, but to study the contact between a liquid and a solid phase for different aims. If soil or waste was classified as hazardous by batch tests, there was little choice but to abandon the possibility of using it, even for overstated results that would only entail a minimal risk of contamination that is fully acceptable. Such an approach is unsustainable, environmentally unfriendly and uneconomic (disposal instead of utilization, minor land reclamation, expensive measures and taxation to manage the waste).

This has led to novel ways of restricting leaching to acceptable levels over long periods of time, as assessed using methods such as column and monolithic tank tests, which help to simulate conditions more representative of actual waste management, in order to predict long-term leaching behaviour. They provide a better understanding of actual liquids infiltration through roads or similar structures and a sealing of contaminants to limit leaching through encapsulation. Ultimately, the narrow focus of "*pass the test*" should be avoided, waste characteristics should be improved and leaching should be reduced under actual use or disposal scenarios. With a view to the circular economy, the design of the new standards should focus on the content of valuable resources and not only on the contaminants mass (Regadío and Binnemans 2018; Johansson and Krook 2021). Generally, the application of a leaching test alone is not sufficient and a combination of different test methods is advised to determine the detailed leaching behaviour. This is because some elements may be leached only in one condition, for example, in the first time-interval of the column tests, or only in the last flush of monolithic tests, or only at high liquid-to-solid ratios with longer periods of time, or only at a specific pH value (Palden et al. 2019).

Leaching tests should next be supported by modelling and model validation (CEN—EN 12920:2006 + A1:2008[17]). The leaching test results are used in fate and transport modelling to evaluate the exposure in risk assessment. Initial concentration (C_0, in mg/L), infiltration rate (q, in mm/h) and subsurface hydrogeological conditions, such as vadose zone depth (m), aquifer thickness (m), soil partitioning coefficients (Kd, in L/Kg) or dilution are typical model inputs. The assumptions used in pollutant fate and transport models are more important than the type of leaching test chosen. Using different model inputs may totally change the final outcome of the risk assessment, while using different leaching tests may show little impact (Townsend et al. 2016). In terms of field testing and validation, much remains to be done to develop a better understanding and promote greater awareness among risk assessors.

References

Agrela F, Díaz-López JL, Rosales J, Cuenca-Moyano GM, Cano H, Cabrera M (2021) Environmental assessment, mechanical behavior and new leaching impact proposal of mixed recycled aggregates to be used in road construction. J Clean Prod 280 124362.https://doi.org/10.1016/j.jclepro.2020.124362

Al Sayed MH, Madany IM, Buali ARM (1995) Use of sewage sludge ash in asphaltic paving mixes in hot regions. Constr Build Mater 9:19–23. https://doi.org/10.1016/0950-0618(95)92856-C

Alleman JE, Berman NA (1984) Constructive sludge management: biobrick. J Environ Eng 110:301–311. https://doi.org/10.1061/(ASCE)0733-9372(1984)110:2(301)

[17] CEN—EN 12920:**2006 +** A1:**2008**, Characterization of waste—Methodology for the determination of the leaching behaviour of waste under specified conditions. pp. 1–12.

Almeida J, Ribeiro AB, Silva AS, Faria P (2020) Overview of mining residues incorporation in construction materials and barriers for full-scale application. J Build Eng 29:101215.https://doi.org/10.1016/j.jobe.2020.101215

Alvarenga P, Rodrigues D, Mourinha C, Palma P, de Varennes A, Cruz N, Tarelho LAC, Rodrigues S (2019) Use of wastes from the pulp and paper industry for the remediation of soils degraded by mining activities: chemical, biochemical and ecotoxicological effects. Sci Total Environ 686:1152–1163. https://doi.org/10.1016/j.scitotenv.2019.06.038

Anderson M (2002) Encouraging prospects for recycling incinerated sewage sludge ash (ISSA) into clay-based building products. J Chem Technol Biotechnol 77:352–360. https://doi.org/10.1002/jctb.586

Aubert JE, Husson B, Vaquier A (2004) Use of municipal solid waste incineration fly ash in concrete. Cem Concr Res 34:957–963. https://doi.org/10.1016/j.cemconres.2003.11.002

Azad NM, Samarakoon SMSMK (2021) Utilization of industrial By-Products/Waste to Manufacture Geopolymer Cement/Concrete. Sustainability 13:873. https://doi.org/10.3390/su13020873

Berg ER, Neal JA (1998) Municipal solid waste bottom ash as portland cement concrete ingredient. J Mater Civ Eng 10:168–173. https://doi.org/10.1061/(ASCE)0899-1561(1998)10:3(168)

Cabrera M, Agrela F, Ayuso J, Galvín AP, Rosales J (2016) Feasible use of biomass bottom ash in the manufacture of cement treated recycled materials. Mater Struct 49:3227–3238. https://doi.org/10.1617/s11527-015-0715-2

Cabrera M, Galvín AP, Agrela F (2019) Leaching issues in recycled aggregate concrete. In: New trends in eco-efficient and recycled concrete. Elsevier, pp 329–356. https://doi.org/10.1016/B978-0-08-102480-5.00012-9

Cabrera M, Galvín AP, Agrela F, Beltrán MG, Ayuso J (2016) Reduction of leaching impacts by applying biomass bottom ash and recycled mixed aggregates in structural layers of roads. Materials 9:228. https://doi.org/10.3390/ma9040228

Camberato JJ, Gagnon B, Angers DA, Chantigny MH, Pan WL (2006) Pulp and paper mill by-products as soil amendments and plant nutrient sources. Can J Soil Sci 86:641–653. https://doi.org/10.4141/S05-120

Capasso I, Lirer S, Flora A, Ferone C, Cioffi R, Caputo D, Liguori B (2019) Reuse of mining waste as aggregates in fly ash-based geopolymers. J Clean Prod 220:65–73. https://doi.org/10.1016/j.jclepro.2019.02.164

Carmona MO, González Paules J, Sánchez Catalán JC, Fernández Pousa L, Ade Beltrán R, Quero Sanz F (2010) Reciclado de escorias de fondo de central térmica para su uso como áridos en la elaboración de componentes prefabricados de hormigón. Mater Constr 60:99–113. https://doi.org/10.3989/mc.2010.52109

Castro F, Vilarinho C, Trancoso D, Ferreira P, Nunes F, Miragaia A (2009) Utilisation of pulp and paper industry wastes as raw materials in cement clinker production. Int J Mater Eng Innov 1:74. https://doi.org/10.1504/IJMATEI.2009.024028

Chakraborty S, Jo BW, Jo JH, Baloch Z (2017) Effectiveness of sewage sludge ash combined with waste pozzolanic minerals in developing sustainable construction material: an alternative approach for waste management. J Clean Prod 153:253–263. https://doi.org/10.1016/j.jclepro.2017.03.059

Chang J, Lin T, Ko M, Liaw D (1999) Stabilization/solidification of sludges containing heavy metals by using cement and waste pozzolans. J Environ Sci Health Part A 34:1143–1160. https://doi.org/10.1080/10934529909376887

Chen J-S, Chu P-Y, Chang J-E, Lu H-C, Wu Z-H, Lin K-Y (2008) Engineering and environmental characterization of municipal solid waste bottom ash as an aggregate substitute utilized for asphalt concrete. J Mater Civ Eng 20:432–439. https://doi.org/10.1061/(ASCE)0899-1561(2008)20:6(432)

Chesner WH, Collins RJ, MacKay MH, Emery J (2002) User guidelines for waste and by-product materials in pavement construction, Recycled Materials Resource Center, April 1998, p 683. Geographical Coverage: New Hampshire; USA | Publication/Report Number: FHWA-RD-97–148, Guideline Manual, Rept No. 480017

Coelho A, de Brito J (2012) Influence of construction and demolition waste management on the environmental impact of buildings. Waste Manag 32:532–541. https://doi.org/10.1016/j.wasman.2011.11.011

Donatello S, Cheeseman CR (2013) Recycling and recovery routes for incinerated sewage sludge ash (ISSA): a review. Waste Manag 33:2328–2340. https://doi.org/10.1016/j.wasman.2013.05.024

Esquinas AR, Álvarez JI, Jiménez JR, Fernández JM (2018) Durability of self-compacting concrete made from non-conforming fly ash from coal-fired power plants. Constr Build Mater 189:993–1006. https://doi.org/10.1016/j.conbuildmat.2018.09.056

Ferreira C, Ribeiro A, Ottosen L (2003) Possible applications for municipal solid waste fly ash. J Hazard Mater 96:201–216. https://doi.org/10.1016/S0304-3894(02)00201-7

Forteza R, Far M, Seguı C, Cerdá V (2004) Characterization of bottom ash in municipal solid waste incinerators for its use in road base. Waste Manag 24:899–909. https://doi.org/10.1016/j.wasman.2004.07.004

Frías M, Rodríguez O, Sánchez de Rojas MI (2015) Paper sludge, an environmentally sound alternative source of MK-based cementitious materials. A Review. Constr Build Mater 74:37–48. https://doi.org/10.1016/j.conbuildmat.2014.10.007

Gao D, Wang F-P, Wang Y-T, Zeng Y-N (2020) Sustainable utilization of steel slag from traditional industry and agriculture to catalysis. Sustainability 12:9295. https://doi.org/10.3390/su12219295

Guo X, Shi H, Hu W, Wu K (2014) Durability and microstructure of CSA cement-based materials from MSWI fly ash. Cem Concr Compos 46:26–31. https://doi.org/10.1016/j.cemconcomp.2013.10.015

Jamshidi A, White G (2019) Evaluation of performance and challenges of use of waste materials in pavement construction: a critical review. Appl Sci 10:226. https://doi.org/10.3390/app10010226

Jayaranjan MLD, van Hullebusch ED, Annachhatre AP (2014) Reuse options for coal fired power plant bottom ash and fly ash. Rev Environ Sci Biotechnol 13:467–486. https://doi.org/10.1007/s11157-014-9336-4

Johansson N, Krook J (2021) How to handle the policy conflict between resource circulation and hazardous substances in the use of waste? J Ind Ecol 25:994–1008. https://doi.org/10.1111/jiec.13103

Jurič B, Hanžič L, Ilić R, Samec N (2006) Utilization of municipal solid waste bottom ash and recycled aggregate in concrete. Waste Manag 26:1436–1442. https://doi.org/10.1016/j.wasman.2005.10.016

Khudyakova LI, Kislov EV, Paleev PL, IYu, Kotova (2020) Nephrite-bearing mining waste as a promising mineral additive in the production of new cement types. Minerals 10:394. https://doi.org/10.3390/min10050394

Kiventerä J, Lancellotti I, Catauro M, Poggetto FD, Leonelli C, Illikainen M (2018) Alkali activation as new option for gold mine tailings inertization. J Clean Prod 187:76–84. https://doi.org/10.1016/j.jclepro.2018.03.182

Kiventerä J, Sreenivasan H, Cheeseman C, Kinnunen P, Illikainen M (2018) Immobilization of sulfates and heavy metals in gold mine tailings by sodium silicate and hydrated lime. J Environ Chem Eng 6:6530–6536. https://doi.org/10.1016/j.jece.2018.10.012

Kosson DS, van der Sloot HA, Sanchez F, Garrabrants AC (2002) An integrated framework for evaluating leaching in waste management and utilization of secondary materials. Environ Eng Sci 19:159–204. https://doi.org/10.1089/109287502760079188

Kuranchie FA, Shukla SK, Habibi D (2013) Mine wastes in Western Australia and their suitability for embankment construction. In: Geo-Congress 2013. Presented at the Geo-Congress 2013, American Society of Civil Engineers, San Diego, California, United States, pp 1443–1452. https://doi.org/10.1061/9780784412787.145

Lederer J, Trinkel V, Fellner J (2017) Wide-scale utilization of MSWI fly ashes in cement production and its impact on average heavy metal contents in cements: the case of Austria. Waste Manag 60:247–258. https://doi.org/10.1016/j.wasman.2016.10.022

Lemeshev VG, Gubin IK, Savel'ev, Yu A, Tumanov DV, Lemeshev DO (2004) Utilization of coal-mining waste in the production of building ceramic materials. Glass Ceram 61:308–311. https://doi.org/10.1023/B:GLAC.0000048698.58664.97

Lin C-L, Weng M-C, Chang C-H (2012) Effect of incinerator bottom-ash composition on the mechanical behavior of backfill material. J Environ Manage 113:377–382. https://doi.org/10.1016/j.jenvman.2012.09.013

Liu J, Yu B, Wang Q (2020) Application of steel slag in cement treated aggregate base course. J Clean Prod 269 121733.https://doi.org/10.1016/j.jclepro.2020.121733

Liu Y-N, Guo Z-H, Xiao X-Y, Wang S, Jiang Z-C, Zeng P (2017) Phytostabilisation potential of giant reed for metals contaminated soil modified with complex organic fertiliser and fly ash: a field experiment. Sci Total Environ 576:292–302. https://doi.org/10.1016/j.scitotenv.2016.10.065

Malagavelli V, Patura PN (2011) Strength characteristics of concrete using solid waste an experimental investigation. Int J Earth Sci Eng 4:937–940

Martínez C, Cotes T, Corpas FA (2012) Recovering wastes from the paper industry: development of ceramic materials. Fuel Process Technol 103:117–124. https://doi.org/10.1016/j.fuproc.2011.10.017

Maschio S, Tonello G, Piani L, Furlani E (2011) Fly and bottom ashes from biomass combustion as cement replacing components in mortars production: rheological behaviour of the pastes and materials compression strength. Chemosphere 85:666–671. https://doi.org/10.1016/j.chemosphere.2011.06.070

Modolo RCE, Ferreira VM, Machado LM, Rodrigues M, Coelho I (2011) Construction materials as a waste management solution for cellulose sludge. Waste Manag 31:370–377. https://doi.org/10.1016/j.wasman.2010.09.017

Modolo RCE, Ferreira VM, Tarelho LA, Labrincha JA, Senff L, Silva L (2013) Mortar formulations with bottom ash from biomass combustion. Constr Build Mater 45:275–281. https://doi.org/10.1016/j.conbuildmat.2013.03.093

Modolo RCE, Labrincha JA, Ferreira VM, Machado LM (2007) Use of cellulose sludge in the production of fiber-cement building materials. In: Bragança L (ed) Portugal SB07: Sustainable construction, materials and practices: challenge of the industry for the New Millenium. Presented at the Portugal SB07: sustainable construction, materials and practices: challenge of the industry for the new millenium, IOS Press, Lisbon, pp 918–923

Monte MC, Fuente E, Blanco A, Negro C (2009) Waste management from pulp and paper production in the European Union. Waste Manag 29:293–308. https://doi.org/10.1016/j.wasman.2008.02.002

Monzó J, Payá J, Borrachero MV, Morenilla J, Bonilla M, Calderon P (2004) Some strategies for reusing residues from waste water treatment plants: preparation of building materials. In: Vázquez E, Hendriks CF, Janssen GMT (eds) International RILEM Conference on the use of recycled materials in buildings and structures, Proceedings. Presented at the International RILEM conference on the use of recycled materials in buildings and structures, RILEM publications (Réunion internationale des laboratoires d'essais et de recherches sur les matériaux et les constructions), Barcelona (Spain), pp 814–823

Motz H, Geiseler J (2001) Products of steel slags an opportunity to save natural resources. Waste Manag 21:285–293. https://doi.org/10.1016/S0956-053X(00)00102-1

Muñoz P, Morales MP, Letelier V, Mendivil MA (2016) Fired clay bricks made by adding wastes: assessment of the impact on physical, mechanical and thermal properties. Constr Build Mater 125:241–252. https://doi.org/10.1016/j.conbuildmat.2016.08.024

Naik TR, Kraus RN (2003) Recycled materials in concrete industry center for By-Products Utilization (No. CBU-2003–08 REP-503), Dept. of Civil Engineering and Mechanics, College of Engineering and Applied Science. The University of Wisconsin, Milwaukee

Ondrasek G, Kranjčec F, Filipović L, Filipović V, Bubalo Kovačić M, Badovinac IJ, Peter R, Petravić M, Macan J, Rengel Z (2021) Biomass bottom ash & dolomite similarly ameliorate an acidic low-nutrient soil, improve phytonutrition and growth, but increase Cd accumulation in radish. Sci Total Environ 753 141902.https://doi.org/10.1016/j.scitotenv.2020.141902

Palden T, Regadío M, Onghena B, Binnemans K (2019) Selective metal recovery from Jarosite residue by leaching with acid-equilibrated ionic liquids and precipitation-stripping. ACS Sustain Chem Eng 7:4239–4246. https://doi.org/10.1021/acssuschemeng.8b05938

Pan S-C, Tseng D-H, Lee C-C, Lee C (2003) Influence of the fineness of sewage sludge ash on the mortar properties. Cem Concr Res 33:1749–1754. https://doi.org/10.1016/S0008-8846(03)00165-0

Pandeline DA, Cosentino PJ, Kalajian EH, Chavez MF (1997) Shear and deformation characteristics of municipal waste combustor bottom ash for highway applications. Transp Res Rec J Transp Res Board 1577:101–108. https://doi.org/10.3141/1577-13

Ram LC, Masto RE (2014) Fly ash for soil amelioration: a review on the influence of ash blending with inorganic and organic amendments. Earth-Sci Rev 128:52–74. https://doi.org/10.1016/j.earscirev.2013.10.003

Rashid RA, Frantz GC (1992) MSW incinerator ash as aggregate in concrete and masonry. J Mater Civ Eng 4:353–368. https://doi.org/10.1061/(ASCE)0899-1561(1992)4:4(353)

Regadío M, Binnemans K (2018) Solvoleaching of (landfilled) industrial residues and a low-grade laterite ore with diluted HCl in the ionic liquid Aliquat 336. In: Jones PT, Machiels L (eds) 4th International Symposium on Enhanced Landfill Mining (ELFM IV). Mechelen (Belgium), pp 121–126. ISBN: 9789082825909

Regadío M, Black JA, Thornton SF (2020) The role of natural clays in the sustainability of landfill liners. Detritus 100–113. https://doi.org/10.31025/2611-4135/2020.13946

Rosales J, Beltrán MG, Cabrera M, Velasco A, Agrela F (2016) Feasible use of biomass bottom ash as addition in the manufacture of lightweight recycled concrete. Waste Biomass Valorization 7:953–963. https://doi.org/10.1007/s12649-016-9522-4

Shen H, Forssberg E, Nordström U (2004) Physicochemical and mineralogical properties of stainless steel slags oriented to metal recovery. Resour Conserv Recycl 40:245–271. https://doi.org/10.1016/S0921-3449(03)00072-7

Srinivasan R, Sathiya K, Palanisamy M (2010) Experimental investigation in developing low cost concrete from paper industry waste. Bull Polytech Inst Jassy Constr Arquit LVI (LX):43–56

Sutcu M, Akkurt S (2009) The use of recycled paper processing residues in making porous brick with reduced thermal conductivity. Ceram Int 35:2625–2631. https://doi.org/10.1016/j.ceramint.2009.02.027

Tang P, Florea MVA, Spiesz P, Brouwers HJH (2015) Characteristics and application potential of municipal solid waste incineration (MSWI) bottom ashes from two waste-to-energy plants. Constr Build Mater 83:77–94. https://doi.org/10.1016/j.conbuildmat.2015.02.033

Tay J (1987) Sludge ash as filler for portland cement concrete. J Environ Eng 113:345–351. https://doi.org/10.1061/(ASCE)0733-9372(1987)113:2(345)

Tay J, Yip W, Show K (1991) Clay-blended sludge as lightweight aggregate concrete material. J Environ Eng 117:834–844. https://doi.org/10.1061/(ASCE)0733-9372(1991)117:6(834)

Townsend TG, Hofmeister M, Monroy-Sarmiento L, Blaisi N (2016) Application of new leaching protocols for assessing beneficial use of solid wastes in Florida (No. Final project report). Hinkley Center for Solid and Hazardous Waste Management

van der Sloot HA, Comans RNJ, Hjelmar O (1996) Similarities in the leaching behaviour of trace contaminants from waste, stabilized waste, construction materials and soils. Sci Total Environ 178:111–126. https://doi.org/10.1016/0048-9697(95)04803-0

van der Sloot HA, Hoede D, Bonouvrie P (1991) Comparison of different regulatory leaching test procedures for waste materials and construction materials (No. No. ECN-C--91–082). Netherlands Energy Research Foundation (ECN) Technical report first published online: 13 May 2001

van der Sloot HA, Seignette PFAB, Meeussen JCL, Hjelmar O, Kosson DS (2008) A database, speciation modeling and decision support tool for soil, sludge, sediments, wastes and construction products: LeachXS^{TM}- ORCHESTRA. In: 2nd International Symposium on Energy from Biomass and Waste. Fondazione Cini, Venice, Italy
Wyrzykowski M, Ghourchian S, Sinthupinyo S, Chitvoranund N, Chintana T, Lura P (2016) Internal curing of high performance mortars with bottom ash. Cem Concr Compos 71:1–9. https://doi.org/10.1016/j.cemconcomp.2016.04.009

Chapter 24
Behavior of Laterally Loaded Mono-Piled Raft Foundation in Sloping Ground

Ayush Kumar, Sonu Kumar, and Ashutosh Kumar

24.1 Introduction

The steep increase in the population of urban areas has now become a challenging task for civil engineers to accommodate people in the limited available land and provide sustainable and resilient infrastructure such as high-rise buildings, high-speed trains, wind farms, etc. As a core domain of civil engineering, geotechnical engineering has the responsibility to provide sustainable and economic foundation solutions. Hence, it is necessary to develop a cost-effective foundation solution for a given set of geotechnical and wind conditions. The foundation system should show high performance in terms of shear strength and deformation characteristics to satisfy their capacity and serviceability requirements. In recent years, piled raft foundation system is considered as a widely accepted foundation system due to the use of a limited number of piles below the raft which leads to considerable savings in cost without compromising the bearing capacity demand and serviceability requirement (Poulos 2001; Katzenbach et al. 2000; Kumar et al. 2015a, 2017; Choudhury et al. 2019). It is a hybrid foundation system that utilizes the capacities of both shallow and deep foundations for carrying the superstructure load through the complex soil-pile-raft interactions. The successful use of piled raft having 64 piles beneath 256 m high Messeturm Tower of Germany proved it as an alternative and economical foundation system by saving approximately 5.9 million USD, over the conventional group pile foundation (316 piles) (Katzenbach et al. 2000; Kumar et al. 2016). Several of such applications of this rational foundation system built in Germany and other European countries and Asian Countries are documented in the literature where piled raft

A. Kumar · S. Kumar · A. Kumar (✉)
Indian Institute of Technology Mandi, Kamand, Mandi 175005, India
e-mail: ashutosh@iitmandi.ac.in

S. Kumar
e-mail: d20015@students.iitmandi.ac.in

K. R. Reddy et al. (eds.), *Advances in Sustainable Materials and Resilient Infrastructure*, Springer Transactions in Civil and Environmental Engineering,
https://doi.org/10.1007/978-981-16-9744-9_24

has been used below high-rise buildings, high-speed railways, industrial structures such as oil tanks and very heavy and dynamically loaded structures such as nuclear power plants (Yamashita et al. 2012; Katzenbach et al. 2016; Russo et al. 2013; Patil et al. 2021). Several researchers have given different design approaches for this foundation subjected to vertical loading conditions. Clancy and Randolph (1996) proposed an equivalent pier raft approach to analyze piled rafts subjected to vertical loading conditions. Poulos (2001) proposed a three-stage design solution of CPRF considering piles as settlement reducers. de Sanctis and Mandolini (2006) proposed a simplified expression for obtaining the bearing capacity of CPRF resting on soft clayey soils. Kumar and Choudhury (2018) proposed a settlement-based capacity expression for a piled raft foundation system subjected to static vertical loading conditions where piles are embedded in the sand. Currently, the design solution for piled raft under vertical loading conditions for predominantly static loading resting on the leveled ground is robust and well in place. In addition, foundations usually are subjected to the combination of vertical, lateral, and moment loading which may get induced especially in the event of an earthquake and wind. These forces can also be generated if the center of gravity of the foundation system lies above the foundation level as in the case of the wind turbine. The combined action of both pile and raft in lateral load sharing, which is unlike the case of mono-pile foundation or pile group where only piles carry the entire load, interests several researchers and intrigued them towards the use of this improved and economical foundation in seismically vulnerable areas (Dash et al. 2009; Kumar and Choudhury 2016, 2017; Roy et al. 2018, 2020; Bhaduri and Choudhury 2021). However, all of these studies mainly focus on the behavior of piled raft foundations resting on the leveled ground and it should be noted that applying the leveled ground design procedure on the sloped ground might be difficult as the deformation and failure mechanism is different in case of foundation resting near the slopes.

The behavior of structures constructed on slopes is different which is mainly because of the lesser confining pressure present in piles leading to a lesser load-carrying capacity along the sloped ground and the pile foundation may tend to induce slope failure, particularly at shallow depth (Rowe and Poulos 1979; Gabrand and Borden 1990; Chae et al. 2004; Almas et al. 2008). Several researchers have studied the lateral capacity of the pile near the sloped ground (Poulos 1976; Chae et al. 2004; Georgiadis and Georgiadis 2010; Sawant and Shukla 2012; Peng et al. 2019; Sivapriya and Ramanathan 2019; Sivapriya and Gandhi 2013, 2020; Jiang et al. 2020; Peng et al. 2020). Poulos (1976) reported that the lateral deflection in the pile embedded in the sloped ground can be around 1.6 times more than the similar pile embedded in the leveled ground. Finite element analysis for lateral loading condition performed by Trochanis et al. (1991) and Brown and Shea (1991) outlined that the slope angle has a significant influence on the load-carrying capacity mobilized as shear resistance is reduced when a slope angle is increased. It was also reported that the lateral resistance along the rear side of the pile reduced when the pile moved towards the slope. Compared with the leveled ground, laboratory studies overestimated the pressure difference generated at the back of the pile. If the active force was considered, the passive resistance would be reduced by 500% (Davidson 1982;

Gabrand and Borden 1990). When the piles are placed on a ridge in loose soil, the passive wedge is formed directly on the surface of the pile (Muthukkumaran 2004, 2014; Barker 2012). The failure of pile resting in sloped ground and subjected to lateral loading in the case of a long pile is due to the excessive deformation of the pile, and for the short pile, lateral soil resistance was fully mobilized to cause failure (Georgiadis et al. 2013).

However, the studies related to understanding the performance of piled raft foundation resting near slope is limited in the available literature. To develop the load-bearing mechanism of piled raft foundations existing near the sloped ground, this study focuses on understanding the behavior of a mono-piled raft foundation existing near the slope and subjected to lateral loading conditions. Mono-pile raft is the foundation system having concentrically aligned single pile and a raft component over it. Vertical load transfer in a mono-pile raft foundation is possible through the combined action of load-bearing by the pile and raft. However, for lateral loading, the pile develops the lateral resistance due to the passive resistance provided by the soil opposite to the direction of loading and the raft provides the resistance due to the frictional resistance developed along the sides and the base of the raft. The presence of slope may tend to reduce the passive resistance developed against the lateral deformation near the crest of the slope. Figure 24.1 illustrates the schematic representation of load-bearing in a mono-piled raft foundation existing near the slope where the resistance against the lateral load is jointly taken by the pile and the raft.

It is to be noted that the mono piled-raft can be used as a foundation solution to the structures like wind turbines and power transmission towers on sloping grounds and are required to provide safety and serviceability under the combined vertical, horizontal, and moment (V-H-M) loading. For the condition of combined V-H-M loading and sloped ground topography, the load-bearing mechanism and the interaction behavior of mono-piled raft with the surrounding soils are of great engineering significance and not fully understood yet.

This study presents the load-bearing mechanism in a mono-piled raft foundation embedded in slopes and subjected to lateral loading conditions. After successful

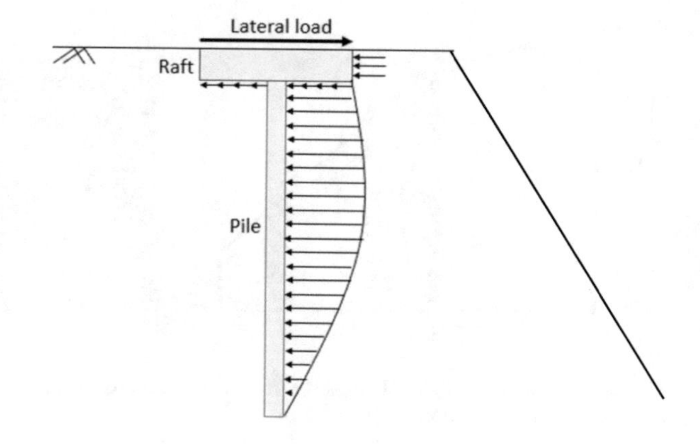

Fig. 24.1 Schematic illustration of lateral load-bearing in case of mono-pile raft near slope

validation of the developed numerical model with a case of a single pile, two cases of raft attached to a pile were studied. First being the raft resting on the soil surface and the other being the embedded raft. The influence of changing slenderness ratio of the pile and the varying slope angle was also studied. Finally, the effect of loading combination was studied.

24.2 Numerical Modeling of Single Pile and Validation

To understand the behavior of a mono-piled raft foundation on sloping ground, firstly the behavior of a single pile embedded on the crest of the slope was studied, and then two conditions were modeled: (i) a mono-piled raft where the raft is resting on the ground and (ii) a mono-piled raft where the raft is embedded just below the ground.

To this end, the experimental study reported by Sivapriya and Gandhi (2020) on the behavior of a single pile embedded in slope 1 V: 2H and subjected to lateral loading was used for the validation of the developed numerical model, and thereafter raft was modeled in a similar geometrical configuration to perform further analysis. Sivapriya and Gandhi (2020) developed a laboratory-scale experimental setup where a single pile was embedded in the sloping ground. The foundation material was clay having a plasticity index of 38% collected from the Chennai region of India. The foundation bed was made with having cohesive strength of 30 kPa. Pile was made of Aluminum having a diameter of 16 mm and length of 450 mm. To understand the behavior of pile in the lateral direction, the pile base was touched with a model base simulating the attainment of vertical equilibrium before applying the lateral loading. The lateral displacement to 5 mm was applied at the pile head and the response of the pile was obtained.

Similar conditions were simulated using the developed numerical model. Figure 24.2 shows the schematic representation and the discretized finite element mesh with its dimensions where the slope of 1 V:2H was kept. It can be seen that a single pile was placed just at the crest of the slope (Fig. 24.2a). The soil model which

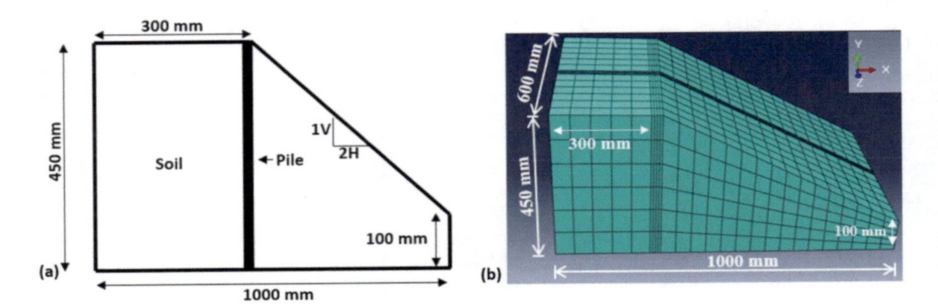

Fig. 24.2 Model of single pile in slope: **a** Schematic representation and **b** Discretized mesh of the developed numerical model

Table 24.1 Soil properties used for finite element model (Sivapriya and Gandhi 2020)

SI	Parameter	Symbol (Units)	Soil	Pile
1	Material model	–	Mohr–Coulomb	Linear elastic
2	Youngs modulus	E (kN/m^2)	8025	70,000,000
3	Unit weight	γ(kN/m^3)	17.84	27
4	Poisson's ratio	μ	0.495	0.2
5	Cohesion	c (kN/m^2)	30	–
6	Friction angle	Ø (°)	0	–

has dimensions of 1000 mm base, 300 mm top, and 450 mm height was developed using an 8-noded linear brick element. A conventional Mohr–Coulomb elastoplastic constitutive model was used to model the clay. Standard fixities were assigned where the vertical axes were restricted to move along the x-axis, i.e., $U_x = 0$ and the base of the soil model was restricted to move along all three directions $U_x = 0$, $U_y = 0$, and $U_z = 0$. The pile of length 450 mm and diameter 16 mm was modeled using a linear elastic material. The geotechnical properties of soil and mechanical properties of pile are mentioned in Table 24.1 as per Sivapriya and Gandhi (2020). The interaction between pile and soil was established using the contact behavior where a normal interaction was set as hard contact and tangential interaction was set in the form of penalty by assigning a friction factor.

The friction coefficient of 0.7 was chosen as per IS 2911 (Part 1/Section 4) (BIS 2010). This would allow the interaction of two different materials having an interface property. Figure 24.2b shows the developed three-dimensional finite element model with its dimensions. It can be observed that the finer mesh was chosen along the pile compared to the relatively coarser mesh away from it. The choice of the mesh size would have the capability to achieve accuracy in the simulation result and control the computational cost of the numerical analysis, discussed in the following section.

24.2.1 *Mesh Convergence and Validation of the Numerical Model*

The mesh size in the finite element study dictates the accuracy of the results indicating finer the mesh size more accurate would be the simulation results. The coarser mesh may result in improper interactions between the pile and soil, however, a very fine mesh would increase the computation cost of the analysis. A mesh size selected may be called convergent if two successive mesh sizes have approximately similar response under the influence of simulated loading conditions. In this study, a mesh optimization was carried out by assigning different meshes ranging from coarse to fine, and the analysis was performed in stages by establishing the equilibrium condition in the first instance, and then pile was activated. Also, to achieve proper

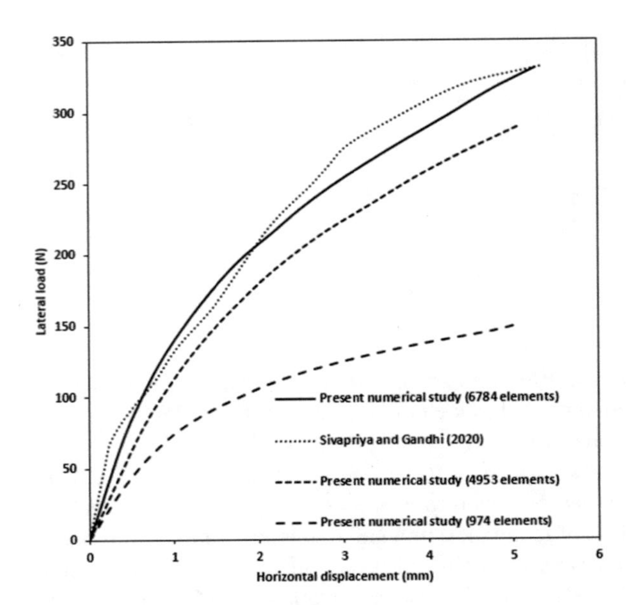

Fig. 24.3 Comparison of lateral load versus horizontal displacement between Sivapriya and Gandhi (2020) and present numerical study

interaction between pile and soil elements, the mesh size selected near to the interface of pile and soil was finer than the mesh size for outer parts of the foundation footprint. Thereafter, the lateral load was applied towards the slope direction and the response of the pile in terms of load-deformation was noted. Figure 24.3 shows the load–displacement response of a single pile having different numbers of elements generated during the finite element study. It can be observed that the accuracy of the results increased with an increase in the number of elements from 974 to 6784.

The load–displacement results obtained using a numerical model having 6784 number of elements matches closely with the experimental results reported by Sivapriya and Gandhi (2020) which validated the developed numerical model both qualitatively and quantitatively. Figure 24.4 shows the deformation contour under a lateral load of 330 N which corresponds to a maximum displacement of around 5 mm in pile. The test results and deformation contour obtained using a numerical model were comparable and hence can be considered as a validation of the developed numerical model. The same model can be now used for further analysis.

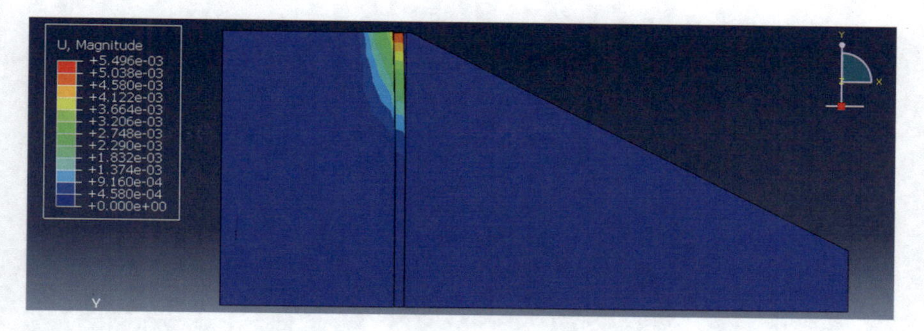

Fig. 24.4 Deformation contours of the finite element model for single pile under a lateral load of 330 N

24.3 Mono Piled Raft Foundation

This section provides the analysis and results of two types of mono-piled raft foundations: (i) a mono-piled raft where the raft is resting on the ground and (ii) a mono-piled raft where the raft is embedded just below the ground. Here, the raft of diameter 48 mm and thickness 20 mm is modeled using linear elastic material and the properties of the raft are given in Table 24.1. The dimension of the raft is chosen to provide rigid nature to the raft foundation as per Horikoshi and Randolph (1997). It is to be noted that the rigid connection between pile and raft was modeled here which is similar to the cases modeled by (Kumar et al. 2016, 2017), (Kumar and Choudhury 2017, 2018). This was achieved by modeling both raft and pile as a single volume. The interaction between raft and soil was modeled similar to that of the pile having a friction coefficient of 0.7. The lateral load was applied at the center of gravity of the raft. This study considers the presence of a mono-piled raft at a distance of 94 mm from the slope indicating the presence of the foundation system near the sloped ground.

24.3.1 Comparison Between On-Surface Raft and Embedded raft- Mono-Piled Raft Foundation

Figure 24.5 shows the developed numerical model for a mono-piled raft foundation in a sloping ground where Fig. 24.5a represents the case of raft resting on the soil surface and Fig. 24.5b represents the case of raft embedded just below the soil surface that means the top of the raft is matching with the ground level. The slope of the soil bed was taken as 1 V:2H. Figure 24.6 shows the horizontal displacement contour of a mono-piled raft foundation subjected to 400 N of lateral load. The results indicate that a maximum lateral deformation of around 5.5 mm can be observed in the on-surface raft (Fig. 24.6a) compared to 2.26 mm in the embedded raft mono-piled

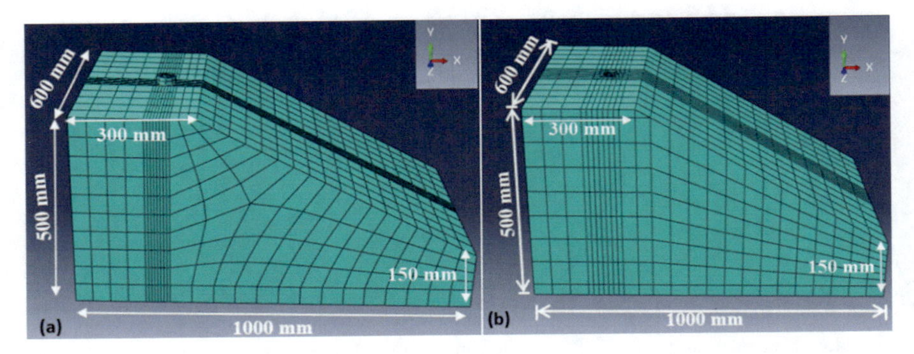

Fig. 24.5 Numerical model for mono-piled raft foundation in sloping ground. **a** Raft on the soil surface; **b** Raft embedded

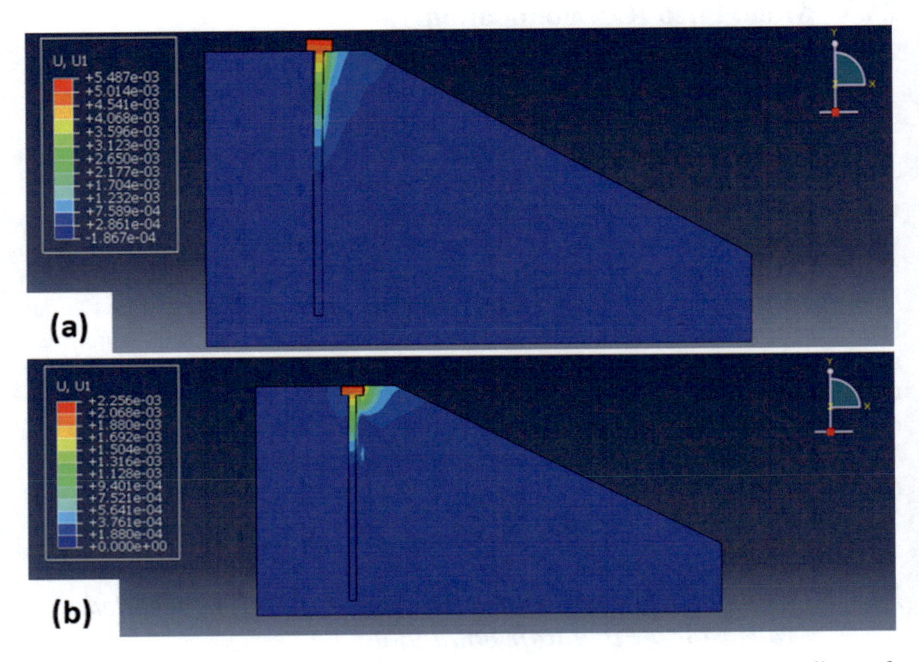

Fig. 24.6 Horizontal displacement contours under lateral load of 400 N for mono-pile **a** raft: on-surface raft; **b** embedded raft

raft foundation (Fig. 24.6b) resulting in a 138% reduction in the displacement at the same load level. The deformation contour illustration indicates the deformation distribution only towards the slope in the case of raft on surface mono-piled raft foundation whereas there is a deformation distribution towards the slope and away from the slope in case of embedded raft mono-piled raft foundation. This is mainly because of the predominant impact of raft-soil interaction in the case of embedded

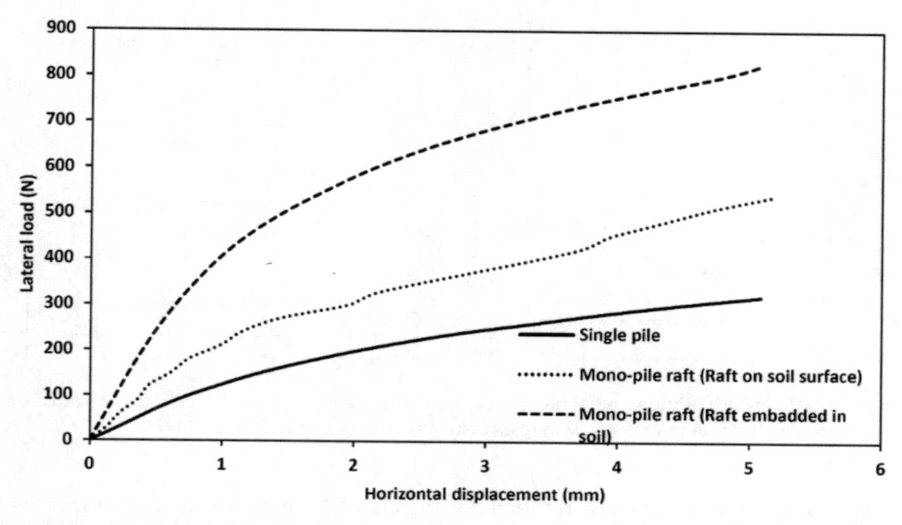

Fig. 24.7 Load–displacement response of single pile, mono-piled raft (raft on surface and embedded raft)

raft mono-piled raft foundations. This is also because of the higher surface area available in the embedded raft case compared to the on-surface raft case leading to the generation of higher passive resistance. The active length of the pile which contribute significantly towards the deformation distribution is 0.53 times the pile length in the case of on-surface mono-piled raft foundation compared to 0.4 times the pile length in the case of embedded raft mono-piled raft foundation. This is mainly because of the contribution of the sides of the raft in sharing the lateral load.

Figure 24.7 shows the load–displacement curves for single pile and mono-piled raft foundations. It can be observed that the capacity of the foundation to deform reduced with the presence of the raft where maximum capacity can be observed in the case of embedded raft foundation compared to a minimum in the case of a single pile. For example, the capacity at the displacement level of 5 mm in the single pile is 300 N whereas in a mono-piled raft foundation the capacity increased to 530 N (on-surface raft case) and 825 N (embedded raft case), respectively. The capacity of the foundation increased by 76% with the introduction of the raft and the capacity further increased 55% if the raft is embedded. This increase is mainly due to the presence of a raft which is providing lateral resistance against the displacement. Higher the raft contributing area in providing resistance, higher would be capable of the mono-piled raft foundation system. Here, the analysis has been shown up to 5 mm of lateral deformation as per the recommendation of IS 2911-part IV (1985).

Figure 24.8 shows the comparison between the horizontal displacement and bending moment obtained along the pile length for mono-piled raft foundation under a lateral load of 400 N. The lateral displacement reduced with the soil depth due to increasing resistance of the soil along the depth. The deformation reduced

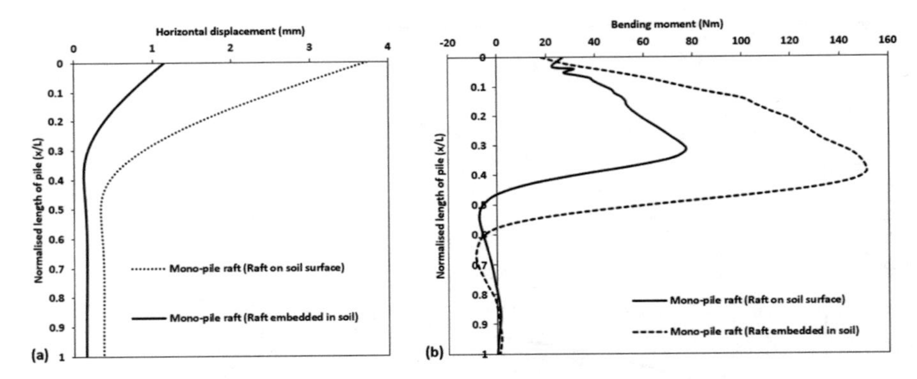

Fig. 24.8 Response of mono-piled raft with raft on soil surface and embedded raft. **a** Horizontal displacement. **b** Bending moment with respect to normalized length of pile

considerably for the case of the embedded raft (1.15 mm) compared to the on-surface raft where a maximum displacement of 3.75 mm was observed (Fig. 24.8a). Figure 24.8b shows the bending moment developed along the pile length where the bending moment of 27 N.m and 18 N.m was obtained at the top of the pile in case of embedded raft and on-surface raft, respectively. This is mainly because of the rigid connection between pile and raft. The maximum bending moment was obtained at a depth below the pile head which was mainly because of the effect of slope in changing the bending moment response which is unlike the case of level ground where the pile develops maximum bending moment at the pile head compared to elsewhere in the pile. Poulos and Davis (1980) and Kumar and Choudhury (2017) reported similar response in case a rigidly connected pile when embedded in the leveled ground. The magnitude of bending moment at the pile tip reached close to zero. The cross-over point, i.e., a point where bending moment changes its signs from a positive value to a negative value can be observed in both cases which is around 0.4 times the pile length for the embedded raft case and 0.56 times the pile length for the on-surface raft case. This is mainly because of an increase in the overall foundation depth due to the raft embedment. The maximum bending moment in the case of an embedded raft mono-piled raft is 94% more than the maximum bending moment in the case of on-surface raft mono-piled raft foundation.

24.3.2 Paramteric Study

In this section, the result of parametric study is presented which was performed by varying the slope angle of the soil and length to diameter ratio of pile for the case of embedded raft mono-piled raft and the load-deformation response was analyzed.

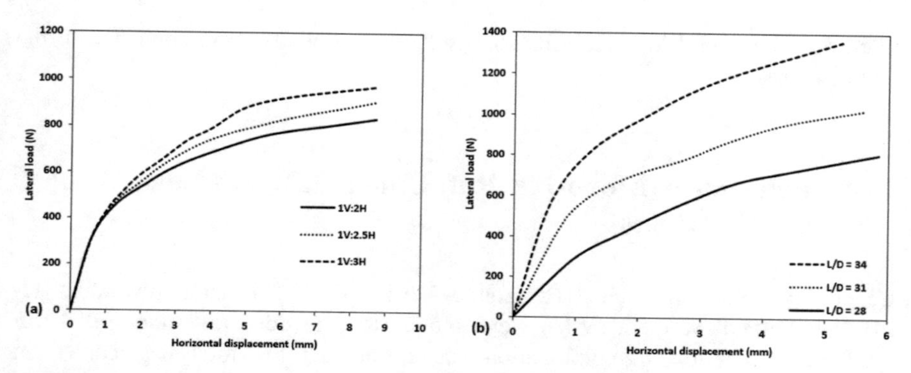

Fig. 24.9 Lateral load versus horizontal displacement curves for **a** variable slopes **b** and different slenderness ratio

24.3.3 Varying the Slope Angle

The influence of slope angle on the load-bearing mechanism of a mono-piled raft foundation embedded was studied by varying the slope from 1 V:2H, 1 V:2.5H, and 1 V:3H. Figure 24.9a shows the lateral load–displacement response of the pile, and it can be observed that the capacity of mono-piled raft foundation increases with a reduction in the slope angle that means lower is the slope angle higher is the resistance offered by the soil. The lateral load-carrying capacity increased about 16% with a change in slope from 1 V:2H to 1 V:3H and an increase of about 8% was observed for change in slope from 1 V:2H to 1 V:2.5H. This is mainly because of an increase in the passive resistance of the soil with a reduction in the slope angle. It can also be concluded that the lateral load-bearing capacity of a mono-piled raft embedded in soil has an inverse relationship with the steepness of the slope. This result also emphasizes that an increment in the slope angle reduced confinement around the pile which subsequently reduced the capacity of the foundation system to carry the lateral load.

24.3.4 Varying the L/D *Ratio*

The slenderness ratio of the pile was varied from 28 to 34 and the lateral load–displacement response was obtained for 1 V: 2H case of the slope. Figure 24.9b shows the load–displacement response of mono-piled raft foundation where an increase in the capacity of the foundation was observed with an increase in the slenderness ratio of the pile from 28 to 34. The lateral capacity of the pile at 5 mm displacement increased by around 88% with an increment in the slenderness ratio from 28 to 34. This increasing trend in the lateral load with the horizontal displacement can be attributed to the fact that with an increase in the length to diameter ratio of mono piled-raft, the lateral load mobilization by the mono piled-raft in contact with the soil increases.

24.4 Behavior of Mono-Pile Raft Under Different Load Combinations

Loading combinations such as Horizontal–Vertical (H-V), Horizontal-Moment (H-M), and Vertical-Moment (V-M) were applied to embedded raft mono-piled raft foundations. The loading combination was chosen such that total displacement in the foundation corresponds to 5 mm and an interaction curve is obtained.

24.4.1 Results in Terms of H-V, H-M, and V-M Loading Combinations

To obtain H-V interaction curve, the horizontal load was applied at the center of gravity of the raft whereas the vertical load was applied concentrically to the axis of pile and raft as a point load. The soil boundary at the base was extended to 5 times the pile diameter to avoid the boundary effect. The loading direction for lateral force in the horizontal direction was applied towards the slope. The load combination was chosen such that the resultant displacement achieved is 5 mm. Figure 24.10a shows the deformation contour when a foundation system is subjected to combined horizontal and vertical loads. The deformation progresses along the entire length

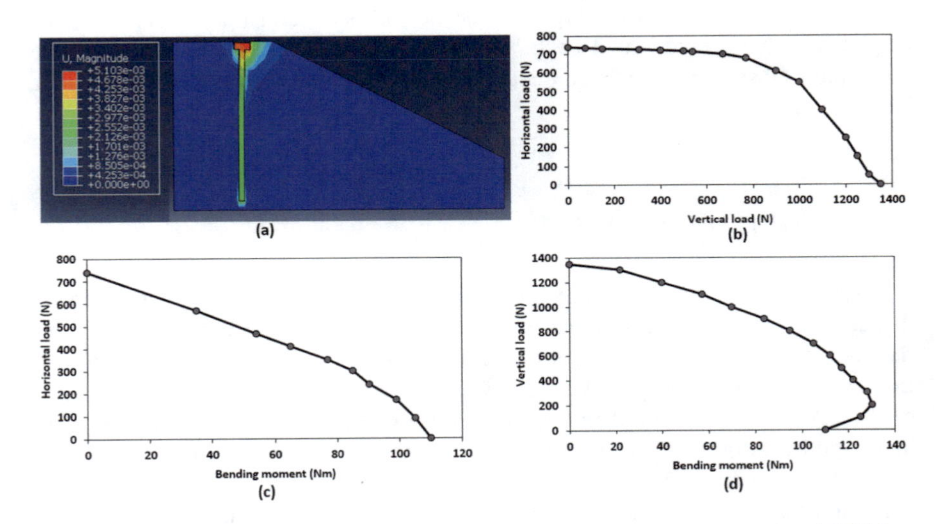

Fig. 24.10 Results of interaction analysis. **a** Deformation contours, **b** H-V, **c** H-M, and **d** V-M

of the pile which is unlike the case of only lateral loading as can be seen from Fig. 24.6. Figure 24.10b shows the interaction curve for the H-V combination of loads indicating an enveloping curve. All the values below the interaction curve represent the safe loading for the combination of horizontal and vertical loadings for an embedded raft mono-piled raft foundation resting on slope 1 V: 2H.

To obtain H-M interaction curve, the moment and the horizontal load were applied at the center of the gravity of the raft where zero moment means a case of pure horizontal load application and zero horizontal load means pure moment application. The direction of the applied horizontal load was towards the slope and the moment was in an anti-clockwise direction which has a tendency to rotate the foundation system towards the slope. Figure 24.10c shows the interaction curve for the H-M combination of loads indicating an enveloping curve. All the values below the interaction curve represent the safe loading for the combination of horizontal load and moment for an embedded raft mono-piled raft foundation resting on slope 1 V: 2H.

To obtain V-M interaction curve, the vertical loading and the moment in combination were applied at the center of the gravity of the raft where zero moment means a case of pure vertical load and vice-versa. The direction of the applied moment was in an anti-clockwise direction which has a tendency to rotate the foundation system towards the slope. Figure 24.10d shows the interaction curve for the H-M combination of loads indicating an enveloping curve. All the values below the interaction curve represent the safe loading for the combination of vertical load and the moment for an embedded raft mono-piled raft foundation resting on slope 1 V: 2H.

24.5 Conclusions

The present study discusses the behavior of on-surface and embedded raft mono-piled raft foundations resting near the slope where the foundation system was subjected to lateral loading conditions. The analyses were carried out using the three-dimensional finite element-based program Abaqus where the numerical analysis results simulated the experimental results both qualitatively and quantitatively thereby validating the numerical simulation. Following conclusions are drawn from the study.

1. The mono-piled raft foundation system is an improvement over the conventional pile foundation due to the additional resistance provided by the raft component. In addition, it has been observed that the embedded raft performed better in reducing the displacement and the bending moment in piles because of the additional resistance offered by the thickness of the raft.
2. The active length of the pile was dependent on the presence of raft either on the surface or embedded where higher active length was observed in the case of on-surface raft mono-piled raft foundation.
3. The bending moment response in the pile was maximum at a depth below the pile head because of the effect of the slope which reduced the development of

the passive resistance in the pile which is unlike that of leveled ground where the maximum bending moment is observed at the pile head.

4. The lateral load-carrying capacity of the foundation system was reduced with an increase in the slope angle which was mainly because of the tendency of slope in reducing the lateral confinement.

References

Almas BN, Seethalakshmi P, Muthukkumaran K (2008) Lateral capacity of single pile located at slope crest. Indian Geotech J 38(3):278–294

Barker PD (2012) Effects of soil slope on the lateral capacity of piles in cohesionless soils. Master of Science thesis, Oregon State University, Corvallis, USA

Bhaduri A, Choudhury D (2021) Steady-state response of flexible combined pile-raft foundation under dynamic loading. Soil Dynamics and Earthquake Engineering, Elsevier, 145, 106664. https://doi.org/10.1016/j.soildyn.2021.106664

Brown DA, Shie CF (1991) Some numerical experiments with a three-dimensional finite element model of a laterally loaded pile. Comput Geotech 12(2):149–162. https://doi.org/10.1016/0266-352X(91)90004-Y.

Chae KS, Ugai K, Wakai A (2004) Lateral resistance of short single piles and pile groups located near slopes. Int J Geomech 4(2):93–104. https://doi.org/10.1061/(ASCE)1532–3641(2004)

Choudhury D, Kumar A, Patil M, Rao VD, Bhaduri A, Singbal P, Shukla J (2019) Sustainable foundation solutions for industrial structures under earthquake conditions theory to practice. In: Proceedings of 16th Asian Regional Conference on Soil Mechanics and Geotechnical Engineering (16ARC), Taipei, Taiwan

Clancy P, Randolph MF (1996) Simple tools for pile raft foundations. Geotechnique 46(2):313–328. https://doi.org/10.1680/geot.1996.46.2.313

Dash S, Govindaraju L, Bhattacharya S (2009) A case study of damages of the Kandla Port and Customs Office tower supported on a mat–pile foundation in liquefied soils under the 2001 Bhuj earthquake. Soil Dyn Earthquake Eng 29(2):333–346. https://doi.org/10.1016/j.soildyn.2008.03.004

Davidson HL (1982) Laterally loaded drilled pier research. Final report prepared for electrical power research institute. Volume 1: Design methodology, DE82901901

De Sanctis L, Mandolini A (2006) Bearing capacity of piled rafts on soft clay soils. J Geotech Geoenviron Eng 132(12):1600–1610. https://doi.org/10.1061/(ASCE)1090-0241(2006)132:12(1600)

Gabrand MA, Borden RH (1990) Lateral analysis of piers constructed on slopes. J Geotech Eng, ASCE 116(12):1831–1850. https://doi.org/10.1061/(ASCE)0733-9410(1990)

Georgiadis K, Georgiadis M (2010) Undrained lateral pile response in sloping ground. J Geotech Geoenviron Eng ASCE 136(11):1489–1501. https://doi.org/10.1061/(ASCE)GT.1943–5606.0000373

Georgiadis K, Georgiadis M, Anagnostopoulos C (2013) Lateral bearing capacity of rigid piles near clay slopes. Soils Found 53(1):144–154. https://doi.org/10.1016/j.sandf.2012.12.010

Horikoshi K, Randolph MF (1997) On the definition of raft-soil stiffness ratio for rectangular rafts. Geotechnique 47(5):1055–1061

IS 2911-Part IV (1985) Indian standard code of practice for design and construction of pile foundations, Part 4 Load tests on pile, Bureau of Indian standards (BIS), New Delhi 110002

IS 2911- Part 1/Section 4 (2010) Indian standard code of practice for design and construction of pile foundations, Part 1 Concrete piles, Section 4 Precast concrete piles in prebored holes. Bureau of Indian standards (BIS), New Delhi 110002

Jiang C, Liu L, He JL, Xie HS (2020) Effect of the proximity of slope and pile shape on lateral capacity of piles in clay slopes. Eur J Environ Civ Eng 1–15. https://doi.org/10.1080/19648189.2020.1858452

Katzenbach R, Arslan U, Moormann C (2000) Piled raft foundation projects in Germany. In: Hemsley JA (ed) Design applications of raft foundations. Thomas Telford, London, pp 323–392

Katzenbach R, Leppla S, Choudhury D (2016) Foundation systems for high-rise structures. CRC Press, Taylor and Francis Group, UK, pp 1–298. (ISBN: 978-1-4978-4477-5). https://doi.org/10.1201/9781315368870

Kumar A, Choudhury D, Shukla J, Shah DL (2015a) Seismic design of pile foundation for oil tank by using PLAXIS3D. Disaster Adv 8(6):33–42

Kumar A, Choudhury D, Katzenbach R (2015b) Behaviour of combined pile-raft foundation (CPRF) under static and pseudo-static conditions using PLAXIS3D. In: Proceedings of 6th international conference on earthquake geotechnical engineering (6ICEGE), Christchurch, New Zealand, paper ID-140

Kumar A, Choudhury D, Katzenbach R (2016) Effect of earthquake on combined pile-raft foundation. Int J Geomech ASCE 16(5):04016013:1–16. https://doi.org/10.1061/(ASCE)GM.1943-5622.0000637

Kumar A, Choudhury D (2016) DSSI analysis of pile foundations for an oil tank in Iraq. In: Proceedings of Institution of Civil Engineers-Geotechnical Engineering. 169(2):129–138. https://doi.org/10.1680/jgeen.15.00025

Kumar A, Choudhury D (2017) Load sharing mechanism of combined pile-raft foundation (CPRF) under seismic loads. Geotech Eng J Southeast Asian Geotechnical Society (SEAGS). Association of Geotechnical Society Southeast Asia (AGSSEA) 48(3):95–10

Kumar A, Patil M, Choudhury D (2017) Soil-structure interaction in a combined pile-raft foundation-a case study. In: Proceedings of Institution of Civil Engineers-Geotechnical Engineering. 170(2):117–128. https://doi.org/10.1680/jgeen.16.00075

Kumar A, Choudhury D (2018) Development of new prediction model for capacity of combined pile-raft foundations. Comput Geotech 97:62–68. https://doi.org/10.1016/j.compgeo.2017.12.008

Mezazigh S, Levacher D (1998) Laterally loaded piles in sand: slope effect on p-y reaction curves. Can Geotech J 35(3):433–441. https://doi.org/10.1139/t98-016

Muthukkumaran K (2004) Non-linear soil–structure interaction of piles on sloping ground, Doctoral of Philosophy Thesis, IIT-Madras, Chennai

Muthukkumaran K (2014) Effect of slope and loading direction on laterally loaded piles in cohesionless soil. Int J Geomech 14(1):1–7. https://doi.org/10.1061/(ASCE)GM.1943-5622.0000293

Ng CWW, Zhang LM (2001) Three-dimensional analysis of performance of laterally loaded sleeved piles in sloping ground. J Geotech Geoenviron Eng 127(6):499–509. https://doi.org/10.1061/(ASCE)1090-0241(2001)127:6(499).

Patil G, Choudhury D, Mondal A (2021) Three-Dimensional soil–foundation–superstructure interaction analysis of nuclear building supported by combined piled–raft system. Int J Geomech 21(4):04021029. https://doi.org/10.1061/(ASCE)GM.1943-5622.0001956

Poulos HG, Davis EH (1980) Pile foundation analysis and design. Wiley, New York

Peng W, Zhao M, Xiao Y, Yang C, Zhao H (2019) Analysis of laterally loaded piles in sloping ground using a modified strain wedge model. Comput Geotech 107:163–175. https://doi.org/10.1016/j.compgeo.2018.12.007

Peng W, Zhao M, Zhao H, Yang C (2020) Behaviors of a laterally loaded pile located in a mountainside. Int J Geomech 20(8):04020123. https://doi.org/10.1061/(ASCE)GM.1943-5622.0001745

Poulos HG (1976) Behaviour of laterally loaded piles near a cut or slope. Aust Geomech J G6(1):6–12

Poulos HG, Davis EH (1980) Pile foundation analysis and design. Wiley, New York.

Poulos HG (2001) Piled raft foundations: design and applications. Geotechnique 51(2):95–113

Rathod D, Muthukkumaran K, Thallak SG (2019) Experimental investigation on behavior of a laterally loaded single pile located on sloping ground. Int J Geomech 19(5):04019021. https://doi.org/10.1061/(ASCE)GM.1943-5622.0001381

Rowe RK, Poulos HG (1979) A method for predicting the effect of piles on slope behaviour. In: 3rd International conference on numerical methods in geomechanics, pp 261–262

Roy J, Kumar A, Choudhury D (2018) Natural frequencies of piled raft foundation including superstructure effect. Soil Dyn Earthquake Eng 112:69–75. https://doi.org/10.1016/j.soildyn.2018.04.048

Roy J, Kumar A, Choudhury D (2020) Pseudostatic approach to analyze combined pile-raft foundation. Int J Geomech 20(10):06020028. https://doi.org/10.1061/(ASCE)GM.1943-5622.0001806

Russo G, Abagnara V, Poulos HG, Small JC (2013) Re-assessment of foundation settlements for the Burj Khalifa, Dubai. Acta Geotechnica 8(1):3–15

Sawant V, Shukla SK (2012) Finite element analysis of laterally loaded piles in sloping ground. Coupled Syst Mech 1(1):59–78. https://doi.org/10.1007/s40098-012-0022-6

Sivapriya SV, Gandhi SR (2013) Experimental and numerical behaviour of single pile subjected to lateral load. Indian Geotech J 43(1):105–114. https://doi.org/10.1007/s40098-012-0037-z

Sivapriya SV, Ramanathan R (2019) Load–displacement behaviour of a pile on a sloping ground for various L/D ratios. Slovak J Civ Eng 27(1):1–6. https://doi.org/10.2478/sjce-2019-0001

Sivapriya SV, Gandhi SR (2020) Soil–structure interaction of pile in a sloping ground under different loading conditions. Geotech Geol Eng 38:1185–1194. https://doi.org/10.1007/s10706-019-01080-z

Trochanis AM, Bielak J, Christiano P (1991) Three-dimensional nonlinear study on piles. J Geotech Eng ASCE 117(3):429–447. https://doi.org/10.1061/(ASCE)0733-9410(1991)117:3(429)

Yamashita K, Hamada J, Onimaru S, Higashino M (2012) Seismic behaviour of piled raft with ground improvement supporting a base-isolated building on soft ground in Tokyo. Soils Found 52:1000–1015. https://doi.org/10.1016/j.sandf.2012.11.017

Chapter 25
The Role of Civil Engineering in Achieving UN Sustainable Development Goals

Lavanya Addagada, Srikrishnaperumal T. Ramesh, Dwarika N. Ratha, Rajan Gandhimathi, and Prangya Ranjan Rout

25.1 Introduction

In 2015, member states of the United Nations (UN) adopted a set of universal, comprehensive, transformative goals and targets to address the greatest economic, social, and environmental barriers that exist within and among the countries by 2030 (UNDG 2017). The major goal of this 2030 agenda is to build a better future and provide a good quality of life for humans and life on the planet. Moreover, the 2030 agenda stimulates the action in critical areas of importance such as people, planet, prosperity, peace, and partnership to improve lives and profoundly transform the world for a better future. The 2030 agenda consists of 17 sustainable development goals (SDGs) with 169 targets, which are integrated and indivisible in nature (Moyer and Hedden 2020; UNGA 2015). These goals and targets are mainly envisaged to establish a path toward sustainable development. The SDGs are majorly targeted to protect the environment, minimize poverty and hunger, and provide peace and human prosperity. In addition to that, these SDGs are established to address the twenty-first

L. Addagada
CSIR- National Environmental Engineering Research Institute Nagpur, Nagpur, Maharashtra 440020, India

S. T. Ramesh · R. Gandhimathi
Department of Civil Engineering, National Institute of Technology, Tiruchirappalli, Tamilnadu 620015, India

D. N. Ratha
Department of Civil Engineering, Thapar Institute of Engineering and Technology, Patiala, Punjab 147004, India

P. R. Rout (✉)
Department of Biotechnology, Thapar Institute of Engineering and Technology, Patiala, Punjab 147004, India
e-mail: prr10@iitbbs.ac.in

K. R. Reddy et al. (eds.), *Advances in Sustainable Materials and Resilient Infrastructure*, Springer Transactions in Civil and Environmental Engineering,
https://doi.org/10.1007/978-981-16-9744-9_25

Fig. 25.1 Sustainable development goals and their icons (Summarized from United Nations website, https://sdgs.un.org/goals)

century's most significant challenges that include natural resource depletion, climate change, threats due to natural disasters, and rapid urbanization (Griggs et al. 2013). The 17 SDG's along with logos are shown in Fig. 25.1.

All the 17 SDGs are interrelated and interlinked, therefore prioritizing one over the other leads to conflicts. Hence, all the SDGs need to be tackled simultaneously rather than individually. SDG 1 is focused on ending extreme poverty throughout the world, while SDG 2 is targeted to eradicate hunger by promoting sustainable agriculture. Similarly, SDG 3 is directed to ensure a healthy life and promote well-being, whereas SDG 4 is bound to provide universal education, SDG 5 aimed to achieve gender equality and empower women and girls. SDG 6 deals with sustainable management of water and sanitation for all, SDG 7 is focused on ensuring reliable, affordable, sustainable energy for all, SDG 8 is targeted to promote sustainable economic growth and full and productive employment for all, SDG 9 aimed to develop resilient infrastructure and foster innovation. Likewise, SDG 10 intended to minimize inequality within and among the countries, SDG 11 targeted to develop sustainable cities, SDG 12 directed to ensure sustainable production and consumption patterns, SDG 13 deals with necessary actions required to combat climate change and its impacts. Furthermore, SDG 14 and 15 aimed to conserve the life below water and life on land, respectively, SDG 16 directed to foster peace at all levels, and SDG 17 is aimed to provide a brief guidance on effective implementation of all the above-mentioned 16 SDGs and it highlights the necessity of partnerships at local, regional, and global level to promote sustainable development. All the SDGs have specific targets and need to address various issues that are interconnected. For example, SDG 1 (eradicating poverty) is interconnected with SDG 10 (reducing inequality), which is again interrelated with SDG 5 (gender equality), which in turn connected with SDG 8 (decent work and economic growth) and SDG 4 (quality education). Therefore, it is extremely necessary to understand the linkages within and among 17 SDGs to meet the envisioned goals and targets.

To meet the SDGs by 2030, an integrated and sustainable approach is necessary to fulfill the needs of all people ensuring equitable opportunities and economic prosperity for all, while reducing its detrimental impacts on the planet. Furthermore, the integrated and sustainable approach seeks scientific evidence to accomplish societal needs through its developmental actions. In this regard, it highlights the role of innovation in addressing societal challenges in line with the SDGs. Engineers play a critical role in developing innovative approaches and sustainable technologies to achieve SDGs and ultimately contributing to enhance the quality of life of each individual. Among the engineers, especially civil engineers, are at the front line in delivering SDGs. To attain any of the above-mentioned 17 SDGs, infrastructure development throughout the globe is extremely important. Infrastructure development particularly in transportation, energy, water, sanitation, and other sectors profoundly influences delivering the SDG 3, 6, 7, 9, and 11 and can be achieved only with the help of civil engineers. Civil engineering professionals must build resilient and sustainable infrastructure by employing eco-friendly and green materials in the construction and the design phase of projects, thus SDGs can be accomplished by ensuring prosperity to people and life on the planet. Therefore, the present study highlights the critical role of civil engineers in delivering SDGs. In addition, the present work also discusses the linkage between civil engineering and 17 SDGs, concerning social, economic, and environmental factors in building a resilient society. Moreover, the current study also illustrated the leadership role of civil engineers in delivering the SDGs by 2030. Additionally, the present study also illustrates the necessity of partnerships to implement the 2030 agenda successfully.

25.2 Background for SDG'S

The concept of SDGs started at the UN conference on sustainable development held in Rio de Janeiro in 2012. These SDGs supersede Millennium Development Goals (MDGs), which aimed to address poverty, hunger, disease, and providing primary education to all the children are among the significant priorities (World Bank 2015). The world showed significant progress with the implementation of MDGs, but still, there is a lot to do to achieve uniformity throughout the world. Furthermore, MDGs mainly focused on developing countries and these MDGs highlighted the necessity of fundamental transformations within the society. Therefore, SDGs were framed with an ambition to discourse the social, economic, and environmental barriers that existed globally. SDGs are applicable globally irrespective of the nation's level of development, capacities, and policies. Additionally, SDGs emphasized the necessity of sustainable development worldwide to ensure adequate resources for future generation's endurance. Thus, in 2015 UN assembly passed a resolution, "Transforming our World: The 2030 agenda for Sustainable Development," which is an agreement among UN member states that is envisaged to promote sustainable development throughout the globe. The 2030 agenda consists of 17 SDGs and 169 targets.

25.3 Sustainable Development

Sustainable development is defined as "Development that meets the needs of the present without compromising the ability of future generations to meet their own needs" (Rout et al. 2020b). The major pillars of sustainable development are social, economic, and environmental components (Fig. 25.2) (WCED 1987). The social component of sustainable development intends to enhance the quality of life of an individual and society. The social component unveiled the inevitability of uniform distribution of goods and services to all the people to encourage overall society development. The economic component recommends to offer incentives to the organizations that follow sustainable practices and maximize the system's efficiency despite limited natural resources. Environmental protection is our utmost priority and to achieve that, harmonious balance between social and economic components is imperative. The core idea of the environmental component is to conserve and protect the ecosystems and natural resources without damaging the stability of the physical and biological systems.

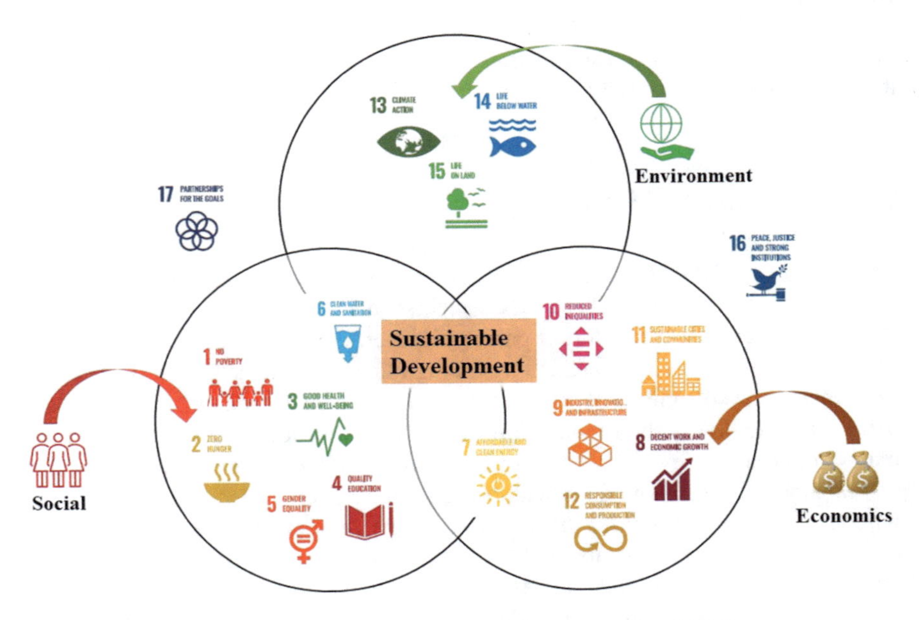

Fig. 25.2 Components of sustainable development along with associated SDGs

25.4 Engineering for Sustainable Development

The 17 SDGs indicate that an integrated approach is vital to discourse poverty, to fulfill basic amenities requirements in education, sanitation, and the health sector, to mitigate the influence of climate change on life, agriculture, and natural resources. To resolve the above-mentioned global challenges, building a smarter world with innovations, inventions, technologies, and the development of infrastructure committed for sustainable development is desirable (UNESCO 2021). Science, technology, and engineering are the heart of sustainable development and play a crucial role in progressing the SDGs in an integrated manner (UN 2019; 2020). The engineering profession would help to accomplish all these needs through a scientific approach by developing solutions, technologies, and infrastructure sustainably. Engineering is the art, discipline, or field that utilize technical, mathematical knowledge, and design principles that hamper the development of sustainable structures, systems, processes, and new materials for specific purposes. Among the various engineering fields, civil engineering is the oldest and imperative to fulfill the day-to-day needs of society. People who practice engineering are regarded as engineers. Engineers should practice and encourage sustainable, innovative, eco-friendly, and safer methods for advancement in the SDGs and ensure prosperity for all and a healthy planet. Overall, these challenges provide opportunities for engineers globally (Kelly et al. 2016).

25.5 Role of Civil Engineering to Achieve SDGS

The role of civil engineers in developing society is remarkable. Their tremendous work enhanced the lever of life standards from the ancient era to the modern twenty-first century. Infrastructure development is another key component that needs to be developed by civil engineers to accomplish all the SDGs for a fair future for the people and the planet. Civil engineers contribution toward SDGs entails various skills such as a high level of creativity, ability to work with diversified groups to develop a road map for sustainable practices, broad knowledge over social, ecological, and environmental aspects, and promote circular economy approach while formulating strategies. The vital role of civil engineering and civil engineers to attain each of the SDGs is mentioned below.

25.5.1 SDG 1

SDG 1 is no poverty, which means that end poverty in all forms from everywhere. To achieve this, providing basic amenities such as nutrition, clean water, sanitation, employment, health care, transportation facilities, power supplies, etc., to each corner of the world is essential, which is the job of civil engineers. In addition to that, poverty alleviation also entails consideration of social, economic, and political aspects. Civil engineers need to develop basic infrastructure facilities in remote geographical areas by providing connectivity either with railways, roadways, and waterways and further provide access to basic essentials. Moreover, traditional methods of infrastructure development deteriorate the environment, which is not a sustainable approach. The sustainable approach to eradicate poverty requires multi-disciplinary, cross-country, and inter-cultural solutions that are affordable and enhance access to all the resources. Moreover, the inclusion of local community people in decision-making and policy-making would help to understand the conditions and thus deliver effective and efficient solutions. The solutions also should have a positive and neutral effect on natural resource consumption and utilization. Creating employment for poor and local community people would push them from below the poverty line to above the poverty line and contribute to poverty reduction. In addition to that, construction of cost-effective housing units for locals is possible only with the involvement of civil engineers and their role in fulfilling SDG 1 is irreplaceable.

25.5.2 SDG 2

SDG 2 is mainly for no hunger. Ending hunger in all forms and food security for all need to be accomplished through sustainable agricultural practices. It requires inter-linkages within and among small and medium-scale farmers to tackle the impact of climate change on agriculture. To alleviate hunger and to provide food to the growing population, crop production needs to be augmented. Extensive utilization of pesticides and fertilizers increases the crop production; however, it has a significant negative impact on the environment. Similarly, natural disasters like floods and cyclones and irrigation practices influence the crop productivity rate. Therefore, civil engineers need to take a leadership role during the design and construct phase of flood management systems, irrigation infrastructures, and develop soil erosion control systems to protect the crops from natural disasters that significantly aid to enhance the crop yield. Additionally, civil engineers must extend their services to build resilient infrastructure such as roads, crop storage facilities, etc., and promote a farm-to-market approach, thereby empowering the farmers to increase yields and encourage agricultural-related businesses (UNESCO 2016). Consequently, providing all these facilities directly or indirectly contribute to attain global food security.

25.5.3 SDG 3

SDG 3 is aimed to ensure healthy lives and promote the well-being of people of all ages globally. The main objective of this goal is to reduce maternal mortality, eradicate epidemics like tuberculosis, malaria, and various communicable diseases, minimize the fatality rate due to traffic accidents, illness, and air pollution, and provide access to healthcare facilities for all the people of all ages throughout the world. In order to control the spreading of transmissible diseases, designing and building basic sanitation and wastewater treatment systems is the utmost importance and is possible only with the involvement of civil engineers. Air pollution is another global issue, strategies necessary to mitigate its effect on the environment need to be framed, adopted, and implemented for people's overall well-being. Moreover, it is the role of civil engineers to frame strict and stringent norms to address air pollution-related problems and prevent associated ailments. Building healthcare amenities and providing access to those amenities is possible only with the laying of roads, which need to be done by civil engineers. Therefore, civil engineers must consider social, economic, and environmental aspects during the design stage of the projects to foster sustainable development. Supplying clean water to the poor and needy people is essential to suppress water transmissible and infectious illnesses for healthy life and well-being of people.

25.5.4 SDG 4

The main objective of SDG 4 is to ensure quality education for all and facilitate lifelong learning opportunities. Education plays a key role across all the SDGs and nurtures sustainable development. Providing education at primary and secondary levels is essential irrespective of their gender for effective and efficient outcomes (UNESCO 2017). Enable equal access to both women and men for technical and tertiary education and ensure their fit for employment, job, and entrepreneurship. In addition to that, technical education must provide an opportunity for all the learners to acquire the necessary skills and knowledge indispensable to practice sustainable development. For achieving all the above-stated objectives, building schools and the necessary infrastructure is the most imperative component. Most importantly, SDGs need to be incorporated as part of the curriculum and encourage the students to practice sustainable ways in day-to-day life. Interdisciplinary courses should be implemented in the engineering curriculum so that the technical knowledge essential to address global challenges through new technologies and processes would be acquired. Therefore, to achieve SDGs holistic approach and basic infrastructure vital for that must be developed by civil engineers along with safeguarding the environment.

25.5.5 SDG 5

SDG 5 is to achieve gender equality and empower all women and girls. Girls and women represent nearly half of the world's population and hence contributing half of the potential. Gender inequality persisted in society is stagnating the social progress. Therefore, gender equality is the key driver for development, and it allows women to cross their social barriers and involve them in decision-making and policy-making processes. During the beginning stage of the project itself, engineers must identify the challenges that need to be addressed to minimize gender inequality and social inclusion problems. Then these objectives should be embedded in the action plans and provide a route map, which guides the planning before and during the project development. Infrastructure must widen the access to education, employment, and primary health services, which in turn enhance the economic growth of the nation by making women economically active and develop careers for them.

25.5.6 SDG 6

The aim of this goal is to provide clean water and sanitation facilities universally. This SDG is closely associated with SDG 1, SDG 2, SDG 3, SDG 4, SDG 5, SDG 7, SDG 11, SDG 12, SDG 13, and SDG 15. All the SDGs require a combined approach rather than an individual approach to achieve them successfully in a sustainable way. Ensuring clean water and sanitation facilities for all is the core of sustainable development. In recent years, pollution, overexploitation, and climate change impacted the quantity and quality of water resources and led to serious water stress conditions (WHO 2017). Climate change further worsens the water stress situation due to the increased frequency of natural disasters such as floods and droughts. Pollution is another major threat to the freshwater resources since wastewaters and wastes are being discharged into these resources without sufficient treatment (Rout et al. 2021a; Rout et al. 2017). Civil engineers saved billions of lives through various effective, innovative, and efficient technologies developed to treat the wastewater and supply clean water to society (Rout et al. 2020a; Akshay et al. 2016; Rout et al. 2021b; Shahid et al. 2020; Kwak et al. 2020; Rout et al. 2018). Consequently, the spreading of waterborne diseases is eradicated completely. Civil engineers designed and constructed efficient water distribution systems to ensure a clean and clear water supply for all. In addition to that, various low-cost waste treatment systems like anaerobic treatment systems and composting technologies promoted the production of methane and manure, respectively, from waste, thereby effectively contributed for environmental protection. For example, meeting the global phosphorus demand with limited resources is impossible, therefore to overcome this issue, alternatives such as the recovery of phosphorous from rich domestic and industrial wastewaters are advisable for sustainable utilization (Lavanya et al. 2019; Lavanya and Ramesh 2020; Lavanya and Ramesh 2021). Rainwater harvesting systems need to be implemented in each and every household to conserve water resources. Promote reduce, reuse, recycle, and recovery approach in treating the wastewaters and wastes. Basic sanitation facilities, including handwashing with soap and water, need to be ensured globally (UNDESA 2017). Sanitation practices not only help maintain personnel hygiene but also supports overall well-being. Cost-effective sanitation facilities need to be built by utilizing waste materials and making sanitation a circular economy (Rout et al. 2016b). Decentralized treatment options need to explore for effective management of waste that is being generated from sanitation facilities. Additionally, nature-based treatment solutions should be preferred over conventional energy-consuming treatment systems to promote sustainable development (Rout et al. 2015b).

25.5.7 SDG 7

The major intention of this goal is to provide affordable, reliable, and modern energy to all the people in the world. SDG 7 is the heart for the remaining SDGs and for sustainable development as well. Energy is the basis to improve living standards and economic growth (Mohanty et al. 2021; Rout et al. 2021c). Still, billions of people were lack of essential electricity sources. Promoting the utilization of renewable energy sources such as wind, solar, and geothermal energy contributes to reduced greenhouse gas emissions into the atmosphere. The role of civil engineering in progressing this SDG is not significant compared tó their role in other SDGs. New, affordable, and innovative technology development require greater involvement of electrical and mechanical engineers. However, civil engineers would play a small role in delivering those technologies to the people. Installation of electric poles, design, and construction of houses and multistory buildings with different renewable energy sources not only promote sustainable development but also contribute to environmental protection, which is possible only with the involvement of civil engineers. Advancement in SDG 7 spur the progress in remaining SDGs. Therefore, cross-sectorial coordination is required at all levels between policymakers and executive members.

25.5.8 SDG 8

The purpose of this goal is to promote sustained, inclusive, and sustainable economic growth, employment, and decent work for all. Railways, roadways, water supply, and electricity are the basic amenities underpinning the economic growth. Except electricity, remaining facilities design, construction, and maintenance would be done by civil engineers. Basic amenities like water, housing, and energy needs are also fulfilled by civil engineers and enable people to live healthy and productive lives. Additionally, civil engineers play an imperative role in utilizing locally available material and waste by-products from various industries; for example, usage of ash generated from the coal industry for construction and other purposes greatly reduces the negative impact on environment. Extensive research would help to utilize industrial waste in various applications (construction) and not only help to minimize waste handling issues and promote the circular economy concept and create new employment opportunities (Rout et al. 2015a) for all. Incorporating occupational health and safety measures at the workplace is utmost important to promote decent work for all.

In addition to that, promote tourism, which provides job opportunities to the local people and encourages them to showcase their culture and products to the tourists.

25.5.9 SDG 9

SDG 9 highlights the importance of resilient infrastructure, sustainable industrialization, and innovation. Industrialization is the underpinning for economic growth. Industrialization creates employment and job opportunities, thereby providing income to the individuals, mitigating poverty and hunger, delivering decent job opportunities for youth, and promoting well-being of people and the planet. Industrialization ensures the application of science, technology, and innovation and enhances the skills and education necessary to meet sustainable developmental goals. New innovative approaches must be affordable and offer easy access to every individual. In addition to that, retrofitting the existing industries with sustainable approaches would minimize the release of emissions into the atmosphere. On the other hand, transportation is the driving component for sustainable development. Access to all the locations is possible only with transportation, but conventional transportation techniques released a significant amount of greenhouse gasses into the atmosphere. However, sustainable transportation practices like utilization of electric vehicles instead of fossil fuel-based vehicles, etc., would help to overcome the shortcomings of conventional methods. To advance in SDG 9, incorporation of green engineering concept in the early stages of projects, usage of life cycle assessment approaches in infrastructure development is necessary.

25.5.10 SDG 10

This is aimed to eliminate the inequalities existed within and among the countries. Elimination of inequalities persisted in education, health, and income is possible only with economic growth. Economic growth enables opportunities for everyone. Additionally, civil engineers need to promote urbanization, thereby local people would be benefitted through employment. In addition to that, population explosion in the South Asia region offers dire warning of social and human costs in terms of housing, water supply, and sanitation. Thus, to overcome the water and sanitation issues, the development of low-cost water purification systems, conservation of existing water resources, and affordable solid waste management facilities is imperative and would uplift the level of people, which reduce the inequalities in the society and offer decent living for all. Moreover, minimizing the inequalities empowers the women and makes them participate in the workforce to address the issues related to inequalities. Various awareness programs highlighting the negative consequences of inequality must be conducted to reduce the inequalities.

25.5.11 SDG 11

The objective of this goal is to build sustainable cities and communities. To achieve the goal, it is absolutely necessary to provide access to affordable, adequate housing facilities, public transport, and upgrade the slums through various developmental programs. In addition to that building a new infrastructure that is resilient to natural disasters, planning and managing integrated systems for human settlements contribute to delivering this SDG efficiently. Furthermore, the impacts of human settlements on the environment need to be minimized by controlling air pollution and effective management of waste and wastewaters (Rout et al. 2016a; Lee et al. 2021). Civil engineers are part of this goal, and their presence is very much needed to advance this SDG. Civil engineers must utilize locally available options in terms of material, employment, and resources during the design and development phase of projects to raise the living standard of people. Sustainable cities' concept must ensure the incorporation of smart energy options, efficient water conservation systems, and the utilization of renewable energy sources for lighting and other purposes. Therefore, civil engineers' role is very much needed to build sustainable cities. Likewise, transportation and mobility are central to sustainable development and are possible only with involvement of civil engineers. Transportation facilities would promote economic growth and accessibility. Moreover, sustainable transportation advances health, rural productivity rate, urbanization, rural–urbanization linkages, and social equity by protecting the environment.

25.5.12 SDG 12

The objective of this goal is to ensure responsible consumption and protection of natural resources. This goal highlights the importance of environmental management of chemicals and waste using life cycle assessment by adopting reduce, reuse, recycling, and recovery approach. Sustainable tourism is another aspect that allows people to explore various biodiversity-rich areas, available products, and cultures in those areas. Moreover, circular economy approach from cradle to crave of material flow minimizes the waste accumulation and encourages its effective management. Adequate and proper intervention measures need to be adopted to conserve the natural resources for future generations. Civil engineers need to develop different technologies that enhance the utilization of waste materials in the infrastructure and transportation sector to address waste management issues. In addition, incentives must be given to those organizations that follow sustainable approaches to promote the sustainability concept.

25.5.13 SDG 13

This goal deals with the measures that need to be followed to combat climate change and its effects. Civil engineers are at the forefront to mitigate the impacts of climate change through the development of resilient infrastructure, transportation systems, and smart cities (Albino et al. 2015). Systems must be built in such a way that they must withstand extreme climate and weather conditions. In addition to that, developmental activities must emit the least amount of greenhouse gasses into the atmosphere to foster sustainability. Moreover, the utilization of fossil fuels needs to be minimized and the usage of renewable energy sources should be maximized for a safer environment (Lee et al. 2020). Moreover, integration of climate change policies in a national framework, raising education and awareness about climate change mitigation, adaptation, and risk minimization are imperative to deliver the SDGs and to attain food security for the growing population. Sustainable agricultural practices need to be implemented to protect the crops from extremely harsh conditions and also to meet the global food production demand. Most importantly a coordinated approach is necessary to achieve progress in this SDG.

25.5.14 SDG 14

This goal describes the necessity of conserving oceans, seas, and marine resources to achieve sustainable development. Oceans and seas are an integral part of the earth's ecosystem, hence it is our responsibility to protect and conserve them. Oceans and seas are the huge sources for marine foods and water transportation. Preserving the oceans, seas, and life within them is the responsibility of civil engineers. Additionally, preventing marine pollution from various anthropogenic activities is also the duty of civil engineers. Oceans and seas are the major source of sinks for greenhouse gasses and regulate the climate. Oceans and seas through various marine foods give life to many people. Stringent and stringent norms should be implemented by government bodies for disposing the waste into oceans and seas to protect the life within them.

25.5.15 SDG 15

This SDG's aim is to protect the life on land. Protecting biodiversity, including flora and fauna, controlling desertification, and supplying adequate food supplies to the needy people is essential for delivering this SDG (Verma et al. 2020). Sustainable solutions development is required to control deforestation and safeguarding the habitats of wildlife, which is one of the primary roles of civil engineers. In addition to that, protecting indigenous fauna is also a major responsibility since they are part of the ecosystem. The new innovative technologies developed by civil engineers must be in such a way that they provide an opportunity to live in harmony with nature and restore the ecosystem. For example, promoting urban agriculture above the roofs of buildings, thereby creating a habitat for several bird species. The severe loss of biodiversity has a significant negative impact on food security, the health of people, and nutrition. In addition to that, encourage partnerships to restore land resources and assessing the extent of desertification is extremely important to accomplish sustainability.

25.5.16 SDG 16

The objective of this goal is to ensure peace and justice for all and build effective, accountable institutions at all levels. The promotion of peace and inclusive societies through good governance and institutions is a priority for all engineers (UNDP and UN Environment 2018). Civil engineering professionals are partnering with educational institutions for regulation and accreditation and it is imperative for ensuring the competency among civil engineers of different countries. Institutions must ensure civil engineering education standards for the forthcoming graduates reflect ethical, inclusive, and sustainable engineering values. Organizations must develop frameworks that prevent corruption in civil engineering to maximize infrastructure benefits that promote sustainable development. For example, American society of civil engineers is an organization that provides guidelines, ethics of civil engineering, and, most importantly, provides the best human resources for fostering SDGs. Protecting children from all sorts of violence is one of the primary aims of this SDG. Moreover, eradicating child labor, female genital mutilation, and child marriages is essential throughout the globe for peace. Adequate intervention measures need to be implemented to protect the children from all sorts of violence.

25.5.17 SDG 17

This goal highlights the necessity of global partnerships to strengthen the implementation and revitalize sustainable development. This goal is majorly for advancing the remaining 16 SDGs through partnerships and a multi-stakeholder approach. Partnerships among various engineering disciplines or across nations and with international organizations would significantly contribute to developing innovative, novel, and efficient technologies and lays a roadmap for successful implementation of these technologies for sustainable development. Knowledge transfer within and among countries, capacity development for inclusive approaches, financial inclusion and multi-stakeholder partnership, and voluntary commitment is essential to discourse today and future issues. Capacity-building activities are determined to minimize the negative impacts of present social, economic, and environmental challenges on SDGs progress. Capacity-building actions promote integrated, coordinated, and cohesive sustainable development practices by sharing lessons learned and good practices by organizing various workshops. Financial assistance is very much needed to the developing countries for practicing sustainable strategies. Partnerships through collaborative network share the best practices and approaches that nurture the SDGs progress.

25.6 Civil Engineering Tools for Sustainable Development

Civil engineers can bring a significant change in natural sources, water, and energy usage patterns, thereby reducing the adverse impacts on the environment. Several tools have been developed to assess and mitigate the impacts of anthropogenic activities on the environment.

25.6.1 Environmental Impact Assessment

Initial environment examinations for the upcoming and proposed projects are necessary to guarantee that the project's planning, design, and implementation do not adversely affect the environment. Therefore, performing initial environmental studies or feasibility studies at the early stage of the project would provide an opportunity for the engineers to verify different possible alternatives to avoid or reduce the negative social and environmental effects, thus making it sustainable.

25.6.2 Strategic Environmental Assessment

This is another tool used by environment regulating agencies and guides decision-making over project developmental activities toward sustainability. It is mainly performed for policies and programs before the project has been sanctioned or approved by a competent authority. This assessment is performed at the beginning stage of the project to analyze sensitivity to environment, availability, and utilization of alternate resources, financial status, investment, etc. Proper and effective investigation at the early stage provides a brief idea about necessary measures that need to be taken to make the project sustainable.

25.6.3 Technology Modification and Green Building Approach

Life cycle assessment for newly developed alternative materials and technologies need to be assessed, thus this analysis provides choices for builders and consumers. The sustainable construction of buildings requires more focus on the energy aspect rather than its lifespan. Civil engineers must develop effective and efficient infrastructure with a greater lifespan. In addition to that, civil engineers must ensure that the builders follow the urban and architectural standards by taking social and cultural concerns into account. As far as possible, utilization of locally available materials needs to be preferred over importing to avoid transportation costs and enable opportunities for local people, thus promoting sustainable approaches.

Geographical information systems are effective in identifying sensitive ecosystems, soil types, land uses, watersheds, and existing infrastructure in the proposed project location. In addition to that, possible future scenarios with alternative solutions can be shown pictorially and offer a better understanding of the limitations and growth of the proposed site.

Green building is an approach to enhance the efficiency of resource (material, energy, and water) usage patterns to prevent adverse impacts on humans and the environment throughout the life of a building. This is possible only with adequate planning, designing, implementation, operation, maintenance, and efficient waste management. The Green building concept would significantly contribute to minimize operating costs, enhance livability, productivity, and, most importantly, health. Development using the green building method would provide harmony with nature and promote sustainable practices.

25.7 Conclusions

Sustainable development is consistent only with a combined and cohesive systems approach, which nurtures to resolve the issues pertaining to social, economic, and environmental components. This study highlights and indicate a clear direction for civil engineering professionals to collaborate, participate, and engage themselves in different infrastructure developmental projects for effective and efficient greater quality sustainable outcomes. Moreover, civil engineers should offer exceptional services through strategies, advancement in technologies, and innovative solutions beyond their minimum compliance to foster sustainable development. Furthermore, higher education institutions must incorporate SDGs in the civil engineering curriculum for the forthcoming graduates to create a sustainable world. The demand for civil engineering expertise and skills would escalate when compared to past years to accomplish the SDGs by 2030.

References

Akshaya VK, Prangya RR, Puspendu B, Rajesh DR (2016) Anaerobic treatment of wastewater. Green Technologies for Sustainable Water Management 297–336

Albino V, Berardi U, Dangelico RM (2015) Smart cities: definitions, dimensions, and performance. J Urban Technol 22:3–21. https://doi.org/10.1080/10630732.2014.942092

Griggs D, Stafford-Smith M, Gaffney O, Rockström J, Öhman MC, Shyamsundar P et al (2013) Policy: sustainable development goals for people and planet. Nature 495:305–307. https://doi.org/10.1038/495305a

https://sdgs.un.org/goals

Kelly WE, Mohsen JP, Haselbach LH (2016) Engineering the UN Post-2015 sustainable development goals. In: ASEE's 123rd Annual Conference & Exposition, New Orleans, USA

Kwak W, Rout PR, Lee E, Bae J (2020) Influence of hydraulic retention time and temperature on the performance of an anaerobic ammonium oxidation fluidized bed membrane bioreactor for low-strength ammonia wastewater treatment. Chem Eng J 386:123992

Lavanya A, Ramesh SKT (2020) Effective removal of phosphorous from dairy wastewater by struvite precipitation: process optimization using response surface methodology and chemical equilibrium modeling. Sep Sci Technol 56:395–410

Lavanya A, Ramesh SKT (2021) Crystal seed-enhanced ammonia nitrogen and phosphate recovery from landfill leachate using struvite precipitation technique. Environ Sci Pollut Res 287

Lavanya A, Ramesh ST, Nandhini S (2019) Phosphate recovery from swine wastewater by struvite precipitation and process optimization using response surface methodology. Desalin Water Treat 164:134–143

Lee E, Rout PR, Bae J (2021) The applicability of anaerobically treated domestic wastewater as a nutrient medium in hydroponic lettuce cultivation: Nitrogen toxicity and health risk assessment. Sci Total Environ 780:146482

Lee E, Rout PR, Kyun Y, Bae J (2020) Process optimization and energy analysis of vacuum degasifier systems for the simultaneous removal of dissolved methane and hydrogen sulfide from anaerobically treated wastewater. Water Res 182:115965

Mohanty A, Rout PR, Dubey B, Meena SS, Pal P, Goel M (2021) A critical review on biogas production from edible and non-edible oil cakes. Biomass Convers Biorefinery 1–18

Moyer JD, Hedden S (2020) Are we on the right path to achieve the sustainable development goals? World Dev 127. https://doi.org/10.1016/j.worlddev.2019.104749

Rout PR, Bhunia P, Dash RR (2015a) Effective utilization of a sponge iron industry by-product for phosphate removal from aqueous solution: a statistical and kinetic modelling approach. J Taiwan Inst Chem Eng 46:98–108

Rout PR, Bhunia P, Dash RR (2015b) A mechanistic approach to evaluate the effectiveness of red soil as a natural adsorbent for phosphate removal from wastewater. Desalin Water Treat 54(2):358–373

Rout PR, Bhunia P, Dash RR (2017) Simultaneous removal of nitrogen and phosphorous from domestic wastewater using Bacillus cereus GS-5 strain exhibiting heterotrophic nitrification, aerobic denitrification and denitrifying phosphorous removal. Biores Technol 244:484–495

Rout PR, Bhunia P, Lee E, Bae J (2021a) Microbial Electrochemical Systems (MESs): promising alternatives for energy sustainability. The Way to a Sustainable Modern Society. In: Alternative Energy Resources. pp 223–251

Rout PR, Bhunia P, Ramakrishnan A, Surampalli RY, Zhang TC, Tyagi RD (2016a) Sustainable hazardous waste management/treatment: framework and adjustments to meet grand challenges. In: Sustainable Solid Waste Management. pp 319–364

Rout PR, Dash RR, Bhunia P (2016b) Nutrient removal from binary aqueous phase by dolochar: highlighting optimization, single and binary adsorption isotherms and nutrient release. Process Saf Environ Prot 100:91–107

Rout PR, Dash RR, Bhunia P (2020a) Insight into a waste material-based bioreactor for nutrient removal from domestic wastewater. In: Recent Developments in Waste Management. Springer, Singapore, pp. 397–407
Rout PR, Dash RR, Bhunia P, Rao S (2018) Role of Bacillus cereus GS-5 strain on simultaneous nitrogen and phosphorous removal from domestic wastewater in an inventive single unit multi-layer packed bed bioreactor. Biores Technol 262:251–260
Rout PR, Shahid MK, Dash RR, Bhunia P, Liu D, Varjani S, Zhang TC, Surampalli RY (2021b) Nutrient removal from domestic wastewater: a comprehensive review on conventional and advanced technologies. J Environ Manage 296:113246
Rout PR, Verma AK, Bhunia P, Surampalli RY, Zhang TC, Tyagi RD, Brar SK, Goyal MK (2020b) Introduction to sustainability and sustainable development. Sustain Fundam Appl 1–19
Rout PR, Zhang TC, Bhunia P, Surampalli RY (2021c) Treatment technologies for emerging contaminants in wastewater treatment plants: a review. Sci Total Environ 753:141990
Shahid MK, Kashif A, Rout PR, Aslam M, Fuwad A, Choi Y, Park JH, Kumar G (2020) A brief review of anaerobic membrane bioreactors emphasizing recent advancements, fouling issues and future perspectives. J Environ Manage 270:110909
UN (2019) The sustainable development goals progress report 2019, United Nations Economic and Social Council. https://unstats.un.org/sdgs/files/report/2019/secretary-general-sdg-report-2019--EN.pdf
UN (2020) The sustainable development goals progress report 2020, United Nations Economic and Social Council. https://unstats.un.org/sdgs/files/report/2020/secretary-general-sdg-report-2020--EN.pdf
UNDESA. (2017) SDG indicators metadata repository. https://unstats.un.org/sdgs/metadata/
UNDG (2017) Mainstreaming the 2030 agenda for sustainable development: reference guide to UN country teams, United Nations Development Group. https://undg.org/wp-content/uploads/2017/03/UNDG-Mainstreaming-the-2030-AgendaReference-Guide-2017.pdf
UNDP and UN Environment. (2018) Managing mining for sustainable development: A sourcebook. United Nations Development Programme, Bangkok
UNESCO (2016) Towards resilient non-engineered construction: guide for risk-informed policy making, Paris, UNESCO Publishing. https://unesdoc.unesco.org/ark:/48223/pf0000246077
UNESCO (2017) Cracking the code: Girls' and women's education in science technology, engineering, and mathematics (STEM), Paris: UNESCO Publishing. https://unesdoc.unesco.org/ark:/48223/pf0000253479
UNESCO (2021) Engineering for sustainable development, United Nations Educational, Scientific and Cultural Organization
UNGA (2015) Transforming our world: The 2030 agenda for sustainable development, A/RES/70/1, United Nations General Assembly
Verma AK, Rout PR, Lee E, Bhunia P, Bae J, Surampalli RY, Tyagi RD, Lin P, Chen Y (2020) Biodiversity and sustainability. Sustain Fundam Appl 255–275
WCED (1987) Our common future. World Commission on Environment and Development
WHO (2017) Progress on drinking water, sanitation, and hygiene: 2017 update and SDG baselines, World Health Organization
World Bank (2015) Global monitoring report 2015/2016: development goals in an era of demographic change. The World Bank. https://doi.org/10.1596/978-1-4648-0669-8

Chapter 26
Towards a Sustainable and Resilient Infrastructure Through Interdependency Among Performance Indicators

Suchith Reddy Arukala

26.1 Introduction

to United Nations (UN), by the end of the year 2050 globally around 6.3 billion people are expected to live in the cities (Tathagat and Dod 2015) due to which the urban inhabitants increase rapidly with huge requirement of infrastructural facilities for transportation, housing, health, and education. This unintended population growth suffers with natural resource, energy consumption, and pollution leading to environmental degradation. Today's cities energy consumption accounts to 70% of Greenhouse Gas emissions (GHG) (Hossain et al. 2019; McCormick et al. 2013). This major swift from rural to urban occupants demands a huge amount of energy and resource consumption by emitting pollution and waste. In addition, the building operating consequences further ruin the environmental ecology. In comparison to developed countries, the transformation of traditional construction to sustainable construction is vigorous in developing countries, which have got new trend in accepting the sustainable building guidelines (Aghimien and Aigbavboa 2019; Magent et al. 2009; Reddy et al. 2019). The challenging solution for these emerging problems is a paradigm swift to sustainable construction (Fig. 26.1). At this growth rate in the next 10 years, India will be using huge material resources at a much faster rate than they have ever been used. Along with growth rate, depletion of the environment and natural resources will be very high (Dhanjode et al. 2013). Recently, Government of India has implemented the development plans and concept of smart cities, which can contribute to ecological imbalance and carbon footprint either directly or indirectly. The viable solution to tackle this enormous growth in urban transformation is to shift the current scenario to a sustainable urban development model. Many researchers have defined sustainable development according to their scope of research, where

S. R. Arukala (✉)
Department of Civil Engineering, Kakatiya Institute of Technology & Science, Warangal, Telangana 506010, India

K. R. Reddy et al. (eds.), *Advances in Sustainable Materials and Resilient Infrastructure*, Springer Transactions in Civil and Environmental Engineering,
https://doi.org/10.1007/978-981-16-9744-9_26

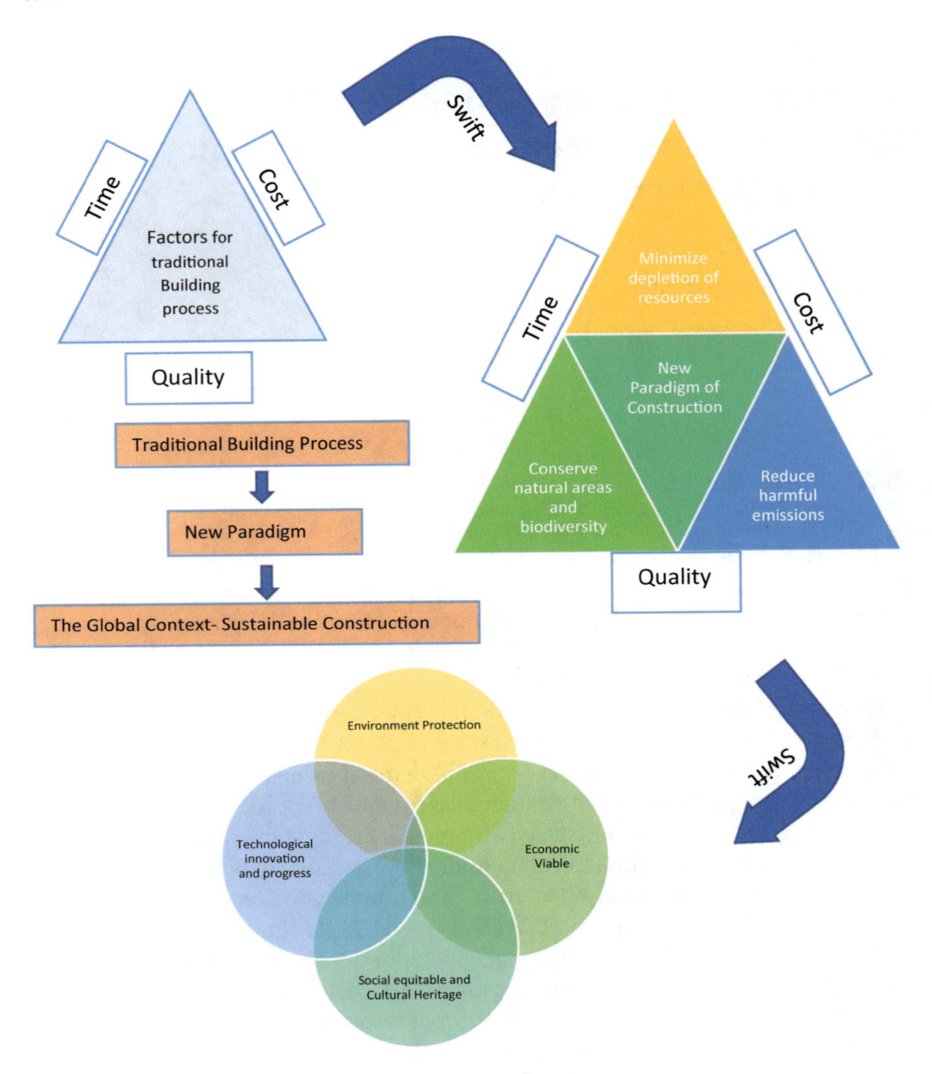

Fig. 26.1 Challenges of construction industry towards sustainability

the whole idea revolves around the main definition which was given by the Brundtland Commission report (Keeble 1988). According to Gesellschaft für Internationale Zusammenarbeit (GIZ), under the Economic Policy Forum (EPF), it was recognized that India needs further action and development in three broad areas: (1) Policy and regulation, (2) Capacity and skills, and (3) Awareness and understanding of benefits (Economic Policy Forum, 2014). The aspects of sustainability change with factors as location, climate variations, local context, topographical, culture, and heritage (Ding, 2008). Adopting sustainable principles in the construction sector can reduce

GHG emissions and carbon footprint (Jain et. al. 2013). To enhance the likelihood of sustainable construction, the first step has to be taken to identify substantial indicators for all the dimensions of sustainability (Zhou and Castro 2011).

26.1.1 *Approach to Building Sustainability*

The sustainable building excites the improvement of global ecology in the construction industry in all the phases of construction, and enables the monitoring of sustainable performance by indicators and criteria. There are several assessment tools that have failed to implement in developing countries considering only environmental aspects as a whole to assess the building performance. Most of the methods contain prescriptive requirement that mainly focussed on environment and economy (Rosa 2013). The building is said to be sustainable only when it obeys the principles of sustainability throughout its life-cycle. The assessment of building through sustainable indicators and criteria involves multi-dimensional aspects (Bamgbade et al. 2019; Li et al. 2005). The present study is focussed on Environmental, Social, Economic, and Technological criteria in all the phases of building life-cycle. The main objective of the study is to identify, evaluate, analyse, and measure the interdependency of sustainable indicators on sustainable criteria using Multicriteria Decision-Making Methods. The purpose of specific sustainable indicators and criteria is to gather the information for decision-making throughout the life-cycle thinking of the building, material, and equipment in order to develop a building sustainable assessment tool, etc.

26.2 Methodology

Approaches based on the performance of building-related process can meet the design objective of the building. The hierarchical model facilitates to provide a common platform for assessing the building in all disciplines of sustainability and also serves for the development of design and technical solution for achieving sustainability (Ubarte and Kaplinski 2016). The sustainable indicators performance can be assessed by considering multi-dimensional criteria (Banani et al. 2016). Scientific evidence proposes that the assessment of significant performance indicator can be performed by a consensus-based process which best suits the comprehensive analysis (Alyami and Rezgui 2012). The proposed methodology as shown in.

Figure 26.2, and has been developed based on how sustainability in construction has to be achieved in developing countries like India with varied local and regional context, climate changes, construction procedures, topographical and cultural changes, keeping in mind the four dimensions of sustainable criteria, the questionnaire survey was drawn out with 7-point Likert scale, and experts from

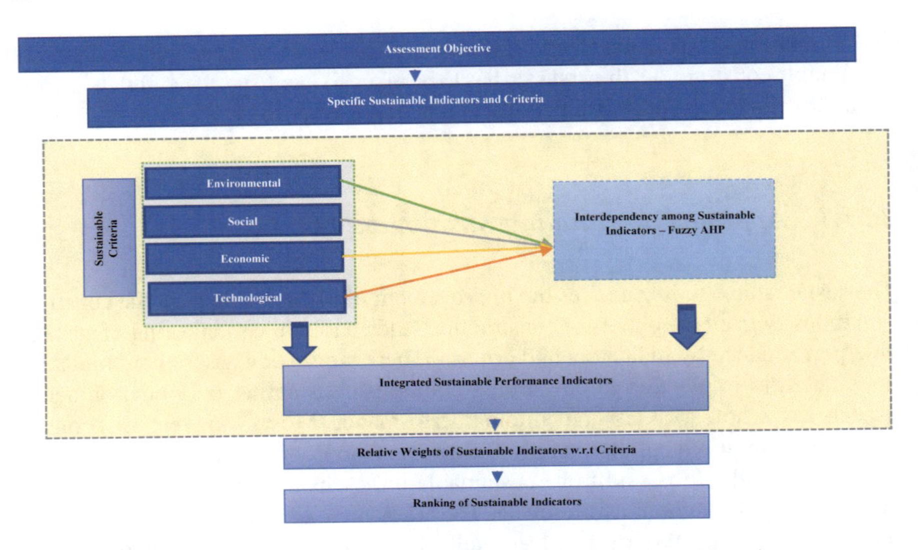

Fig. 26.2 Methodology for assessing the specific sustainable indicators and criteria

public and private organizations participated. The collected survey results were segregated among academicians, designers, architects, consultants, clients, contractors, and others to analyse their significance towards sustainability. The data extracted was observed to be consistent using Cronbach's alpha coefficient for four dimensions of sustainability are found to be above 0.85. They are then processed, analysed, and interpreted using statistical techniques to extract the required information as shown in Fig. 26.2. This information is then analysed using Fuzzy Analytical Hierarchy Process, a pairwise comparison Multi-Criteria Decision-Making (MCDM) method to establish interrelationship among criteria and indicators to rank the overall sustainable performance-based indicator with respect to sustainable criteria.

26.3 Results and Discussions

Considering only the environmental aspects existing in the literature and ignoring socio-economic and technological impacts of construction is not adequate to assess the sustainability accurately. Most of the performance indicators related to environmental aspect are ranked high Among all indicators, about 50% of them were aligned towards environmental criteria and the remaining were almost equally distributed to other criteria with respect to their ranking as per Fig. 26.3. Considering equal importance for all indicators, the average weights attained for Environmental, Technological, Social, and Economic criteria were 27.21%, 25.61%, 24.74, and 22.43%, respectively. It is noticed that among all stakeholders, Engineers, Academicians, and Consultants have given higher importance towards sustainability aspects. Among all

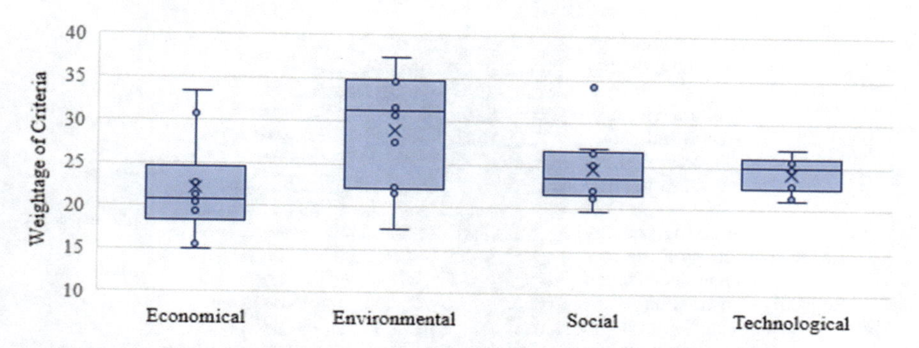

Fig. 26.3 Non-parametric observations of sustainable performance of criteria for all indicator

indicators, it was observed that the cost indicator has been rated high. This might be due to high initial investment to gain the benefits during Serviceability and Functionality. From Table 26.1, it is evident that the GHG emission and controlling pollution were ranked high on the environmental criteria. Similarly, cost and profitability indicators have been ranked high in the economic criteria, social welfare, and cultural heritage for social and innovative technology and design process in the technological aspect. Fig. 26.4 exhibits the level of importance of each indicator. The proposed criteria and indicators play a vital role in achieving sustainable construction. By integrating several stakeholders from various disciplines in the process of assessment

Table 26.1 Overall sustainable ranking of specific sustainable indicators

Specific sustainable indicator	Criteria	Rank
Reducing GHG emission	Environmental	1
Innovative technology	Technological	2
Controlling pollution	Environmental	3
Preserving ecology	Environmental	4
Social welfare	Social	5
Cost	Economical	6
Cultural heritage	Social	7
Water reducible and conservation	Environmental	8
Renewable energy resources	Environmental	9
Health and safety	Social	10
Design process	Technological	11
Reducing construction waste	Environmental	12
Profitability	Economical	13
Functionality and usability	Technological	14
Human satisfaction	Environmental	15

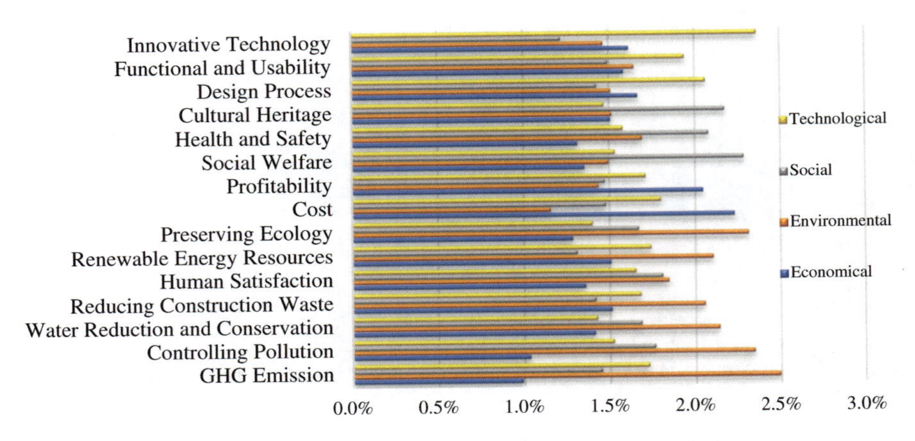

Fig. 26.4 Normalized weightage % of sustainable criteria with respect to indicators

provides creative sustainable measures, strategies, and initiatives. The key sustainable performance indicators analyzed can be utilized to further develop a sustainable assessment rating tool and also aids to frame the guidelines, statutory rules, and regulations. Further, these findings will assist the decision/policymakers, client, contractors, architects, engineers, and designers to implement sustainability aspects in the construction industry and optimize the output for achieving a sustainable construction.

26.4 Conclusions

Fast developing countries like India are typically facing the paradigm shift to examine and implement sustainable construction. The survey findings show that India is facing a shortage of Technological aspects in achieving sustainability. For a comprehensive evaluation, one must consider all aspects: Environment, Social, Economic, and Technological rather than solely focussing on environmental aspects. In view of reducing waste, resource, and energy consumption, it is recommended to adopt the technological aspects in every stage of construction activity. The proposed criteria and indicators play a vital role in achieving sustainable construction. The findings reveal that the determined 15 indicators in relation to 4 criteria are most relevant to assess the sustainable performance instead of assessing with numerous indicators. The analyses undertaken in the present study using Fuzzy AHP method determine the efficiency of the opted criteria in achieving sustainability. The average weights attained for Environmental, Technological, Social, and Economic criteria were 27.21%, 25.61%, 24.74, and 22.43%, respectively. The study has built its strength by integrating stakeholder responses from various disciplines in finding the interdependency among indicators. The key sustainable performance indicators analysed can be utilized to further develop a sustainable assessment rating tool and also aids to frame the guidelines, statutory rules, and regulations. Further, these findings will assist the decision/policymakers, client, contractors, architects, engineers, and designers to implement sustainability aspects in the construction industry and optimize the output for achieving a sustainable construction.

References

Aghimien DO, Aigbavboa CO (2019). Microscoping the Challenges of Sustainable Construction in Developing Countries. https://doi.org/10.1108/JEDT-01-2019-0002

Alyami SH, Rezgui Y (2012) Sustainable building assessment tool development approach. Sustain Cities Soc 5(1):52–62. https://doi.org/10.1016/j.scs.2012.05.004

Bamgbade JA, Kamaruddeen AM, Nawi MNM, Adeleke AQ, Salimon MG, Ajibike WA (2019) Analysis of some factors driving ecological sustainability in construction firms. J Clean Prod 208:1537–1545. https://doi.org/10.1016/j.jclepro.2018.10.229

Banani R, Vahdati MM, Shahrestani M, Clements- D (2016) The development of building assessment criteria framework for sustainable non-residential buildings in Saudi Arabia. Sustain Cities Soc 26:289–305. https://doi.org/10.1016/j.scs.2016.07.007

Dhanjode, C. S., Ralegaonkar, R. V, & Dakwale, V. A. (2013). Design and Development of Sustainable Construction Strategy for Residential Buildings: a Case Study for Composite Climate. *International Journal of Sustainable Construction Engineering & Technology*, *4*(1), 2180–3242. http://penerbit.uthm.edu.my/ojs/index.php/IJSCET

Ding GKC (2008) Sustainable construction-The role of environmental assessment tools. J Environ Manage 86(3):451–464. https://doi.org/10.1016/j.jenvman.2006.12.025

Economic Policy Forum. (2014). *Promoting sustainable and inclusive growth in emerging economies : Green Buildings.* 1–62. https://economic-policy-forum.org/wp-content/uploads/2016/02/Sustainable-and-Inclusive-Growth-Green-Buildings.pdf

Hossain U, Sohail A, Ng ST (2019) Resources, Conservation & Recycling Developing a GHG-based methodological approach to support the sourcing of sustainable construction materials and products. Resour Conserv Recycl 145(March):160–169. https://doi.org/10.1016/j.resconrec.2019.02.030

Jain M, Mital M, Syal M (2013) Obstacles and Catalysts Associated with Implementation of LEED-EB?? India. *Environment and Urbanization Asia* 4(2):349–363. https://doi.org/10.1177/0975425313511164

Keeble BR (1988) The Brundtland Report: "Our Common Future." Med War 4(1):17–25. https://doi.org/10.1080/07488008808408783

Li Z, Chau CK, Zhou X (2005) Accelerated Assessment and Fuzzy Evaluation of Concrete Durability. J Mater Civ Eng 17(3):257–263. https://doi.org/10.1061/(ASCE)0899-1561(2005)17:3(257)

Magent CS, Korkmaz S, Klotz LE, Riley DR (2009) A design process evaluation method for sustainable buildings. Architectural Engineering and Design Management 5(1–2):62–74. https://doi.org/10.3763/aedm.2009.0907

McCormick K, Anderberg S, Coenen L, Neij L (2013) Advancing sustainable urban transformation. J Clean Prod 50:1–11. https://doi.org/10.1016/j.jclepro.2013.01.003

Reddy, A. S., Kumar, P. R., & Raj, A. (2019). Quantitative Assessment of Sustainable Performance Criteria for Developing a Sustainable Building Assessment Tool (SBAT). *International Conference on Sustainable Infrastructure 2019: Leading Resilient Communities through the 21st Century. Reston, VA: American Society of Civil Engineers, 2019.*, 689–702. https://doi.org/10.1061/9780784482650.073

Rosa, L. V. (2013). Assessing the Sustainability of Existing Buildings Using the Analytic Hierarchy Process. *American Journal of Civil Engineering*, *1*(1), 24. https://doi.org/10.11648/j.ajce.20130101.14

Tathagat, D., & Dod, R. D. (2015). Role of Green Buildings in Sustainable Construction-Need, Challenges and Scope in the Indian Scenario. *IOSR Journal of Mechanical and Civil Engineering Ver. II*, *12*(2), 2320–2334. https://doi.org/10.9790/1684-12220109

Ubarte I, Kaplinski O (2016) Review of the sustainable built environment in 1998–2015. Engineering Structures and Technologies 8(2):41–51. https://doi.org/10.3846/2029882X.2016.1189363

Zhou H, Castro D (2011) Key Performance Indicators for Infrastructure Sustainability - A Comparative Study between China and the United States. Advanced Materials Research 250–253:2984–2992. https://doi.org/10.4028/www.scientific.net/AMR.250-253.2984

Chapter 27
Imbibing Energy Efficiency in Buildings Through Sustainable Materials—A Review

P. Mani Rathnam and Shashi Ram

27.1 Introduction

A large portion of the world's total energy consumption and carbon dioxide emissions are attributed to buildings. In the construction industry, 35% is total energy consumption and accounts for 38% carbon dioxide emission. Also, around 55% of the world's electricity is consumed by buildings (United Nations Environment Programme 2020). Although progress towards more environment-friendly buildings and construction is being made, advancements are not keeping up with the increasing construction industry and escalating demand for energy services. To fulfil the Paris Agreement's global climate goals, the worldwide buildings sector's energy intensity per square metre must improve by 30% by 2030 compared to 2015 (United Nations Environment Programme 2020). India being a part of Paris Agreement, makes it important for the Indian construction industry to identify solutions for reducing the energy demand of buildings.

United Nations Habitat estimates that 65% of the 169 targets that form the basis for the 17 Sustainable Development Goals (SDGs) are related to regional and urban development (Oborn and Walters 2020), and the built environment professionals play a crucial role in the design of cities and human settlements that meet the requirements of 11th SDG goal for inclusive, safe, resilient, and sustainable development.

To combat climate change and drive sustainable development, there is an urgent need to progress towards a pollution-free globe by using sustainable materials and practises. Significant action in this area will be helpful to achieve SDG objectives. Solutions that are both technologically and financially feasible exist, but stronger

P. M. Rathnam · S. Ram (✉)
Department of Civil Engineering, N.I.T. Warangal, Telangana 506004, India
e-mail: rshashi@nitw.ac.in

P. M. Rathnam
e-mail: pmanirathnam@student.nitw.ac.in

K. R. Reddy et al. (eds.), *Advances in Sustainable Materials and Resilient Infrastructure*, Springer Transactions in Civil and Environmental Engineering,
https://doi.org/10.1007/978-981-16-9744-9_27

regulations and collaborations for effective implementation are needed. So that significant reduction of carbon footprint can be achieved to make our built environment sustainable and energy efficient for the long-term survival of humankind (Harrison 2020).

27.2 Government Policies

Energy policy refers to how a certain organization (typically a government) has chosen to approach challenges of energy development, such as energy conversion, distribution, and usage. Examples of energy policy features include legislation, international treaties, investment incentives, energy conservation recommendations, taxes, and other public policy approaches. Since modern economies rely heavily on energy, a functional economy requires not just labour and capital but also energy for production, transportation, communication, agriculture, and other reasons. These energy policies also apply to the building and construction sector which are highly energy intensive. The building sector consumes 20% of the total energy generated globally, with residential and commercial end users accounting for the majority of it (Zabalza Bribián et al. 2009). Building energy regulations are usually promulgated by the government to control the design and operation of buildings to minimize or manage their energy use. Several variables influence the energy consumption of buildings, including the building type and construction as well as the efficiency of cooling systems, heating, and ventilation during operational stage of a building.

27.2.1 Building Energy Codes

Building energy regulations are being revised all around the world. Buildings global survey report, survey respondents, and broader literature searches have highlighted recent trends as shown in Table 27.1.

27.2.2 Minimum Energy Performance Standards (MEPS)

Minimum energy performance standard (MEPS) is a set of performance standards for energy-consuming equipment that limits the maximum amount of energy that a product can use while performing its intended activity. A government energy-efficiency authority will generally make a MEPS mandatory. It may contain standards that aren't directly connected to energy, in order to guarantee that increasing energy efficiency doesn't have a negative impact on overall performance and user happiness and for both freshly constructed and refurbished buildings, energy-saving solutions are being created. However, in addition to fundamental energy-efficiency measures,

Table 27.1 Buildings energy codes updates (United Nations Environment Programme 2020)

SI	Country	Updates
1	India	The 2018 Energy-efficient building Code, which aims to reduce building energy consumption by 30%, will be incorporated into the laws of Himachal Pradesh
2	Azerbaijan	In 2019, new energy-saving and efficiency legislation came into effect
3	Oman	Starting in 2021, the British Standards Institute in the UK will draft Oman's new national building code
4	Philippines	New specifications are being drafted, which will include energy-efficiency standards
5	Canada	The new national new building energy code model for the 2020 code cycle is approaching completion. These restrictions will apply to small buildings, residential buildings, as well as commercial and institutional structures. As part of the new standards, all grades or hierarchical performance targets must be implemented by 2030 in order for jurisdictions to be "net zero power ready"

established renewable energy technologies (Chwieduk 2003), as well as sustainable building materials, should be used in conjunction with passive building designs to achieve a significant reduction in energy usage. The European Union will incorporate all these features to achieve almost zero energy buildings for operational energy savings by 2020 in all the new buildings (Chel and Kaushik 2018). There are four primary ways of lowering building energy use, which finally leads to a reduction in carbon dioxide emissions (Chel and Kaushik 2018). These characteristics are as follows:

1. Solar energy harvesting for comfort passive temperature through building design and orientation.
2. Building construction with materials of low embodied energy is preferred.
3. Reducing the amount of energy used during building operations using energy-efficient household appliances.
4. Buildings that use renewable energy technology for fulfilling their building energy demand.

Modernization of buildings to improve their poor energy efficiency is also one of the methods to minimize energy usage. While building energy rehabilitation is one of the United Nations benchmarks and 2030 objectives for sustainable development, it isn't the only one. Due to the stringent rules put on the construction industry in recent years, the most significant improvements in energy efficiency happened after the 1990s. It was found that the energy use of the residential structure built in 2002 was reduced by 24% (Patiño-Cambeiro et al. 2019).

Therefore, there is a huge opportunity for energy conservation in buildings not only from the construction side but also on the regulatory side, i.e. the role of government is very critical in setting minimum energy performance standards and making them mandatory to follow.

27.2.3 Current Global Scenario

Extraction and use of non-renewable fossil fuels are plagued by excessive energy consumption and greenhouse gas emissions, which have a major impact on the environment. About 75% of greenhouse gas emissions are attributed to energy generation. Climate change mitigation has been a topic of discussion in recent decades. According to European Union targets announced in 2012, energy consumption and greenhouse gas emissions should be reduced by 20% by the year 2020 compared with the year 1990 levels. However, more current statistics in 2017 indicate that this target will not be achieved, despite the fact that the 2030 target is proposed to increase energy consumption by about 27% and greenhouse gas emissions by 40% (Patiño-Cambeiro et al. 2019). In this strategy, more effort is needed to achieve these goals. In 2019, 73 countries enacted mandatory or optional energy regulations, of which 4 countries indicated that they are enacting regulations for the year 2021 and beyond. Indian states like Madhya Pradesh and Telangana have implemented building energy codes voluntarily for part of the building sector from the year 2020 (United Nations Environment Programme 2020).

27.3 Sustainable Building

Assessing sustainability is increasingly important for long-term growth, especially in the global construction industry. Prevention of environmental degradation caused by facilities and infrastructure throughout their life cycle is the primary goal of sustainable design. It also aims to provide a safe and efficient building environment by utilization of solar and water energy.

Burdova and Vilcekova (2015) suggested and implemented a study of the Building Environmental Assessment System (BEAS) tool in Slovakia to fully protect resources (Burdova and Vilcekova 2015). The purpose of this tool is to determine the importance weights of the major field of assessment and evaluation of residential structures. Table 27.2 shows that each primary field in the BEAS system contains a number of indicators that indicate the aim of the assessment as well as the scope of the assessment. BEAS system methodology aims to improve the design, construction, operation, and maintenance of sustainable residential structures. Therefore, these indicators can be targeted for energy conservation and also reduce the harmful effects caused by buildings throughout their life.

Through the use of nature-based solutions in building, we can help restore nature to human projects, enhance the sectors towards environmentally friendly and sustainable development, and give a range of health care and pliable benefits (Enzi et al. 2017). Solutions that rely on natural processes and ecosystems are known as nature-based solutions. As a result of the advantages, ecosystem services are being incorporated into the built environment, the link between humans and nature may be re-established while also contributing to the resolution of modern urban issues. Urban

Table 27.2 The environmental assessment method for buildings (BEAS) at Slovakia (Burdova and Vilcekova 2015)

SI	Field	Sub-field	Indicators
1	Project planning and site selection	I. Site selection II. Site development	Distance between commercial and cultural organizations, distance to public green space, and distance to road transport infrastructure are all factors that should be considered while using formerly sensitive or important biological property This includes densification, the potential of changing a building's purpose, the impact of design on an existing urban landscape, and rules to limit the usage of personal cars. There should be a sufficient quantity of public green areas, as well as trees with shade potential, compatibility of urban design with local values of culture
2	Building construction	I. Materials II. Life Cycle Assessment (LCA)	Use of locally produced materials, material economy of structural components and building envelopes, radioactivity of building materials, Reusable, recyclable, and easy to deconstruct Global warming and acidification potential of construction materials and primary energy incorporated in building materials
3	Indoor environment	–	Thermal comfort, humidity, acoustics, natural lighting, total volatile organic compounds, indoor air quality, radon, nitrogen oxides, PM10, and microorganisms are some of the factors to consider

(continued)

Table 27.2 (continued)

SI	Field	Sub-field	Indicators
4	Energy performance	I. Operational energy II. Systems using renewable energy sources III. Energy Management	Lighting power, heating, hot water, ventilation, and cooling energy consumption; home appliances energy consumption Solar system; heat pump; photovoltaic technology; heat recovery Energy management system; operation and maintenance; the amount of local control of the lighting system; the amount of personal control of the technical system by the tenant
5	Water management	–	Water flow reduction and control; Surface water runoff; Supply of potable water; Using "grey water" filtering
6	Waste management	–	Measures to reduce waste generated by building operations; Measures to reduce emissions generated by building construction, operations, and destruction; Hazardous waste risk stemming from facility activities

heat island effect (UHI) may be reduced by using plants on and around buildings (Energy Sector Management Assistance Program (ESMAP) 2020). Several towns and developers across the world have taken strong steps to promote and implement natural solutions in order to restore nature to human settlements and structures, despite the fact that the advantages are not new.

Table 27.3 provides data about various nature-based solutions implemented in the cities around the world and also green infrastructure rules are prevalent in many areas across the world and encourage adoption of solutions based on nature in the sector, despite the fact that they are not (yet) mainstream. Hence sustainable, circular, and liveable buildings may be achieved through a combination of R&D expenditures, incentives, and consumer awareness.

Apart from the above-mentioned strategies, material selection also plays a major role in the reduction of the total energy demand of a building. Construction sector uses a huge amount of materials and pollutes both manmade and natural environments. Out of the total available sustainable building materials, few are solid concrete blocks, stabilized mud blocks, Autoclaved Aerated Concrete (AAC) blocks, fly ash

Table 27.3 Solutions based on nature (United Nations Environment Programme 2020)

SI	Cities around the world	Examples of solutions based on nature
1	City of Berlin	One thousand green roofs programme was announced, which will reimburse green roof installation expenses up to 60 euros per square metre and 60,000 euros per building up to 2023 totalling 2.7 million
2	City of Toronto	According to the Bylaw of green roof, buildings larger than 2,000 square feet must have between 20% and 60% of their usable roof area covered with vegetation or covered with plantations
3	Shanghai city	A target of 2 million square metres of green roofs by 2020 as part of the "Sponge City Initiative" to combat heat, pollution, and floods was set [48]. To achieve this goal, the city employed a combination of incentives and limitations, as well as green roof and wall construction subsidies of up to US$29/m^2 for qualified buildings
4	City of Melbourne	An evaluation tool for green infrastructure, the Green Factor Tool was created as part of the City's Greening the City programme to help in the design of ecologically sustainable, green infrastructure-integrated development
6	Thailand	Earlier this year, the Urban Rooftop Farm at Thammasat University became Asia's largest urban rooftop farm. An outdoor classroom for pupils and a green space for building users, the living roof is 22,000 m^2 in size (236,806 sq ft) with water management system, this green roof collects 11,718 cubic metres of rain and reduces runoff, while also responding to runoff issues and climate predictions in Thailand

lime gypsum blocks, and fly ash concrete blocks which have a low environmental effect. Blocks which have low thermal conductivity compared to conventional burnt clay (BC) bricks which help in reduction in operational energy of the building which in turn contributes to the energy efficiency of the building. Fly ash is a waste product that is utilized in significant amounts in Autoclaved aerated concrete blocks, fly ash concrete blocks, and fly ash lime gypsum blocks. Stone dust from stone crushing equipment is also used in materials, which has a cheaper cost and less influence on the environment (Tanuj and Khitoliya Ar Singh 2016). In comparison to hollow concrete or brick masonry, the use of local materials such as adobe, cow dung, straw, mud, and low energy-intensive materials results in lower CO_2 emissions. Mud Brick (Adobe) houses feature thick walls and a lot of thermal mass which significantly inhibits heat transmission. Therefore in the summer, mud-brick walls provide the best thermal comfort. Using locally available materials in building construction, embodied energy may be decreased by 37% also when energy-intensive components like cement, steel, and glass are substituted with local alternative materials (Cow dung, Straw, and Mud), less embodied energy is required (Mishra and Usmani 2013). Adobe has a low thermal conductivity, which means it maintains a more consistent temperature within a dwelling and prevents heat loss. It is cool during the day and warm at night in an adobe home, which was built in a hot and dry region in the American Southwest. Outside temperature fluctuations are delayed by 12 h due to the effect of the wall's filter (Duffin and Knowles 1981). Nanotechnology use will also improve the functioning of conventional materials while also lowering carbon emissions, energy usage, and raw material consumption (Oke et al. 2017). A detailed discussion of sustainable materials developed using waste is provided in Sect. 27.4.2 of this book chapter.

Hence, sustainable building materials are all related to the choice of materials produced from resource-efficient processes, for example, the selection of materials that produce a low content of energy, are supplied from local vicinity and use renewable sources of energy, and the selection of materials which contribute less GHG emissions to the atmosphere at the end of their lives.

27.4 Construction Materials

The built environment places a great strain on the world's resources, and the manufacturing of building materials is the primary source of greenhouse gas emissions during the life cycle of structures (Eddy 2019). Construction materials are anticipated to account for one-third of the growth in global material consumption which is estimated to be tripled by 2060. By 2060, it is estimated that concrete alone will account for 12% of global greenhouse gas emissions (OECD 2018). Housing, construction, and infrastructure use 40% to 50% of the resources mined globally (De Wit et al 2018). Homes, buildings, and infrastructure use between 40% and 50% of the world's extracted resources.

The world consumption of construction minerals is expected to be around 43 gigatonnes (Gt) per year, accounting for almost half of total resource production (Krausmann et al. 2018), including more than 4 Gt of cement (Andrew 2021), as well as aggregates, asphalt, brick, plaster, stone, and glass. An additional 0.6 Gt of steel, along with other polluting and energy-consuming metals such as aluminium and copper, will be mined and produced (Cullen et al. 2012). Likewise, the construction sector is heavily dependent on pristine resources. For sand and non-binding materials (such as crushed stone, slag or crushed concrete, or shale), 30 billion tonnes of sand must be removed annually, or around 4 tonnes per person, primarily from rivers and coastlines, which might aggravate current coastal area problems (World Business Council for Sustainable 2019). Each year, about 0.9 billion m^3 of wood products are consumed, necessitating huge amounts of forest and the limited usage of forest certification standards demonstrating sustainable harvesting. There are also many other materials used in large quantities that have disproportionate environmental impacts compared to the quantities consumed. These materials include asbestos, plastic window frames and architectural profiles, floor coverings, and wall coverings as well as pipes and thermal and electrical insulation.

Waste materials make for about 40% of mass after completion of building construction as well as the demolition of the building. Concrete, bricks, gypsum, tiles, ceramics, and excavated soil accounts for up to 30% of total waste generated in the European countries, and many of these materials are reusable (Iyer-Raniga and Huovila 2021). Metals are easily collected from construction and demolition debris in many nations due to their high salvage value (European Enviroment Agency 2020). As a result of downcycling, which reduces the quality and worth of materials while ignoring their manufacturing and processing consequences, a large portion of this waste is generated (Construction Leadership Council 2020).

27.4.1 *Circular Materials Flows in Construction*

As it is understood from the above section that materials play an important part in the energy consumption of a building. In order to reduce this energy consumption, it is important to understand material flow in a construction activity of a building. An understanding of the natural and built surroundings as well as the construction industry's consumption and flow of building materials is necessary throughout the building materials entire life cycle. This material flow in construction is depicted in Fig. 27.1 (Winterstetter et al. 2021). Data on the technical and other attributes of building materials, as well as their distribution across time and place, and their environmental, social, and economic consequences, must be supplemented with data on their amounts. This data will be used to:

(a) Assess supply and demand, linear approaches that have failed, and obstacles and circularity potential in order to determine which flows or processes require intervention.
(b) Replacement of materials as well as policy and technological changes are examined through scenario analysis.
(c) The discovery of links to other systems, such as energy and water.

In Fig. 27.1, we can observe that all the processes in the life cycle of a material, right from the excavation of raw materials from natural resources to the treatment or recycling, are generally a norm which is being followed, but after the disposal through anthropogenic resource mining the materials are reintroduced into the system as new

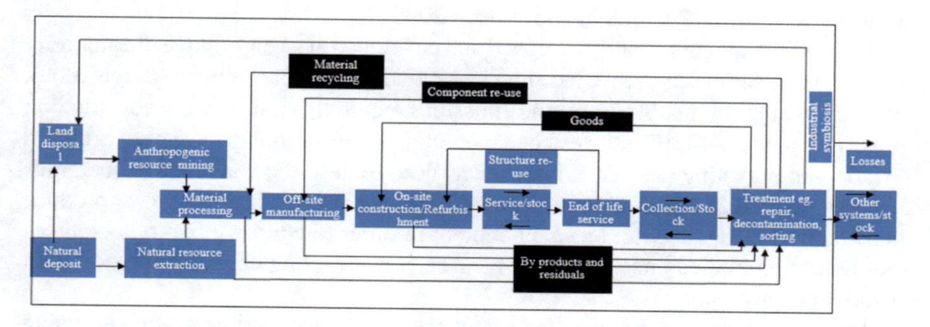

Fig. 27.1 Circular materials flows in construction (Winterstetter et al. 2021)

material. Also after the treatment stage the materials formed are being used in various processing stages as recycled material for material processing, component re-use for off-site manufacturing, and goods on-site construction. After the end of service, the materials lead to structural re-use for on-site construction. All the by-products that are generated in each of the processes are reintroduced into the flow from the treatment stage. After the treatment stage, the materials that cannot be introduced into the flow are introduced into other systems based on the usage.

27.4.2 Sustainable Building Materials

For a sustainable future, the selection of materials must take into account a number of factors. Materials that are resource and energy efficient, pollutant-free, and have no negative health impacts are considered sustainable building materials (Esin 2007), (Franzoni 2011). According to the following criteria, environmentally friendly building materials should be used:

1. Pollution prevention (including indoor air quality)
2. Efficiency in resources
3. Efficiency in energy (including initial and repeating embodied energy as well as greenhouse gas emissions)

The recovery and manufacture of raw materials, construction energy on-site, and transportation all contribute to the energy consumption of building materials. The embodied energy intensity of construction materials varies from region to region and from plant to plant, depending on energy sources, technologies, and production processes. The estimate of embodied energy includes both the initial and recurrent embodied energy of building materials. The initial embodied energy is accounted for by the materials used in construction, whereas the recurring embodied energy is needed during operation. As soon as the building is inhabited, and for the remainder of its economic life, its value is estimated. Whenever it comes to building materials energy efficiency, both initial and ongoing embedded energy play a major role. The amount of energy embodied in a building varies from 10 to 60% (Huberman and Pearlmutter 2008), (Puettmann and Wilson 2005).

If the buildings are properly designed and developed at an early stage, the majority of the rare materials can be replaced by less rare or renewable sources (Berge 2009). To decrease the amount of energy and emissions required in transporting construction materials, it is (Esin 2007) suggested that building supplies should have been acquired locally and recycling and reusing are terms used to describe construction materials that can be readily removed at the end of their useful lives. It is crucial to produce construction materials with recycled content in order to conserve natural resources and reduce embodied energy (Gao et al. 2001). If recycling creates more pollution it should be avoided.

Table 27.4 provides information about the building materials developed using waste. It is observed that the use of rice husk, cigarette butts, Recycled paper mill

Table 27.4 Reuse of industrial waste for development of building materials (Ram et al. 2018)

SI	Industrial waste	Observations
1	Rice husk	Sand and cement are used to make construction blocks, which have a density of 904 kg/m^3, a compressive strength of 3.3 MPa, a water absorption of 26%, and a thermal conductivity of 0.3 W/m K (Lertsatitthanakorn et al. 2009)
2	Cigarette butts	Fire bricks are developed using brown silty clayey sand as raw materials and the properties of the bricks are: density 1482–1949 kg/m^3, compressive strength 3–12 MPa, water absorption 9–18 wt%, and thermal conductivity is 0.4–0.8 W/m K (Kadir et al. 2009)
3	a. Recycle paper from mill waste b. wastage from Cotton	Bricks are developed using cement as raw material and the properties of the bricks are: density 560 kg/m^3, compressive strength 22 MPa, water absorption 100 wt%, and thermal conductivity is 0.2–0.3 W/m K (Rajput et al. 2012)
4	a. Olive pomace b. Agricultural waste (cleared underbrush, olive and fruit tree trimmings and energy crops	Building blocks are developed with cement as raw materials and the properties of the blocks are: density 1490–2037 kg/m^3, compressive strength 61 MPa, water absorption 9–19 wt%, and thermal conductivity is 0.7 W/m K (Carrasco Hurtado et al. 2014)
5	Sugarcane bagasse ash	Bricks are developed with quarry dust and lime as raw materials and the properties of the bricks are: density 1050–1409 kg/m^3, compressive strength 6–3 MPa, water absorption 19.7–20 wt%, and thermal conductivity is 0.4–0.5 W/m K (Madurwar et al. 2015)
6	Bio-briquette ash	Bricks are developed with sand and cement as raw materials and the properties of the bricks are: density 1340 kg/m^3, compressive strength 4 MPa, water absorption 19 wt%, and thermal conductivity is 0.5 W/m K (Sakhare and Ralegaonkar 2016)

(continued)

waste, cotton waste, agricultural waste, sugarcane bagasse ash, and cofired blended ash were helpful in developing walling materials. This research work promotes the concept of utilizing industrial waste for the development of building materials. Also from the table, it is observed that these bricks have thermal conductivity values ranging from 0.3 to 0.7 W/mK. The thermal condutivity of fly ash bricks is

Table 27.4 (continued)

SI	Industrial waste	Observations
7	Cofired blended ash	Bricks are developed with sand, cement, and ash as raw materials and the properties of the bricks are: density 1600 kg/m^3, compressive strength 5.86 MPa, water absorption 16.93 wt%, and thermal conductivity is 0.4 W/m K (Ram and Ralegaonkar 2018)

1.05 W/mK (Ram and Ralegaonkar 2018) and burnt clay bricks is 0.4 W/mK (Dondi et al. 2004). This range of thermal conductivity value 0.3–0.7 W/mK is lower than the burnt clay bricks and fly ash bricks. The achieved lower thermal conductivity value is advantageous in maintaining the comfortable temperature within a building. Which helps in reducing the operational energy demand of a building.

27.4.2.1 Bio-Based Materials

Biomaterials are often biodegradable alternative construction materials that assist to regenerate natural systems by mitigating the demand for excavation for building minerals and excavation operations. Carbon dioxide may be stored in some biobased materials and the soil can be replenished over the life by using construction biomaterials (Food and Agriculture Organization of the united nations 2019). Recycling post-agricultural waste into biomaterials helps to establish a circular economy by diverting rubbish from waste streams. They provide ground-breaking and cutting-edge solutions to severe shortage of construction materials. Construction of buildings using biomaterials is compatible with modern inventiveness, digital, and productive techniques such as prefabrication, modular construction, and 3-D printing.

In local applications, biomaterials provide a wide range of advantages. Climate-specific local biobased renewable materials offer favourable performance characteristics in climate-responsive design, while also giving additional money and jobs to local economies (Collet et al. 2017).

27.4.2.2 Phase Changing Materials (PCM)

Energy demands for space cooling might triple in the next 30 years (International Energy Agency (IEA) 2018), notably in hot and tropical nations, with residential structures accounting for more than two-thirds of this rise. Since 1990, annual worldwide sales of air conditioning systems have nearly tripled. A mere 8% of the world's 2.8 billion residents who experience daily temperatures over 25 °C throughout the year have air conditioners in their homes (International Energy Agency (IEA) 2018), (United Nations Environment Programme 2019). Also the increased trend in pay can be attributed to a variety of factors, including increasing incomes, predicted doubling of building floor space by 2060 (Agency 2019), and a warming planet with higher temperatures and more frequent heatwaves.

To tackle this crisis, sustainable materials like organic phase changing materials (PCM) can be used for better thermal comfort. It is common knowledge that all materials interact with their surroundings. The majority of materials, on the other hand, are unable to modify their qualities in response to the features of the environment in which they are used. As a result, the PCM's state changes according to the temperature of the surrounding area. It is at this time that PCMs collect and store energy by converting it to liquid form. Its ability to release previously stored energy is revealed as the temperature drops, resulting in the material's transition from liquid to solid form (da Cunha and de Aguiar 2020). Chemical and thermal stability, non-reactivity and corrosion resistance, compatibility with construction materials, and recyclable nature are the main advantages of organic phase change materials. They are available in a wide temperature range, undergo congruent phase change, and have a high latent heat of fusion and a low liquid phase undercooling capability. Other disadvantages of these materials include a poor enthalpy and heat conductivity as well as a low density and flammability and high cost (Sharma et al. 2009), (Cabeza et al. 2011).

27.4.3 Operational Energy

The energy needed for a structure, including lighting, heating, cooling, ventilation systems, and operating building devices, over their full service life is operational energy. Operational energy generally accounts for longer portion of service life of the infrastructure and can account for 80%–90% of the total structural energy. However, the operating energy of buildings has decreased considerably with the development of energy-efficient building systems and appliances. Knowing the R value, thermal conductivity, thermal transfer, reflectivity, thermal absorption, and thermal efficiency of the materials to be used and the combined energy effect of these materials on energy efficiencies may save a lot of money in the short and long term and reduce the energy demand. R value is used to assess how effectively heat cannot travel through a particular substance by a material. The higher the rating, the better the resistance and the thermal insulation power. The highest R value, about R-45, is for vacuum insulated panels. The thermal conductivity is the capacity of a material to transfer heat and lower the value better thermal comfort in the dwelling. At ambient temperature and atmospheric pressure, Aerogel has the least thermal conductivity and has a value of 0.013 W/mK. The rate of heat transfer through matter is known as thermal transmittance; the lower the value, the greater the insulating ability. As a result, low thermal conductivity and transmittance materials must be used for energy efficiency throughout the operating phase of a building. Thermal efficiency is a dimensionless performance measure of a device that consumes thermal energy; the greater the thermal efficiency of appliances in the dwelling, the less energy is consumed. Reflectivity is the reflecting quality or power of a surface or substance. A good reflective surface is smooth, glossy surfaces such as mirrors and polished metals that reflect light well. Dull and dark surfaces, such as dark textiles, do not reflect light well. Adopting materials with all of the aforementioned qualities will result in reduced operational energy and demand, contributing to a building's overall energy efficiency.

27.5 Conclusion

Sustainable building materials are generally thought of as natural materials that give specific benefits to users such as low maintenance, energy efficiency, enhanced health and comfort of people living in the house, and improved performance while being less harmful to the environment. The use of locally available material reduces the amount of energy required for the transportation of material to the site. Reuse of the industrial waste and also the waste generated by agriculture in development of materials for constructions contribute to the energy reduction involved in raw material extraction from the natural resources. These wastes can also be used for developing low thermal conductivity sustainable materials which help in reducing the operational energy requirement and enhancing the thermal comfort inside the built environment. More research has been conducted to develop sustainable materials to mitigate the global environmental crisis. Also adopting more measures in providing awareness towards sustainable materials in the construction sector will be a key factor towards sustainable development.

References

Agency IE (2019) Energy efficiency. Iea 110

Andrew RM (2021) Global CO 2 emissions from cement production Introduction to previous estimates of global cement emissions. 10(4):1–20. https://doi.org/10.5194/essd-10-2213-2018

Berge B (2009) The ecology of building materials. 2nd edn. Architectural Press

Burdova EK, Vilcekova S (2015) Sustainable building assessment tool in Slovakia. Energy Procedia 78:1829–1834. https://doi.org/10.1016/j.egypro.2015.11.323

Cabeza LF et al (2011) Materials used as PCM in thermal energy storage in buildings: a review. Renew Sustain Energy Rev 15:1675–1695. https://doi.org/10.1016/j.rser.2010.11.018

Carrasco Hurtado B et al (2014) An evaluation of bottom ash from plant biomass as a replacement for cement in building blocks. Fuel 118:272–280. https://doi.org/10.1016/j.fuel.2013.10.077

Chel A, Kaushik G (2018) Renewable energy technologies for sustainable development of energy efficient building. Alex Eng J 57(2):655–669. https://doi.org/10.1016/j.aej.2017.02.027

Chwieduk D (2003) Towards sustainable-energy buildings. Appl Energy 76(1–3):211–217. https://doi.org/10.1016/S0306-2619(03)00059-X

Collet F, Prétot S, Lanos C (2017) Hemp-Straw composites: thermal and hygric performances. Energy Procedia 139:294–300. https://doi.org/10.1016/j.egypro.2017.11.211

Construction Leadership Council (2020) Zero avoidable waste in construction. p 6

Cullen JM, Allwood JM, Bambach MD (2012) Mapping the global flow of steel: from steelmaking to end-use goods. Environ Sci Technol 46(24):13048–13055. https://doi.org/10.1021/es302433p

da Cunha SRL, de Aguiar JLB (2020) Phase change materials and energy efficiency of buildings: a review of knowledge. J Energy Storage 27. https://doi.org/10.1016/j.est.2019.101083

Dondi M et al (2004) Thermal conductivity of clay bricks. J Material Civil Eng 16. https://doi.org/10.1061/(ASCE)0899-1561(2004)16:1(8)

Duffin RJ, Knowles G (1981) Temperature control of buildings by adobe wall design. Sol Energy 27(3):241–249. https://doi.org/10.1016/0038-092X(81)90125-0

Eddy R (2019) The circular economy. Building Engineer 94(11):24–26. https://doi.org/10.36661/2596-142x.2019v1i1.10902

Energy Sector Management Assistance Program (ESMAP) (2020) Primer for cool cities: reducing excessive urban heat

Enzi V et al (2017) Nature-based solutions and buildings—the power of surfaces to help cities adapt to climate change and to deliver biodiversity BT—nature-based solutions to climate change adaptation in urban areas: linkages between science, policy and practice. In: Kabisch N et al (eds) Cham: Springer International Publishing, pp 159–183. https://doi.org/10.1007/978-3-319-56091-5_10

Esin T (2007) A study regarding the environmental impact analysis of the building materials production process (in Turkey). Build Environ 42:3860–3871. https://doi.org/10.1016/j.buildenv.2006.11.011

European Enviroment Agency (2020) Construction and demolition waste: challenges and opportunities in a circular economy. Briefing no. 14/2019, p 8

Food and Agriculture Organiszation of the united nations (2019) Forest products annual market review, 2018–2019. Unece

Franzoni E (2011) Materials selection for green buildings: which tools for engineers and architects? Procedia Engineering 21:883–890. https://doi.org/10.1016/j.proeng.2011.11.2090

Gao W et al (2001) Energy impacts of recycling disassembly material in residential buildings. Energy Buildings 33(6):553–562. https://doi.org/10.1016/S0378-7788(00)00096-7

Harrison J (2020) The role of materials in sustainable construction. ISOS Conference, pp 1–8

Huberman N, Pearlmutter D (2008) A life-cycle energy analysis of building materials in the Negev desert. Energy Buildings 40:837–848. https://doi.org/10.1016/j.enbuild.2007.06.002

International Energy Agency (IEA) (2018) The future of cooling opportunities for energy-efficient air conditioning. Int Energy Agency. www.iea.org

Iyer-Raniga U, Huovila P (2021) Global state of play for circular built environment

Kadir AA et al (2009) Density, strength, thermal conductivity and leachate characteristics of light-weight fired clay bricks incorporating cigarette butts. World Acad Sci Eng Technol 53(5):1035–1040

Krausmann F et al (2018) From resource extraction to outflows of wastes and emissions: the socioeconomic metabolism of the global economy 1900–2015'. Glob Environ Chang 52:131–140. https://doi.org/10.1016/j.gloenvcha.2018.07.003

Lertsatitthanakorn C, Atthajariyakul S, Soponronnarit S (2009) Techno-economical evaluation of a rice husk ash (RHA) based sand–cement block for reducing solar conduction heat gain to a building. Constr Build Mater 23:364–369. https://doi.org/10.1016/j.conbuildmat.2007.11.017

Madurwar MV, Mandavgane SA, Ralegaonkar RV (2015) Development and feasibility analysis of Bagasse ash bricks. J Energy Eng 141(3):04014022. https://doi.org/10.1061/(asce)ey.1943-7897.0000200

Mishra S, Usmani DJA (2013) Comparison of embodied energy in different masonary wall material. Int J Adv Eng Technol 4(4):1–3

Oborn P, Walters J (2020) Survey of the built environment professions in the Commonwealth

OECD (2018) Global material resources outlook to 2060: economic drivers and environmental consequences. p 212. https://doi.org/10.1787/9789264307452-en

Oke AE, Aigbavboa CO, Semenya K (2017) Energy savings and sustainable construction: examining the advantages of nanotechnology. Energy Procedia 142:3839–3843. https://doi.org/10.1016/j.egypro.2017.12.285

Patiño-Cambeiro F et al (2019) Economic appraisal of energy efficiency renovations in tertiary buildings. Sustain Cities Soc 47. https://doi.org/10.1016/j.scs.2019.101503

Puettmann ME, Wilson JB (2005) Life-cycle analysis of wood products: Cradle-to-gate LCI of residential wood building materials. Wood Fiber Sci 37(2001):18–29

Rajput D et al (2012) Reuse of cotton and recycle paper mill waste as building material. Constr Build Mater 34:470–475. https://doi.org/10.1016/j.conbuildmat.2012.02.035

Ram S, Ralegaonkar RV (2018) Development of low thermal conductivity walling material using industrial by-product. J Clean Prod 204:767–777. https://doi.org/10.1016/j.jclepro.2018.08.338

Ram S, Ralegaonkar RV, Gavali HR (2018) Assessment of energy efficiency in buildings using synergistic walling material. In: Proceedings of Institution of Civil Engineers: Energy 171(4):182–189. https://doi.org/10.1680/jener.17.00029

Sakhare VV, Ralegaonkar RV (2016) Use of bio-briquette ash for the development of bricks. J Cleaner Prod 112:pp 6–689. https://doi.org/10.1016/j.jclepro.2015.07.088

Sharma A et al (2009) Review on thermal energy storage with phase change materials and applications. Renew Sustain Energy Rev 13(2):318–345. https://doi.org/10.1016/j.rser.2007.10.005

Tanuj V, Singh KA, R. J. (2016) An evaluative study on energy efficient building materials. Int J Innovative Res Adv Eng (IJIRAE) 3(08):98–103

United Nations Environment Programme (2019) Global status report for buildings and construction. UN Enviroment programme

United Nations Environment Programme (2020) 2020 Global status report for buildings and construction—Towards a zero-emissions, efficient and resilient buildings and construction sector. Glob Status Rep p 80

Winterstetter A et al (2021) The role of anthropogenic resource classification in supporting the transition to a circular economy. J Clean Prod 297. https://doi.org/10.1016/j.jclepro.2021.126753

De Wit M, Hoogzaad J, Ramkumar S, Friedl H, Douma A (2018) The circularity gap report. an analysis of the circular state of the global economy. Circle Econ 1–36

World Business Council for Sustainable (2019) The building system carbon framework WBCSD

Zabalza Bribián I, Aranda Usón A, Scarpellini S (2009) Life cycle assessment in buildings: state-of-the-art and simplified LCA methodology as a complement for building certification. Build Environ 44(12):2510–2520. https://doi.org/10.1016/j.buildenv.2009.05.001

Printed in the United States
by Baker & Taylor Publisher Services